T0212930

Lecture Notes in Computer Science　　9210

Commenced Publication in 1973
Founding and Former Series Editors:
Gerhard Goos, Juris Hartmanis, and Jan van Leeuwen

Editorial Board

More information about this series at http://www.springer.com/series/7407

Adrian Kosowski · Igor Walukiewicz (Eds.)

Fundamentals of Computation Theory

20th International Symposium, FCT 2015
Gdańsk, Poland, August 17–19, 2015
Proceedings

Editors
Adrian Kosowski
Université Paris Diderot
Paris
France

Igor Walukiewicz
Université Bordeaux 1
Talence
France

ISSN 0302-9743   ISSN 1611-3349   (electronic)
Lecture Notes in Computer Science
ISBN 978-3-319-22176-2   ISBN 978-3-319-22177-9   (eBook)
DOI 10.1007/978-3-319-22177-9

Library of Congress Control Number: 2015944500

LNCS Sublibrary: SL1 – Theoretical Computer Science and General Issues

Springer Cham Heidelberg New York Dordrecht London
© Springer International Publishing Switzerland 2015

Printed on acid-free paper

Springer International Publishing AG Switzerland is part of Springer Science+Business Media
(www.springer.com)

Preface

The Program Committee also conferred a special Test-of-Time Award to the FCT 1991 paper *Lattice basis reduction: Improved practical algorithms and solving subset sum problems*, co-authored by Claus-Peter Schnorr and Martin Euchner. This volume contains the papers presented at FCT 2015: The 20th International Symposium on Fundamentals of Computation Theory, which was held during August 17–19, 2015, in Gdańsk, Poland. The Symposium on Fundamentals of Computation Theory was established in 1977 for researchers interested in all aspects of theoretical computer science, in particular in algorithms, complexity, and formal and logical methods. It is a biennial conference, which has previously been held in Poznań (1977), Wendisch-Rietz (1979), Szeged (1981), Borgholm (1983), Cottbus (1985), Kazan (1987), Szeged (1989), Gosen-Berlin (1991), Szeged (1993), Dresden (1995), Kraków (1997), Iasi (1999), Riga (2001), Malmö (2003), Lübeck (2005), Budapest (2007), Wrocław (2009), Oslo (2011), and Liverpool (2013). The suggested topics of FCT 2015 included three main areas:

- Algorithms (algorithm design and optimization; approximation, randomized, and heuristic methods; circuits and Boolean functions; combinatorics and analysis of algorithms; computational algebra; computational complexity; computational geometry; online algorithms; streaming algorithms; distributed and parallel computing)
- Formal methods (algebraic and categorical methods; automata and formal languages; computability and nonstandard computing models; database theory; foundations of concurrency and distributed systems; logics and model checking; models of reactive, hybrid, and stochastic systems; principles of programming languages; program analysis and transformation; specification, refinement, and verification; security; type systems)
- Emerging fields (ad hoc, dynamic, and evolving systems; algorithmic game theory; computational biology; foundations of cloud computing and ubiquitous systems; quantum information and quantum computing)

This year we received 60 submissions in response to the call for papers, of which 27 were accepted by the Program Committee after a careful review and discussion process. The conference program included three invited talks, by Marek Karpinski (University of Bonn), Antonín Kučera (Masaryk University), and Peter Widmayer (ETH Zürich), as well as a special session devoted to some of the most influential papers presented at FCT in the 40 years of the history of the conference. This volume contains the accepted papers and abstracts of the invited talks.

We would like to thank the members of the Program Committee for the evaluation of the submissions and the additional reviewers for their excellent cooperation in this work. We are grateful to all the contributors of the conference, in particular to the invited speakers for their willingness to share their insights on interesting new developments. Furthermore, we thank the Organizing Committee chaired by Paweł Żyliński and Łukasz Kuszner for their invaluable help.

We are especially grateful to the sponsors of the conference: the Mayor of the City of Gdańsk, the University of Gdańsk, the Gdańsk University of Technology, and Springer, for their financial support. Finally, we acknowledge the use of the EasyChair system in managing paper submissions, the refereeing process, and preparation of the conference proceedings.

August 2015   Adrian Kosowski
                                                               Igor Walukiewicz

Conference Organization

Program Committee

Per Austrin	KTH Royal Institute of Technology, Stockholm, Sweden
Christel Baier	Technische Universität Dresden, Germany
Marcin Bieńkowski	University of Wrocław, Poland
Tomás Brázdil	Masaryk University, Brno, Czech Republic
Luis Caires	Universidade Nova de Lisboa, Portugal
Thomas Colcombet	CNRS and Université Paris Diderot, France
Marek Cygan	University of Warsaw, Poland
Stéphane Demri	CNRS and ENS Cachan, France
Dariusz Dereniowski	Gdańsk University of Technology, Poland
Konstantinos Georgiou	University of Waterloo, Canada
Radu Grosu	Vienna University of Technology, Austria
Rolf Klein	University of Bonn, Germany
Barbara König	University of Duisburg-Essen, Germany
Adrian Kosowski	Inria and Université Paris Diderot, France
Dan Král	University of Warwick, UK
Leonid Libkin	University of Edinburgh, UK
Andrzej Murawski	University of Warwick, UK
Jelani Nelson	Harvard University, USA
Gennaro Parlato	University of Southampton, UK
Andrzej Pelc	Université du Québec en Outaouais, Canada
Guido Proietti	University of L'Aquila, Italy
Andrzej Proskurowski	University of Oregon, USA
Stanisław Radziszowski	Rochester Institute of Technology, USA
Davide Sangiorgi	Inria and University of Bologna, Italy
Thomas Sauerwald	University of Cambridge, UK
Pawel Sobocinski	University of Southampton, UK
Andrzej Szepietowski	University of Gdańsk, Poland
Wojciech Szpankowski	Purdue University, USA
Igor Walukiewicz	CNRS and Université de Bordeaux, France
Paweł Żyliński	University of Gdańsk, Poland

Steering Committee

Bogdan Chlebus	University of Colorado, USA
Zoltan Esik	University of Szeged, Hungary
Marek Karpinski	University of Bonn, Germany

Andrzej Lingas	Lund University, Sweden
Miklos Santha	CNRS and Université Paris Diderot, France
Eli Upfal	Brown University, USA

Additional Reviewers

Bauwens, Bruno
Bezakova, Ivona
Bhattacharya, Binay
Bilò, Davide
Blasiok, Jaroslaw
Blum, Norbert
Borowiecki, Piotr
Choi, Yongwook
Curticapean, Radu
Damaschke, Peter
Downey, Rod
Dueck, Gerhard
Dybizbanski, Janusz
Elbassioni, Khaled
Enea, Constantin
Farley, Arthur
Fijalkow, Nathanaël
Fitzsimmons, Zack
Gajarský, Jakub
Gleich, David
Golovach, Petr
Gualà, Luciano
Gurvich, Vladimir

Habermehl, Peter
Horn, Florian
Jeż, Łukasz
Jurdzinski, Tomasz
Kaminsky, Alan
Knop, Dusan
Knorr, Matthias
Kolpakov, Roman
Kretinsky, Jan
Krzywkowski, Marcin
Kucera, Petr
La Torre, Salvatore
Laskoś-Grabowski, Paweł
Lonati, Violetta
Magner, Abram
Mamede, Margarida
Marcinkowski, Jan
Mikkelsen, Jesper W.
Narvaez, David
Nickovic, Dejan
Obdrzalek, Jan
Ordyniak, Sebastian
Pajak, Dominik

Place, Thomas
Prusinkiewicz,
  Przemyslaw
Radoszewski, Jakub
Raman, Venkatesh
Rauf, Imran
Sauerhoff, Martin
Saurabh, Saket
Scheffel, Torben
Schlipf, Lena
Schmitz, Sylvain
Silberstein, Natalia
Skrzypczak, Michal
Souto, André
Spyra, Aleksandra
Sreejith, A.V.
Truthe, Bianca
Volec, Jan
Wilson, Christopher
Wood, Christopher
Zenil, Hector
Zhang, Yu
Ziemann, Radosław

Institutional Organizers

University of Gdańsk
Gdańsk University of Technology

Honorary Patronage

Mayor of the City of Gdańsk
Committee on Informatics, Polish Academy of Sciences

Cooperating Institution

Sphere Research Labs Sp. z o.o.

Organizing Committee

Piotr Borowiecki
Zuzanna Kosowska-Stamirowska
Adrian Kosowski
Łukasz Kuszner
Paweł Żyliński

Invited Talks

It is Better to be Vaguely Right Than Exactly Wrong[*]

Peter Widmayer

Institute of Theoretical Computer Science, ETH Zürich, Switzerland
widmayer@inf.ethz.ch

Abstract. In the past few years, more and more attention has been paid to the annoying difference between abstract algorithmic problems and their messy origins in the real world. While algorithms theory is highly developed for a large variety of clean, combinatorial optimization problems, their practical counterparts cannot always be solved by well understood theoretical methods, for a variety of reasons: Practical inputs suffer from uncertainties and inaccuracies, like noisy data, or algorithmic outputs cannot be realized exactly, due to physical inaccuracies. We aim at a better theoretical understanding of how to cope with uncertain input data for real problems. We propose a general approach that tends to lead to substantial algorithmic challenges and inefficiencies on the one hand, but promises on the other hand to deliver good results for practical problems as varied as robust trip planning in public transportation and robust de novo peptide sequencing in computational biology.

This talk is about joint work with Katerina Böhmova, Joachim Buhmann, Matus Mihalak, Tobias Pröger, and Rasto Sramek.

---

[*] A quote going back to Carveth Read, but incorrectly attributed to John Maynard Keynes after his death as "It is better to be roughly right than precisely wrong".

Towards Better Inapproximability Bounds for TSP: A Challenge of Global Dependencies

Marek Karpinski

Department of Computer Science and the Hausdorff Center for Mathematics,
University of Bonn, Bonn, Germany
marek@cs.uni-bonn.de

Abstract. We present in this paper some of the recent techniques and methods for proving best up to now explicit approximation hardness bounds for metric symmetric and asymmetric *Traveling Salesman Problem* (TSP) as well as related problems of Shortest Superstring and Maximum Compression. We attempt to shed some light on the underlying paradigms and insights which lead to the recent improvements as well as some inherent obstacles for further progress on those problems.

Marek Karpinski—Research supported by DFG grants and the Hausdorff grant EXC59-1.

On the Existence and Computability
of Long-Run Average Properties in
Probabilistic VASS

Antonín Kučera

Faculty of Informatics, Masaryk University, Brno, Czech Republic
kucera@fi.muni.cz

Abstract. We present recent results about the long-run average properties of probabilistic vector additions systems with states (pVASS). Interestingly, for probabilistic pVASS with two or more counters, long-run average properties may take several different values with positive probability even if the underlying state space is strongly connected. This contradicts the previous results about stochastic Petri nets established in 80s. For pVASS with three or more counters, it may even happen that the long-run average properties are undefined (i.e., the corresponding limits do not exist) for almost all runs, and this phenomenon is stable under small perturbations in transition probabilities. On the other hand, one can effectively approximate eligible values of long-run average properties and the corresponding probabilities for some sublasses of pVASS. These results are based on new exponential tail bounds achieved by designing and analyzing appropriate martingales. The paper focuses on explaining the main underlying ideas.

Based on a joint work with Tomáš Brázdil, Joost-Pieter Katoen, Stefan Kiefer, and Petr Novotný. The author is supported by the Czech Science Foundation, grant No. 15-17564S.

Contents

Set Algorithms, Covering, and Traversal

Graph Algorithms and Networking Applications

Anonymity and Indistinguishability

Graphs, Automata, and Dynamics

Logic and Games

Invited talks

Towards Better Inapproximability Bounds for TSP: A Challenge of Global Dependencies

Marek Karpinski[✉]

Department of Computer Science and the Hausdorff Center for Mathematics,
University of Bonn, Bonn, Germany
marek@cs.uni-bonn.de

Abstract. We present in this paper some of the recent techniques and methods for proving best up to now explicit approximation hardness bounds for metric symmetric and asymmetric *Traveling Salesman Problem* (TSP) as well as related problems of Shortest Superstring and Maximum Compression. We attempt to shed some light on the underlying paradigms and insights which lead to the recent improvements as well as some inherent obstacles for further progress on those problems.

1 Introduction

The metric *Traveling Salesman Problem* (TSP) is one of the best known and broadly studied combinatorial optimization problems. Nevertheless its approximation status remained surprisingly elusive and very resistant for any new insights even after several decades of research. Basically, there were no improvements of the approximation algorithm of Christofides [C76] for the general metric TSP, and also a very slow improvement of the explicit inapproximability bounds for that problem [PY93, E03, EK06, PV06, L12, KS12, KS13a, KS13b, KLS13, AT15]. The attainable values of the explicit inapproximability bounds, and especially the methods for proving them, could give valuable insights into the algorithmic nature of the problem at hand. Unfortunately, there is still a huge gap between upper and lower approximation bounds for TSP. The best upper bound stands at the moment still firmly at 50 % (approximation ratio 1.5). There were however several improvements of underlying approximation ratios for special cases of metric TSP like $(1, 2)$-metric [BK06] and the Graphic TSP [MS11, SV14]. Also the corresponding inapproximability bounds for those special instances were established in [E03, EK06, KS13a, KS13b].

We discuss also some new results on related problems of the *Shortest Superstring, Maximum Asymmetric TSP* and *Maximum Compression Problem* (cf. [KS13a]).

In this paper we introduce an essentially different method from the earlier work of [PV06] to attack the general problem. This method uses some new ideas on small occurrence optimization. The inspiration for it came from the constructions used for restricted cases of TSP in [E03] and [EK06].

M. Karpinski—Research supported by DFG grants and the Hausdorff grant EXC59-1.

A. Kosowski and I. Walukiewicz (Eds.): FCT 2015, LNCS 9210, pp. 3–11, 2015.
DOI: 10.1007/978-3-319-22177-9_1

2   Underlying Idea

The general idea is to use somehow instances of metric TSP to solve approximately another instance of optimization problems with provable inapproximability bounds. Thus, establishing the possible approximation hardness barriers for the TSP-solver itself. The reduction must be approximation preserving, i.e. validate *goodness* of feasible solutions of a given problem and also validating *goodness* of a corresponding tour. The direction from the tour to a feasible solution would be crucial for that method. To give a simple illustration, we start from that optimization problem and construct an instance of the TSP. We have to establish now the correspondence between the solutions of the problem and the tours of TSP. That correspondence must satisfy the crucial property that the problem has a *good* solution if and only if TSP has a *short* tour. The second direction from TSP to the optimization problem seems conceptually at the first glance more difficult. That intuition is correct and the main effort has been devoted to that issue. The suitable optimization problem will be a specially tailored bounded occurrence optimization problem of Sects. 4, 5 and 6.

3   TSP and Related Problems

We are going to define now main optimization problems of the paper.

- Metric TSP (here TSP for short): Given a metric space (V, d) (usually given by a complete weighted graph or a connected weighted graph). Construct a shortest tour visiting every vertex exactly once.
- Asymmetric Metric TSP (ATSP): Given an asymmetric metric space (V, d), d may be asymmetric. Construct a shortest tour visiting every vertex exactly once.
- Graphic TSP: Given a connected graph $G = (V, E)$. Construct a shortest tour in the *shortest path* metric completion of G or equivalently construct a smallest Eulerian spanning multi-subgraph of G.
- Shortest Superstring Problem (SSP): Given a finite set of strings S. Construct a shortest superstring such that every string in S is a substring of it.
- Maximum Compression Problem (MCP): Given a finite set of strings S. Construct a superstring of S with maximum *compression* which is the difference between the sum of the lengths of the strings in S and the length of the superstring.
- Maximum Asymmetric Traveling Salesman Problem (MAX-ATSP): Given a complete directed graph with nonnegative weights. Construct a tour of maximum weight visiting every vertex exactly once.

4   Bounded Occurrence Optimization Problems

We introduce here a notion of bounded occurrence optimization playing important roles in our construction.

- MAX-E3-LIN2: Given a set of equations mod 2 with exactly 3 variables per equation. Construct an assignment maximizing the number of equations satisfied.
- 3-Occ-MAX-HYBRID-LIN2: Given a set of equations mod 2 with exactly 2 or 3 variables per equation and the number of occurrences of each variable being bounded by 3.

The approximation hardness of 3-Occ-MAX-HYBRID-LIN2 problem was proven for the first time by Berman and Karpinski [BK99] (see also [K01, BK03]) by randomized reduction f from MAX-E3-LIN2 and the result of Håstad [H01] on that problem, f : MAX-E3-LIN2 → 3-Occ-MAX-HYBRID-LIN2.

Theorem 1 ([BK99]). *For every $0 < \varepsilon < \frac{1}{2}$, it is NP-hard to decide whether an instance of $f(MAX\text{-}E3\text{-}LIN2) \in$ 3-Occ-MAX-HYBRID-LIN2 with $60n$ equations with exactly two variables and $2n$ equations with exactly three variables has its optimum value above $(62 - \varepsilon)n$ or below $(61 + \varepsilon)n$.*

The above result will be used in the simulational approximation reduction g : 3-Occ-MAX-HYBRID-LIN2 → TSP to the instances of the metric TSP.

5 Bi-wheel Amplifier Graphs

We shortly describe here one of the main concepts of our construction and proofs, that is a concept of a bi-wheel amplifier introduced in [KLS13].

The construction extends the notion of a wheel amplifier of [BK99, BK01] (we refer to [BK99] for the notions of *contact* and *checker* vertices).

A bi-wheel amplifier with $2n$ contact vertices is constructed in the following way. First we construct two disjoint cycles with each $7n$ vertices and we number the vertices by $0, 1, \ldots 7n - 1$. The *contacts* will be the vertices with the numbers being a multiple of 7, while the remaining vertices will be *checkers*. We complete the construction by selecting at random a perfect matching from the checkers of one cycle to the checkers of the other cycle (see Fig. 1).

We use a bi-wheel amplifier in a similar way to the standard wheel amplifier. The crucial difference is that the cycle edges will correspond to the equality constraints, and matching edges will correspond to inequality constraints. The contacts of one cycle will represent the positive appearance of the original variable, and the contacts of the others the negative ones. The main reason is that encoding inequality constraints will be more efficient then encoding equalities with TSP gadgets.

We can prove the following crucial result on bi-wheels.

Lemma 1 ([KLS13]). *With high probability, a bi-wheel is a 3-regular amplifier.*

We are going to apply Lemma 1 in the next section.

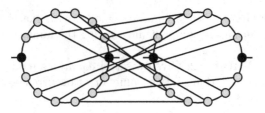

Fig. 1. A bi-wheel amplifier with $n = 2$. The vertices ○ denote checkers and the vertices ● denote contacts.

6   Preparation Lemma

We need the following preparation lemma to simplify our constructions.

We will call in the sequel the problem 3-Occ-MAX-HYBRID-LIN2 simply the HYBRID problem. Lemma 1 will be used now to prove the following lemma.

Lemma 2 ([KLS13]). *For any constant $\varepsilon > 0$ and $b \in \{0, 1\}$, there exists an instance I of the HYBRID problem with all variables occurring exactly three times and having $21m$ equations of the form $x \oplus y = 0$, $9m$ equations of the form $x \oplus y = 1$ and m equations of the form $x \oplus y \oplus z = b$, such that it is NP-hard to decide whether there is an assignment to variables which leaves at most $\varepsilon \cdot m$ equations unsatisfied, or every assignment to variables leaves at least $(0.5 - \varepsilon)m$ equations unsatisfied.*

Because of the above lemma we can assume, among other things, that equations with three variables in I are of the form, say, $x \oplus y \oplus z = 0$. For explicit constructions of simulating gadgets especially the bi-wheel amplifiers and improved gadgets for size-three equations we refer to [KLS13].

7   Instances of Metric TSP

We describe first an underlying idea of the construction of g. We start with instances of the HYBRID problem by constructing a graph with special gadgets structures representing equations. Here we use a new tool of *bi-wheel amplifier* graphs [KLS13], sketched shortly in Sect. 5.

Given an instance I of the HYBRID problem. Let us denote the corresponding graph by G_S. We analyse first the process of constructing a tour in G_S for a given assignment a to the variables of a HYBRID instance I.

Lemma 3 ([KLS13]). *If there is an assignment to the variables of an instance I of the HYBRID problem with $31m$ equations and ν bi-wheels which makes k equations unsatisfied, then there exists a tour in G_S which costs at most $61m + 2\nu + k + 2$.*

We have to prove also corresponding bounds for the opposite direction. Given a tour in G_S, we construct an assignment to the variables of the *associated* instance of the HYBRID problem.

Lemma 4 ([KLS13]). *If there exists a tour in G_S with cost $61m + k - 2$, then there exists an assignment to the variables of the corresponding instance of the HYBRID problem which makes at most k equations unsatisfied.*

Lemmas 3 and 4 entail now straightforwardly our main result.

Theorem 2 ([KLS13]). *The TSP problem is NP-hard to approximate to within any approximation ratio less than* **123/122**.

On the upper approximation bound of this problem, the best approximation algorithm after more than three decades research is still Christofides [C76] algorithm with approximation ratio 1.50. This leaves a curious huge gap between upper bound 50 % and lower bound of about 1 % wide open.

8   Instances of Asymmetric TSP (ATSP)

We consider now asymmetric metric instances of TSP. There is no polynomial time constant approximation ratio algorithm known for that problem. The best known approximation algorithm achieves an approximation ratio of $O(\log n / \log \log n)$ [AGM+10]. It motivates a strong interest on inapproximability bounds for this problem. We establish here the best-of-now explicit inapproximability bounds, still constant and very far from the best upper approximation bound.

The plan of our attack is similar to the case of symmetric TSP. We construct here for an instance I of the HYBRID problem a directed, for this case, graph G_A using the constructions of bi-directed edges. The corresponding lemmas describe the opposite directions of the reductions: assignment to tour and tour to assignment.

Lemma 5 ([KLS13]). *Given an assignment to the variables of an instance I of the HYBRID problem with ν bi-wheels which makes k equations of I unsatisfied, then there exists a tour in G_A which costs at most $37m + 5\nu + 2m\lambda + 2\nu\lambda + k$ for a fixed constant $\lambda > 0$.*

Lemma 6 ([KLS13]). *If there exists a tour in G_A with cost $37m + k + 2\lambda m$, then there exists an assignment to the variables of the corresponding instance of the HYBRID problem which leaves at most k equations unsatisfied.*

Lemmas 5 and 6 entail now our main result of this section.

Theorem 3 ([KLS13]). *The ATSP problem is NP-hard to approximate to within any approximation ratio less than* **75/74**.

9   A Role of Weights

A natural question arises about the role of the magnitudes of weights necessary to carry good approximation reductions from the HYBRID problem. The bounded metric situations were studied for the first time in [EK06]. We define $(1, B)$-TSP ($(1, B)$-ATSP) as the TSP (ATSP) problem taking values from the set of integers $\{1, \ldots, B\}$. The case of $(1, 2)$-TSP is important for its connection to the Graphic TSP. Using a specialization of the methods of Sects. 7 and 8 we obtain the following explicit inapproximability bounds.

Theorem 4 ([KS12]). *It is NP-hard to approximate*

(1)   *the $(1, 2)$-TSP to within any factor less than 535/534;*
(2)   *the $(1, 4)$-TSP to within any factor less than 337/336;*
(3)   *the $(1, 2)$-ATSP to within any factor less than 207/206;*
(4)   *the $(1, 4)$-ATSP to within any factor less than 141/140.*

   For the most restrictive case of $(1, 2)$-TSP, there are better algorithms known [BK06] than the Christofides algorithm.

10   Graphic TSP

We consider now another restricted case of TSP called the Graphic TSP and an interesting for generic reasons case of Graphic TSP on cubic graphs. There was a significant progress recently in designing improved approximation algorithms for the above problems, cf. [MS11, SV14]. Taking an inspiration from the technique on bounded metric TSP, we prove the following

Theorem 5 ([KS13b]). *The Graphic TSP is NP-hard to approximate to within any factor less than 535/534.*

Theorem 6 ([KS13b]).   *The Graphic TSP on cubic graphs is NP-hard to approximate to within any factor less than 1153/1152.*

   The result of Theorem 6 is also the first inapproximability result for the cubic graph instances (cf. [BSS+11]).

11   Some Applications

As a further application of our method, we prove new explicit inapproximability bounds for the Shortest Superstring Problem (SSP) and Maximum Compression Problem (MCP) improving the previous bounds by an order of magnitude [KS13a].

Theorem 7 ([KS13a]). *The SSP is NP-hard to approximate to within any factor less than 333/332.*

Theorem 8 ([KS13a]). *The MCP is NP-hard to approximate to within any factor less than* 204/203.

The best approximation algorithm for MCP reduces that problem to MAX-ATSP (cf., e.g. [KLS+05]). On the other hand, MCP can be seen as a restricted version of the MAX-ATSP. This entails the NP-hardness of approximating MAX-ATSP to within any factor less than 204/203.

12 Summary of Main Results

We summarize here the best known explicit inapproximability results on the TSP problems (Table 1) and the applications (Table 2).

Table 1. Inapproximability results on the TSP problems

Problem	Approximation hardness bound	
TSP	123/122	[KLS13]
Asymmetric TSP	75/74	[KLS13]
Graphic TSP	535/534	[KS13b]
Graphic TSP on cubic graphs	1153/1152	[KS13b]
$(1,2)$-TSP	535/534	[KS12]
$(1,4)$-TSP	337/336	[KS12]
$(1,2)$-ATSP	207/206	[KS12]
$(1,4)$-ATSP	141/140	[KS12]

Table 2. Inapproximability results on the related problems

Problem	Approximation hardness bound	
SSP	333/332	[KS13a]
MCP	204/203	[KS13a]
MAX-ATSP	204/203	[KS13a]

13 Further Research

We presented a method for improving the best known explicit inapproximability bounds for TSP and some related problems. The method depends essentially on a new construction of bounded degree amplifiers. It is still a sensible direction to go for improving design of our new amplifiers and perhaps discovering new proof methods. Also, improving an explicit lower bound for the HYBRID problem would have immediate consequences toward inapproximability bound of the TSP.

Here again the best upper approximation bounds are much higher (but within a couple of percentage points) from the currently provable lower approximation bounds. The TSP problem lacks good definability properties and is definitionally *globally dependent* on all its variables. Our methods used in the proofs were however local in that sense.

In order to get essentially better lower approximation bounds (if such are in fact possible) one should perhaps try to design some more global and perhaps some more weight dependent methods. One possible way would be perhaps to design directly a new *global* PCP construction for the TSP. This seems for the moment to be a very difficult undertaking because of the before mentioned definability properties of the problem.

References

[AT15] Approximation Taxonomy of Metric TSP (2015). http://theory.cs. uni-bonn.de/info5/tsp

[AGM+10] Asadpour, A., Goemans, M., Mądry, A., Gharan, S., Saberi, A.: An $O(\log n/\log\log n)$-approximation algorithm for the asymmetric traveling salesman problem. In: Proceedings of the 21st SODA, pp. 379–389 (2010)

[BK99] Berman, P., Karpinski, M.: On some tighter inapproximability results. In: Wiedermann, J., Van Emde Boas, P., Nielsen, M. (eds.) ICALP 1999. LNCS, vol. 1644, p. 200. Springer, Heidelberg (1999)

[BK01] Berman, P., Karpinski, M.: Efficient amplifiers and bounded degree optimization. ECCC TR01-053 (2001)

[BK03] Berman, P., Karpinski, M.: Improved approximation lower bounds on small occurrence optimization. ECCC TR03-008 (2003)

[BK06] Berman, P., Karpinski, M.: 8/7-approximation algorithm for (1, 2)-TSP. In: Proceedings of the 17th SODA, pp. 641–648 (2006)

[BSS+11] Boyd, S., Sitters, R., van der Ster, S., Stougie, L.: TSP on cubic and subcubic graphs. In: Günlük, O., Woeginger, G.J. (eds.) IPCO 2011. LNCS, vol. 6655, pp. 65–77. Springer, Heidelberg (2011)

[C76] Christofides, N.: Worst-Case Analysis of a New Heuristic for the Traveling Salesman Problem, Technical Report CS-93-13. Carnegie Mellon University, Pittsburgh (1976)

[E03] Engebretsen, L.: An explicit lower bound for TSP with distances one and two. Algorithmica **35**, 301–318 (2003)

[EK06] Engebretsen, L., Karpinski, M.: TSP with bounded metrics. J. Comput. Syst. Sci. **72**, 509–546 (2006)

[H01] Håstad, J.: Some optimal inapproximability results. J. ACM **48**, 798–859 (2001)

[KLS+05] Kaplan, H., Lewenstein, M., Shafrir, N., Sviridenko, M.: Approximation algorithms for asymmetric TSP by decomposing directed regular multigraphs. J. ACM **52**, 602–626 (2005)

[K01] Karpinski, M.: Approximating bounded degree instances of NP-hard problems. In: Freivalds, R. (ed.) FCT 2001. LNCS, vol. 2138, p. 24. Springer, Heidelberg (2001)

[KLS13] Karpinski, M., Lampis, M., Schmied, R.: New inapproximability bounds for TSP. In: Cai, L., Cheng, S.-W., Lam, T.-W. (eds.) Algorithms and Computation. LNCS, vol. 8283, pp. 568–578. Springer, Heidelberg (2013)

[KS12] Karpinski, M., Schmied, R.: On approximation lower bounds for TSP with bounded metrics. CoRR abs/1201.5821 (2012)

[KS13a] Karpinski, M., Schmied, R.: Improved lower bounds for the shortest superstring and related problems. In: Proceedings 19th CATS, CRPIT 141, pp. 27–36 (2013)

[KS13b] Karpinski, M., Schmied, R.: Approximation hardness of graphic TSP on cubic graphs, CoRR abs/1304.6800, 2013. Journal version RAIRO-Operations Research 49, pp. 651–668 (2015)

[L12] Lampis, M.: Improved inapproximability for TSP. In: Gupta, A., Jansen, K., Rolim, J., Servedio, R. (eds.) APPROX 2012 and RANDOM 2012. LNCS, vol. 7408, pp. 243–253. Springer, Heidelberg (2012)

[MS11] Mömke, T., Svensson, O.: Approximating graphic TSP by matchings. In: Proceedings of the IEEE 52nd FOCS, pp. 560–569 (2011)

[PV06] Papadimitriou, C., Vempala, S.: On the approximability of the traveling salesman problem. Combinatorica **26**, 101–120 (2006)

[PY93] Papadimitriou, C., Yannakakis, M.: The traveling salesman problem with distances one and two. Math. Oper. Res. **18**, 1–11 (1993)

[SV14] Sebő, A., Vygen, J.: Shorter tours by nicer ears: 7/5-approximation for the Graph-TSP, 3/2 for the Path Version, and 4/3 for two-edge-connected subgraphs. Combinatorica **34**, 1–34 (2014)

On the Existence and Computability of Long-Run Average Properties in Probabilistic VASS

Antonín Kučera[✉]

Faculty of Informatics, Masaryk University, Brno, Czech Republic
kucera@fi.muni.cz

Abstract. We present recent results about the long-run average properties of probabilistic vector additions systems with states (pVASS). Interestingly, for probabilistic pVASS with two or more counters, long-run average properties may take several different values with positive probability even if the underlying state space is strongly connected. This contradicts the previous results about stochastic Petri nets established in 80s. For pVASS with three or more counters, it may even happen that the long-run average properties are undefined (i.e., the corresponding limits do not exist) for almost all runs, and this phenomenon is stable under small perturbations in transition probabilities. On the other hand, one can effectively approximate eligible values of long-run average properties and the corresponding probabilities for some sublasses of pVASS. These results are based on new exponential tail bounds achieved by designing and analyzing appropriate martingales. The paper focuses on explaining the main underlying ideas.

1 Introduction

Probabilistic vector addition systems with states (pVASS) are a stochastic extension of ordinary VASS obtained by assigning a positive integer weight to every rule. Every pVASS determines an infinite-state Markov chain where the states are pVASS configurations and the probability of a transition generated by a rule with weight ℓ is equal to ℓ/T, where T is the total weight of all enabled rules. A closely related model of stochastic Petri nets (SPN) has been studied since early 80s [2,10] and the discrete-time variant of SPN is expressively equivalent to pVASS.

In this paper we give a summary of recent results about the long-run average properties of runs in pVASS achieved in [4,5]. We show that long-run average properties may take several different values with positive probability even if the state-space of a given pVASS is strongly connected. It may even happen that these properties are undefined (i.e., the corresponding limits do not exist)

A. Kučera—Based on a joint work with Tomáš Brázdil, Joost-Pieter Katoen, Stefan Kiefer, and Petr Novotný. The author is supported by the Czech Science Foundation, grant No. 15-17564S.

A. Kosowski and I. Walukiewicz (Eds.): FCT 2015, LNCS 9210, pp. 12–24, 2015.
DOI: 10.1007/978-3-319-22177-9_2

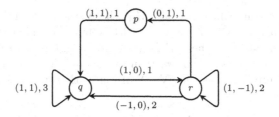

Fig. 1. An example of a two-dimensional pVASS.

for almost all runs. These results contradict the corresponding claims about SPNs published in 80s (see Sect. 2 for more comments). On the other hand, we show that long-run average properties of runs in one-counter pVASS are defined almost surely and can be approximated up to an arbitrarily small relative error in *polynomial time*. This result is obtained by applying several deep observations about one-counter probabilistic automata that were achieved only recently. Further, we show that long-run average properties of runs in two-counter pVASS can also be effectively approximated under some technical (and effectively checkable) assumption about the underlying pVASS which prohibits some phenomena related to (possible) null-recurrency of the analyzed Markov chains.

2   Preliminaries

We use $\mathbb{Z}$, $\mathbb{N}$, $\mathbb{N}^+$, and $\mathbb{R}$ to denote the set of all integers, non-negative integers, positive integers, and real numbers, respectively. We assume familiarity with basic notions of probability theory (probability space, random variable, expected value, etc.). In particular, *Markov chains* are formally understood as pairs of the form $M = (S, \rightarrow)$, where S is a finite or countably infinite set of states and $\rightarrow \subseteq S \times (0,1] \times S$ is a transition relation such that for every $s \in S$ we have that $\sum_{s \rightarrow t} x$ is equal to one. Every state $s \in S$ determines the associated probability space over all runs (infinite paths) initiated in s in the standard way.

2.1   Probabilistic Vector Addition Systems with States

A *probabilistic Vector Addition System with States (pVASS)* with $d \geq 1$ counters is a finite directed graph whose edges are labeled by pairs κ, ℓ, where $\kappa \in \mathbb{Z}^d$ is a vector of *counter updates* and $\ell \in \mathbb{N}$ is a *weight*. A simple example of a two-counter pVASS is shown in Fig. 1. Formally, a pVASS is a triple $\mathcal{A} = (Q, \gamma, W)$, where Q is a finite set of *control states*, $\gamma \subseteq Q \times \mathbb{Z}^d \times Q$ is a set of *rules*, and $W : \gamma \rightarrow \mathbb{N}^+$ is a *weight assignment*. In the following, we write $p \xrightarrow{\kappa} q$ to denote that $(p, \kappa, q) \in \gamma$, and $p \xrightarrow{\kappa,\ell} q$ to denote that $(p, \kappa, q) \in \gamma$ and $W((p, \kappa, q)) = \ell$.

A *configuration* of a pVASS $\mathcal{A}$ is a pair pv where $p \in Q$ is the current control state and $v \in \mathbb{N}^d$ is the vector of current counter values. A rule $p \xrightarrow{\kappa} q$ is *enabled* in a configuration pv iff $v + \kappa \in \mathbb{N}^d$, i.e., the counters remain non-negative when applying the counter change κ to v. The semantics of $\mathcal{A}$ is defined by its associated infinite-state Markov chain $M_{\mathcal{A}}$ whose states are the configurations of $\mathcal{A}$ and $pv \xrightarrow{x} qu$ if there is a rule $p \xrightarrow{\kappa} q$ with weight ℓ enabled in pv such that $u = v + \kappa$ and $x = \ell/T$, where T is the total weight of all rules enabled in pv. If there is no rule enabled in pv, then pv has only one outgoing transition $pv \xrightarrow{1} pv$. For example, if $\mathcal{A}$ is the pVASS of Fig. 1, then $r(3,0) \xrightarrow{1/3} p(3,1)$.

2.2   Patterns and Pattern Frequencies

Let $\mathcal{A} = (Q, \gamma, W)$ be a pVASS, and let $Pat_{\mathcal{A}}$ be the set of all *patterns* of $\mathcal{A}$, i.e., pairs of the form $p\alpha$ where $p \in Q$ and $\alpha \in \{0,+\}^d$. To every configuration pv we associate the pattern $p\alpha$ such that $\alpha_i = +$ iff $v_i > 0$. Thus, every run $w = p_0v_0, p_1v_1, p_2v_2, \ldots$ in the Markov chain $M_{\mathcal{A}}$ determines the unique sequence of patterns $p_0\alpha_0, p_1\alpha_1, p_2\alpha_2, \ldots$ For every $n \geq 1$, let $\mathcal{F}^n(w) : Pat_{\mathcal{A}} \to \mathbb{R}$ be the pattern frequency vector computed for the first n configurations of w, i.e., $\mathcal{F}^n(w)(p\alpha) = \#_{p\alpha}^n(w)/n$, where $\#_{p\alpha}^n(w)$ is the total number of all $0 \leq j < n$ such that $p_j\alpha_j = p\alpha$. The *limit* pattern frequency vector, denoted by $\mathcal{F}(w)$, is defined by $\mathcal{F}(w) = \lim_{n\to\infty} \mathcal{F}^n(w)$. If this limit does not exist, we put $\mathcal{F}(w) = \bot$.

Note that $\mathcal{F}$ is a random variable over $Run(pv)$. The very basic questions about $\mathcal{F}$ include the following:

- Do we have that $\mathcal{P}[\mathcal{F}{=}\bot] = 0$?
- Is $\mathcal{F}$ a discrete random variable?
- If so, is the set of values taken by $\mathcal{F}$ with positive probability finite?
- Can we compute these values and the associated probabilities?

Since the set of rules enabled in a configuration pv is fully determined by the associated pattern $p\alpha$, the frequency of patterns also determines the frequency of rules. More precisely, almost all runs that share the same pattern frequency also share the same frequency of rules performed along these runs, and the rule frequency is easily computable from the pattern frequency.

The above problems have been studied already in 80 s for a closely related model of stochastic Petri nets (SPN). In [8], Sect. 4.B, is stated that if the state-space of a given SPN (with arbitrarily many unbounded places) is strongly connected, then the firing process is ergodic. In the setting of discrete-time probabilistic Petri nets, this means that for almost all runs, the limit frequency of transitions performed along a run is defined and takes the same value. This result is closely related to the questions formulated above. Unfortunately, this claim is invalid. In Fig. 2, there is an example of a SPN (with weighted transitions) with two counters (places) and strongly connected state space where the limit frequency of transitions takes two eligible values (each with probability 1/2). Intuitively, if both places/counters are positive, then both of them have a tendency to decrease, i.e., a configuration where one of the counters is empty is reached almost surely. When we reach a configuration where, e.g., the first place/counter is zero and the second place/counter is positive, then the second place/counter starts to *increase*, i.e., it never becomes zero again with some positive probability. The first place/counter stays zero for most of the time, because when it

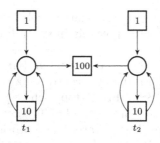

Fig. 2. A discrete-time SPN $\mathcal{N}$.

becomes positive, it is immediately emptied with a very large probability. This means that the frequency of firing t_2 will be much higher than the frequency of firing t_1. When we reach a configuration where the first place/counter is positive and the second place/counter is zero, the situation is symmetric, i.e., the frequency of firing t_1 becomes much higher than the frequency of firing t_2. Further, almost every run eventually behaves according to one of these two scenarios, and therefore there are two limit frequencies of transitions, each of which is taken with probability $1/2$. This possibility of reversing the "global" trend of the counters after hitting zero in some counter was not considered in [8]. Further, there exists a three-counter pVASS $\mathcal{A}$ with strongly connected state-space where the limit frequency of transitions is undefined for almost all runs, and this property is preserved for all ε-perturbations in transition weights for some fixed $\varepsilon > 0$ (see [4]). So, we must unfortunately conclude that the results of [8] are invalid for fundamental reasons.

In the next sections, we briefly summarize the results of [4] about pattern frequency vector in pVASS of dimension one and two. From now on, we assume that

- every counter is changed at most by one when performing a single rule, i.e., the vector of counter updates ranges over $\{-1, 0, 1\}^d$;
- for every pair of control states p, q, there is at most one rule of the form $p \xrightarrow{\kappa} q$.

These assumptions are not restrictive, but they have some impact on complexity, particularly when the counter updates are encoded in binary.

3   Pattern Frequency in One-Counter pVASS

For one-counter pVASS, we have the following result [4]:

Theorem 1. *Let $p(1)$ be an initial configuration of a one-counter pVASS $\mathcal{A}$. Then*

- $\mathcal{P}[\mathcal{F} = \bot] = 0$;
- $\mathcal{F}$ *is a discrete random variable;*
- *there are at most $2|Q| - 1$ pairwise different vectors F such that $\mathcal{P}(\mathcal{F} = F) > 0$;*
- *these vectors and the associated probabilities can be approximated up to an arbitrarily small relative error $\varepsilon > 0$ in polynomial time.*

Since pattern frequencies and the associated probabilities may take irrational values, they cannot be computed precisely in general; in this sense, Theorem 1 is the best result achievable.

A proof of Theorem 1 is not too complicated, but it builds on several deep results that have been established only recently. As a running example, consider the simple one-counter pVASS of Fig. 3 (top), where $p(1)$ is the initial configuration. The first important step in the proof of Theorem 1 is to classify the runs initiated in $p(1)$ according to their *footprints*. A footprint of a run w initiated in $p(1)$ is obtained from w by deleting all intermediate configurations in all maximal subpaths that start in a configuration with counter equal to one, end in a configuration with counter equal to zero, and the counter stays positive in all intermediate configurations (here, the first/last configurations of a finite path are not considered as intermediate). For example, let w be a run of the form

$$\underline{p(1)}, p(2), r(2), r(1), s(1), \underline{s(0)}, \underline{r(0)}, \underline{s(0)}, \underline{s(1)}, r(1), s(1), r(1), \underline{r(0)}, \ldots$$

Then the footprint of w starts with the underlined configurations

$$p(1), s(0), r(0), s(0), s(1), r(0), \ldots$$

Note that a configuration $q(\ell)$, where $\ell > 1$, is preserved in the footprint of w iff all configurations after $q(\ell)$ have positive counter value. Further, for all $p, q \in Q$, let

- $[p{\downarrow}q]$ be the probability of all runs that start with a finite path from $p(1)$ to $q(0)$ where the counter stays positive in all intermediate configurations;
- $[p{\uparrow}] = 1 - \sum_{q \in Q} [p{\downarrow}q]$.

Almost every footprint can be seen as a run in a *finite-state* Markov chain $X_{\mathcal{A}}$ where the set of states is $\{q_0, q_1, q_{\uparrow} \mid q \in Q\}$ and the transitions are determined as follows:

- $p_0 \xrightarrow{x} q_\ell$ in $X_{\mathcal{A}}$ if $x > 0$ and $p(0) \xrightarrow{x} q(\ell)$ in $M_{\mathcal{A}}$;
- $p_1 \xrightarrow{x} q_0$ in $X_{\mathcal{A}}$ if $x = [p{\downarrow}q] > 0$;
- $p_1 \xrightarrow{x} p_{\uparrow}$ in $X_{\mathcal{A}}$ if $x = [p{\uparrow}] > 0$;
- $p_{\uparrow} \xrightarrow{1} p_{\uparrow}$.

The structure of $X_{\mathcal{A}}$ for the one-counter pVASS of Fig. 3 (top) is shown in Fig. 3 (down). In particular, note that since $r_{\uparrow} = s_{\uparrow} = 0$, there are no transitions $r_1 \to r_{\uparrow}$ and $s_1 \to s_{\uparrow}$ in $X_{\mathcal{A}}$.

For almost all runs w initiated in $p(1)$, the footprint of w determines a run in $X_{\mathcal{A}}$ initiated in p_1 in the natural way. In particular, if w contains only finitely many configurations with zero counter, then the footprint of w takes the form $u, s(0), r(1), v$, where $s(0)$ is the last configuration of w with zero counter and $r(1), v$ is an infinite suffix of w. This footprint corresponds to a run $u, s(0), r(1), r_{\uparrow}, r_{\uparrow}, \ldots$ of $X_{\mathcal{A}}$. In general, it may also happen that the footprint of w *cannot* be interpreted as a run in $X_{\mathcal{A}}$, but the total probability of all such w is equal to zero. As a concrete example, consider the run

$$p(1), r(1), s(0), s(1), s(2), s(3), s(4), \ldots$$

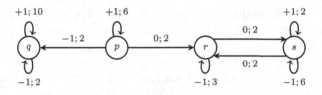

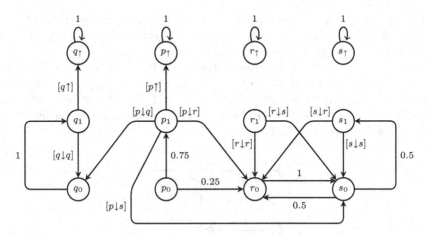

Fig. 3. A one-counter pVASS $\mathcal{A}$ (top) and its associated finite-state Markov chain $X_{\mathcal{A}}$ (down).

in $M_{\mathcal{A}}$ where the counter never reaches zero after the third configuration. The footprint of this run cannot be seen as a run in $X_{\mathcal{A}}$ (note that the infinite sequence $p_1, s_0, s_1, s_{\uparrow}, s_{\uparrow}, s_{\uparrow}, \ldots$ is *not* a run in $X_{\mathcal{A}}$).

Since almost every run in $M_{\mathcal{A}}$ initiated in $p(1)$ determines a run in $X_{\mathcal{A}}$ (via its footprint), we obtain that almost every run in $M_{\mathcal{A}}$ initiated in $p(1)$ visits a bottom strongly connected component (BSCC) of $X_{\mathcal{A}}$. Formally, for every BSCC B of $X_{\mathcal{A}}$ we define $Run(p(1), B)$ as the set of all runs w in $M_{\mathcal{A}}$ initiated in $p(1)$ such that the footprint of w determines a run in $X_{\mathcal{A}}$ that visits B. One is tempted to expect that almost all runs of $Run(p(1), B)$ share the same pattern frequency vector. This is true if (the underlying graph of) $\mathcal{A}$ has at most one *diverging* BSCC. To explain this, let us fix a BSCC D of $\mathcal{A}$ and define a Markov chain $\mathcal{D} = (D, \rightarrow)$ such that $s \xrightarrow{x} t$ in $\mathcal{D}$ iff $s \xrightarrow{\kappa, \ell} t$ and $x = \ell/T_s$, where T_s is the sum of the weights of all outgoing rules of s in $\mathcal{A}$. Further, we define the *trend of D* as follows:

$$t_d = \sum_{s \in D} \mu_{\mathcal{D}}(s) \cdot \sum_{s \xrightarrow{\kappa, \ell} t} \ell/T_s \cdot \kappa \qquad (1)$$

Here, $\mu_{\mathcal{D}}$ is the invariant distribution of $\mathcal{D}$. Intuitively, the trend t_D corresponds to the expected change of the counter value per transition (mean payoff) in D when the counter updates are interpreted as payoffs. If t_D is positive/negative, then the counter has a tendency to increase/decrease. If $t_D = 0$, the situation is more subtle. One possibility is that the counter can never be emptied to zero when it reaches a sufficiently high value (in this case, we say that D is *bounded*). The other possibility is that the counter can always be emptied, but then the frequency of visits to a configuration with zero counter is equal to zero almost surely. We say that D is *diverging* if either $t_D > 0$, or $t_D = 0$ and D is bounded. For the one-counter pVASS of Fig. 3, we have that the BSCC $\{q\}$ is diverging, because its trend is positive. The other BSCC $\{r, s\}$ is not diverging, because its trend is negative.

Let us suppose that $\mathcal{A}$ has at most one diverging BSCC, and let B be a BSCC of $X_\mathcal{A}$. If $B = \{q_\uparrow\}$ for some $q \in Q$, then almost all runs of $Run(p(1), \{q_\uparrow\})$ share the same pattern frequency vector F where $F(s(+)) = \mu_D(s)$ for all $s \in D$, and $F(pat) = 0$ for all of the remaining patterns pat. In the example of Fig. 3, we have that almost all runs of $Run(p(1), \{q_\uparrow\})$ and $Run(p(1), \{p_\uparrow\})$ share the same pattern frequency vector F such that $F(q(+)) = 1$. Now let B be a BSCC of $X_\mathcal{A}$ which is *not* of the form $\{q_\uparrow\}$. As an example, consider the following BSCC of the chain $X_\mathcal{A}$ given in Fig. 3:

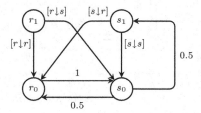

For all $r_1, s_0 \in B$ and all $t \in S$, let

- $E\langle r{\downarrow}s\rangle$ be the conditional expected length of a path from $r(1)$ to $s(0)$ under the condition that $s(0)$ is reached from $r(1)$ via a finite path where the counter stays positive in all configurations preceding $s(0)$;
- $E_{\#_t}\langle r{\downarrow}s\rangle$ be the conditional expected number of visits to a configuration with control state t along a path from $r(1)$ to $s(0)$ (where the visit to $s(0)$ does not count) under the condition that $s(0)$ is reached from $r(1)$ via a finite path where the counter stays positive in all configurations preceding $s(0)$;

If some $E\langle r{\downarrow}s\rangle$ is infinite, then the frequency of visits to configurations with zero counter is equal to zero for almost all runs of $Run(p(1), B)$. Further, there is a non-bounded BSCC D of $\mathcal{A}$ with zero trend such that $r \in D$, and one can easily show that almost all runs of $Run(p(1), B)$ share the same pattern frequency vector F where $F(s(+)) = \mu_D(s)$ for all $s \in D$, and $F(pat) = 0$ for all of the remaining patterns pat.

Now suppose that all $E\langle r{\downarrow}s\rangle$ are finite (which implies that all $E_{\#_t}\langle r{\downarrow}s\rangle$ are also finite). Recall that every transition of $X_\mathcal{A}$ represents a finite subpath of a run in $M_\mathcal{A}$. The expected length of a subpath represented by a transition of B is given by

$$E[L] = \sum_{s_0 \in B} \mu_B(s_0) \cdot 1 + \sum_{r_1 \in B} \mu_B(r_1) \cdot \sum_{r_1 \xrightarrow{x} s_0} x \cdot E\langle r{\downarrow}s\rangle$$

where μ_B is the invariant distribution of B. Similarly, we can define the expected number of visits to a configuration $t(k)$, where $k > 0$, along a subpath represented by a transition of B by

$$E[t(+)] \;=\; \sum_{r_1 \in B} \mu(r_1) \cdot \sum_{r_1 \overset{x}{\to} s_0} x \cdot E_{\#_t}\langle r \downarrow s \rangle \,.$$

The expected number of visits to a configuration $t(0)$ along a subpath represented by a transition of B (where the last configuration of a subpath does not count) is given by $E[t(0)] \;=\; \mu_B(t_0)$. It follows that almost all runs of $Run(p(1), B)$ share the same pattern frequency vector F where

$$F(t(+)) = \frac{E[t(+)]}{E[L]}, \qquad F(t(0)) = \frac{E[t(0)]}{E[L]}$$

for all $t \in S$.

To sum up, for each BSCC B of $X_{\mathcal{A}}$ we need to approximate the probability of $Run(p(1), B)$ and the associated pattern frequency vector up to a given relative error $\varepsilon > 0$. To achieve that, we need to approximate the following numbers up to a sufficient relative precision, which is determined by a straightforward error propagation analysis:

– the probabilities of the form $[p \downarrow q]$ and $[p \uparrow]$;
– the conditional expectations $E\langle r \downarrow s \rangle$ and $E_{\#_t}\langle r \downarrow s \rangle$

The algorithms that approximate the above values are non-trivial and have been obtained only recently. More detailed comments are given in the next subsections.

Finally, let us note that the general case when $\mathcal{A}$ has more than one diverging BSCC does not cause any major difficulties; for each diverging BSCC D, we construct a one-counter pVASS $\mathcal{A}_D$ where the other diverging BSCCs of $\mathcal{A}$ are modified so that their trend becomes negative. The analysis of $\mathcal{A}$ is thus reduced to the analysis of several one-counter pVASS where the above discussed method applies.

3.1 Approximating $[p \downarrow q]$ and $[p \uparrow]$

In this subsection we briefly indicate how to approximate the probabilities of the form $[p \downarrow q]$ and $[p \uparrow]$ up to a given relative error $\varepsilon > 0$.

Let $x_{\min}$ be the least positive transition probability in $M_{\mathcal{A}}$. It is easy to show that if $[p \downarrow q] > 0$, then $[p \downarrow q] > x_{\min}^{|Q|^3}$ (one can easily bound the length of a path from $p(1)$ to $q(0)$). Hence, it suffices to show how to approximate $[p \downarrow q]$ up to a given *absolute* error $\varepsilon > 0$.

The vector of all probabilities of the form $[p \downarrow q]$ is the least solution (in $[0, 1]^{|Q|^2}$) of a simple system of recursive non-linear equations constructed as follows:

$$[p \downarrow q] \;=\; \sum_{p(1) \overset{x}{\to} q(0)} x \;+\; \sum_{p(1) \overset{x}{\to} t(1)} x \cdot [t \downarrow q] \;+\; \sum_{p(1) \overset{x}{\to} t(2)} x \cdot \sum_{s} [t \downarrow s] \cdot [s \downarrow r]$$

These equations are intuitive, and can be seen as a special case of the equations designed for a more general model of probabilistic pushdown automata [6,7]. A solution to this system (even for pPDA) can be approximated by a decomposed Newton method [7] which produces one bit of precision per iteration after exponentially many initial iterations [9]. For one-counter pVASS, this method produces one bit of precision per iteration after *polynomially* many iterations. By implementing a careful rounding, a *polynomial time approximation algorithm* for $[p{\downarrow}q]$ was designed in [11].

Since $[p{\uparrow}] = 1 - \sum_{q \in Q}[p{\downarrow}q]$, we can easily approximate $[p{\uparrow}]$ up to an arbitrarily small *absolute* error in polynomial time. To approximate $[p{\uparrow}]$ up to a given *relative* error, we need to establish a reasonably large lower bound for a positive $[p{\uparrow}]$. Such a bound was obtained in [3] by designing and analyzing an appropriate martingale. More precisely, in [3] it was shown that if $[p{\uparrow}] > 0$, then one of the following possibilities holds:

- There is $q \in Q$ such that $[q{\uparrow}] = 1$ and $p(1)$ can reach a configuration $q(k)$ for some $k > 0$. In this case, $[p{\uparrow}] \geq x_{\min}^{|Q|^2}$.
- There is a BSCC D of $\mathcal{A}$ such that $t_D > 0$ and

$$[p{\uparrow}] \geq \frac{x_{\min}^{4|Q|^2} \cdot t_D^3}{7000 \cdot |Q|^3}.$$

3.2 Approximating $E\langle R{\downarrow}s \rangle$ and $E_{\#_t}\langle r{\downarrow}s \rangle$

The conditional expectations of the form $E\langle r{\downarrow}s \rangle$ satisfy a simple system of *linear* recursive equations constructed in the following way:

$$E\langle q{\downarrow}r \rangle = \sum_{q(1)\xrightarrow{x}r(0)} \frac{x}{[q{\downarrow}r]} + \sum_{q(1)\xrightarrow{x}t(1)} \frac{x \cdot [t{\downarrow}r]}{[q{\downarrow}r]} \cdot (1 + E\langle t{\downarrow}r \rangle)$$

$$+ \sum_{q(1)\xrightarrow{x}t(2)} x \cdot \sum_s \frac{[t{\downarrow}s] \cdot [s{\downarrow}r]}{[q{\downarrow}r]} \cdot (1 + E\langle t{\downarrow}s \rangle + E\langle s{\downarrow}r \rangle)$$

The only problem is that the coefficients are fractions of probabilities of the form $[p{\downarrow}q]$, which may take irrational values and cannot be computed precisely in general. Still, we can approximate these coefficients up to an arbitrarily small relative error in polynomial time by applying the results of the previous subsection. Hence, the very core of the problem is to determine a sufficient precision for these coefficients such that the approximated linear system still has a unique solution which approximates the vector of conditional expectations up to a given relative error. This was achieved in [3] by developing an upper bound on $E\langle r{\downarrow}s \rangle$, which was then used to analyze the condition number of the matrix of the linear system.

The same method is applicable also to $E_{\#_t}\langle r{\downarrow}s \rangle$ (the system of linear equation presented above must be slightly modified).

4   Pattern Frequency in Two-Counter pVASS

The analysis of pattern frequencies in two-counter pVASS imposes new difficulties that cannot be solved by the methods presented in Sect. 3. Still, the results achieved for one-counter pVASS are indispensable, because in certain situations, one of the two counters becomes "irrelevant", and then we proceed by constructing and analyzing an appropriate one-counter pVASS.

The results achieved in [4] for two-counter pVASS are formulated in the next theorem.

Theorem 2. *Let pv be an initial configuration of a stable two-counter pVASS $\mathcal{A}$. Then*

- *$\mathcal{P}[\mathcal{F}{=}{\perp}] = 0$;*
- *$\mathcal{F}$ is a discrete random variable;*
- *there are only finitely many vectors F such that $\mathcal{P}(\mathcal{F}{=}F) > 0$;*
- *these vectors and the associated probabilities can be effectively approximated up to an arbitrarily small absolute/relative error $\varepsilon > 0$.*

The condition of *stability* (explained below) can be checked in exponential time and guarantees that certain infinite-state Markov chains that are used to analyze the pattern frequencies of $\mathcal{A}$ are not null-recurrent.

Let $p(1,1)$ be an initial configuration of $\mathcal{A}$, and let $\mathcal{A}_1$ and $\mathcal{A}_2$ be one-counter pVASS obtained from $\mathcal{A}$ by preserving the first and the second counter, and abstracting the other counter into a *payoff* which is assigned to the respective rule of $\mathcal{A}_1$ and $\mathcal{A}_2$, respectively. The analysis of runs initiated in $p(1,1)$ must take into account the cases when one or both counters become bounded, one or both counters cannot be emptied to zero anymore, etc. For simplicity, let us assume that

(1) the set of all configurations reachable from $p(1,1)$ is a strongly connected component of $M_{\mathcal{A}}$, and for every $k \in \mathbb{N}$ there is a configuration reachable from $p(1,1)$ where both counters are larger than k;
(2) the set of all configuration reachable from $p(1)$ in $\mathcal{A}_1$ and $\mathcal{A}_2$ is infinite and forms a strongly connected component of $M_{\mathcal{A}_1}$ and $M_{\mathcal{A}_2}$, respectively.

Note that Assumption (1) implies that the graph of $\mathcal{A}$ is strongly connected. An example of a two-counter pVASS satisfying these assumptions is given in Fig. 4. Now we define

- the *global trend* $t = (t_1, t_2)$, where t_i is the trend of $\mathcal{A}_i$ as defined by Eq. (1) where $D = Q$ (recall that the graph of $\mathcal{A}_i$ is strongly connected);
- the *expected payoff* τ_i of a run initiated in a configuration of $\mathcal{A}_i$ reachable from $p(1)$. Since the set of all configurations reachable from $p(1)$ in $\mathcal{A}_i$ is strongly connected, it is not hard to show that τ_i is independent of the initial configuration and the mean payoff of a given run is equal to τ_i almost surely.

Intuitively, the global trend t specifies the average change of the counter values per transition if *both* counters are abstracted into payoffs. Note that if t is positive in both components, then almost all runs initiated in $p(1,1)$ "diverge",

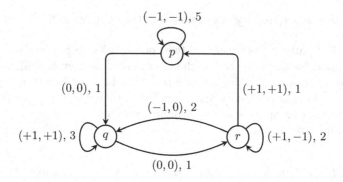

Fig. 4. A two-counter pVASS $\mathcal{A}$.

i.e., both counters remain positive from certain point on (here we need Assumption (1)). This means that almost all runs initiated in $p(1,1)$ share the same pattern frequency vector F where $F(q(+,+)) = \mu_{\mathcal{A}}$ and $F(pat) = 0$ for all of the remaining patterns pat (here $\mu_{\mathcal{A}}$ is the invariant distribution of $\mathcal{A}$; see the definition of μ_D in Eq. (1) and recall that $\mathcal{A}$ is strongly connected).

Now suppose that t_2 is negative, and consider a configuration $q(k,0)$ reachable from $p(1,1)$, where k is "large". Obviously, a run initiated in $q(k,0)$ hits a configuration of the form $q'(k',0)$ without visiting a configuration with zero in the first counter with very high probability. Further, if $\tau_2 > 0$, then k' is larger than k "on average". Hence, the runs initiated in $q(k,0)$ will have a tendency to "diverge along the x-axis". If $\tau_2 < 0$, then k' is smaller than k on average, and the runs initiated in $q(k,0)$ will be moving towards the y-axis. A symmetric observation can be made when t_1 is negative. Hence, if *both* t_1 and t_2 are negative, we can distinguish three possibilities:

- $\tau_1 > 0$ and $\tau_2 > 0$. Then, almost all runs initiated in $p(1,1)$ will eventually diverge either along the x-axis or along the y-axis. That is, one of the counters eventually becomes irrelevant almost surely, and the pattern frequencies for $\mathcal{A}$ can be determined from the ones for $\mathcal{A}_1$ and $\mathcal{A}_2$ (since $\mathcal{A}_1$ and $\mathcal{A}_2$ are one-counter pVASS, we can apply the results of Sect. 3). The SPN of Fig. 2 is one concrete example of this scenario.
- $\tau_1 < 0$ and $\tau_2 > 0$. Then almost all runs initiated in $p(1,1)$ will eventually diverge along the x-axis. The case when $\tau_1 > 0$ and $\tau_2 < 0$ is symmetric.
- $\tau_1 < 0$ and $\tau_2 < 0$. In this case, there is a computable m such that the set of all configurations of the form $q(k,0)$ and $q(0,k)$, where $q \in Q$ and $k \leq m$, is a *finite eager attractor*. That is, this set of configurations is visited infinitely often by almost all runs initiated in $p(1,1)$, and the probability of revisiting this set in ℓ transitions decays (sub)exponentially in ℓ. The pattern frequencies for the runs initiated in $p(1,1)$ can be analyzed be generic methods for systems with a finite eager attractor developed in [1].

The cases when t_1, t_2, τ_1, or τ_2 is equal to zero are disregarded in [4], because the behaviour of $\mathcal{A}$, $\mathcal{A}_1$, or $\mathcal{A}_2$ can then exhibit strange effects caused by (possible) null recurrency of the underlying Markov chains, which requires different analytical methods (the *stability* condition in Theorem 2 requires that t_1, t_2, τ_1, and τ_2 are non-zero). We have not discussed the case when t_1 is negative and t_2 positive (or vice versa), because the underlying analysis is similar to the one presented above.

To capture the above explained intuition precisely, we need to develop an explicit lower bound for the probability of "diverging along the x-axis from a configuration $q(k,0)$" when $\tau_2 > 0$, examine the expected value of the second counter when hitting the y-axis by a run initiated in $q(k,0)$ when $\tau_2 < 0$, etc. These bounds are established in [4] by generalizing the martingales designed in [3] for one-counter pVASS.

5 Future Research

One open problem is to extend Theorem 2 so that it also covers two-counter pVASS that are not necessarily stable. Since one can easily construct a (non-stable) two-counter pVASS such that $\mathcal{P}(\mathcal{F}{=}F) > 0$ for infinitely many pairwise different vectors F, and there even exists a (non-stable) two-counter pVASS such that $\mathcal{P}(\mathcal{F}{=}\bot) = 1$, this task does not seem trivial.

Another challenge is to design algorithms for the analysis of long-run average properties in reasonably large subclasses of multi-dimensional pVASS. Such algorithms might be obtained by generalizing the ideas used for two-counter pVASS.

References

1. Abdulla, P., Henda, N.B., Mayr, R., Sandberg, S.: Limiting behavior of Markov chains with eager attractors. In: Proceedings of 3rd International Conference on Quantitative Evaluation of Systems (QEST 2006), pp. 253–264. IEEE Computer Society Press (2006)
2. Marsan, M.A., Conte, G., Balbo, G.: A class of generalized stochastic Petri nets for the performance evaluation of multiprocessor systems. ACM Trans. Comput. Syst. **2**(2), 93–122 (1984)
3. Brázdil, T., Kiefer, S., Kučera, A.: Efficient analysis of probabilistic programs with an unbounded counter. J. ACM **61**(6), 41:1–41:35 (2014)
4. Brázdil, T., Kiefer, S., Kučera, A., Novotný, P.: Long-run average behaviour of probabilistic vector addition systems. In: Proceedings of LICS 2015 (2015)
5. Brázdil, T., Kiefer, S., Kučera, A., Novotný, P., Katoen, J.-P.: Zero-reachability in probabilistic multi-counter automata. In: Proceedings of CSL-LICS 2014, pp. 22:1–22:10. ACM Press (2014)
6. Esparza, J., Kučera, A., Mayr, R.: Model-checking probabilistic pushdown automata. In: Proceedings of LICS 2004, pp. 12–21. IEEE Computer Society Press (2004)
7. Etessami, K., Yannakakis, M.: Recursive Markov chains, stochastic grammars, and monotone systems of nonlinear equations. J. ACM **56**(1), 1–66 (2009)
8. Florin, G., Natkin, S.: Necessary and sufficient ergodicity condition for open synchronized queueing networks. IEEE Trans. Softw. Eng. **15**(4), 367–380 (1989)

9. Kiefer, S., Luttenberger, M., Esparza, J.: On the convergence of Newton's method for monotone systems of polynomial equations. In: Proceedings of STOC 2007, pp. 217–226. ACM Press (2007)
10. Molloy, M.K.: Performance analysis using stochastic Petri nets. IEEE Trans. Comput. **31**(9), 913–917 (1982)
11. Stewart, A., Etessami, K., Yannakakis, M.: Upper bounds for newton's method on monotone polynomial systems, and P-time model checking of probabilistic one-counter automata. In: Sharygina, N., Veith, H. (eds.) CAV 2013. LNCS, vol. 8044, pp. 495–510. Springer, Heidelberg (2013)

Geometry, Combinatorics, Text Algorithms

Longest α-Gapped Repeat and Palindrome

Paweł Gawrychowski[1] and Florin Manea[2]([⊠])

[1] Institute of Informatics, University of Warsaw, Warsaw, Poland
gawry@mimuw.edu.pl
[2] Department of Computer Science, Kiel University, Kiel, Germany
flm@informatik.uni-kiel.de

Abstract. We propose an efficient algorithm finding, for a word w and an integer $\alpha > 0$, the longest word u such that w has a factor uvu, with $|uv| \leq \alpha|u|$ (i.e., the longest α-gapped repeat of w). Our algorithm runs in $\mathcal{O}(\alpha n)$ time. Moreover, it can be easily adapted to find the longest u such that w has a factor $u^R vu$, with $|uv| \leq \alpha|u|$ (i.e., the longest α-gapped palindrome), again in $\mathcal{O}(\alpha n)$ time.

1 Introduction

Gapped repeats and palindromes have been investigated for a long time (see, e.g., [2,4,6,8,12–14] and the references therein), with motivation coming especially from the analysis of DNA and RNA structures, where tandem and interspersed repeats as well as hairpin structures play important roles in revealing structural and functional information of the analysed genetic sequence (see, e.g., [2,8,12] and the references therein).

Following [12,13], we analyse gapped repeats uvu or palindromes $u^R vu$ where the length of the gap v is upper bounded by the length of the arm u multiplied by some factor. More precisely, in [13], the authors investigate α-*gapped repeats*: words uvu with $|uv| \leq \alpha|u|$. Similarly, [12] deals with α-*gapped palindromes*, i.e., words $u^R vu$ with $|uv| \leq \alpha|u|$. For $\alpha = 2$, these structures are called long armed repeats (or pairs) and palindromes, respectively; for $\alpha = 1$, they are squares and palindromes of even length, respectively. Intuitively, one is interested in repeats or palindromes whose arms are roughly close one to the other; therefore, the study of α-gapped repeats and palindromes was rather focused on the cases with small α. Here, we address the general case, of searching in a word w α-gapped repeats or palindromes for $\alpha \leq |w|$.

In [12] the authors propose an algorithm that, given a word of length n, finds the set S of all its factors which are maximal α-gapped palindromes (i.e., the arms cannot be extended to the right or to the left) in $\mathcal{O}(\alpha^2 n + |S|)$ time. No upper bound on the possible size of the set S was given in [12], but, following the ideas of [13], we conjecture it is $\mathcal{O}(\alpha^2 n)$.

The algorithms of [2] can be directly used to find in $\mathcal{O}(n \log n + |S|)$ time the set S of all the factors of a word of length n which are maximal α-gapped repeats.

The work of Florin Manea was supported by the DFG grant 596676.

A. Kosowski and I. Walukiewicz (Eds.): FCT 2015, LNCS 9210, pp. 27–40, 2015.
DOI: 10.1007/978-3-319-22177-9_3

Note that all the square factors of a word are maximal α-gapped repeats, with empty gap; thus, there are words for which $|S|$ is $\Omega(n \log n)$ (see, e.g., [3]). In [13] the size of the set of maximal α-gapped repeats with non-empty gap is shown to be $\mathcal{O}(\alpha^2 n)$, and can be computed in $\mathcal{O}(\alpha^2 n)$ time for integer alphabets.

A classical problem for palindromes asks to find the longest palindromic factor of a word [15]. Inspired by this, we address the following problems.

Problem 1. Given a word w and $\alpha \leq |w|$, find the longest word u such that w has a factor uvu with $|uv| \leq \alpha|u|$ (the longest α-gapped repeat of w).

Problem 2. Given a word w and $\alpha \leq |w|$, find the longest word u such that w has a factor $u^R v u$ with $|uv| \leq \alpha|u|$ (the longest α-gapped palindrome of w).

In this paper, we present a solution of Problem 1, working in $\mathcal{O}(\alpha n)$ time, and explain briefly how our algorithms can be adapted to solve Problem 2 within the same time complexity Our approach is more efficient than producing first the list of all maximal α-gapped repeats or palindromes and then returning the one with the longest arms; the algorithms of [13] (for repeats) and [12] (for palindromes), which produce such lists, are slower by an α factor than ours. Our solutions are based on a careful combinatorial analysis (e.g., in each solution we find separately the repeats or palindromes with periodic arms and those with aperiodic arms, respectively) as well as on the usage of several word-processing data-structures (e.g., suffix arrays, dictionary of basic factors). They are essentially different from the approaches of [2] (which is based on an efficient processing of the suffix tree of the input word) and from those in [12–14] (which have as crucial idea the construction and analysis of the LZ-factorisation of the word).

This extended abstract is structured as follows. After giving a series of basic facts regarding combinatorics on words and data structures, we develop the tools we need in our solutions. We then give a solution for Problem 1 and briefly point how it can be adapted to solve Problem 2.

2   Preliminaries

Definitions. The computational model we use to design and analyse our algorithms is the standard unit-cost RAM (Random Access Machine) with logarithmic word size, which is generally used in the analysis of algorithms. In the upcoming algorithmic problems, we assume that the words we process are sequences of integers (called letters, for simplicity). In general, if the input word has length n then we assume its letters are in $\{1, \ldots, n\}$, so each letter fits in a single memory-word. This is a common assumption in stringology (see, e.g., the discussion in [9]). Also, all logarithms appearing here are in base 2; we denote by $\log n$ the value $\lfloor \log_2 n \rfloor$. While we do not assume to be able to compute the value of $\log x$ in constant time, for some $x \leq n$, we note that one can compute in $\mathcal{O}(n)$ time all the values $\log x$ with $x \leq n$; therefore, we assume that in the algorithms addressed by Remark 2, or Lemmas 6 and 7, on an input of length n, we implicitly compute all the values $\log x$ with $x \leq n$.

Let V be a finite alphabet; V^* is the set of all finite words over V. The *length* of a word $w \in V^*$ is denoted by $|w|$. The *empty word* is denoted by λ. A word $u \in V^*$ is a *factor* of $v \in V^*$ if $v = xuy$, for some $x, y \in V^*$; we say that u is a *prefix* of v, if $x = \lambda$, and a *suffix* of v, if $y = \lambda$. We denote by $w[i]$ the symbol at position i in w, and by $w[i..j]$ the factor of w starting at position i and ending at position j, consisting of the catenation of the symbols $w[i], \ldots, w[j]$, where $1 \le i \le j \le n$; we define $w[i..j] = \lambda$ if $i > j$. The powers of a word w are defined recursively by $w^0 = \lambda$ and $w^n = ww^{n-1}$ for $n \ge 1$. If w cannot be expressed as a nontrivial power (i.e., w is not a repetition) of another word, then w is *primitive*. A *period* of a word w over V is a positive integer p such that $w[i] = w[j]$ for all i and j with $i \equiv j \pmod{p}$; if p is a period of w, then w is called p-*periodic*. Let $per(w)$ be the smallest period of w. A word w with $per(w) \le \frac{|w|}{2}$ is called run; a run $w[i..j]$ (so, $p = per(w[i..j]) \le \frac{j-i+1}{2}$) is maximal if and only if it cannot be extended to the left or right to get a word with period p, i.e., $i = 1$ or $w[i-1] \ne w[i+p-1]$, and, $j = n$ or $w[j+1] \ne w[j-p+1]$. In [11] it is shown that the number of maximal runs of a word is linear and their list (with a run $w[i..j]$ represented as the triple $(i, j, per(w[i..j]))$) can be computed in linear time (see [1] for an algorithm constructing all maximal runs of a word, without producing its LZ-factorisation).

We now give the basic definitions of the data structures we use. For a word u, $|u| = n$, over $V \subseteq \{1, \ldots, n\}$ we build in $\mathcal{O}(n)$ time its suffix tree and suffix array, as well as LCP-data structures, allowing us to retrieve in constant time the length of the longest common prefix of any two suffixes $u[i..n]$ and $u[j..n]$ of u, denoted $LCP_u(i, j)$ (the subscript u is omitted when there is no danger of confusion). See, e.g., [8,9], and the references therein.

Note that, given a word w of length n and $\ell < n$ we can use one LCP query to compute the longest prefix w' of w which is ℓ-periodic: $|w'| = \ell + LCP(1, \ell+1)$. In our solution for Problem 2, we construct LCP data structures for the word $v = ww^R$; this takes $\mathcal{O}(|w|)$ time. To check whether $w[i..j]$ occurs at position ℓ in w (respectively, $w[i..j]^R$ occurs at position ℓ in w) we check whether $\ell + (j-i) \le n$ and $LCP(i, \ell) \ge j - i + 1$ (respectively, $LCP(\ell, 2|w| - j + 1) \ge j - i + 1$).

Given a word w, its dictionary of basic factors (introduced in [5]) is a structure that labels the factors of the form $w[i..i+2^k - 1]$ (called basic factors), for $k \ge 0$ and $1 \le i \le n - 2^k + 1$, such that every two equal basic factors get the same label and the label of a basic factor can be retrieved in $\mathcal{O}(1)$ time. The dictionary of basic factors of a word of length n is constructed in $\mathcal{O}(n \log n)$ time.

Preliminary Results and Basic Tools. In the following we introduce our basic tools, of both algorithmic and combinatorial nature.

The first lemma concerns overlapping maximal runs in a word.

Lemma 1. *The overlap of two maximal runs u and u' with the same period (i.e., $per(u)=per(u'))$ is shorter than the period.*

A consequence of this lemma is that no position of w is contained in more than two maximal runs having the same period. Also, for $\ell, p \ge 1$, a factor of length ℓp of w contains at most $\ell - 1$ maximal runs of period p; this holds because the overlap of each two consecutive (when ordered w.r.t. their starting position) runs of that factor is shorter than $p - 1$.

The following lemma can be shown using standard tools.

Lemma 2. *Let w be a word of length n. We can preprocess the word w in $\mathcal{O}(n \log n)$ time to construct data structures that allow us to answer in $\mathcal{O}(\log n)$ time queries asking for the length of the period of the factors of w.*

For a word u, let $lper(u)$ be the lexicographically minimal factor x of u of length equal to $per(u)$; clearly, $lper(u)$ is primitive for all u. For a word w and a factor x of w, let L_x be the list of the maximal runs $v = w[i..j]$ of w such that $lper(v) = x$, ordered with respect to the starting positions of these runs.

Lemma 3. *Let w be a word of length n. We can compute in $\mathcal{O}(n)$ time the lists L_x for all $x \in H_w$, where H_w is the set of the factors x of w with $L_x \neq \emptyset$.*

During the computation of the lists L_x we can also compute for each $v \in L_x$ the position where x occurs firstly in v. Accordingly, each list L_x is stored in a structure where it is identified by the starting position of the first occurrence of x in the first run of L_x and by $|x|$. However, for simplicity, we keep writing L_x.

Remark 1. Assume that w is a word and let v be a factor of w with $per(v) = p \leq \frac{|v|}{2}$. Let z be a factor of length $\ell|v|$ of w. Each occurrence of v in z is part of a maximal run of period p; moreover, if v occurs in a maximal run of period p at the position i of that run, it will also occur at positions $i - p$ and $i + p$, provided that $i - p$ and $i + p + |v| - 1$ fall inside the run, respectively. So, we can represent succinctly the occurrences of v in z by returning succinct representations of the maximal runs u contained in z, of length at least $|v|$, with $lper(v) = lper(u)$; for each of these runs we also store the first occurrence of v in the run. Finally, note that there are at most $2\ell - 1$ such runs contained in z.

Similarly, for a word w of length n, we note that a basic factor $w[i..i + 2^k - 1]$ occurs either at most twice in a factor $w[j..j + 2^{k+1} - 1]$ or its occurrences are part of a run of period $per(w[i..i + 2^k - 1]) \leq 2^{k-1}$ (so, the positions where $w[i..i + 2^k - 1]$ occurs in $w[j..j + 2^{k+1} - 1]$ form an arithmetic progression of ratio $per(w[i..i + 2^k - 1])$, see [10]). Hence, the occurrences of $w[i..i + 2^k - 1]$ in $w[j..j + 2^{k+1} - 1]$ can be represented in a compact manner, just like before. To the same end, for an integer $c \geq 2$, the occurrences of the basic factor $w[i..i + 2^k - 1]$ in $w[j..j + c2^k - 1]$ can be presented in a compact manner: the positions (at most c) of the separate occurrences of $w[i..i + 2^k - 1]$ (that is, occurrences that do not form a run) and/or at most c maximal runs determined by the overlapping occurrences of $w[i..i + 2^k - 1]$. We can show the following result.

Lemma 4. *Given a word w of length n and a number $c \geq 2$, we can preprocess w in time $\mathcal{O}(n \log n)$ such that given any basic factor $y = w[i..i + 2^k - 1]$ and any factor $z = w[j..j + c2^k - 1]$, with $k \geq 0$, we can compute in $\mathcal{O}(\log \log n + c)$ time a succinct representation of all the occurrences of y in z.*

The following results are shown using tools developed in [7].

Lemma 5. *Given a word v, $|v| = \alpha \log n$, we can process v in time $\mathcal{O}(\alpha \log n)$ time such that given any basic factor $y = v[j \cdot 2^k + 1..(j + 1)2^k]$, with $j, k \geq 0$ and $j2^k + 1 > (\alpha - 1) \log n$, we can find in $\mathcal{O}(\alpha)$ time $\mathcal{O}(\alpha)$ bit sets, each storing $\mathcal{O}(\log n)$ bits, characterising all the occurrences of y in v.*

Note that each of the bit sets produced in the above lemma can be stored in a constant number of memory words in our model of computation. Essentially, this lemma states that we can obtain in $\mathcal{O}(\alpha \log n)$ time a representation of size $\mathcal{O}(\alpha)$ of all the occurrences of y in v.

Remark 2. By the previous lemma, given a word v, $|v| = \alpha \log n$, and a basic factor $y = v[j \cdot 2^k + 1..(j + 1)2^k]$, with $j, k \geq 0$ and $j2^k + 1 > (\alpha - 1) \log n$, we can produce $\mathcal{O}(\alpha)$ bit sets, each containing exactly $\mathcal{O}(\log n)$ bits, characterising all the occurrences of y in v. Let us also assume that we have access to all values $\log x$ with $x \leq n$. Now, using the bit-sets encoding the occurrences of y in v and given a factor z of v, $|z| = c|y|$ for some $c \geq 1$, we can obtain in $\mathcal{O}(c)$ time the occurrences of y in z: the positions (at most c) where y occurs outside a run and/or at most c maximal runs containing the occurrences of y. Indeed, the main idea is to select by bitwise operations on the bit-sets encoding the factors of v that overlap z the positions where y occurs (so the positions with an 1). For each two consecutive such occurrences of y we detect whether they are part of a run in v (by LCP-queries on v) and then skip over all the occurrences of y from that run (and the corresponding parts of the bit-sets) before looking again for the 1-bits in the bit-sets.

3   Our Solution

We now give the solution to Problem 1. Our solution has two major steps. In the first one, described in Lemma 6, we compute the longest α-gapped repeat uvu with u periodic, i.e., $per(u) \leq \frac{|u|}{2}$. In the second one, described in Lemma 7, we compute the longest α-gapped repeat uvu with u aperiodic, i.e., $per(u) > \frac{|u|}{2}$. Then, we return the one repeat of these two that has the longest arms.

We first show how to find the longest α-gapped uvu repeat with u periodic.

Lemma 6. *Given a word w of length n and $\alpha \leq n$, the longest α-gapped repeat uvu with u periodic contained in w can be found in $\mathcal{O}(\alpha n)$ time.*

Proof. For the simplicity of the exposure, let $c = \alpha - 1$. We want to find a repeat $u_1 v u_2 = uvu$ with $u_1 = u_2 = u$ periodic and $|v| \leq c|u|$. As u is periodic, both its occurrences u_1 and u_2 must be contained in maximal runs of w having the same $lper$. Accordingly, let y_1 and y_2 denote the maximal runs containing u_1 and u_2; we have, $lper(y_1) = lper(y_2) = x$.

The outline of our algorithm is the following. For each x, we have three cases to analyse. In the first y_1 and y_2 denote the same run, and in the second y_1 and y_2 are overlapping runs with the same period. In both cases, we can easily find the α-gapped repeat $u_1 v u_2$ with u_1 contained in y_1 and u_2 contained in y_2.

The last case occurs when y_1 and y_2 do not overlap. Then, it is harder to find the repeat u_1vu_2 we look for. In this case, we try each possibility for y_2 and, at a very intuitive level, the restriction on the gap suggests that we do not have to look for y_1 among too many runs occurring to the left of y_2, and having the same *lper* as y_2: either they are not long enough to be worth considering as a possible place for the left arm of the longest α-gapped repeat with the right arm in y_2 or the gap between them and y_2 is too large. We exploit this and are able to find efficiently the repeat u_1vu_2 with the longest arms. With care, the analysis of the first two cases takes constant time per run, respectively, per pair of overlapping runs, while in the last case, for each run y_2 we can check whether it is the place of the right arm of the repeat we look for in amortised $\mathcal{O}(c)$ time. This adds up to $\mathcal{O}(cn) = \mathcal{O}(\alpha n)$ time. We now give more details for each case.

In the first case (see Fig. 1), u_1 and u_2 occur in the same run, so $y_1 = y_2 = y$. Let us first assume that in u_1vu_2 we have $v \neq \lambda$. Then u_1 must be a prefix of y and u_2 a suffix of y. Otherwise, we could extend both u_1 and u_2 with at least one symbol to get an α-gapped repeat with longer arm. For instance, when u_1 is not a prefix of y, then we extend both u_1 and u_2 with a letter to the left, the last letter of v becoming now part of the new u_2. This contradicts the fact that u_1vu_2 was the longest α-gapped repeat. Now, if $y = t^\ell t'$ is the maximal run, we get that $u = t^{\ell'}t'$ where $\ell' = \lfloor \frac{\ell-1}{2} \rfloor$. Further, assume that $v = \lambda$, so u_1vu_2 is in fact a square uu. Like before, we have that uu is either the longest square prefix of y or its longest square suffix; these can be immediately computed, and we just return the longest of them. This way, we can compute the longest α-gapped repeat uvu with both occurrences of u contained in the same maximal run y of w.

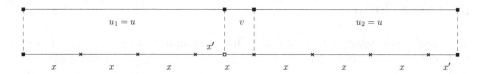

Fig. 1. An α-gapped repeat $u_1vu_2 = uvu$ inside a run $x^k x'$.

In the second case (see Fig. 2), y_1 and y_2 are distinct runs that overlap. By Lemma 1, the length of the overlap between these two runs is at most $|x|$; also, y_1 and y_2 should occur on consecutive positions in the list L_x. We first assume that $v \neq \lambda$. If u_1 is not a prefix of y_1, then u_2 should be a prefix of y_2, or we could extend both u_1 and u_2 to the left and get a longer α-gapped repeat (with a shorter gap v). So assume that u_2 is a prefix of y_2, and let x' be the prefix of length $|x|$ of y_2. It is clear that u_2 is obtained by taking the longest word of period $|x|$ that starts with x' occurring both in y_2 and in y_1, such that it ends in y_1 before the position where y_2 starts. From the computation of L_x we have the first occurrence of x in both y_1 and y_2, so we can also obtain the first occurrence of x' in y_1; consequently, we can also compute the aforementioned longest word. This concludes the case when u_1 is not a prefix of y_1. The case when u_1 is a prefix of y_1 can be treated in a very similar fashion. This way, we get the longest

α-gapped repeat $u_1 v u_2$ with u_1 contained in y_1, u_2 contained in y_2, and $v \neq \lambda$. Further, we consider the case when $v = \lambda$, so $u_1 u_2 = uu$ is a square. If neither u_1 nor u_2 is a prefix of y_1 and y_2, respectively, then we can shift the entire square $u_1 u_2$ to the left with one symbol, to get a new square with the same length of the arm. We can repeat this process until one of u_1 or u_2 becomes a prefix of its corresponding maximal run. Now, just like before, we can determine the longest square uu such that either uu starts on the same position as y_1 or the second u starts on the same position as y_2, and this gives us one of the longest squares uu such that the first u is contained in y_1 and the second in y_2. Hence, we obtained the longest α-gapped repeat $u_1 u_2$ with u_1 contained in y_1 and u_2 in y_2, where y_1 and y_2 overlap.

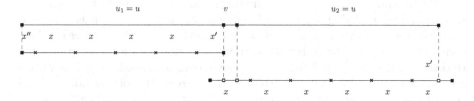

Fig. 2. An α-gapped repeat $u_1 v u_2 = uvu$ with u_1 in $y_1 = x'' x^{k_1} x'$, u_2 in $y_2 = x^{k_2}$, and y_1 and y_2 overlap.

In the final case (see Fig. 3), the runs y_1 and y_2 do not overlap. We first analyse the ways the arms of the longest α-gapped repeat uvu, namely u_1 and u_2, may occur in the runs y_1 and y_2, respectively.

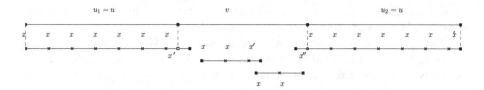

Fig. 3. An α-gapped repeat $u_1 v u_2 = uvu$ with u_1 in $y_1 = x^{k_1}$ and u_2 in $y_2 = x'' x^{k_2} x'$. Between y_1 and y_2 there are some runs with the same $lper$: $y_3 = x^2 x'$, $y_4 = x^2$.

Firstly, assume that y_1 equals some factor of y_2. Then, either $u_1 = y_1 (= u_2)$ or u_1 is a suffix of y_1 and u_2 is a prefix of y_2. Indeed, if we cannot construct an α-gapped repeat $u_1 v u_2$ with $u_1 = u_2 = y_1$ because the gap is too long, then any repeat $u_1 v u_2$ with u_1 prefix of y_1 and u_2 suffix of y_2 is not α-gapped either. Moreover, unless $u_1 = y_1$, then u_1 should be a suffix of y_1 and u_2 a prefix of y_2; otherwise, a longer repeat could be constructed. Finally, if y_1 equals a factor of y_2 but we cannot construct an α-gapped repeat $u_1 v u_2$ as above with $u_1 = u_2 = y_1$, then we will not be able to construct an α-gapped repeat $u_1 v u_2$ with u_1 being the run y_1 and u_2 contained in some run occurring to the right of y_2.

Secondly, the case when y_2 equals a factor of y_1 is similar. Either $u_1 = u_2 = y_2$ or u_1 is a suffix of y_1 and u_2 is a prefix of y_2. Moreover, if y_2 is contained in y_1 but we cannot construct an α-gapped repeat u_1vu_2 as above with $u_1 = u_2 = y_2$, then we will not be able to construct an α-gapped repeat $u_1'vu_2$ with u_2 being part of the run y_2 and $u_1' = y_2$ contained in any run occurring to the left of y_1.

Finally, assume that neither y_1 equals a factor of y_2, nor y_2 a factor of y_1. This means that the difference between the $|y_1|$ and $|y_2|$ is rather small; anyway, smaller than $|x|$. Now, there are two possibilities. The first one is when u_1 is a prefix of y_1 with $|y_1| - |u_1| < |x|$ and u_2 a suffix of y_2 with $|y_2| - |u_2| < |x|$. The second one is when u_1 is a suffix of y_1 with $|y_1| - |u_1| < |x|$ and u_2 a prefix of y_2 with $|y_2| - |u_2| < |x|$. In both cases, we get $\max\{|y_1|, |y_2|\} \leq |u| + |x|$ and $\min\{|y_1|, |y_2|\} \geq |u|$.

Now we can describe how to compute the longest α-gapped repeat u_1vu_2, with u_1 and u_2 occurring in non-overlapping runs. Unlike the previous case, here it may be the case that y_1 and y_2 are not consecutive runs of L_x. So, we consider each run $y_2 \in L_x$ at a time. For the current y_2 we maintain a set M of possible runs that may contain the left arm u_1 of the longest α-gapped repeat. The main point of our solution is that, at each step, the current M (corresponding to the current run y_2) is computed efficiently from the set M corresponding to the run of L_x previously considered in the role of y_2. Then, we use M to find the longest α-gapped repeat we can construct with the right arm in y_2, and, finally, eliminate from M the runs that are useless for the rest of the computation. That is, when we reached y_2, we know already that some of the runs occurring before it are too short to be able to contain the left arm of the longest α-gapped repeat; also, some runs may not be too short to produce the repeat with the longest arms, but are not long enough to ensure that any repeat with its left arm in them and the right arm in one of the runs occurring to the right of y_2 would be α-gapped. These runs should not be considered further in the computation, so they are discarded. We execute this process for all $y_2 \in L_x$.

More precisely, our algorithm runs as follows. We assume that before analysing $y_2 \in L_x$ we found the longest α-gapped repeat u_1vu_2, with both u_1 and u_2 contained in two non-overlapping runs which occur before y_2 in L_x, so we processed them already. Let $\ell = |u_1|$ and $k = \left\lfloor \frac{\ell}{|x|} \right\rfloor$. We now consider the run $y_2 \in L_x$, and assume that $|y_2| > \ell$; clearly, shorter runs could not contain the right arm of a repeat with arms longer than u_1. By the same reason, we do not have to consider any run that contains at most $k - 2$ full occurrences of x as a place for the left arm of the repeat: the length of this arm would be less than $k|x| \leq \ell$. Also, we do not have to consider a run y_1 with at most $k + 1$ occurrences of x as a possible place for the left arm of the repeat unless it is one of the rightmost $6c$ runs occurring in L_x that end before y_2 begins and have at least $k - 1$ occurrences of x. Indeed, if y_1 is not one of these $6c$ runs, then the gap between the end of y_1 and the start of y_2 would be, if $k \geq 3$, longer than $6c(k-2)|x|$, or, if $k = 2$, longer than $6c|x|$. This is greater than $(k+3)c|x| \geq c|y_1|$. Therefore, any repeat with the left arm in such a run y_1 and right arm in the considered run y_2 would not be α-gapped.

So, we only need to consider for some y_2 the rightmost $6c$ runs of L_x which end before y_2 and have at least $k-1$ occurrences of x as well as the runs occurring before (i.e., to the left of) these $6c$ runs, that contain at least $k+2$ occurrences of x. These runs are stored in the structure M. First, we check which is the α-gapped repeat with longest arm that can be constructed with the left arm in one of the aforementioned $6c$ runs and the right one in y_2. Then we try to do the same thing for the other runs of M. Let y_1 be a run with at least $k+2$ occurrences of x.

It might be the case that y_1 equals a factor of y_2. We produce the longest α-gapped repeat with left arm in y_1 and right arm in y_2. In the case when this repeat is $y_1 v y_1$, we store it if its arm y_1 is longer than the arm of the current longest α-gapped repeat, and then discard y_1: we already obtained the longest repeat whose left arm could be in y_1, so it bears no importance for the rest of the computation. We then check the next run that occurs to the left of y_1 and is contained in M, and so on. If we cannot obtain $y_1 v y_1$, we discard y_1 because it will never lead to another α-gapped repeat again, as explained in the discussion on how an α-gapped repeat with arms in non-overlapping runs may be placed inside these runs, and, again, continue our search. If y_2 is contained in y_1, we either obtain the α-gapped repeat $y_2 v y_2$ and stop analysing y_2, as it already produced the longest α-gapped repeat it can ever produce, or we obtain only a shorter α-gapped repeat and we stop analysing y_2 because it will never produce another α-gapped repeat with a run occurring to the left of y_1: the rest of the runs in M are simply to far away from y_2.

If neither y_1 equals a factor of y_2 nor y_2 equals a factor of y_1, we proceed as follows. Let $u_1 v u_2$ be the longest α-gapped repeat with the arms contained in y_1 and y_2, respectively. Assume that $\left\lceil \frac{|u|}{|x|} \right\rceil = m$; then $\max\{|y_1|, |y_2|\} \le (m+1)|x|$ and $\min\{|y_1|, |y_2|\} \ge (m-1)|x|$. It follows that we do not have to consider many more runs to the left of y_1 as a possible place for the left arm of the longest α-gapped repeat with the right arm in y_2. Indeed, it is enough to consider, besides y_1, the next (rightmost) $8c$ runs occurring to the left of y_1 in L_x and having at least $m-2$ occurrences of x. All the other runs we meet while traversing L_x to select these $8c$ runs are discarded: they are too short to produce a repeat with a longer arm than what we already have found. The runs occurring to the left of these $8c$ runs could not produce an α-gapped repeat with the right arm in y_2, as the gap between them and y_2 would too long. Indeed, the gap between any of them and y_2 is at least $8c(m-3)|x| + 6c|x|$, if $m \ge 4$, or $14c|x|$, otherwise; this is, anyway, greater than $c|y_2|$. Therefore, after checking the selected $8c$ runs as a possible place for the left arm of the longest α-gapped repeat, we are sure to have obtained the longest α-gapped repeat with the right arm in y_2.

Further, we have to update M so that it becomes ready to be used when considering a new run of L_x in the role of y_2. Assume the current longest α-gapped repeat has the arm length ℓ' and let $k' = \left\lfloor \frac{\ell'}{|x|} \right\rfloor$. From the runs of M (so, excluding the ones we already discarded), we first store the rightmost $6c$ runs with at least $k'-1$ occurrences of x. Then we also discard from the rest of M all the runs with at most $k'+1$ occurrences of x, occurring to the right of the leftmost run of L_x considered in our search. Once M is cleaned up like this, if the next run we need to consider in the role of y_2 (so, which has length greater than ℓ') is y', we add to M, in the order of their starting positions, the runs with at least $k' - 1$ occurrences of x and ending between the starting position of y_2 and the starting position of y'. These runs are added one by one, such that the set of the rightmost $6c$ runs of M with at least $k' - 1$ occurrences of x is correctly maintained. When a run with at most $k' + 1$ occurrences of x is no longer among the last $6c$ runs of M, we just discard it. We then process the updated M with y' in the role of y_2.

In this way, we compute for each x the longest α-gapped repeat $u_1 v u_2$ with $u_1 = u_2$ periodic, contained in non-overlapping runs with $lper = x$. We do this for all possible values of x, and, alongside the analysis from the cases discussed at the beginning of this proof, we get the longest α-gapped repeat $u_1 v u_2$ with $u_1 = u_2$ periodic.

The complexity of this algorithm is $\mathcal{O}(\alpha n)$. Indeed, in the first case we spend constant time per each run. In the second case we need constant time to analyse each pair of consecutive runs from the list L_x, for each x. This adds up to $\mathcal{O}(n)$ as the number of runs in a word is linear. In the last case, the time needed to process a run $y_2 \in L_x$ is proportional to the number of elements discarded from the structure M during this processing plus the number of elements inserted in M before considering the next run of L_x, to which we add an extra $\mathcal{O}(c) = \mathcal{O}(\alpha)$ processing time. As each run of L_x is added at most once in M, and then removed once, the total number of deletions and insertions we make is $\mathcal{O}(n)$. In total, the analysis of the third case takes, as claimed, $\mathcal{O}(\alpha n)$. This concludes our proof. □

Next we show how to find the longest α-gapped repeats uvu with u aperiodic.

Lemma 7. *Given a word w of length n and an integer $\alpha \leq n$, the longest α-gapped repeat uvu with u aperiodic, contained in w, can be found in $\mathcal{O}(\alpha n)$ time.*

Proof. Here we cannot use the runs structure of the input word to guide our search for the arms of the longest α-gapped repeat. So we need a new approach.

Informally, this new approach works as follows (see also Fig. 4). For each k, we try to find the longest α-gapped repeat $u_1 v u_2 = uvu$, with $u_1 = u_2 = u$ aperiodic, and $2^{k+1} \log n \leq |u| \leq 2^{k+2} \log n$. In each such repeat, the right arm u_2 must contain a factor (called k-block) z, of length $2^k \log n$, starting on a position of the form $j2^k \log n + 1$. So, we try each such factor z, fixing in this way a range of the input word where u_2 could appear. Now, u_1 must also contain a copy of z. However, it is not mandatory that this copy of z occurs nicely aligned to its original occurrence; from our point of view, this means the copy of z does not necessarily occur on a position of the form $i \log n + 1$. But, it is not hard to see that z has a factor y of length $2^{k-1} \log n$, starting in its first $\log n$ positions and whose corresponding occurrence in u_1 should start on a position of the form $i \log n + 1$. Further, we can use the fact that $u_1 v u_2$ is α-gapped and apply Lemma 4 to a suitable encoding of the input word to locate in constant time for each y starting in the first $\log n$ positions of z all possible occurrences of y on a position of the form $i \log n + 1$, occurring not more than $(8\alpha + 2)|y|$ positions to the left of z. Intuitively, each occurrence of y found in this way fixes a range where u_1 might occur in w, such that $u_1 v u_2$ is α-gapped. So, around each such occurrence of y (supposedly, in the range corresponding to u_1) and around the y from the original occurrence of z we try to effectively construct the arms u_1 and u_2, respectively, and see if we get the α-gapped repeat. In the end, we just return the longest repeat we obtained, going through all the possible choices for z and the corresponding y's. We describe in the following an $\mathcal{O}(\alpha n)$ time implementation of this approach.

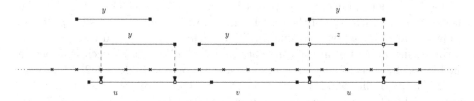

Fig. 4. Segment of w, split into blocks of length $\log n$. In this segment, z is a k-block of length $2^k \log n$. For each factor y, of length $2^{k-1} \log n$, occurring in the first $\log n$ symbols of z (not necessarily a sequence of blocks), we find the occurrences of y that correspond to sequences of 2^{k-1} blocks, and start at most $(8\alpha+2)|y| = (4\alpha+1)\cdot 2^k \log n$ symbols (or, alternatively, $(4\alpha + 1) \cdot 2^k$ blocks) to the left of the considered z. These y factors may appear as runs or as separate occurrences. Some of them can be extended to form an α-gapped repeat $u_1 v u_2 = u v u$ such that the respective occurrence of y has the same offset in u_1 as the initial y in u_2.

The first step of the algorithm is to construct a word w', of length $\frac{n}{\log n}$, whose symbols, called *blocks*, encode $\log n$ consecutive symbols of w grouped together. Basically, now we have two versions of the word w: the original one, and the one where it is split in blocks. Clearly, the blocks can be encoded into numbers between 1 and n in linear time, so we can construct in $\mathcal{O}(n)$ time the suffix arrays and LCP-data structures for both w and w'. We can also build in $\mathcal{O}(n)$ time the data structures of Lemma 4 for the word w'.

Now, we try to find the longest α-gapped repeat $u_1 v u_2 = u v u$ of w, with $u_1 = u_2 = u$ aperiodic, and $2^{k+1} \log n \leq |u| \leq 2^{k+2} \log n$, for each $k \geq 1$ if $\alpha > \log \log n$ or $k \geq \log \log n$, otherwise. Let us consider now one such k. We split again the word w, this time in factors of length $2^k \log n$, called k-blocks. For simplicity, assume that each split is exact.

Clearly, if an α-gapped repeat $u_1 v u_2$ like above exists, then u_2 contains at least one of the k-blocks. Consider such a k-block z and assume it is the leftmost k-block of u_2. On the other hand, u_1 contains at least $2^{k+1} - 1$ consecutive blocks from w', so there should be a factor y of w corresponding to 2^{k-1} of these $(2^{k+1} - 1)$ blocks which is also a factor of z, and starts on one of the first $\log n$ positions of z. Now, for each k-block z and each y, with $|y| = 2^{k-1} \log n$ and starting in its prefix of length $\log n$, we check whether there are occurrences of y in w ending before z that correspond to exactly 2^{k-1} consecutive blocks of

w' (one of them should be the occurrence of y in u_1); note that the occurrence of y in z may not necessarily correspond to a group of 2^{k-1} consecutive blocks, but the one from u_1 should. As u_1vu_2 is α-gapped and $|u_1| \leq 2^{k+2}\log n$, then the occurrence of y from u_1 starts at most $(4\alpha + 1)2^k \log n$ symbols before z (as $|u_1v| \leq \alpha|u_2| \leq \alpha2^{k+2}\log n$, and z occurs with an offset of at most $2^k \log n$ symbols in u_2). So, the block-encoding of the occurrence of the factor y from the left arm u_1 should occur in a factor of $(4\alpha + 1)2^k$ blocks of w', to the left of the blocks corresponding to z.

For the current z and an y as above, we check whether there exists a factor y' of w' whose blocks correspond to y, by binary searching the suffix array of w' (using LCP-queries on w to compare the factors of $\log n$ symbols of y and the blocks of w', at each step of the search). If not, we try another possible y. If yes, using Lemma 4 for w', we retrieve (in $\mathcal{O}(\log\log|w'| + \alpha)$ time) a representation of the occurrences of y' in the range of $(4\alpha + 1)2^k$ blocks of w' occurring before the blocks of z; this range corresponds to a range of length $(4\alpha + 1)2^k \log n$ of w.

If y' is aperiodic then there are only $\mathcal{O}(\alpha)$ such occurrences. Each factor of w corresponding to one of these occurrences might be the occurrence of y from u_1, so we try to extend both the factor corresponding to the respective occurrence of y' from w' and the factor y from z in a similar way to the left and right to see whether we obtain the longest α-gapped repeat. If y' is periodic (so, y is periodic as well), Remark 1 shows that the representation of its occurrences consists of $\mathcal{O}(\alpha)$ separate occurrences and $\mathcal{O}(\alpha)$ runs in which y' occurs. The separate occurrences are treated as above. Each run r' of w' where y' occurs is treated differently, depending on whether its corresponding run r from w (made of the blocks corresponding to r') supposedly starts inside u_1, ends inside u_1, or both starts and ends inside u_1. We can check each of these three cases separately, each time trying to establish a correspondence between r and the run containing the occurrence of y from z, which should also start, end, or both start or end inside u_2, respectively. Then we define u_1 and u_2 as the longest equal factors containing these matching runs on matching positions. Hence, for each separate occurrence of y' or run of such occurrences, we may find an α-gapped repeat in w; we just store the longest. This whole process takes $\mathcal{O}(\alpha)$ time.

If $\alpha > \log\log n$, we run this algorithm for all $k \geq 1$ and find the longest α-gapped repeat uvu, with u aperiodic, and $4\log n \leq |u|$, in $\mathcal{O}(\alpha n)$ time.

If $\alpha \leq \log\log n$, we run this algorithm for all $k \geq \log\log n$ and find the longest α-gapped repeat uvu, with u aperiodic, and $2^{\log\log n+1}\log n \leq |u|$, in $\mathcal{O}(\alpha n)$ time. If our algorithm did not find such a repeat, we should look for α-gapped repeats with shorter arm. Now, $|u|$ is upper bounded by $2^{\log\log n+1}\log n = 2(\log n)^2$, so $|uvu| \leq \ell_0$, for $\ell_0 = \alpha \cdot 2(\log n)^2 + 2(\log n)^2 = (2\alpha + 2)(\log n)^2$. Such an α-gapped repeat uvu is, thus, contained in (at least) one factor of length $2\ell_0$ of w, starting on a position of the form $1 + m\ell_0$ for $m \geq 0$. So, we take the factors $w[1 + m\ell_0..(m + 2)\ell_0]$ of w, for $m \geq 0$, and apply for each such factor, separately, the same strategy as above. The total time needed to do that is $\mathcal{O}\left(\alpha\ell_0\frac{n}{\ell_0}\right) = \mathcal{O}(\alpha n)$. Hence, we found the longest α-gapped repeats uvu, with u aperiodic, and $2^{\log\log(2\ell_0)+1}\log(2\ell_0) \leq |u|$. If our search was still fruitless, we need to search α-gapped repeats with $|u| \leq 2^{\log \log(2\ell_0)+1} \log(2\ell_0) \leq 16 \log n$ (a rough estimation, based on the fact that $\alpha \leq \log \log n$).

So, in both cases, $\alpha > \log \log n$ or $\alpha \leq \log \log n$, it is enough to find the longest α-gapped repeats with $|u| \leq 16 \log n$. The right arm u_2 of such a repeat is contained in a factor $w[m \log n+1..(m+17) \log n]$ of w, while u_1 surely occurs in a factor $x = w[m \log n - 16\alpha \log n+1..(m+17) \log n]$ (or, if $m \log n - 16\alpha \log n+1 \leq 0$, then in a factor $x = w[1..(m+17) \log n]$); in total, there are $\mathcal{O}(n/\log n)$ such x factors. In each of these factors, we look for α-gapped repeats $u_1 v u_2 = uvu$ with $2^{k+1} \leq |u| \leq 2^{k+2}$, where $0 \leq k \leq \log \log n + 2$ (the case $|u| < 2$ is trivial), and u_2 occurs in the suffix of length $17 \log n$ of this factor. Moreover, u_2 contains a factor y of the form $x[j2^k + 1..(j + 1)2^k]$. Using Lemma 5 and Remark 2, for each such possible y occurring in the suffix of length $17 \log n$ of x, we assume it is the one contained in u_2 and we produce in $\mathcal{O}(\alpha)$ time a representation of the $\mathcal{O}(\alpha)$ occurrences of y in the factor of length $(4\alpha + 1)|y|$ preceding y. One of these should be the occurrence of y from u_1. Similarly to the previous cases, we check in $\mathcal{O}(\alpha)$ time which is the longest α-gapped repeat obtained by pairing one of these occurrences to y, and extending them similarly to the left and right. The time needed for this is $\mathcal{O}(\alpha \log n)$ per each of the $\mathcal{O}(\frac{n}{\log n})$ factors x defined above. This adds up to an overall complexity of $\mathcal{O}(\alpha n)$, again.

This was the last case we needed to consider. In conclusion, we can find the longest α-gapped repeat uvu, with u aperiodic, in $\mathcal{O}(\alpha n)$ time.   □

Lemmas 6 and 7 lead to the following theorem.

Theorem 1. *Problem 1 can be solved in $\mathcal{O}(\alpha n)$ time.*

To solve Problem 2 we construct LCP-structures for ww^R (allowing us to test efficiently whether a factor $w[i..j]^R$ occurs at some position ℓ in w) and a mapping connecting the lists L_x, of maximal runs y with $lper(y) = x$, to the lists L'_x of maximal runs y in w^R with $lper(y) = x$. Using the same strategy as in the case of repeats we can solve Problem 2 in $\mathcal{O}(\alpha n)$ time.

We first look for α-gapped palindromes with periodic arm, and then for such palindromes with aperiodic arms. The main difference is that, when looking for α-gapped palindrome $u^R vu$ with u contained in or containing a part of a run from L_x, for some x, we get that u^R is in contained in or contains part of a run from L'_x, respectively.

Basically, when we search $u^R vu$ with the longest periodic u and $|uv| \leq \alpha|u|$, we choose a maximal run y, with $lper(y) = x$, as the possible place for the right arm of the gapped palindrome; then, we only have to check (counting in an amortised setting) the rightmost $\mathcal{O}(\alpha)$ runs of L'_x that end before y as the possible place of u^R. When we search the longest α-gapped palindrome $u^R vu$ with u aperiodic, we split again w in blocks and k-blocks, for each $k \leq \log |w|$, to check in whether there exists such an $u^R vu$ with $2^k \leq |u| \leq 2^{k+1}$. This search is conducted pretty much as in the case of repeats, only that now when we fix some factor y of u, we have to look for the occurrences of y^R in the factor of length $\mathcal{O}(\alpha|y|)$ preceding it; the LCP-structures for ww^R are useful for this. The following results follows.

Theorem 2. *Problem 2 can be solved in $\mathcal{O}(\alpha n)$ time.*

References

1. Bannai, H., Tomohiro, I., Inenaga, S., Nakashima, Y., Takeda, M., Tsuruta, K.: A new characterization of maximal repetitions by lyndon trees. In: Proceedings of the SODA, pp. 562–571 (2015)

2. Brodal, G.S., Lyngsø, R.B., Pedersen, C.N.S., Stoye, J.: Finding maximal pairs with bounded gap. In: Crochemore, M., Paterson, M. (eds.) CPM 1999. LNCS, vol. 1645, pp. 134–149. Springer, Heidelberg (1999)

3. Crochemore, M.: An optimal algorithm for computing the repetitions in a word. Inf. Process. Lett. **12**(5), 244–250 (1981)

4. Crochemore, M., Iliopoulos, C.S., Kubica, M., Rytter, W., Waleń, T.: Efficient algorithms for two extensions of LPF table: the power of suffix arrays. In: van Leeuwen, J., Muscholl, A., Peleg, D., Pokorný, J., Rumpe, B. (eds.) SOFSEM 2010. LNCS, vol. 5901, pp. 296–307. Springer, Heidelberg (2010)

5. Crochemore, M., Rytter, W.: Usefulness of the Karp-Miller-Rosenberg algorithm in parallel computations on strings and arrays. Theoret. Comput. Sci. **88**(1), 59–82 (1991). http://dx.doi.org/10.1016/0304-3975(91)90073-B

6. Crochemore, M., Tischler, G.: Computing longest previous non-overlapping factors. Inf. Process. Lett. **111**(6), 291–295 (2011)

7. Gawrychowski, P.: Pattern matching in Lempel-Ziv compressed strings: fast, simple, and deterministic. In: Demetrescu, C., Halldórsson, M.M. (eds.) ESA 2011. LNCS, vol. 6942, pp. 421–432. Springer, Heidelberg (2011)

8. Gusfield, D.: Algorithms on strings, trees, and sequences: computer science and computational biology. Cambridge University Press, New York (1997)

9. Kärkkäinen, J., Sanders, P., Burkhardt, S.: Linear work suffix array construction. J. ACM **53**, 918–936 (2006)

10. Kociumaka, T., Radoszewski, J., Rytter, W., Waleń, T.: Efficient data structures for the factor periodicity problem. In: Calderón-Benavides, L., González-Caro, C., Chávez, E., Ziviani, N. (eds.) SPIRE 2012. LNCS, vol. 7608, pp. 284–294. Springer, Heidelberg (2012)

11. Kolpakov, R., Kucherov, G.: Finding maximal repetitions in a word in linear time. In: Proceedings of the FOCS, pp. 596–604 (1999)

12. Kolpakov, R., Kucherov, G.: Searching for gapped palindromes. Theor. Comput. Sci. **410**(51), 5365–5373 (2009)

13. Kolpakov, R., Podolskiy, M., Posypkin, M., Khrapov, N.: Searching of gapped repeats and subrepetitions in a word. In: Kulikov, A.S., Kuznetsov, S.O., Pevzner, P. (eds.) CPM 2014. LNCS, vol. 8486, pp. 212–221. Springer, Heidelberg (2014)

14. Kolpakov, R.M., Kucherov, G.: Finding repeats with fixed gap. In: Proceedings of the SPIRE, pp. 162–168 (2000)

15. Manacher, G.K.: A new linear-time on-line algorithm for finding the smallest initial palindrome of a string. J. ACM **22**(3), 346–351 (1975)

On the Enumeration of Permutominoes

Ana Paula Tomás[(✉)]

DCC & CMUP, Faculdade de Ciências, Universidade do Porto, Porto, Portugal
apt@dcc.fc.up.pt

Abstract. Although the exact counting and enumeration of polyominoes remain challenging open problems, several positive results were achieved for special classes of polyominoes. We give an algorithm for direct enumeration of permutominoes [13] by size, or, equivalently, for the enumeration of grid orthogonal polygons [23]. We show how the construction technique allows us to derive a simple characterization of the class of convex permutominoes, which has been extensively investigated [5]. The approach extends to other classes, such as the row convex and the directed convex permutominoes.

1 Introduction

The generation of geometric objects has applications to the experimental evaluation and testing of geometric algorithms. No polynomial time algorithm is known for generating polygons uniformly on a given set of vertices. Some generators employ heuristics [1,7] or restrict to certain classes of polygons, e.g., monotone, convex or star-shaped polygons [22,24]. Numerous related problems have also been extensively investigated, as the exact counting or enumeration of polyominoes [10]. These remain challenging open problems in computational geometry and enumerative combinatorics. A *polyomino* is an edge-connected set of unit squares on a regular square lattice (grid). Polyominoes are defined up to translations. In this paper, we give an algorithm for the enumeration of *permutominoes* by *size*, or, equivalently, for the enumeration of *grid orthogonal polygons* by the number of vertices [23]. Research on permutominoes has focused the enumeration of some subclasses of permutominoes according to the size and the charaterization of pairs of permutations defining various classes of permutominoes [20]. Polyominoes are usually enumerated by area (i.e., number of cells). The direct enumeration of polyominoes is a computational problem of exponential complexity. An overview of the main developments concerning direct and indirect approaches is given in [3]. Jensen's transfer-matrix algorithm [14] – an indirect method – is currently the most powerful algorithm for counting fixed polyominoes. Exact counts are known for polyominoes that have up to 56 cells [3,15]. As far as we can see, Jensen's algorithm cannot be adapted for

Partially supported by CMUP (UID/MAT/00144/2013), which is funded by FCT (Portugal) with national (MEC) and European structural funds through the programs FEDER, under the partnership agreement PT2020.

© Springer International Publishing Switzerland 2015
A. Kosowski and I. Walukiewicz (Eds.): FCT 2015, LNCS 9210, pp. 41–52, 2015.
DOI: 10.1007/978-3-319-22177-9_4

counting permutominoes. Our algorithm for direct enumeration of permutominoes is based on INFLATE-PASTE, a construction technique we developed in [23].

The rest of the paper is organized as follows. Sections 2 and 3 introduce fundamental background, in particular, the INFLATE-PASTE technique. Section 4 describes our enumeration algorithm for generic permutominoes. In Sect. 5, we see how to tailor the approach to count some specific classes (or to create instances in such sets), such as the convex and the row-convex permutominoes. Section 6 concludes the paper.

2   Preliminaries

A polygon is called orthogonal if its edges meet at right angles (of $3\pi/2$ radians at reflex vertices and $\pi/2$ at convex vertices). If r is the number of reflex vertices of an n-vertex orthogonal polygon, then $n = 2r + 4$ (e.g. [19]). The grid orthogonal polygons (*grid ogons*) were introduced in [23] as a relevant class for generation. A *grid ogon* is an orthogonal polygon without collinear edges, embedded in a regular square grid and that has exactly one edge in each line of its minimal bounding square. The grid ogons correspond to the *permutominoes* introduced in [5,9,13]. A permutomino is a polyomino that is given by two suitable permutations of $\{1, 2, \ldots, r + 2\}$, for $r \geq 0$, which define the sequences of the x and y coordinates of its vertices. Its *size* is $r + 1$ and represents the width of its minimal bounding square. We adopt the notion of size given in [9], which is slightly different from [5] (where it is defined as one plus the width of the minimal bounding square). The topological border of a permutomino of size $r + 1$ is a grid ogon with r reflex vertices, and so, it has $2r + 4$ vertices in total. Conversely, the region delimited by a n-vertex grid ogon is a permutomino of size $r + 1$. All polyominoes we consider are simply-connected and, similarly, all polygons are simple and without holes.

3   The Inflate-Paste Technique

The INFLATE-PASTE technique[1] was proposed in [23] for creating n-vertex grid ogons (i.e., permutominoes), at random, as sketched in Fig. 1.

The algorithm yields an n-vertex grid ogon in $O(n^2)$ time. It exploits the fact that every n-vertex grid ogon results from a unit square by applying INFLATE-PASTE $r = (n - 4)/2$ times. INFLATE-PASTE glues a new rectangle to a grid ogon to obtain a new one with 1 more reflex vertex. The rectangle is glued by PASTE to an horizontal edge incident to a convex vertex v, is fixed at v and must be in a region that we called *the free neighbourhood of v* (Fig. 2b). This region is denoted by $FSN(v)$ and consists of the external points that are rectangularly visible from v in the quadrant with origin v that contains the horizontal edge $e_H(v)$ and the inversion of the vertical edge $e_V(v)$, incident to v. Here, the inversion of $e_V(v)$ is its reflection with respect to v. Two points z

---

[1] Demos at http://www.dcc.fc.up.pt/~apt/genpoly.

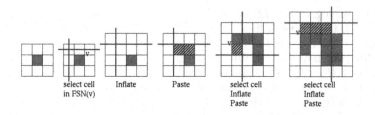

Fig. 1. Using INFLATE-PASTE to create a permutomino with 3 reflex vertices (size 4).

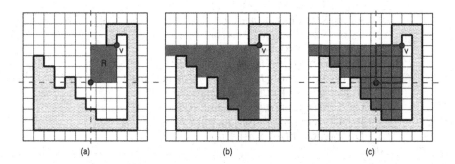

Fig. 2. INFLATE-PASTE: (a) gluing a rectangle to v (b) $FSN(v)$ is the dark shaded region (c) the rectangle is defined by v and the center of a cell of $FSN(v)$.

and w are *rectangularly visible* if the axis-aligned rectangle that has z and w as opposite corners does not intersect the interior of the polygon [18]. At each step, the algorithm selects a convex vertex v and a cell c in $FSN(v)$. The center of c and v define the rectangle that is glued, but the cell is *inflated* first. The INFLATE operation keeps the grid regular: the grid lines are shifted to insert two new lines – one horizontal and one vertical line – for the new edges, that will meet at the center of c. We assume that the starting unit square is inside a 3×3 square, whose boundary contains no edges of P (see Fig. 1). The boundary is kept free along the method. In this way, $FSN(v)$ is always a bounded region and a Ferrers diagram, with origin at v. Hence it can be defined by a sequence of integers, representing the number of cells that form each row of the diagram. Any cell in $FSN(v)$ can be used for growing the polygon using v. Hence, for the instance shown in Fig. 2c), we can make $9 + 7 + 6 + 4 + 4 + 3 + 2 = 35$ distinct grid ogons using the selected vertex v.

4   Direct Enumeration of Permutominoes

ECO was introduced in [2] as a construction paradigm for the *enumeration of combinatorial objects* of a given class, by performing local transformations that increase a certain parameter (the size) of the objects. In this section we propose a direct enumeration procedure for the grid ogons (i.e., permutominoes) using INFLATE-PASTE. The enumeration algorithm is as follows.

```
PermutominoEnum(P,S,G,n₀,r)
    if r = 0 then return fi
    MakeEmptyStack(T)
    while not IsEmpty(S) do
        v := Pop(S)    /* v has coordinates (vₓ, v_y) */
        e_H(v) := the horizontal edge of P that contains v
        if IsConvex(v,P) then
            for C in FreeNeighbourhood(v, P, G) do
                (p, q) := the southwest corner of C
                InflateGrid(p,q,G)
                w₁ := NewVertex(p + 1,v_y)
                w₂ := NewVertex(p + 1,q + 1)
                w₃ := NewVertex(vₓ,q + 1)
                PasteRectangle(v, [w₁, w₂, w₃], P)
                PushConvex([w₁, w₂, w₃],e_H(v),S,P)
                OutputPolygon(P,n₀ + 2)    /* or increment a counter */
                PermutominoEnum(P,S,G,n₀ + 2,r − 1)
                CutRectangle(v, [w₁, w₂, w₃], P)
                PopConvex([w₁, w₂, w₃],e_H(v),S,P)
                DeflateGrid(p + 1,q + 1,G)
            done
        fi
        Push(v,T);
    done
    while not IsEmpty(T) do
        Push(Pop(T),S)
    done
```

Here, P is the initial polygon, G a representation of the grid lines, S a stack that contains the convex vertices of P that are available for expansion, n_0 the number of vertices of P and r the maximum number of reflex vertices of the polygons. PermutominoEnum enumerates recursively all descendants of P that have up to $n_0 + 2r$ vertices. If initially $P := \{(1,1),(2,1),(2,2),(1,2)\}$ (w.r.t. the standard 2D cartesian coordinate system and given in CCW-order), $S := \{(1,2),(2,2)\}$, and $n_0 = 4$, then the algorithm will enumerate (or count) all grid ogons that have up to $2r + 4$ vertices. Nevertheless, in our description of the algorithm in pseudocode, we assume that P is represented by a doubly-linked circular list and that the vertices are linked by pointers to the grid lines that contain them (do not keep their coordinates explicitly). In the same way, the stack S contains pointers to the vertices that are in the stack and the new vertices w_1, w_2 and w_3 keep a similar representation. This means that, in the pseudocode, p, q, $p + 1$, v_y, $q + 1$, v_x refer to the pointers to the corresponding horizontal and vertical grid lines (not simple coordinates). InflateGrid(p,q,G) shifts the vertical grid lines $x > p$, one position to the right and the horizontal grid lines $y > q$ one position upwards, and inserts two new lines $x = p + 1$ and $y = q + 1$. DeflateGrid($p + 1,q + 1,G$) does the reverse operation, removing the lines $x = p + 1$ and $y = q + 1$ from the grid. This implies that the coordinates of the vertices of P change accordingly. A trail-stack T is used to restore the contents of the stack S in the end of the function (we can make a copy of S at start and use it, instead). This is important for exploring the higher branches of the enumeration search tree. The call to PUSHCONVEX puts the new convex vertices on the top of the stack: w_2 and w_3 are convex vertices always and w_1 is convex if it does not lie in the interior of $e_H(v)$. For the correctness of the enumeration algorithm, it does not matter how the algorithm sorts these new vertices to push them onto the stack, but we can adopt a particular order, as we discuss below. After the recursive call to PERMUTOMINOENUM, we restore the polygon, the stack and grid before we proceed to the next cell in $FSN(v)$. This is done by the calls to CUTRECTANGLE, POPCONVEX and DEFLATEGRID. During the enumeration procedure, the bottom horizontal edge cannot move upwards, which is ensured by placing the initial unit square in a 2×3 grid.

The correctness and completeness of the INFLATE-PASTE construction are proved in [23] and follow from the analysis of the horizontal partition of orthogonal polygons (see Fig. 3). The *horizontal partition* of an n-vertex grid ogon P consists of $r + 1$ rectangles and is obtained by adding all horizontal chords incident to the reflex vertices of P. Each face of the horizontal partition, which is a rectangle, gives a node of its *dual graph* and two nodes are connected by an edge in the graph if the two corresponding rectangles are adjacent. Now, our enumeration procedure is based on the existence of a unique depth-first generating tree for each polygon P, once we fix an order for visiting the dual graph of its horizontal partition. One possibility is to define the order as the one induced by a clockwise walk around the polygon, starting at the lowest rectangle, from its SW-vertex. In Fig. 3, we used numbers to indicate the order in which the rectangles (i.e., the nodes of the dual graph) are found.

Figure 4 shows all permutominoes with at most 8 vertices and their construction using the enumeration algorithm. In the figure, we used crosses to indicate convex vertices that can no longer be used for expanding a polygon (to ensure

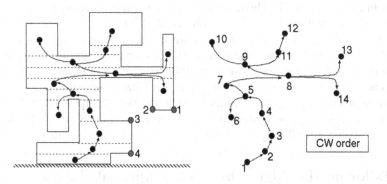

Fig. 3. The horizontal partition of a grid ogon and the unique tree induced by a depth-first visit of its dual graph, if we start from the SW-vertex and walk around the polygon in clockwise order. If we fix this order, only the vertices with labels 1, 2, 3, and 4 remain active for expansion of this instance in our method (with 1 on the top of the stack). The bottom horizontal edge never moves.

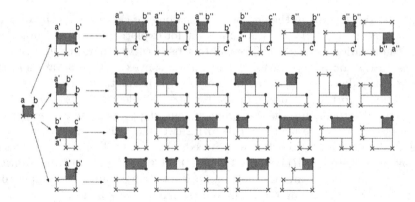

Fig. 4. Permutominoes of size 1, 2, and 3 (i.e., with 4, 6, and 8 vertices) and their horizontal partitions. In each instance, an arrow is attached to the vertex v used in the last step to create the instance (v is no longer a vertex of the polygon).

uniqueness). They will not be in the stack at that stage if the enumeration algorithm proceeds to expand such instance. Since the enumeration is in depth-first, IsConvex checks if a vertex is still active, as Paste can render one vertex reflex.

In contrast to other existing methods for the enumeration of polyominoes, PermutominoEnum, for permutominoes, does not need to keep an exponential number of state configurations in order to count them correctly. Each permutomino is generated exactly once and, hence, there is no need to check for repetitions. Nevertheless, the running time of the algorithm is dominated by the number of permutominoes generated (and thus it is exponential).

By restricting the set of convex vertices that are active for expansion and the notion of free neighbourhood we can design enumerators for particular subclasses, based on the Inflate-Paste construction. In particular, for the convex permutominoes [5,11], the directed convex permutominoes, the row-convex permutominoes, which, by $\pi/2$-rotation, yield the column-convex permutominoes [4], as well as, for the thin, spiral, and min-area permutominoes [17].

Indeed, an algorithm for enumerating the *convex* permutominoes by size was published in [11]. Its running cost is proportional to the number of permutominoes generated. It is quite easy to design a specialized version of our algorithm for enumerating convex permutominoes with identical complexity. Actually, as we will see, for convex permutominoes the free neighbourhoods are linear (rectangles of width 1) and only the two topmost convex vertices can be active.

5   Tailoring the Algorithm to Specific Subclasses

Although the exact counting and enumeration of polyominoes remain challenging open problems, several positive results were achieved for special classes of polyominoes [6,8,16], namely for the class of convex polyominoes and some

of its subfamilies (e.g., directed-convex polyominoes, parallelogram polyominoes, stack polyominoes, and Ferrers diagrams). The larger class of row-convex
(resp. column-convex) polyominoes was considered also [12]. A polyomino is
said to be *row-convex* (resp. *column-convex*) if all its rows (resp. *columns*) are
connected, i.e., the associated orthogonal polygon is y-monotone (x-monotone).
A polyomino is *convex* if it is both row-convex and column-convex (see Fig. 5).

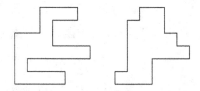

Fig. 5. A row-convex and a convex permutomino.

These classes, which satisfy convexity and/or directness conditions, have been
studied using different approaches and are fairly well characterized, for some
parameters, e.g., area and perimeter [6]. The corresponding classes of permutominoes have been addressed too [4, 5, 9].

5.1 Convex Permutominoes

The analysis of the transformations performed by INFLATE-PASTE during the
application of PERMUTOMINOENUM allow us to derive simple characterizations and exact countings for such classes of permutominoes. Figure 6 shows
all n-vertex convex permutominoes for $n = 4, 6, 8$, each one embedded on a grid.
Only the two topmost convex vertices can be active for INFLATE-PASTE (so,
L and R stand for left or right). Crossed vertices are inactive in the following
transformation steps: "u" means that the vertex would be discarded in PERMU-
TOMINOENUM as well (due to uniqueness conditions) and "c" means that the
resulting permutomino would not be convex. The sequence of $\{0, 1, 2\}^*$ displayed
on the grid top row is the *expansion key* of the corresponding permutomino. Each
element of the key gives the number of active convex vertices that see a certain
grid cell (in Fig. 6, each counter is in its cell). Here, *see* means that the cell
belongs to the free neighborhood of the vertex, already restricted to account
for the convexity condition. For all the remaining empty cells, the counter is 0
and, thus, we omitted it. If we add up the elements of the expansion key of a
given convex permutomino, we get the number of convex permutominoes that
it yields immediately in PERMUTOMINOENUM. In this way, the expansion keys
provide an exact encoding of the structural features that are relevant for counting convex permutominoes according to the number of vertices. Actually, it is
the key as a whole that matters but not the particular cells associated to each
counter. By analysing INFLATE-PASTE in the scope of PERMUTOMINOENUM, we

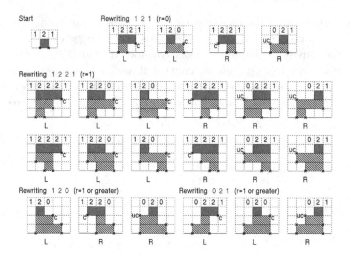

Fig. 6. Enumerating convex permutominoes by size.

may conclude that the *expansion key* of any convex permutomino with $r \geq 0$ reflex vertices must be of one of the following forms:

$$1\,2^{r+1}\,1$$
$$1\,2^j\,0, \quad \text{for } 1 \leq j \leq r-1,$$
$$0\,2^j\,1, \quad \text{for } 1 \leq j \leq r-1, \text{ and}$$
$$0\,2^j\,0, \quad \text{for } 1 \leq j \leq r-2.$$

INFLATE-PASTE operations acting on convex permutominoes can be seen as rewrite rules. Each rule rewrites the key of a convex permutomino with $r - 1$ reflex vertices to the key of one of the convex permutominoes derived from it, having one more reflex vertex, for $r \geq 1$. The rewrite rules are:

$$12^r1 \;\to^{L,R}\; 12^{r+1}1$$
$$12^r1 \;\to^L\; 12^j0, \qquad \text{for } 1 \leq j \leq r$$
$$12^r1 \;\to^R\; 02^j1, \qquad \text{for } 1 \leq j \leq r$$
$$12^{j'}0 \;\to^L\; 12^j0, \qquad \text{for } 1 \leq j \leq j' \leq r-1$$
$$12^{j'}0 \;\to^R\; 12^{j'+1}0, \qquad \text{for } 1 \leq j' \leq r-1$$
$$12^{j'}0 \;\to^R\; 02^j0, \qquad \text{for } 1 \leq j \leq j' \leq r-1$$
$$02^{j'}1 \;\to^R\; 02^j1, \qquad \text{for } 1 \leq j \leq j' \leq r-1$$
$$02^{j'}1 \;\to^L\; 02^{j'+1}1, \qquad \text{for } 1 \leq j' \leq r-1$$
$$02^{j'}1 \;\to^L\; 02^j0, \qquad \text{for } 1 \leq j \leq j' \leq r-1$$
$$02^{j'}0 \;\to^{L,R}\; 02^j0, \qquad \text{for } 1 \leq j \leq j' \leq r-2$$

where L (left) and R (right) identify the topmost vertex selected. For rules with annotation "L, R", both vertices can be selected, one at a time. Figures 7, 8 and 9 illustrate the idea underlying these rules. The correctness and completeness of this rewrite system can be checked easily by case-analysis, taking into account the conditions on convexity.

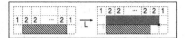 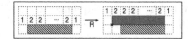

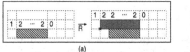 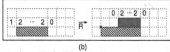

Fig. 7. Rewriting $12^r 1$ using the rewrite rules $12^r 1 \to^L 12^{r+1} 1$ and $12^r 1 \to^R 12^{r+1} 1$.

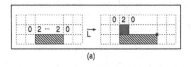 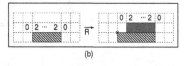

(a)   (b)

Fig. 8. Rewriting $12^{j'} 0$ using (a) $12^{j'} 0 \to^R 12^{j'+1} 0$ and (b) a rule $12^{j'} 0 \to^R 02^j 0$, for $1 \le j \le j'$.

(a)   (b)

Fig. 9. Rewriting $02^{j'} 0$ using (a) $02^{j'} 0 \to^L 02^j 0$ and (b) $02^{j'} 0 \to^R 02^j 0$, for some $1 \le j \le j'$.

Proposition 1. *Let $C_{\alpha,j,\beta}^{(r)}$ be the number of convex permutominoes of the class $\alpha\, 2^j\, \beta$ with r reflex vertices, for $\alpha, \beta \in \{0, 1\}$, $1 \le j \le r+1$ and $r \ge 0$. Then, $C_{0,j,1}^{(r)} = C_{1,j,0}^{(r)}$, for all r and j (symmetry by reflection w.r.t. V-axis) and $C_{\alpha,j,\beta}^{(r)}$ is inductively defined as follows.*

$$C_{1,1,1}^{(0)} = 1$$

$$C_{1,r+1,1}^{(r)} = 2C_{1,r,1}^{(r-1)}, \text{ for } r \ge 1$$

$$C_{1,j,0}^{(r)} = C_{1,r,1}^{(r-1)} + \sum_{j'=\max(1,j-1)}^{r-1} C_{1,j',0}^{(r-1)}, \text{ for } 1 \le j \le r$$

$$C_{0,j,0}^{(r)} = 2\sum_{j'=j}^{r-1} C_{1,j',0}^{(r-1)} + 2\sum_{j'=j}^{r-2} C_{0,j',0}^{(r-1)}, \text{ for } 1 \le j \le r-1$$

The number of convex permutominoes with r reflex vertices (size $r+1$) is given by $C(r) = C_{1,r+1,1}^{(r)} + 2\sum_{j=1}^{r} C_{1,j,0}^{(r)} + \sum_{j=1}^{r-1} C_{0,j,0}^{(r)}$, for $r \ge 0$.

Therefore, using this recurrence, $C(r)$ can be evaluated efficiently using dynamic programming, at least for small values of r, since $C(r)$ grows exponentially. Even for very small values of r, we need to handle big integers (either explicitly or by means of some clever representation). Nevertheless, from [5,9], we know the following closed form for $C(r)$.

$$C(r) = 2(r+4)4^{r-1} - \frac{r+1}{2}\binom{2(r+1)}{r+1}$$

The first terms of the sequence are listed in [21] (ref. A126020): 1, 4, 18, 84, 394, 1836, 8468, 38632, 174426, 780156, ...In a similar way, we can deduce a recurrence for counting the row-convex permutominoes. In both case, the rewrite rules were useful for deducing the recurrences for counting the polygons.

5.2   Row-Convex Permutominoes

The possible forms of the expansion keys of row-convex permutominoes with r reflex vertices are $1^a 2^b 1^c$, with $a + b + c = r + 3$, and $a, b, c \geq 1$ (see Fig. 10).

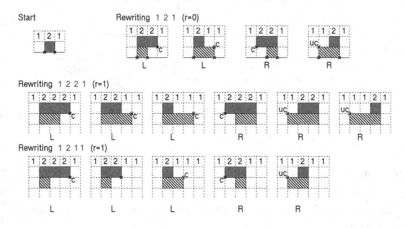

Fig. 10. Creating row-convex permutominoes. To focus on the distinguishing features, for polygons resulting from 1221 and 1211, only the two top rows are shown.

Each rule rewrites the key of a row-convex permutomino with $r - 1$ reflex vertices to the key of one of the row-convex permutominoes derived from it, having one more reflex vertex, for $r \geq 1$. The rewrite rules are:

$$1^a 2^b 1^c \rightarrow^L 1^a 2^{b'} 1^{c'}$$
$$1^c 2^b 1^a \rightarrow^R 1^{c'} 2^{b'} 1^a$$

for all $a, b, c \geq 1$, such that $a + b + c = r + 2$, and for all $b', c' \geq 1$ such that $a + b' + c' = r + 3$.

Let $B_{r,p,k}$ be the number of row-convex permutominoes with r reflex vertices and expansion key $1^p 2^{r+3-(p+k)} 1^k$, for $p, k \geq 1$, $p + k \leq r + 2$. By symmetry, we have $B_{r,p,k} = B_{r,k,p}$. For the recurrence, it interesting to aggregate further. Let $R_{r,p} = \sum_{k=1}^{r+1} B_{r,p,k}$ count the instances whose expansion key starts by 1^p. Then, $R_{0,1} = 1$ and, for $r \geq 1$, and we have

$$R_{r,p} = (r + 2 - p)R_{r-1,p} + \sum_{k=1}^{r+2-p} R_{r-1,k}$$

with $R_{r-1,r+1} = 0$. The first term results from the L-rule and the second from the R-rule (exploiting symmetry). The number of row-convex permutominoes with r vertices is $R(r) = \sum_{p=1}^{r+1} R_{r,p}$, for $r \geq 1$, with $R(0) = 1$. It is not difficult to see that

$$R(r) = 4R(r-1) + 2 \sum_{k=1}^{r-1} (r-k)R_{r-1,k}.$$

Again, we can use these recurrences to compute $R(r)$, by dynamic programming (with big integers), for small values of r. In [4], the authors conjecture that $R(r)$ can be defined asymptotically by $R(r) \sim k(r+2)!h^{r+1}$, with $k = 0.3419111$ and $h = 1.385933$, but the conjecture remains open.

6 Conclusion

In this paper we proposed a direct enumeration algorithm for generic permutominoes, based on the INFLATE-PASTE construction [23]. We developed tailored versions of the method to generate convex and row-convex permutominoes, from which we derived simple recurrences for counting these subclasses. It is worth noting that some of the constructions proposed by other authors for convex and row-convex permutominoes can be seen as instances of the INFLATE-PASTE method.

Acknowledgments. This paper is an extended version of the work presented at the XV Spanish Meeting on Computational Geometry (EGC 2013). The author would like to thank anonymous reviewers for insightful comments.

References

1. Auer, T., Held, M.: Heuristics for the generation of random polygons. In: Proeedings CCCG 1996, pp. 38–43 (1996)
2. Barcucci, E., Del Lungo, A., Pergola, E., Pinzani, R.: ECO: a methodology for the enumeration of combinatorial objects. J. Differ. Equ. Appl. **5**, 435–490 (1999)
3. Barequet, G., Moffie, M.: On the complexity of Jensen's algorithm for counting fixed polyominoes. J. Discrete Algorithms **5**, 348–355 (2007)
4. Beaton, N., Disanto, F., Guttmann, A.J., Rinaldi, S.: On the enumeration of column-convex permutominoes. In: Proceedings of FPSAC 2011, Iceland (2011)
5. Boldi, B., Lonati, V., Radicioni, R., Santini, M.: The number of convex permutominoes. Inf. Comput. **206**, 1074–1083 (2008)
6. Bousquet-Mélou, M.: Bijection of convex polyominoes and equations for enumerating them according to area. Discrete Appl. Math. **48**, 21–43 (1994)
7. Damian, M., Flatland, R., ORourke, J., Ramaswami, S.: Connecting polygonizations via stretches and twangs. Theor. Comp. Syst. **47**, 674–695 (2010)
8. Deutsch, E.: Enumerating symmetric directed convex polyominoes. Discrete Math. **280**, 225–231 (2004)
9. Disanto, F., Frosini, A., Pinzani, R., Rinaldi, S.: A closed formula for the number of convex permutominoes. Electron. J. Combin. **14**, R57 (2007)

10. Golomb, S.: Polyominoes. Princeton U. Press, Princeton (1994)
11. Grazzini, E., Pergola, E., Poneti, M.: On the exhaustive generation of convex per-mutominoes. Pure Math. Appl. **19**, 93–104 (2008)
12. Hickerson, D.: Counting horizontally convex polyominoes. J. Integer Sequences 2, Article 99.1.8 (1999)
13. Insitti, F.: Permutation diagrams, fixed points and Kazhdan-Lusztig R-polynomials. Ann. Comb. **10**, 369–387 (2006)
14. Jensen, I.: Enumerations of lattice animals and trees. J. Stat. Phys. **102**, 865–881 (2001)
15. Jensen, I.: Counting polyominoes: a parallel implementation for cluster computing. In: Sloot, P.M.A., Abramson, D., Bogdanov, A.V., Gorbachev, Y.E., Dongarra, J., Zomaya, A.Y. (eds.) ICCS 2003, Part III. LNCS, vol. 2659, pp. 203–212. Springer, Heidelberg (2003)
16. Del Lungo, A., Duchi, E., Frosini, A., Rinaldi, S.: On the generation and enu-meration of some classes of convex polyominoes. Electron. J. Combin. **11**, R60 (2004)
17. Martins, A.M., Bajuelos, A.: Vertex guards in a subclass of orthogonal polygons. Int. J. Comput. Sci. Netw. Secur. **6**, 102–108 (2006)
18. Overmars, M., Wood, D.: On rectangular visibility. J. Algorithms **9**(3), 372–390 (1998)
19. O'Rourke, J.: An alternate proof of the rectilinear art gallery theorem. J. Geom. **21**, 118–130 (1983)
20. Rinaldi, S., Socci, S.: About half permutations. Electr. J. Comb. **21**(1), P1.35 (2014)
21. Sloane, N.J.A.: The On-Line encyclopedia of integer sequences. OEIS Foundation. http://oeis.org/
22. Sohler, C.: Generating random star-shaped polygons. In: Proceedings CCCG 1999 (1999)
23. Tomás, A.P., Bajuelos, A.: Quadratic-time linear-space algorithms for generating orthogonal polygons with a given number of vertices. In: Laganá, A., Gavrilova, M.L., Kumar, V., Mun, Y., Tan, C.J.K., Gervasi, O. (eds.) ICCSA 2004. LNCS, vol. 3045, pp. 117–126. Springer, Heidelberg (2004). doi:10.1007/978-3-540-24767-8_13
24. Zhu, C., Sundaram, G., Snoeyink, J., Mitchell, J.S.B.: Generating random polygons with given vertices. Comput. Geom. **6**, 277–290 (1996)

Stabbing Segments with Rectilinear Objects

Mercè Claverol[1], Delia Garijo[2], Matias Korman[3,4], Carlos Seara[1],
and Rodrigo I. Silveira[1 (✉)]

[1] Universitat Politècnica de Catalunya, Barcelona, Spain
rodrigo.silveira@upc.edu
[2] Universidad de Sevilla, Seville, Spain
[3] National Institute of Informatics, Tokyo, Japan
[4] JST, ERATO, Kawarabayashi Large Graph Project, Tokyo, Japan

Abstract. We consider *stabbing regions* for a set S of n line segments in
the plane, that is, regions in the plane that contain exactly one endpoint
of each segment of S. Concretely, we provide efficient algorithms for
reporting all combinatorially different stabbing regions for S for regions
that can be described as the intersection of axis-parallel halfplanes; these
are halfplanes, strips, quadrants, 3-sided rectangles, and rectangles. The
running times are $O(n)$ (for the halfplane case), $O(n \log n)$ (for strips,
quadrants, and 3-sided rectangles), and $O(n^2 \log n)$ (for rectangles).

1 Introduction

Let S be a set of n line segments in the plane. We say that a region $\mathcal{R} \subseteq \mathbb{R}^2$ is a
stabbing region for S if $\mathcal{R}$ contains exactly one endpoint of each segment of S; see
Fig. 1(a). Depending on the segment configuration, a stabbing region of a certain
shape may not exist, as shown in Fig. 1(b). Our aim in this paper is to compute
efficiently all stabbing regions $\mathcal{R}$ for a given set of segments S, for regions $\mathcal{R}$ that
can be described as the intersection of axis-parallel (i.e., horizontal or vertical)
halfplanes. Thus, the shapes here considered are halfplanes, strips, quadrants,
3-sided rectangles, and rectangles. This problem fits into the general framework
of *classification or separability problems*, since any stabbing region implicitly
classifies all endpoints of S into two sets: the ones inside $\mathcal{R}$ (including those on its
boundary) are the *red* points, and the ones outside $\mathcal{R}$ are the *blue* points. Thus,
we focus on computing the *combinatorially different* stabbing regions for S,
which are those that provide a different classification of the endpoints of the
segments in S.

Separability and classification problems have been widely investigated and
arise in many diverse problems in computational geometry. In our context, perhaps the simplest stabbing region one can consider is a halfplane, whose boundary is defined by a line, and so it is equivalent to a line that intersects all
segments (thus classifying their endpoints). Edelsbrunner et al. [13] presented
an $O(n \log n)$ time algorithm for solving the problem of constructing a representation of all stabbing lines (with any orientation) of a set S of n line segments.

When no stabbing halfplane exists, it is natural to ask for other types
of stabbing regions. Claverol et al. [10] studied the problem of reporting all

© Springer International Publishing Switzerland 2015
A. Kosowski and I. Walukiewicz (Eds.): FCT 2015, LNCS 9210, pp. 53–64, 2015.
DOI: 10.1007/978-3-319-22177-9_5

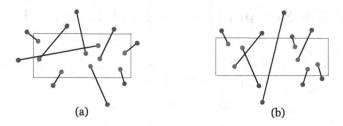

Fig. 1. (a) A set of segments that has a stabbing rectangle. (b) A set of segments for which no stabbing rectangle exists.

combinatorially different stabbing *wedges* (i.e., the stabbing region defined by the intersection of two halfplanes) for a set S of n segments; see also [9]. Their algorithm runs in $O(n^3 \log n)$ time and uses $O(n^2)$ space. They also studied some other stabbers such as double-wedges and zigzags (see [10] for a table comparing the time and space complexities for these stabbers).

Results. Following this line of research, we introduce new shapes of stabbing regions and exploit their geometric structure to obtain efficient algorithms. Concretely, in Sect. 2, we study the case in which the stabbing region is formed by at most two halfplanes, that is, halfplanes, strips, and quadrants. Our approach partitions the plane into three regions: a red one that must be contained in any stabbing region, a blue one that must be avoided by any stabbing region, and a gray region for which we do not have enough information yet. The algorithms are based on iteratively classifying segments and updating the boundaries of these regions, a process that we call *cascading*. The running times obtained are $O(n)$ (for the halfplane case), and $O(n \log n)$ (for strips and quadrants). In Sect. 3 we show that the cascading approach can be extended to 3-sided rectangles, and that the number of combinatorially different solutions is still $O(n)$, resulting in an $O(n \log n)$ algorithm. Finally, we focus on stabbing rectangles in Sect. 4, for which our algorithm runs in $O(n^2 \log n)$ time. This is close to being worst-case optimal, since there can be $\Theta(n^2)$ combinatorially different solutions.

Note that even though we present our algorithms for stabbing regions defined by axis-parallel halfplanes, they extend to any two fixed orientations by making an appropriate affine transformation.

Other Related Work. Díaz-Báñez et al. [12] considered a similar stabbing concept: a region $\mathcal{R}$ stabs a collection of segments if *at least* one endpoint of each segment is in $\mathcal{R}$. In this setting, the endpoints of the segments are not necessarily classified. Moreover, existence of a stabber is always guaranteed (one can always find a large enough region that contains all segments). Thus, all studied problem focus on optimization. For instance, they search for a polygonal stabber with minimum perimeter or area. In their work, they show that the general problem is NP-hard and provide polynomial-time algorithms for some particular cases, like disjoint segments. Other relevant references on stabbing problems that focus on optimization problems in two dimensions are [3,6,7,15,17–20,22]. Variants of these problems have also been studied in three dimensions (e.g. [4,8,14,16,21]).

A more general formulation for the above-mentioned problems is using color-spanning objects. In this case, the input is a set of n colored points, with c colors, and the goal is to find an object (rectangle, circle, etc.) that contains *at least* (or *exactly*) k points of each color. Our setting is the particular case in which $c = n/2$ and we want to contain *exactly* one point of each color class. The color-spanning objects (for the *at least* objective) that have been studied in the literature are strips, axis-parallel rectangles [1,11], and circles [2]. All cases can be solved in roughly $O(n^2 \log c)$ time. Less research has been done for the *exact* objective. Among others, we highlight the research of Barba et al. [5]. In this work they give algorithms that, in $O(n^2 c)$ time, compute disks, squares, and axis-aligned rectangles that contain exactly one element of each of the c color classes. The algorithm that we present in Sect. 4 (axis-aligned rectangle stabber) improves the result of [5] for the particular case in which $c = n/2$ and one looks for an axis-aligned color spanning rectangle. Our algorithm is almost a linear factor faster, and also allows to report all possible solutions (whereas their algorithm only reports one region).

Some Notation. The input consists of a set $S = \{s_1, \ldots, s_n\}$ of n segments. For simplicity, we assume that there is no horizontal or vertical segment in S, and that all segments have non-zero length. The modifications needed to make our algorithms handle these special cases are straightforward, albeit rather tedious. For any $1 \le i \le n$, let p_i and q_i denote the upper and lower endpoint of s_i, respectively. Given a point $p \in \mathbb{R}^2$, we write $x(p)$ (resp. $y(p)$) for its x- (resp. y-) coordinate. Let $y_b = \max_{s_i \in S}\{y(q_i)\}$ and $y_t = \min_{s_i \in S}\{y(p_i)\}$; these values correspond to the y-coordinates of the highest bottom endpoint and the lowest top endpoint, respectively, of the segments in S. The segments attaining those values are denoted by s_b and s_t, respectively (i.e., $y(q_b) = y_b$ and $y(p_t) = y_t$).

2   Stabbing with One or Two Halfplanes

This section deals with stabbing regions that can be described as the intersection of at most two halfplanes. That is, our aim is to obtain a halfplane, strip, or quadrant that contains exactly one endpoint of each segment of S. Note that such stabbing objects do not always exist.

2.1   Stabbing Halfplane

For completeness (since it will be used in the upcoming sections), we explain a straightforward algorithm for determining if a horizontal stabbing halfplane exists. That is, a horizontal line such that one of the (closed) halfplanes defined by the line contains exactly one endpoint of each segment. Observe that such a stabbing halfplane can be *perturbed* so that it has no endpoint of a segment on its boundary. In this case, the complement of a stabbing halfplane is also a stabber. Thus, we are effectively looking for a horizontal line that intersects the interior of all segments. As we are dealing with horizontal stabbers, the problem becomes essentially one-dimensional, and it will ease our presentation to state the problem in that way.

All segments can be projected onto the y-axis, becoming intervals. Considering the set $\overline{S} = \{\overline{s}_1, \ldots, \overline{s}_n\}$ of projected segments, the question is simply whether all the intervals in $\overline{S}$ have a point in common. Clearly, any horizontal line $y := u$ stabbing S must have its y-coordinate between the values y_b and y_t, namely $y_b \leq u \leq y_t$. Therefore, such a line exists if and only if $y_b \leq y_t$. Moreover, whenever this condition happens, both the upper and lower halfplanes will be stabbing halfplanes (and any other horizontal halfplane will be equivalent to one of the two). This simple observation directly leads to a linear-time algorithm.

Observation 1. *All axis-aligned stabbing halfplanes of a set of n segments can be found in $O(n)$ time.*

2.2 Stabbing Strips

We now consider the case in which the stabbing region is a horizontal strip. Note that the existence of stabbing halfplanes directly implies the existence of stabbing strips, but the reverse is not true. Thus, our aim is to compute all *non-trivial* stabbing strips; intuitively speaking, we are not interested in stabbing strips that can be extended to stabbing halfplanes. More formally, we say that a stabbing region $\mathcal{R}$ is *non-trivial* if there is no stabbing region described as the intersection of fewer halfplanes that is combinatorially equivalent to $\mathcal{R}$ (i.e., gives the same classification of the endpoints of the segments in S).

As in Sect. 2.1, we can ignore the x-coordinates of the endpoints, project the points onto the y-axis, and work with the set $\overline{S}$ instead. The endpoints of the classified segments can be seen in the projection onto the y-axis as a set of blue and red points. It follows that there is a separating horizontal strip for them if and only if the red points appear contiguously on the y-axis. More precisely, the points must appear on the y-axis in three contiguous groups, from top to bottom, first a blue group, then a red group, and then another blue group. We refer to the two groups of blue points as the *top* and *bottom* blue points, respectively. We denote the intervals of the y-axis spanned by them by B_t and B_b, respectively. The interval of the y-axis spanned by the group of red points is denoted by R. Since all points above B_t and below B_b must be necessarily blue, we extend B_t and B_b from $+\infty$ and until $-\infty$, respectively. Thus, the y-axis is partitioned into three colored intervals and two uncolored (gray) intervals separating them. See Fig. 2(a).

Observation 2. *Let s_i and s_j be two segments in S such that $\overline{s}_i$ is above $\overline{s}_j$ (in particular, this implies that the projected segments $\overline{s}_i$ and $\overline{s}_j$ are disjoint). Then, any horizontal stabbing strip must contain q_i and p_j.*

Lemma 1. *Any non-trivial horizontal stabbing strip for S contains points q_b and p_t.*

With the previous observations in place, we can present our algorithm. We give an intuitive idea in rather general terms because it will also be used in the upcoming sections. Our algorithm starts by classifying a few segments of S using some geometric observations (say, Lemma 1). As soon as some points are classified, we partition the plane into three regions: the *red* region (a portion of the plane that must be contained by any stabber), the *blue* region (a portion of the plane that cannot be contained in any stabber) and the *gray* region (the complement of the union of the two other regions; that is, regions of the plane for which we still do not know). If a segment of S has an endpoint in either the blue or red regions we can classify it (that is, if an endpoint is in a red region, that endpoint must be red, and its opposite endpoint blue). This coloring may enlarge either the red or blue regions, which may further allow us to classify other segments of S, and so on. We call this process the *cascading procedure*. This approach will continue until either we find a contradiction (say, the red and blue regions overlap), or we have classified all segments of S (and thus we have found a stabber). Thus, at any instant of time we partition S into three sets C, W, and U. A segment is in C if it has been already classified, in W if it is waiting for being classified (there is enough information to classify it, but it has not been done yet), or in U if its classification is still unknown. The algorithm is initialized with $C = W = \emptyset$, and $U = S$.

Regions. In addition to the three sets of segments, the algorithm maintains red and blue candidate regions that are guaranteed to be contained or avoided in any solution, respectively. When we are looking for a horizontal strip, these regions will also be horizontal strips. Thus, it suffices to maintain the projection of the regions on the y-axis. The blue and red regions are represented by the intervals B_t, B_b and R: the *blue region* is $B_t \cup B_b$, and the *red region* is R. The complement of $B_t \cup B_b \cup R$ is called the *gray region*. Note that the gray region, like the blue one, consists of two disjoint components. During the execution of the algorithm the regions will be updated as new segments become classified. See Fig. 2(a).

The algorithm starts by computing s_b and s_t. If a stabbing halfplane exists, one can find a stabbing strip and report it. Otherwise, by Lemma 1, we know how to classify both segments (i.e., q_b and p_t are classified as red, and p_b and q_t as blue). Thus, we move them from U to W. See Fig. 2.

Cascading Procedure. The procedure iteratively classifies segments in W based on the red and blue regions. This is an iterative process in the sense that the classification of one segment can make the blue or red region grow, making other segments move from U to W.

As long as W is not empty, we pick any segment $s \in W$, assign the corresponding colors to its endpoints, and move s from W to C. If a newly assigned endpoint lies outside its corresponding zone, the red or blue area must grow to contain that point. Note that after the red or blue region grows, other segments can change from U to W. The process continues classifying segments of W until either: (i) a contradiction is found (the red region is forced to overlap with the blue region), or (ii) set W becomes empty.

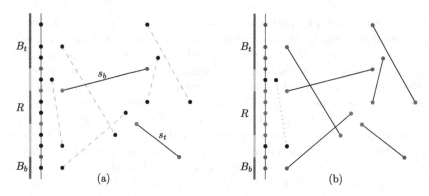

Fig. 2. Computing a stabbing horizontal strip. (a) Result after classifying s_t and s_b. (b) Result after cascading; the red region has grown, and only one segment remains unclassified. In both figures, segments of U are shown dotted, those of W are dashed, and those of C are depicted with a solid line.

Lemma 2. *If the cascading procedure finds no contradiction, each remaining segment in U has an endpoint in each of the connected components of the gray region.*

Proof. By definition, when cascading procedure finishes, all segments still in U must have both endpoints in the gray region. Assume, for the sake of contradiction, that there exists a segment $\bar{s}_i \in U$ whose both endpoints lie in the same gray component, say, the lower one. Recall that, by construction, the red region contains the interval $[y_b, y_t]$. In particular, we have $y(p_i) < y_t$, giving a contradiction with the definition of y_t.   □

Corollary 1. *A horizontal stabbing strip exists for S if and only if the cascading procedure finishes without finding a contradiction.*

Theorem 1. *Determining whether a horizontal stabbing strip exists for a set of n segments can be done in $O(n \log n)$ time and $O(n)$ space.*

Reporting All Horizontal Stabbing Strips. The above algorithm can be modified to report all combinatorially different horizontal stabbing strips without an increase in the running time.

Once the cascading procedure has finished, we define τ as the index of the segment of U whose upper endpoint is lowest (i.e., for any index i such that $s_i \in U$, it holds that $y(p_i) \geq y(p_\tau)$). Likewise, we define β as the index whose lower endpoint is highest. Any stabbing strip will either contain: (*i*) all points in the upper component of the gray region, (*ii*) all points in the lower component of the gray region, or (*iii*) both p_τ and q_β. The first two can be reported in constant time, and for the third case, it suffices to classify the two segments, cascade, and repeat the previous steps.

Theorem 2. *All the combinatorially different horizontal stabbing strips of a set of n segments can be computed in $O(n \log n)$ time.*

2.3 Stabbing Quadrants

We now extend this approach for stabbing quadrants. There are four types of quadrants; without loss of generality, we concentrate on the bottom-right type. Thus throughout this section, the term *quadrant* refers to a bottom-right one. Other types can be handled analogously.

For a segment $s = (p, q)$, let $Q(s)$ denote its *bottom-right quadrant*; that is, the quadrant with apex at $(\max\{x(p), x(q)\}, \min\{y(p), y(q)\})$. See Fig. 3(a).

Observation 3. *Any quadrant classifying a segment $s \in S$ must contain $Q(s)$.*

Given the segment set S, the *bottom-right quadrant* of S, denoted by $Q(S)$, is the (inclusion-wise) smallest quadrant that contains $\cup_{s \in S} Q(s)$; see Fig. 3(b). We now need the equivalent of Lemma 1 to create the initial partition of the plane into red, blue, and gray regions.

Corollary 2. *Any stabbing quadrant of S must contain $Q(S)$.*

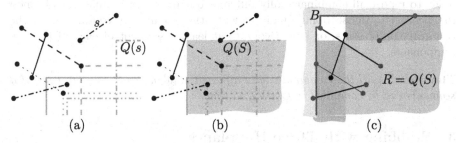

(a)   (b)   (c)

Fig. 3. (a) A set S of five segments and their individual bottom-right quadrants. (b) Bottom-right quadrant $Q(S)$ of the set of segments S. (c) Initial classification given by $Q(S)$, and the red, blue, and gray regions. Note that there is a region (white in the figure) that cannot contain endpoints of S.

Regions. Partition is as follows: the red region R is always defined as the inclusionwise smallest quadrant that contains all classified points and $Q(S)$. At any point in the execution, let $a = (x_R, y_R)$ denote the apex of R. Any blue point b of a classified segment forbids the stabber to include b or any point above and to the left of b (i.e., in the top-left quadrant of b). Moreover, if b satisfies $y(b) \leq y_R$ or $x(b) \geq x_R$, a whole halfplane will be forbidden. Thus, the union of such regions is bounded by a staircase polygonal line (see Fig. 3(c)). Initially, we take the blue region B defined by a point at $(-\infty, \infty)$. We say that a point $p = (x, y)$ is in the gray region if it is not in the red or blue region, and satisfies either $x > x_R$ or $y < y_R$ (see Fig. 3(c)). As in Sect. 2.2, observe that the gray region is the union of two connected components (which we call *right*, and *down* components). Note that, in this case, there is a region, which we call *white*, that is not contained in either the red, blue, or gray regions. However, our first observation is that no endpoint of an unclassified segment can lie in the white region.

Observation 4. *A segment* $s \in S$ *containing an endpoint in the white region must contain its other endpoint in the red region.*

We initially classify all segments that have one endpoint inside $Q(S)$ in W (and all the rest in U). As before, we apply the cascading procedure until a contradiction is found (in which case we conclude that a stabbing quadrant does not exist), or set W eventually becomes empty. In this case we have a very strong characterization of the remaining unclassified segments.

Lemma 3. *If the cascading procedure finishes without finding a contradiction, each remaining segment in* U *has an endpoint in each of the gray components.*

Thus, if no contradiction is found we can extend R until it contains one of the two gray components to obtain a stabber. This can be done because, by construction, each of the connected components of the gray region forms a bottomless rectangle (i.e., the intersection of three axis aligned halfplanes) that shares a corner with the apex of R. As in the strip case, we can use this approach to report all combinatorially different quadrants in an analogous fashion to Theorem 1: any stabbing quadrant will either completely contain one of the two gray components, or it will contain at least a segment of each of the two components.

Theorem 3. *All the combinatorially different stabbing quadrants of a set of* n *segments can be computed in* $O(n \log n)$ *time.*

3   Stabbing with Three Halfplanes

We now consider the case in which the stabber is defined as the intersection of three halfplanes. As in the quadrant case, it suffices to consider those of fixed orientation. Thus, throughout this section, a 3-*rectangle* refers to a rectangle that is missing the lower boundary edge, and that extends infinitely towards the negative y-axis (also called *bottomless rectangle*). Without loss of generality, we may assume that S cannot be stabbed with a halfplane, strip, or quadrant (since any of those regions can be transformed into a bottomless rectangle). As in the previous cases, we solve the problem by partitioning the plane into red-blue-gray regions. However, to generate all stabbing 3-rectangles we need a more involved sweeping phase that is combined with further cascading iterations.

3.1   Number of Different Stabbing 3-Rectangles

First we analyze the number of combinatorially different stabbing 3-rectangles that a set of n segments may have. This analysis will lead to an efficient algorithm to compute them. We start by defining a region that must be included in any stabbing 3-rectangle. Recall that s_b is the segment of S with highest bottom endpoint, and q_b is its bottom endpoint. Analogously, we define s_r and s_ℓ as the segments with leftmost right endpoint and rightmost left endpoint, respectively.

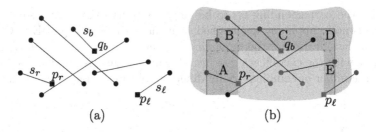

Fig. 4. (a) Example with points q_b, p_r, and p_ℓ highlighted as squares. (b) Red and blue regions after the initial cascading procedure has finished, and partition of the gray region into subregions A, B, C, D, E (Color figure online).

Their corresponding right and left endpoints are denoted by p_r and p_ℓ. See Fig. 4. Finally, we define lines L_ℓ, L_r as the vertical lines passing through p_ℓ and p_r, respectively; analogously, L_b is the horizontal line passing through q_b.

Lemma 4. *Any non-trivial stabbing 3-rectangle $\mathcal{R}$ must contain the intersection points of lines L_ℓ, L_r and L_b.*

The above result allows us to initialize the red region and start the usual cascading procedure. If no contradictions are found, the classified segments define a red and a blue region which must be avoided and included in any solution, respectively. Specifically, the red region is the inclusionwise smallest 3-rectangle that contains all points classified as red. The blue region is the set of points whose inclusion would force a 3-rectangle to contain some point that has already been classified as blue. Finally, the area that is neither red nor blue is called *gray region*. Once the cascading procedure has finished, all remaining unclassified segments must have both of their endpoints in the gray region. It will be convenient to distinguish between different parts of the gray region, depending on their position with respect to the red region. We differentiate between five regions, named A, B, C, D, E, as depicted in Fig. 4(b). These five regions are obtained by drawing horizontal and vertical lines through the two corners of the red region. We say that the *type* of a segment s is XY, for X,Y $\in \{A, B, C, D, E\}$ if s has one endpoint in region X and the other endpoint in region Y. The next lemma shows that, after a cascading, there are only a few possible types.

Lemma 5. *Any unclassified segment after the cascading procedure is of type AC, AD, AE, BE, or CE.*

Now, consider the red and blue regions obtained after a cascading. Let $p_1, \ldots, p_k$ be the endpoints of segments of S inside region A. In addition, we define p_0 as the blue point defining the blue boundary of region A, and p_{k+1} as the red point defining the red boundary of region A. See Fig. 5(a). Let G_i, for $1 \leq i \leq k+1$, be the gray region obtained after classifying points $p_1, \ldots, p_{i-1}$ as blue, $p_i, \ldots, p_k$ as red, and performing a cascading procedure (see Fig. 5(b-d)). If any of those cascading operations results in a contradiction being found, we simply set the corresponding G_i to be empty. Observe that, if for some i, the region G_i is not empty, any unclassified segment must be of type BE or CE. We say that a classification of the remaining segments is *compatible with* G_i if it is compatible with the classification of just classified segments. Next we bound the maximum number of combinatorially different solutions compatible with G_i.

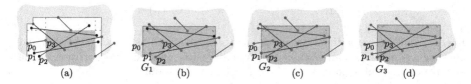

Fig. 5. (a) Initial situation; region A contains points $\{p_1, p_2\}$. (b)-(d): sequence of gray regions G_1, G_2, G_3 (Color figure online).

Lemma 6. *Let G_i be defined as above for some $1 \leq i \leq k+1$ such that $G_i \neq \emptyset$, and let n_i be the number of unclassified segments in G_i, i.e., segments with both endpoints in G_i. Then there are at most n_i combinatorially different solutions compatible with G_i.*

Lemma 7. *For any set S of n segments there are $O(n)$ combinatorially different stabbing 3-rectangles.*

3.2   Algorithm

The previous results give rise to a natural algorithm to generate all combinatorially different stabbing 3-rectangles. The algorithm has two phases:

Initial Cascading. We initialize the red region with Lemma 4 and launch the usual cascading procedure. If this cascading finishes without finding a contradiction, we obtain a red and a blue region that must be included and avoided by any stabbing 3-rectangle, respectively.

Plane Sweep of Region A. We sweep the points in region A from left to right. In the i-th step of the sweep, we classify points $p_1, \ldots, p_{i-1}$ as blue, and points $p_i, \ldots, p_k$ as red. After each such step, we perform a cascading procedure. If the cascading gives no contradiction, we are left with a gray region G_i and a number of unclassified segments that must be of type BE or CE. Then we sweep the endpoints of the unclassified segments in region E from left to right (we call this the *secondary* sweep). At each step of the sweep, we fix those to the left of the sweep line as red, and those to the right of the sweep line as blue, and perform a cascading procedure. From the proof of Lemma 6, we know that each step of this second sweep, after the corresponding cascading procedure, can produce at most one different solution.

Theorem 4. *All combinatorially different stabbing 3-rectangles of a set of n segments can be computed in $O(n \log n)$ time.*

4   Stabbing Rectangles

Applying the cascading approach of the previous sections to rectangles results in a rather involved case distinction, since segments with both endpoints in gray regions can have many different interdependences. Instead, we can use a simple approach based on the algorithm for 3-rectangles, which results in a running time that is close to optimal in the worst case (since it is easy to see that there can be $\Theta(n^2)$ combinatorially different stabbing rectangles).

Any inclusionwise smallest stabbing rectangle R must contain one endpoint on each side, and in particular, an endpoint v of a segment in S must be on its lower boundary segment (otherwise, we could shrink it further). The key observation is that if we fix v (or equivalently, the lower side of a candidate stabbing rectangle), then we can reduce the problem to that of finding a 3-sided rectangle. In particular, by fixing the lower side of the rectangle we are forcing all points below it to be blue, and the point through the fixed side to be red. After a successful cascading procedure, we end up with a certain initial classification that can be completed to a solution to the rectangle if and only if a compatible stabbing 3-sided rectangle exists. Since there are $2n$ candidates for v (each of the endpoints of the segments of S), we have $O(n)$ different instances that are solved independently using Theorem 4.

Theorem 5. *All the combinatorially different stabbing rectangles of a given set S of n segments can be computed in $O(n^2 \log n)$ time.*

Acknowledgments. M. C., C. S., and R.S were partially supported by projects MINECO MTM2012-30951 and Gen. Cat. DGR2014SGR46. D. G. was supported by project PAI FQM-0164. R.S. was partially funded by the Ramón y Cajal program (MINECO).

References

1. Abellanas, M., Hurtado, F., Icking, C., Klein, R., Langetepe, E., Ma, L., Palop, B., Sacristán, V.: Smallest color-spanning objects. In: Meyer auf der Heide, F. (ed.) ESA 2001. LNCS, vol. 2161, pp. 278–292. Springer, Heidelberg (2001)
2. Abellanas, M., Hurtado, F., Icking, C., Klein, R., Langetepe, E., Ma, L., Palop, B., Sacristán, V.: The farthest color Voronoi diagram and related problems. In: Proceedings of the 17th European Workshop on Computational Geometry, pp. 113–116 (2001)
3. Atallah, M., Bajaj, C.: Efficient algorithms for common transversal. Inf. Process. Lett. **25**, 87–91 (1987)
4. Avis, D., Wenger, R.: Polyhedral line transversals in space. Discrete Comput. Geom. **3**, 257–265 (1988)
5. Barba, L., Durocher, S., Fraser, R., Hurtado, F., Mehrabi, S., Mondal, D., Morrison, J., Skala, M., Wahid, M.A.: On k-enclosing objects in a coloured point set. In: Proc. of the 25th Canadian Conference on Computational Geometry, pp. 229–234 (2013)

6. Bhattacharya, B.K., Czyzowicz, J., Egyed, P., Toussaint, G., Stojmenovic, I., Urrutia, J.: Computing shortest transversals of sets. In: Proceedings of the 7th Annual Symposium on Computational Geometry, pp. 71–80 (1991)
7. Bhattacharya, B., Kumar, C., Mukhopadhyay, A.: Computing an area-optimal convex polygonal stabber of a set of parallel line segments. In: Proceedings of the 5th Canadian Conference on Computational Geometry, pp. 169–174 (1993)
8. Brönnimann, H., Everett, H., Lazard, S., Sottile, F., Whitesides, S.: Transversals to line segments in three-dimensional space. Discrete Comput. Geom. **34**, 381–390 (2005)
9. Claverol, M.: Problemas geométricos en morfología computacional. Ph.D. thesis, Universitat Politècnica de Catalunya (2004)
10. Claverol, M., Garijo, D., Grima, C.I., Márquez, A., Seara, C.: Stabbers of line segments in the plane. Comput. Geom. Theor. Appl. **44**(5), 303–318 (2011)
11. Das, S., Goswami, P.P., Nandy, S.C.: Smallest color-spanning objects revisited. Int. J. Comput. Geom. Appl. **19**(5), 457–478 (2009)
12. Díaz-Báñez, J.M., Korman, M., Pérez-Lantero, P., Pilz, A., Seara, C., Silveira, R.I.: New results on stabbing segments with a polygon. Comput. Geom. Theor. Appl. **48**(1), 14–29 (2015)
13. Edelsbrunner, H., Maurer, H.A., Preparata, F.P., Rosenberg, A.L., Welzl, E., Wood, D.: Stabbing line segments. BIT **22**, 274–281 (1982)
14. Fogel, E., Hemmer, M., Porat, A., Halperin, D.: Lines through segments in three dimensional space. In: Proceedings of the 29th European Workshop on Computational Geometry, Assisi, Italy, pp. 113–116 (2012)
15. Goodrich, M.T., Snoeyink, J.S.: Stabbing parallel segments with a convex polygon. In: Proceedings 1st Workshop Algorithms and Data Structures, pp. 231–242 (1989)
16. Kaplan, H., Rubin, N., Sharir, M.: Line transversal of convex polyhedra in $\mathbb{R}^3$. SIAM J. Comput. **39**(7), 3283–3310 (2010)
17. Lyons, K.A., Meijer, H., Rappaport, D.: Minimum polygon stabbers of isothetic line segments. Dept. of Computing and Information Science, Queen's University, Canada (1990)
18. Mukhopadhyay, A., Kumar, C., Greene, E., Bhattacharya, B.: On intersecting a set of parallel line segments with a convex polygon of minimum area. Inf. Process. Lett. **105**, 58–64 (2008)
19. Mukhopadhyay, A., Greene, E., Rao, S.V.: On intersecting a set of isothetic line segments with a convex polygon of minimum area. Int. J. Comput. Geom. Appl. **19**(6), 557–577 (2009)
20. O'Rourke, J.: An on-line algorithm for fitting straight lines between data ranges. Commun. ACM **24**, 574–578 (1981)
21. Pellegrini, M.: Lower bounds on stabbing lines in 3-space. Comput. Geom. Theory Appl. **3**, 53–58 (1993)
22. Rappaport, D.: Minimum polygon transversals of line segments. Int. J. Comput. Geom. Appl. **5**(3), 243–256 (1995)

β-skeletons for a Set of Line Segments in R^2

Mirosław Kowaluk and Gabriela Majewska[✉]

Institute of Informatics, University of Warsaw, Warsaw, Poland
kowaluk@mimuw.edu.pl, gm248309@students.mimuw.edu.pl

Abstract. β-skeletons are well-known neighborhood graphs for a set of points. We extend this notion to sets of line segments in the Euclidean plane and present algorithms computing such skeletons for the entire range of β values. The main reason of such extension is the possibility to study β-skeletons for points moving along given line segments. We show, that relations between β-skeletons for $\beta > 1$, 1-skeleton (Gabriel Graph), and the Delaunay triangulation for sets of points hold also for sets of segments. We present algorithms for computing circle-based and lune-based β-skeletons. We describe an algorithm that for $\beta \geq 1$ computes the β-skeleton for a set S of n segments in the Euclidean plane in $O(n^2\alpha(n)\log n)$ time in the circle-based case and in $O(n^2\lambda_4(n))$ in the lune-based one, where the construction relies on the Delaunay triangulation for S, α is a functional inverse of Ackermann function and $\lambda_4(n)$ denotes the maximum possible length of a $(n, 4)$ Davenport-Schinzel sequence. When $0 < \beta < 1$, the β-skeleton can be constructed in a $O(n^3\lambda_4(n))$ time. In the special case of $\beta = 1$, which is a generalization of Gabriel Graph, the construction can be carried out in a $O(n\log n)$ time.

1 Introduction

β-skeletons in $\mathbb{R}^2$ belong to the family of proximity graphs, geometric graphs in which an edge between two vertices (points) exists if and only if they satisfy particular geometric requirements. In this paper we use the following definitions of the β-skeletons for sets of points in the Euclidean space (β-skeletons are also defined for $\beta \in \{0, \infty\}$ but those cases have no significant influence on our considerations) :

Definition 1. *For a given set of points $V = \{v_1, v_2, \ldots, v_n\}$ in $\mathbb{R}^2$, a distance function d and a parameter $0 < \beta < \infty$ we define a graph*

- *$G_\beta(V)$ – called a lune-based β-skeleton [11] – as follows: two points $v', v'' \in V$ are connected with an edge if and only if no point from $V \setminus \{v', v''\}$ belongs to the set $N(v', v'', \beta)$ (neighborhood, see Fig. 1) where:*
  1. *for $0 < \beta < 1$, $N(v', v'', \beta)$ is the intersection of two discs, each with radius $\frac{d(v', v'')}{2\beta}$ and having the segment $v'v''$ as a chord,*

This research is supported by the ESF EUROCORES program EUROGIGA, CRP VORONOI.

© Springer International Publishing Switzerland 2015
A. Kosowski and I. Walukiewicz (Eds.): FCT 2015, LNCS 9210, pp. 65–78, 2015.
DOI: 10.1007/978-3-319-22177-9_6

2. *for* $1 \le \beta < \infty$, $N(v', v'', \beta)$ *is the intersection of two discs, each with radius* $\frac{\beta d(v', v'')}{2}$, *whose centers are in points* $(\frac{\beta}{2})v' + (1 - \frac{\beta}{2})v''$ *and in* $(1 - \frac{\beta}{2})v' + (\frac{\beta}{2})v''$, *respectively;*

- $G_\beta^c(V)$ – *called a circle-based β-skeleton [5] – as follows: two points* v', v'' *are connected with an edge if and only if no point from* $V \setminus \{v', v''\}$ *belongs to the set* $N^c(v', v'', \beta)$ *(neighborhood, see Fig. 1) where:*
1. *for* $0 < \beta < 1$ *there is* $N^c(v', v'', \beta) = N(v', v'', \beta)$,
2. *for* $1 \le \beta$ *the set* $N^c(v', v'', \beta)$ *is a union of two discs, each with radius* $\frac{\beta d(v', v'')}{2}$ *and having the segment* $v'v''$ *as a chord.*

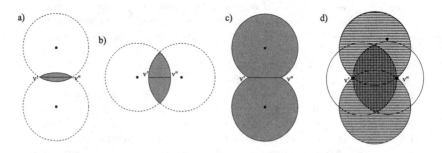

Fig. 1. Neighborhoods of the β-skeleton: (a) for $0 < \beta \le 1$, (b) the lune-based skeleton and (c) the circle-based skeleton for $1 < \beta < \infty$. The relation between neighborhoods: (d) $N(v', v'', \beta) \subseteq N^c(v', v'', \beta)$.

Points $v', v'' \in V$ are called generators of the neighborhood $N(v', v'', \beta)$ ($N^c(v', v'', \beta)$, respectively). The neighborhood $N(v', v'', \beta)$ is called a lune. It follows from the definition that $N(v', v'', 1) = N^c(v', v'', 1)$.

β-skeletons are both important and popular because of many practical applications which span a spectrum of areas from geographic information systems to wireless ad hoc networks and machine learning. For example, they allow us to reconstruct a shape of a two-dimensional object from a given set of sample points and they are also helpful in finding the minimum weight triangulation of a point set.

Hurtado, Liotta and Meijer [9] presented an $O(n^2)$ algorithm for the β-skeleton when $\beta < 1$. Matula and Sokal [14] showed that the lune-based 1-skeleton (Gabriel Graph GG) can be computed from the Delaunay triangulation in a linear time. Supowit [16] described how to construct the lune-based 2-skeleton (Relative Neighborhood Graph RNG) of a set of n points in $O(n \log n)$ time. Jaromczyk and Kowaluk [10] showed how to construct the RNG from the Delaunay triangulation DT for the L_p metric ($1 < p < \infty$) in $O(n\alpha(n))$ time. This result was further improved to $O(n)$ time [13] for β-skeletons where $1 \le \beta \le 2$. For $\beta > 1$, the circle-based β-skeletons can be constructed in $O(n \log n)$ time from the Delaunay triangulation DT with a simple test to filter edges of the DT [5]. On the other hand, so far the fastest algorithm for computing the lune-based β-skeletons for $\beta > 2$ runs in $O(n^{\frac{3}{2}} \log^{\frac{1}{2}} n)$ time [12].

Let us consider the case when we compute the β-skeleton for a set of n points V where every point $v \in V$ is allowed to move along a straight-line segment s_v. Let $S = \{s_v | v \in V\}$. For each pair of segments s_{v_1}, s_{v_2} containing points $v_1, v_2 \in V$, respectively, we want to find such positions of points v_1 and v_2 that $s_v \cap N(v_1, v_2, \beta) = \emptyset$ for any $s_v \in S \setminus \{s_1, s_2\}$. We will attempt to solve this problem by defining a β-skeleton for the set of line segments S as follows.

Definition 2. $G_\beta(S)$ $(G_\beta^c(S)$, respectively) is a graph with n vertices such that there exists a bijection between the set of vertices and the set of segments S, and for $s', s'' \in S$ an edge $s's''$ exists if there are points $v' \in s'$ and $v'' \in s''$ such that $(\bigcup_{s \in S \setminus \{s', s''\}} s) \cap N(v', v'', \beta) = \emptyset$ $((\bigcup_{s \in S \setminus \{s', s''\}} s) \cap N^c(v', v'', \beta) = \emptyset$, respectively).

Note that when segments degenerate to points, we have the standard β-skeleton for a point set.

Geometric structures concerning a set of line segments, e.g. the Voronoi diagram [3,15] or the straight skeleton [1] are well-studied in the literature.

Chew and Kedem [4] defined the Delaunay triangulation for line segments. Their definition was generalized by Brévilliers et al. [2].

However, β-skeletons for a set of line segments were completely unexplored. This paper makes an initial effort to fill this gap.

The paper is organized as follows. In the next section we present some basic facts and we prove that the definition of β-skeletons for a set of line segments preserves inclusions from the theorem of Kirkpatrick and Radke [11] formulated for a set of points. In Sect. 3 we show a general algorithm computing β-skeletons for a set of line segments in Euclidean plane when $0 < \beta < 1$. In Sect. 4 we present a similar algorithm for $\beta \geq 1$ in both cases of lune-based and circle-based β-skeletons. In Sect. 5 we consider an algorithm for Gabriel Graph. The last section contains open problems and conclusions.

2   Preliminaries

Let us consider a two-dimensional plane $\mathbb{R}^2$ with the Euclidean metric and a distance function d.

Let S be a finite set of disjoint closed line segments in the plane. Elements of S are called sites. A circle is tangent to a site s if s intersects the circle but not its interior. We assume that the sites of S are in general position, i.e., no three segment endpoints are collinear and no circle is tangent to four sites.

The Delaunay triangulation for the set of line segments S is defined as follows.

Definition 3. [2] The segment triangulation P of S is a partition of the convex hull $conv(S)$ of S in disjoint sites, edges and faces such that:

– Every face of P is an open triangle whose vertices belong to three distinct sites of S and whose open edges do not intersect S,

– *No face can be added without intersecting another one,*
– *The edges of P are the (possibly two-dimensional) connected components of conv(S) \ (F ∪ S), where F is the set of faces of P.*

The *segment triangulation P such that the interior of the circumcircle of each triangle does not intersect S* is called the segment Delaunay triangulation.

In this paper we will consider a planar graph (a planar multigraph, respectively) $DT(S)$ corresponding to the segment Delaunay triangulation P and its relations with β-skeletons. This graph has a linear number of edges and is dual to the Voronoi Diagram graph for S. It is also possible to study properties of plane partitions generated by β-skeletons for line segments. We will discuss this problem in the last section of this paper.

We can consider open (closed, respectively) neighborhoods $N(v', v'', \beta)$ that lead to *open (closed*, respectively) *β-skeletons*. For example, the *Gabriel Graph GG* [7] is the closed 1-skeleton and the *Relative Neighborhood Graph RNG* [17] is the open 2-skeleton.

Kirkpatrick and Radke [11] showed a following important inclusions connecting β-skeletons for a set of points V with the Delaunay triangulation $DT(V)$ of V : $G_{\beta'}(V) \subseteq G_\beta(V) \subseteq GG(V) \subseteq DT(V)$, where $\beta' > \beta > 1$.

We show that definitions of the β-skeleton and the Delaunay triangulation for a set of line segments S preserve those inclusions. We define $GG(S)$ as a 1-skeleton.

Theorem 1. *Let us assume that line segments in S are in general position and let $G_\beta(S)$ ($G_\beta^c(S)$, respectively) denote the lune-based (circle-based, respectively) β -skeleton for the set S. For $1 \le \beta < \beta'$ following inclusions hold true: $G_{\beta'}(S) \subseteq G_\beta(S) \subseteq GG(S) \subseteq DT(S)$ ($G_{\beta'}^c(S) \subseteq G_\beta^c(S) \subseteq GG(S) \subseteq DT(S)$, respectively).*

Proof. First we prove that $GG(S) \subseteq DT(S)$. Let $v_1 \in s_1, v_2 \in s_2$ be such a pair of points that there exists a disc D with diameter v_1v_2 containing no points belonging to segments from $S \setminus \{s_1, s_2\}$ inside of it. We transform D under a homothety with respect to v_1 so that its image D' is tangent to s_2 in the point t. Then we transform D' under a homothety with respect to t so that its image D'' is tangent to s_1 (see Fig. 2). The disc D'' lies inside of D, i.e., it does not intersect segments from $S \setminus \{s_1, s_2\}$, and it is tangent to s_1 and s_2, so the center of D'' lies on the Voronoi Diagram $VD(S)$ edge. Hence, if the edge s_1s_2 belongs to $GG(S)$ then it also belongs to $DT(S)$.

The last inclusion is based on a fact that for $1 \le \beta < \beta'$ and for any two points v_1, v_2 it is true that $N(v_1, v_2, \beta) \subseteq N(v_1, v_2, \beta')$ (see [11]).

The sequence of inclusions for circle-based β-skeletons is a straightforward consequence of the fact that two different circles intersect in at most two points.

3   Algorithm for Computing β-skeletons for $0 < \beta < 1$

Let us consider a set S of n disjoint line segments in the Euclidean plane. First we show a few geometrical facts concerning β-skeletons $G_\beta(S)$.

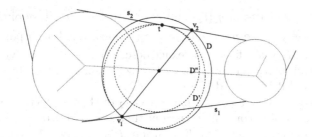

Fig. 2. $GG(S) \subseteq DT(S)$. The dotted line marks a fragment of Voronoi Diagram for the given edges.

The following remark is a straightforward consequence of the inscribed angle theorem.

Remark 1. For a given parameter $0 < \beta \leq 1$ if v is a point on the boundary of $N(v_1, v_2, \beta)$, different from v_1 and v_2, then an angle $\angle v_1 v v_2$ has a constant measure which depends only on β.

Let us consider a set of parametrized lines containing given segments. A line $P(s_i)$ contains a segment $s_i \in S$ and has a parametrization $q_i(t_i) = (x_1^i, y_1^i) + t_i \cdot [x_2^i - x_1^i, y_2^i - y_1^i]$, where (x_1^i, y_1^i) and (x_2^i, y_2^i) are ends of the segment s_i and $t_i \in \mathbb{R}$.

Let respective points from segments s_1 and s_2 be generators of a lune and let an inscribed angle determining a lune for a given value of β be equal to δ. The main idea of the algorithm is as follows. For any point $v_1 \in P(s_1)$ we compute points $v_2 \in P(s_2)$ for which there exists a point $v \in P(s)$, where $s \in S \setminus \{s_1, s_2\}$, such that $\delta \leq \angle v_1 v v_2 \leq 2\pi - \delta$, i.e., $v \in N(v_1, v_2, \beta)$ (see Fig. 3). Then we analyze a union of pairs of neighborhoods generators for all $s \in S \setminus \{s_1, s_2\}$. If this union contains all pairs of points (v_1, v_2), where $v_1 \in s_1$ and $v_2 \in s_2$, then $(s_1, s_2) \notin G_\beta(S)$.

For a given $t_1 \in \mathbb{R}$ and a segment $s \in S \setminus \{s_1, s_2\}$ we shoot rays from a point $v_1 = q_1(t_1) \in P(s_1)$ towards $P(s)$. Let us assume that a given ray intersects $P(s)$ in a point $v = q(t) = (x_1, y_1) + t \cdot [x_2 - x_1, y_2 - y_1]$ for some value of $t \in \mathbb{R}$. Let $w(t) = \overrightarrow{v_1 v}$ be the vector between points v_1 and v. Then $w(t) = [A_1 t + B_1 t_1 + C_1, A_2 t + B_2 t_1 + C_2]$ where coefficients A_i, B_i, C_i for $i = 1, 2$ depend only on endpoints coordinates of segments s_1 and s. The ray refracts in v from $P(s)$ in such a way that the angle between directions of incidence and refraction of the ray is equal to δ. The parametrized equation of the refracted ray is $r(z, t) = v + z \cdot R_\delta w(t)$ for $z \geq 0$ (or $r(z, t) = v + z \cdot R'_\delta w(t)$ for $z \geq 0$, respectively) where R_δ (R'_δ, respectively) denotes a rotation matrix for a clockwise (counter-clockwise, respectively) angle δ. If refracted ray $r(z, t)$ intersects line $P(s_2)$ in a point $q_2(t_2) = r(z, t)$ (it is not always possible - see Fig. 3) then we compare the x-coordinates of $q_2(t_2)$ and $r(z, t)$. As a result we obtain a function containing only parameters t_1 and t_2: $z = \frac{J \cdot t_2 + K \cdot t_1 + L}{D \cdot t + E \cdot t_1 + F}$, where coefficients $J = -(x_2 - x_1), K = x_2^2 - x_1^2, L = x_1^2 - x_1, D = A_1 \cos\delta + A_2 \sin\delta, E = B_1 \cos\delta + B_2 \sin\delta, F = C_1 \cos\delta + C_2 \sin\delta$ are fixed. Since y-coordinates of $q_2(t_2)$ and $r(z,t)$ are also equal we obtain $t_2(t) = \frac{M \cdot t^2 + p_1(t_1) \cdot t + p_2(t_1)}{N \cdot t + p_3(t_1)}$, where p_1, p_2 and p_3 are (at most quadratic) polynomials of variable t_1 and M, N are fixed (the exact description of those polynomials and variables is much more complex than the description of the coefficients in the previous step and it is omitted here).

Let $l_{t_1}^{\delta}(t)$ denote a value of the parameter t_2 of the intersection point of the line $P(s_2)$ and the line containing the ray that starts in $q_1(t_1)$ and refracts in $q(t)$ creating an angle δ. Let $k_{t_1}^{\delta} = l_{t_1}^{\delta}|_I$, where I is a set of values of t such that the ray refracted in $q(t)$ intersects $P(s_2)$. The function $l_{t_1}^{\delta}$ is a hyperbola and the function $k_{t_1}^{\delta}$ is a part of it (see Fig. 3).

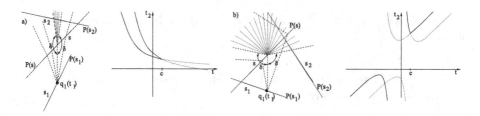

Fig. 3. Examples of correlation between parameters t and t_2 (for a fixed t_1) for a presented composition of segments and (a) a refraction angle near π (dotted lines show refracted rays that are analyzed) and (b) near $\frac{\pi}{2}$. The value c corresponds to the intersection point of lines $P(s)$ and $P(s_2)$. Dotted curves show a case when a line containing a refracted ray intersects $P(s_2)$ but the ray itself does not.

Note that for a given angle δ ($2\pi - \delta$, respectively) extreme points of the function $k_{t_1}^{\delta}$ ($k_{t_1}^{2\pi-\delta}$, respectively) do not have to belong to the set $\{0,1\}$. We can find them by computing a derivative $\frac{dt_2}{dt} = \frac{MN \cdot t^2 + 2Mp_3(t_1) \cdot t + p_1(t_1)p_3(t_1) - Np_2(t_1)}{(N \cdot t + p_3(t_1))^2}$.

Then we can compute the corresponding values of the parameter t_2. This way we obtain the pair (t_1, t_2) such that the segment $q_1(t_1)q_2(t_2)$ is a chord of a circle that is tangent to the analyzed segment s in $q(t)$ and $\angle q_1(t_1)q(t)q_2(t_2) = \delta$ ($\angle q_1(t_1)q(t)q_2(t_2) = 2\pi - \delta$, respectively).

Let $T(t_1, s, s_2) = \bigcup_{\gamma \in [\delta, 2\pi-\delta], t \in [0,1]} k_{t_1}^{\gamma}(t)$, i.e., this is a set of all t_2 such that points $q_1(t_1)$ and $q_2(t_2)$ generate a lune intersected by the analyzed segment s. Let $F(s_1, s, s_2) = \bigcup_{t_1 \in R, x \in T(t_1, s, s_2)} (t_1, x)$ be a set of pairs of parameters (t_1, t_2) such that the segment s intersects a lune generated by points $q_1(t_1)$ and $q_2(t_2)$. The set $F(s_1, s, s_2)$ is an area limited by $O(1)$ algebraic curves of degree at most 3. The curves correspond to the set of values of the parameter t_2 corresponding to extreme points of $k_{t_1}^{\delta}$ ($k_{t_1}^{2\pi-\delta}$, respectively). In particular there are hyperbolas for angles δ and $2\pi - \delta$ corresponding to rays refracted in the ends of the segment s (for parameters $t = 0$ and $t = 1$) - see Fig. 4. In fact, the curves that form the border of the set $F(s_1, s, s_2)$ intersect each other pairwise in at most 4 points, so the length of the Davenport-Schinzel sequence for those curves is $\lambda_4(n) = O(n2^{\alpha(n)})$.

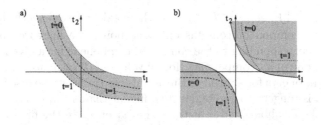

Fig. 4. Examples of sets $F(s_1, s, s_2)$ for β near (a) 0 and (b) 1 (the shape of $F(s_1, s, s_2)$ also depends on the position of the segment s with respect to s_1 and s_2). Dotted (dashed, respectively) curves limit the area corresponding to rays refracted through the segment s and creating the angle δ ($2\pi - \delta$, respectively).

Lemma 1. *The edge s_1, s_2 belongs to the β-skeleton $G_\beta(S)$ if and only if $[0,1] \times [0,1] \setminus \bigcup_{s \in S \setminus \{s_1, s_2\}} F(s_1, s, s_2) \neq \emptyset$.*

Proof. If $[0,1] \times [0,1] \setminus \bigcup_{s \in S \setminus \{s_1, s_2\}} F(s_1, s, s_2) \neq \emptyset$ then there exists a pair of parameters $(t_1, t_2) \in [0,1] \times [0,1]$ such that a lune generated by points $q_1(t_1) \in s_1$ and $q_2(t_2) \in s_2$ is not intersected by any segment $s \in S \setminus \{s_1, s_2\}$, i.e., $(s_1, s_2) \in G_\beta(S)$. The opposite implication can be proved in the same way.

Theorem 2. *For $0 < \beta < 1$ the β-skeleton $G_\beta(S)$ can be found in $O(n^3 \lambda_4(n))$ time.*

Proof. We analyze $O(n^2)$ pairs of line segments. For each pair of segments s_1, s_2 we compute $\bigcup_{s \in S \setminus \{s_1, s_2\}} F(s_1, s, s_2)$. For each $s \in S \setminus \{s_1, s_2\}$ we find a set of pairs of parameters t_1, t_2 such that $N(q_1(t_1), q_2(t_2), \beta) \cap s \neq \emptyset$. The arrangement of $n - 2$ curves in total can be found in $O(n \lambda_4(n))$ time [6]. Then the difference $[0,1] \times [0,1] \setminus \bigcup_{s \in S \setminus \{s_1, s_2\}} F(s_1, s, s_2)$ can be found in $O(n \lambda_4(n))$ time. Therefore we can verify which edges belong to $G_\beta(S)$ in $O(n^3 \lambda_4(n))$ time.

4 Finding β-skeletons for $1 \leq \beta$

Let us first consider the circle-based β-skeletons. According to Theorem 1 for $1 \leq \beta$ there are only $O(n)$ edges which can belong to the β-skeleton for a given set of line segments. We will use this property to compute β-skeletons faster than in the previous section.

Lemma 2. *For $1 \leq \beta$ and the set S of n line segments the number of connected components of the set $[0,1] \times [0,1] \setminus \bigcup_{s \in S \setminus \{s_1, s_2\}} F(s_1, s, s_2)$ is $O(n)$ for any pair $s_1, s_2 \in S$.*

Proof. According to Theorem by Kirkpatrick and Radke [11] for $1 \leq \beta < \beta'$ the following inclusion holds $G_{\beta'}(v) \subseteq G_\beta(V)$. Therefore any neighborhood for β' is included in some neighborhood for β with the same pair of generators. On the other hand, for a given parameter β and a given connected component of the set $[0,1] \times [0,1] \setminus \bigcup_{s \in S \setminus \{s_1, s_2\}} F(s_1, s, s_2)$ there exists a sufficiently big β' such that for β' the component contains only one point (we increase an arbitrary neighborhood corresponding to the connected component for a given β). Hence, the number of one point components (for all values of β) estimates the number of connected components for a given β. But in this case at least one disc forming the neighborhood is tangent to two segments different than s_1 and s_2 or at least one generator of the neighborhood is at the end of s_1 or s_2. In the first case the two segments tangent to the disc and segments s_1, s_2 are the the closest ones to the center of the disc. Therefore the complexity of the set of such components does not exceed the complexity of the 4-order Voronoi diagram for S, i.e., it is $O(n)$ [15]. In the second case there is a constant number of additional components.

Lemma 3. *For any $t_1 \in \mathbb{R}$ and $s_1, s_2 \in S$ there is at most one connected component of the set $[0,1] \times [0,1] \setminus \bigcup_{s \in S \setminus \{s_1, s_2\}} F(s_1, s, s_2)$ that contains points with the same t_1 coordinate.*

Proof. Let the inscribed angle corresponding to $N^c(s_1, s_2, \beta)$ be equal to δ. Let $a = q_1(t_1)$ and $b \in P(s_2)$ ($b' \in P(s_2)$, respectively) be points such that the angle between ab (ab', respectively) and $P(s_2)$ is equal to δ (for $\delta = \frac{\pi}{2}$ we have $b = b'$), see Fig. 5. Boundaries of all neighborhoods $N^c(s_1, s_2, \beta)$ generated by a and a point in s_2 contain either b or b'. There exists the leftmost (rightmost, respectively) position (might be in infinity) of the second neighborhood generator with respect to the direction of t_2. Between those positions no neighborhood intersects segments from $S \setminus \{s_1, s_2\}$. Hence, points corresponding to positions of such generators belong to the same connected component of $[0,1] \times [0,1] \setminus \bigcup_{s \in S \setminus \{s_1, s_2\}} F(s_1, s, s_2)$.

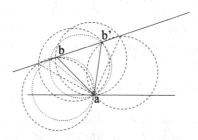

Fig. 5. Neighborhoods that have one common generator.

The algorithm for computing circle-based β-skeletons for $\beta \geq 1$ is almost the same as the algorithm for $\beta < 1$.

Theorem 3. *For $\beta \geq 1$ the circle-based β-skeleton $G_\beta^c(S)$ can be found in $O(n^2 \alpha(n) \log n)$ time.*

Proof. Due to Theorem 1 we have to analyze $O(n)$ edges of $DT(S)$. For $\beta \geq 1$ and for the given segments $s_1, s_2 \in S$ each set $F(s_1, s, s_2)$ can be divided in two sets with respect to the variable t_1. For each t_1 the first set contains part of $F(s_1, s, s_2)$ that is unbound from above with respect to t_2 and the second one contains part of $F(s_1, s, s_2)$ unbound from below (see Fig. 6). The part that contains pairs (t_1, t_2) such that the set of values of t_2 is $\mathbb{R}$ can be divided arbitrarily. We use Hershberger's algorithm [8] to compute unions of sets for $s \in S \setminus \{s_1, s_2\}$ in each group separately. Then, according to Lemma 3 we find an intersection of complements of computed unions. It needs $O(n\alpha(n) \log n)$ time. Hence, the total time complexity of the algorithm is $O(n^2 \alpha(n) \log n)$.

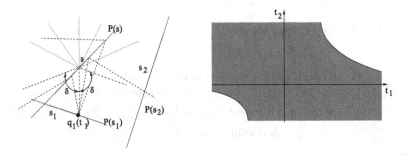

Fig. 6. An example of (a) refracted rays and (b) correlations between variables t_1 and t_2 for circle-based β-skeletons, where $\beta \geq 1$. (the shape of $F(s_1, s, s_2)$ depends on the position of the segment s with respect to s_1 and s_2)

Let us consider the lune-based β-skeletons now. Unfortunately, Lemma 3 does not hold in this case.

According to Theorem 1, in this case we have to consider only $O(n)$ pairs of line segments in S (the pairs corresponding to edges of $DT(S)$). We will analyze pairs of points belonging to given segments $s_1, s_2 \in S$ which generate discs such that each of them is intersected by any segment $s \in S \setminus \{s_1, s_2\}$. We will consider β-skeletons for $\beta > 1$ (a 1-skeleton is the same in the circle-based and lune-based case). Let $q_1(t_1) \in s_1$ and $q_2(t_2) \in s_2$ be generators of a lune $N(q_1(t_1), q_2(t_2), \beta)$ and let $C_1(q_1(t_1), q_2(t_2), \beta)$ be a circle creating a part of its boundary containing point $q_1(t_1)$.

We will shoot a ray from a lune generator and we will compute a possible position of the second generator when the refraction point belongs to the lune. Let an angle between a shot ray and a refracted ray be equal to $\frac{\pi}{2}$ and let $q(t) \in s \cap C_1(q_1(t_1), q_2(t_2), \beta)$. Unfortunately, the ray shot from $q_1(t_1)$ and refracted in $q(t)$ does not intersect the segment s_2 in $q_2(t_2)$. However, we can define a segment s' such that the ray shot from $q_1(t_1)$ refracts in $q(t)$ if and only if the same ray refracted in a point of s' passes through $q_2(t_2)$ (see Fig. 7).

Lemma 4. *Assume that $\beta \geq 1$, $q_1(t_1) \in P(s_1)$ and $q_2(t_2) \in P(s_2)$, where $s_1, s_2 \in S$. Let a point $q(t) \in P(s)$, where $s \in S \setminus \{s_1, s_2\}$, belong to $C_1(q_1(t_1), q_2(t_2), \beta)$. Let l be a line perpendicular to the segment $(q_1(t_1), q(t))$, passing through $q_2(t_2)$ and crossing $(q_1(t_1), q(t))$ in a point w. Then $\frac{d(q_1(t_1), w)}{d(q_1(t_1), q(t))} = \frac{1}{\beta}$.*

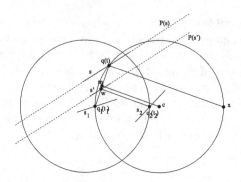

Fig. 7. The auxiliary segment s' and rays refracted in $q(t)$ and w.

Proof. Let x be an opposite to $q_1(t_1)$ end of the diameter of $C_1(q_1(t_1), q_2(t_2), \beta)$. Then $d(q_1(t_1), x) = 2d(q_1(t_1), c)$, where c is the center of $C_1(q_1(t_1), q_2(t_2), \beta)$. From the definition of the β-skeleton follows that $\frac{d(q_1(t_1), q_2(t_2))}{d(q_1(t_1), x)} = \frac{d(q_1(t_1), q_2(t_2))}{2d(q_1(t_1), c)}$. $\frac{2d(q_1(t_1), q_2(t_2))}{2\beta d(q_1(t_1), q_2(t_2))} = \frac{1}{\beta}$. According to Thales' theorem $\frac{d(q_1(t_1), w)}{d(q_1(t_1), q(t))} = \frac{d(q_1(t_1), q_2(t_2))}{d(q_1(t_1), x)} = \frac{1}{\beta}$ (see Fig. 7).

The algorithm computing a lune-based β-skeleton for $\beta \geq 1$ is similar to the previous one. Let $P(s') = h^{\frac{1}{\beta}}_{q_1(t_1)}(P(s))$, where $h^{\frac{1}{\beta}}_{q_1(t_1)}$ is a homothety with respect to a point $q_1(t_1)$ and a ratio $\frac{1}{\beta}$. Like in the case of circle-based β-skeletons we compute pairs of parameters t_1, t_2 such that the ray shot from $q_1(t_1)$ refracts in a point of s' and intersects the segment s_2 in $q_2(t_2)$, i.e., an analyzed segment s intersects a disc limited by the circle $C_1(q_1(t_1), q_2(t_2), \beta)$.

However, in the case of lune-based β-skeletons we analyze only one hyperbola (functions for clockwise and counterclockwise refractions are the same). Moreover, sets $F(s_1, s, s_2)$ and $F(s_2, s, s_1)$ are different. They contain pairs of parameters t_1, t_2 corresponding to points generating discs such that each of them separately is intersected by the segment s. Therefore, we have to intersect those sets to obtain a set of pairs of parameters corresponding to points generating lunes intersected by s (see Fig. 8).

Theorem 4. *For $\beta \geq 1$ the lune-based β-skeleton $G_\beta(S)$ can be found in $O(n^2 \lambda_4(n))$ time.*

Proof. β-skeletons for $\beta \geq 1$ satisfy the inclusions from Theorem 1. Hence, the number of tested edges is linear. For each such pair of segments s_1, s_2 we compute the corresponding sets of pairs of points generating lunes that do not intersect segments from $S \setminus \{s_1, s_2\}$. Similarly as in Theorem 2 we can do it in $O(n\lambda_4(n))$

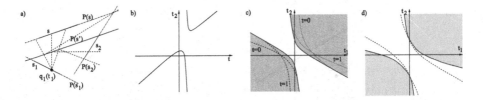

Fig. 8. An example of (a) a composition of three segments s_1, s_2, s, (b) correlations between variables t and t_2 (parametrizing s and s_2, respectively), (c) the set $F(s_1, s, s_2)$ and (d) the intersection $F(s_1, s, s_2) \cap F(s_2, s, s_1)$, where $\beta > 1$.

time. Therefore, the total time complexity of the algorithm (after analysis of $O(n)$ pairs of segments) is $O(n^2 \lambda_4(n))$.

5 Computing Gabriel Graph for Segments

In the previous sections we constructed sets of all pairs of points generating neighborhoods that do not intersect segments other than the segments containing generators. Now we want to find only $O(n)$ pairs of generators (one pair for each edge of a β-skeleton) that define the graph. Let $2 - VR(s_1, s_2)$ denote a region of the 2-order Voronoi diagram for the set S corresponding to s_1, s_2 and $3 - VR(s_1, s_2, s)$ denote a region of the 3-order Voronoi diagram for the set S corresponding to s_1, s_2, s. If an edge $s_1 s_2$, where $s_1, s_2 \in S$ belongs to the Gabriel Graph then there exists a disc $D(p, r)$ centered in p, which does not contain points from $S \setminus \{s_1, s_2\}$ and its diameter is $v_1 v_2$, where $v_1 \in s_1$, $v_2 \in s_2$ and $2r = d(v_1, v_2)$. The disc center p belongs to the set $(2 - VR(s_1, s_2)) \cap (3 - VR(s_1, s_2, s))$ for some $s \in S \setminus \{s_1, s_2\}$.

First, for segments $s_1, s_2 \in S$ we define a set of all middle points of segments with one endpoint on s_1 and one on s_2. This set is a quadrilateral $Q(s_1, s_2)$ (or a segment if $s_1 \parallel s_2$) with vertices in points $\frac{(x_i^1, y_i^1) + (x_j^2, y_j^2)}{2}$, where (x_i^k, y_i^k) for $i = 1, 2$ are endpoints of the segments s_k for $k = 1, 2$ (boundaries of the set are determined by the images of s_1 and s_2 under four homotheties with respect to the ends of those segments and a ratio $\frac{1}{2}$).

Let us analyze a position of a middle point of a segment l whose ends slide along the segments $s_1, s_2 \in S$. Let the length of l be $2r$. We rotate the plane so that segment s_1 lies in the negative part of x-axis and the point of intersection of lines containing segments s_1 and s_2 (if there exists) is $(0, 0)$. Let the segment s_2 lie on the line parametrized by $u \cdot [x_1, y_1]$ for $0 \geq x_1, 0 \leq y_1, 0 \leq u$. Then the middle point of l is (x, y), where $x = -|\sqrt{r^2 - (\frac{uy_1}{2})^2}| + u \cdot x_1, y = \frac{uy_1}{2}$.

Since $(x - 2\frac{x_1}{y_1}y)^2 + (y)^2 = r^2 - (\frac{uy_1}{2})^2 + (\frac{uy_1}{2})^2 = r^2$, then we have $x^2 + y^2(1 + 4(\frac{x_1}{y_1})^2) - 4\frac{x_1}{y_1}xy = r^2$, so all points (x, y) for a given r lie on an ellipse - see Fig. 9.

We want to find a point $p \in 3 - VR(s_1, s_2, s)$ which is a center of a segment $v_1 v_2$, where $v_1 \in s_1$ and $v_2 \in s_2$, and $d(p, s) > \frac{d(v_1, v_2)}{2}$. Then the disc with

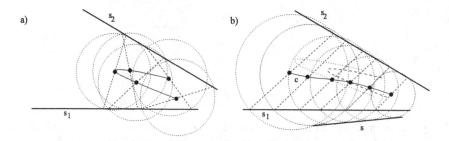

Fig. 9. (a)The set of middle points of segments v_1v_2, where $v_1 \in s_1$ and $v_2 \in s_2$ and (b) the curve c such that the distance between a point on the curve p and the segment s is equal to the length of the radius of a corresponding disc centered in p.

the center in p and the radius $\frac{d(v_1,v_2)}{2}$ intersects only segments s_1, s_2, i.e., there exists an edge of $GG(S)$ between s_1 and s_2.

We need to examine two cases. First, we consider the situation when the closest to p point of a segment s belongs to the interior of s. Let $P(s)$ be the line that contains segment s, which endpoints are (x_1^s, y_1^s) and (x_2^s, y_2^s), and let $q(t_s) = (x_1^s, y_1^s) + t_s \cdot [x_2^s - x_1^s, y_2^s - y_1^s]$ be the parametrization of $P(s)$. Let $L(s, r)$ be a line parallel to $P(s)$ with parametrization $l(t_L) = (x_1^L, y_1^L) + t_L \cdot [x_2^s - x_1^s, y_2^s - y_1^s]$ such that the distance between $P(s)$ and $L(s, r)$ is equal to r. We compute the intersection of the ellipse $x^2 + y^2(1 + 4(\frac{x_1}{y_1})^2) - 4\frac{x_1}{y_1}xy = r^2$ and the line $L(s, r)$. The result is $[x_1^L + t_L(x_2^s - x_1^s)]^2 + [y_1^L + t_L(y_2^s - y_1^s)]^2 - 4\frac{x_1}{y_1}[x_1^L + t_L(x_2^s - x_1^s)][y_1^L + t_L(y_2^s - y_1^s)] = r^2$, so t_L satisfies an equation $At_L^2 + Bt_L + C = r^2$ where coefficients A, B, C are fixed and depend on $x_1, y_1, x_i^s, x_i^L, y_i^s, y_i^L$ for $i = 1, 2$. This equation defines a curve c (see Fig. 9) which intersects corresponding ellipses. A point p which belongs to a part of the ellipse that lies on the opposite side of the curve c than the segment s is a center of a disc which has a diameter v_1v_2, where $v_1 \in S_1$, $v_2 \in S_2$, and does not intersect segment s.

In the second case one of the endpoints of the segment s is the nearest point to p (among the points from s). Let $D_1(r)$ and $D_2(r)$ be discs with diameter r and with centers in corresponding ends of the segment s. We compute the intersection of $D_1(r) = \{(x, y) : (x_1^s - x)^2 + (y_1^s - y)^2 = r^2\}$ $(D_2(r) = \{(x, y) : (x_2^s - x)^2 + (y_2^s - y)^2 = r^2\}$, respectively) and ellipse $x^2 + y^2(1 + 4(\frac{x_1}{y_1})^2) - 4\frac{x_1}{y_1}xy = r^2$. We obtain $x_1^s(x_1^s - 2x) + y_1^s(y_1^s - 2y) - y^2(\frac{x_1}{y_1})^2 + 4(\frac{x_1}{y_1})xy = 0$, so $x = \frac{N_1y^2 + N_2y + N_3}{N_4y + N_5}$ and y satisfies an equation $M_1y^4 + M_2y^3 + M_3y^2 + M_4y + M_5 = 0$ where coefficients N_i and M_j for $i, j = 1, \ldots, 5$ depend on $x_1^s, y_1^s, x_1, y_1, r$ (or on $x_2^s, y_2^s, x_1, y_1, r$, respectively). If there exists a point $p \notin D_1(r) \cup D_2(r)$ that belongs to the part of the ellipse between the segments s_1, s_2, then there also exists a disc with center in p and a diameter $d(v_1, v_2) = 2r$, where $v_1 \in s_1$ and $v_2 \in s_2$, which does not contain ends of the segment s.

In both cases we obtain a curve $c(r)$ dependent on the parameter r - see Fig. 9. We check if a set $Q(s_1, s_2) \cap (2 - VR(s_1, s_2)) \cap (3 - VR(s_1, s_2, s))$ and the segment s are on the same side of the curve c. Otherwise, the segment s_1s_2

belongs to the Gabriel Graph for the set S (i.e., there exists a point p which is the center of a segment $v_1 v_2$, where $v_1 \in s_1$, $v_2 \in s_2$, and $d(p, s) > \frac{d(v_1, v_2)}{2}$ for all $s \in S \setminus \{s_1, s_2\}$).

Theorem 5. *For a set of n segments S the Gabriel Graph $GG(S)$ can be computed in $O(n \log n)$ time.*

Proof. The 2-order Voronoi diagram and the 3-order Voronoi diagram can be found in $O(n \log n)$ time [15]. The number of triples of segments we need to test is linear. For each such triple we can check if there exists an empty 1-skeleton lune in time proportional to the complexity of the set $Q(s_1, s_2) \cap (2 - VR(s_1, s_2)) \cap (3 - VR(s_1, s_2, s))$. The total complexity of those sets is $O(n)$. Hence, the complexity of the algorithm is $O(n) + O(n \log n) = O(n \log n)$.

6   Conclusions

The running time of the presented algorithms for β-skeletons for sets of n line segment ranges between $O(n \log n)$, $O(n^2 \alpha(n) \log n)$ and $O(n^3 \lambda_4(n))$ and depends on the value of β. For $0 < \beta < 1$ the β-skeleton is not related to the Delaunay triangulation of the underlying set of segments. The existence of a relatively efficient algorithm for the Gabriel Graph suggests that it may be possible to find a faster way to compute β-skeletons for other values of β, especially for $1 \leq \beta \leq 2$.

The edges of the Delaunay triangulation for line segments can be represented in the form described in this paper as rectangles contained in $[0, 1] \times [0, 1]$ square in the t_1, t_2-coordinate system. If for each pair of β-skeleton edges the intersection of the corresponding sets for the β-skeleton and the Delaunay triangulation is not empty then there exist a plane partition generated by some pairs of generators of β-skeleton neighborhoods. Unfortunately, it is not always possible. The algorithms shown in this work for each pair of segments find such a position of generators that the corresponding lune does not intersect any other segment. We could consider a problem in which the number of used generators of neighborhoods is n (one generator per each edge). Then the method described in the paper can also be used. We analyze a n-dimensional space and test if $[0, 1]^n \setminus \bigcup_{s_i, s_j \in S, s \in S \setminus \{s_i, s_j\}} F(s_i, s, s_j) \times R^{n-2} \neq \emptyset$, where i and j also define corresponding coordinates in R^n. Unfortunately, such an algorithm is expensive. However, in this case a β-skeleton already generates a plane partition.

The total kinetic problem that can be solved in similar way is a construction β-skeletons for points moving rectilinear but without limitations concerning intersections of neighborhoods with lines defined by the moving points. In this case the form of sets $F(s_i, s, s_j)$ changes and the solution is much more complicated.

Are there any more effective algorithms for those problems?

Additional interesting questions about β-skeletons are related to their connections with k-order Voronoi diagrams for line segments.

Acknowledgments. The authors would like to thank Jerzy W. Jaromczyk for important comments.

References

1. Aichholzer, O., Aurenhammer, F.: Straight skeletons for general polygonal figures in the plane. In: Cai, J.-Y., Wong, C.K. (eds.) COCOON 1996. LNCS, vol. 1090, pp. 117–126. Springer, Heidelberg (1996)
2. Brévilliers, M., Chevallier, N., Schmitt, D.: Triangulations of line segment sets in the plane. In: Arvind, V., Prasad, S. (eds.) FSTTCS 2007. LNCS, vol. 4855, pp. 388–399. Springer, Heidelberg (2007)
3. Burnikel, C., Mehlhorn, K., Schirra, S.: How to compute the voronoi diagram of line segments: theoretical and experimental results. In: van Leeuwen, J. (ed.) ESA 1994. LNCS, vol. 855, pp. 227–239. Springer, Heidelberg (1994)
4. Chew, L.P., Kedem, K.: Placing the largest similar copy of a convex polygon among polygonal obstacles. In: Proceedings of the 5th Annual ACM Symposium on Computational Geometry, pp. 167–174 (1989)
5. Eppstein, D.: β-skeletons have unbounded dilation. Comput. Geom. **23**, 43–52 (2002)
6. Goodman, J.E., O'Rourke, J.: Handbook of Discrete and Computational Geometry. Chapman & Hall/CRC, New York (2004)
7. Gabriel, K.R., Sokal, R.R.: A new statistical approach to geographic variation analysis. Syst. Zool. **18**, 259–278 (1969)
8. Hershberger, J.: Finding the upper envelope of n line segments in O(n log n) time. Inf. Process. Lett. **33**(4), 169–174 (1989)
9. Hurtado, F., Liotta, G., Meijer, H.: Optimal and suboptimal robust algorithms for proximity graphs. Comput. Geom. Theory Appl. **25**(1–2), 35–49 (2003)
10. Jaromczyk, J.W., Kowaluk, M.: A note on relative neighborhood graphs. In: Proceedings of the 3rd Annual Symposium on Computational Geometry, Canada, Waterloo, pp. 233–241. ACM Press (1987)
11. Kirkpatrick, D.G., Radke, J.D.: A framework for computational morphology. In: Computational Geometry, pp. 217–248. North Holland, Amsterdam (1985)
12. Kowaluk, M.: Planar β-skeleton via point location in monotone subdivision of subset of lunes. In: EuroCG, Italy. Assisi 2012, pp. 225–227 (2012)
13. Lingas, A.: A linear-time construction of the relative neighborhood graph from the Delaunay triangulation. Comput. Geom. **4**, 199–208 (1994)
14. Matula, D.W., Sokal, R.R.: Properties of Gabriel graphs relevant to geographical variation research and the clustering of points in plane. Geog. Anal. **12**, 205–222 (1984)
15. Papadopoulou, E., Zavershynskyi, M.: A Sweepline Algorithm for Higher Order Voronoi Diagrams, In: Proceedings of 10th International Symposium on Voronoi Diagrams in Science and Engineering (ISVD), pp. 16–22 (2013)
16. Supowit, K.J.: The relative neighborhood graph, with an application to minimum spanning trees. J. ACM **30**(3), 428–448 (1983)
17. Toussaint, G.T.: The relative neighborhood graph of a finite planar set. Pattern Recognit. **12**, 261–268 (1980)

Complexity and Boolean Functions

Depth, Highness and DNR Degrees

Philippe Moser[1]([✉]) and Frank Stephan[2]

[1] Department of Computer Science, National University of Ireland,
Maynooth, Co Kildare, Ireland
`pmoser@cs.nuim.ie`
[2] Department of Mathematics, The National University of Singapore,
10 Lower Kent Ridge Drive, S17, Singapore 119076,
Republic of Singapore
`fstephan@comp.nus.edu.sg`

Abstract. We study Bennett deep sequences in the context of recursion theory; in particular we investigate the notions of $O(1)$-deep$_K$, $O(1)$-deep$_C$, order-deep$_K$ and order-deep$_C$ sequences. Our main results are that Martin-Löf random sets are not order-deep$_C$, that every many-one degree contains a set which is not $O(1)$-deep$_C$, that $O(1)$-deep$_C$ sets and order-deep$_K$ sets have high or DNR Turing degree and that no K-trivial set is $O(1)$-deep$_K$.

1 Introduction

The concept of logical depth was introduced by C. Bennett [6] to differentiate useful information (such as DNA) from the rest, with the key observation that non-useful information pertains in both very simple structures (for example, a crystal) and completely unstructured data (for example, a random sequence, a gas). Bennett calls data containing useful information logically deep data, whereas both trivial structures and fully random data are called shallow.

The notion of useful information (as defined by logical depth) strongly contrasts with classical information theory, which views random data as having high information content. I.e., according to classical information theory, a random noise signal contains maximal information, whereas from the logical depth point of view, such a signal contains very little useful information.

Bennett's logical depth notion is based on Kolmogorov complexity. Intuitively a logically deep sequence (or equivalently a set) is one for which the more time a compressor is given, the better it can compress the sequence. For example, both on trivial and random sequences, even when given more time, a compressor cannot achieve a better compression ratio. Hence trivial and random sequences are not logically deep.

Several variants of logical depth have been studied in the past [2,9,16,17,20]. As shown in [20], all depth notions proposed so far can be interpreted in the compression framework which says a sequence is deep if given (arbitrarily)more than

---

P. Moser was on Sabbatical Leave to the National University of Singapore, supported in part by SFI Stokes Professorship and Lectureship Programme. F. Stephan was supported in part by NUS grants R146-000-181-112 and R146-000-184-112.

A. Kosowski and I. Walukiewicz (Eds.): FCT 2015, LNCS 9210, pp. 81–94, 2015.
DOI: 10.1007/978-3-319-22177-9_7

$t(n)$ time steps, a compressor can compress the sequence $r(n)$ more bits than if given at most $t(n)$ time steps only. By considering different time bound families for $t(n)$ (e.g. recursive, polynomial time etc.) and the magnitude of compression improvement $r(n)$ - for short: the *depth magnitude* - (e.g. $O(1), O(\log n)$) one can capture all existing depth notions [2,9,16,17,20] in the compression framework [20]. E.g. Bennett's notion is obtained by considering all recursive time bounds t and a constant depth magnitude, i.e., $r(n) = O(1)$. Several authors studied variants of Bennett's notion, by considering different time bounds and/or different depth magnitude from Bennett's original notion [2,3,9,16,20].

In this paper, we study the consequences these changes of different parameters in Bennett's depth notion entail, by investigating the computational power of the deep sets yielded by each of these depth variants.

- We found out that the choice of the depth magnitude has consequences on the computational power of the corresponding deep sets. The fact that computational power implies Bennett depth was noticed in [16], where it was shown that every high degree contains a Bennett deep set (a set is high if, when given as an oracle, its halting problem is at least as powerful as the halting problem relative to the halting problem: A is high iff $A' \geq_T \emptyset''$). We show that the converse also holds, i.e., that depth implies computational power, by proving that if the depth magnitude is chosen to be "large" (i.e., $r(n) = \varepsilon n$), then depth coincides with highness (on the Turing degrees), i.e., a Turing degree is high iff it contains a deep set of magnitude $r(n) = \varepsilon n$.
- For smaller choices of r, for example, if r is any recursive order function, depth still retains some computational power: we show that depth implies either highness or diagonally-non-recursiveness, denoted DNR (a total function is DNR if its image on input e is different from the output of the e-th Turing machine on input e). This implies that if we restrict ourselves to left-r.e. sets, recursive order depth already implies highness. We also show that highness is not necessary by constructing a low order-deep set (a set is low if it is not powerful when given as an oracle).
- As a corollary, our results imply that weakly-useful sets introduced in [16] are either high or DNR (set S is weakly-useful if the class of sets reducible to it within a fixed time bound s does not have measure zero within the class of recursive sets).
- Bennett's depth [6] is defined using prefix-free Kolmogorov complexity. Two key properties of Bennett's notion are the so-called slow growth law, which stipulates that no shallow set can quickly (truth-table) compute a deep set, and the fact that neither Martin-Löf random nor recursive sets are deep. It is natural to ask whether replacing prefix-free with plain complexity in Bennett's formulation yields a meaningful depth notion. We call this notion plain-depth. We show that the random is not deep paradigm also holds in the setting of plain-depth. On the other hand we show that the slow growth law fails for plain-depth: every many-one degree contains a set which is not plain-deep of magnitude $O(1)$.
- A key property of depth is that "easy" sets should not be deep. Bennett [6] showed that no recursive set is deep. We give an improvement to this result

by observing that no K-trivial set is deep (a set is K-trivial if the complexity of its prefixes is as low as possible). Our result is close to optimal, since there exist deep ultracompressible sets [17].

- In most depth notions, the depth magnitude has to be achieved almost everywhere on the set. Some feasible depth notions also considered an infinitely often version [9]. Bennett noticed in [6] that infinitely often depth is meaningless because every recursive set is infinitely often deep. We propose an alternative infinitely often depth notion that doesn't suffer this limitation (called i.o. depth). We show that little computational power is needed to compute i.o. depth, i.e., every hyperimmune degree contains an i.o. deep set of magnitude εn (a degree is hyperimmune if it computes a function that is not bounded almost everywhere by any recursive function), and construct a Π_1^0-class where every member is an i.o. deep set of magnitude εn.

In summary, our results show that the choice of the magnitude for logical depth has consequences on the computational power of the corresponding deep sets, and that larger depth magnitude is not necessarily preferable over smaller magnitude. We conclude with a few open questions regarding the constant magnitude case. Due to lack of space, some proofs are ommitted and will appear in the journal version of this paper.

2 Preliminaries

We use standard computability/algorithmic randomness theory notations see [11,21,23]. We use $\leq^+$ to denote less or equal up to a constant term. We fix a recursive 1-1 pairing function $\langle \cdot \rangle : \mathbb{N} \times \mathbb{N} \to \mathbb{N}$. We use sets and their characteristic sequences interchangeably, we denote the binary strings of length n by 2^n, and 2^ω denotes the set of all infinite binary sequences. The join of two sets A, B is the set $A \oplus B$ whose characteristic sequence is $A(0)B(0)A(1)B(1)\ldots$, that is, $(A \oplus B)(2n) = A(n)$ and $(A \oplus B)(2n + 1) = B(n)$ for all n. An order function is an unbounded non-decreasing function from $\mathbb{N}$ to $\mathbb{N}$. A time bound function is a recursive order t such that there exists a Turing machine Φ such that for every n, $\Phi(n)[t(n)] \downarrow = t(n)$, i.e., $\Phi(n)$ outputs the value $t(n)$ within $t(n)$ steps of computation. Set A is left-r.e. iff the set of dyadic rationals strictly below the real number $0.A$ (a.k.a. the left-cut of A denoted $L(A)$) is recursively enumerable (r.e.), i.e., there is a recursive sequence of non-decreasing rationals whose limit is $0.A$. All r.e. sets are left-r.e., but the converse fails.

We consider standard Turing reductions $\leq_T$, truth-table reductions $\leq_{tt}$ (where all queries are made in advance and the reduction is total on all oracles) and many-one reductions $\leq_m$. Two sets A, B are Turing equivalent ($A \equiv_T B$) if $A \leq_T B$ and $B \leq_T A$. The Turing degree of a set A is the set of sets Turing equivalent to A. Fix a standard enumeration of all oracle Turing machines $\Phi_1, \Phi_2, \ldots$. The jump A' of a set A is the halting problem relative to A, i.e., $A' = \{e : \Phi_e^A(e) \downarrow\}$. The halting problem is denoted $\emptyset'$. A set A is high (that is, has high Turing degree) if its halting problem is as powerful as the halting problem of the halting problem, i.e., $\emptyset'' \leq_T A'$. High sets are equivalent to sets that compute dominating functions (i.e., sets A such that there is a function f with $f \leq_T A$ such that for every computable function g and for almost every n, $f(n) \geq g(n)$), i.e., a set is high iff it computes a dominating function [23]. A set A is low if its halting problem is not more powerful than the halting problem of a recursive set, i.e., $A' \leq_T \emptyset'$. Note that $\emptyset'$ is high relative to every low set.

If one weakens the dominating property of high sets to an infinitely often condition, one obtains hyperimmune degrees. A set is of hyperimmune degree if it computes a function that dominates every recursive function on infinitely many inputs. Otherwise the set is called of hyperimmune-free degree.

Another characterization of computational power used in computability theory is the concept of diagonally non-recursive function (DNR). A total function g is DNR if for every e, $g(e) \neq \Phi_e(e)$, i.e., g can avoid the output of every Turing machine on at least one specified input. A set is of DNR degree, if it computes a DNR function. It is known that every r.e. DNR degree is high, actually even Turing equivalent to $\emptyset'$ [1].

If one requires a DNR function to be Boolean, one obtains the PA-complete degrees: A degree is PA-complete iff it computes a Boolean DNR function. It is known that there exists low PA-complete degrees [23].

Fix a universal prefix free Turing machine U, i.e., such that no halting program of U is a prefix of another halting program. The prefix-free Kolmogorov complexity of string x, denoted $K(x)$, is the length of the length-lexicographically first program x^* such that U on input x^* outputs x. It can be shown that the value of $K(x)$ does not depend on the choice of U up to an additive constant. $K(x, y)$ is the length of a shortest program that outputs the pair (x, y), and $K(x|y)$ is the length of a shortest program such that U outputs x when given y as an advice. We also consider standard time bounded Kolmogorov complexity. Given time bound t (resp. $s \in \mathbb{N}$), $K^t(x)$ (resp. $K_s(x)$) denotes the length of the shortest prefix free program p such that $U(p)$ outputs x within $t(|x|)$ (resp. s) steps. Replacing U above with a plain (i.e., non prefix-free) universal Turing machine yields the notion of plain Kolmogorov complexity, and is denoted $C(x)$. We need the following counting theorem.

Theorem 1 (Chaitin [7]). *There exists $c \in \mathbb{N}$ such that for every $r, n \in \mathbb{N}$, $|\{\sigma \in 2^n : K(\sigma) \leq n + K(n) - r\}| \leq 2^{n-r+c}$.*

A set A is Martin-Löf random (MLR) if none of its prefixes are compressible by more than a constant term, i.e., $\forall n\ K(A \restriction n) \geq n - c$ for some constant c, where $A \restriction n$ denotes the first n bits of the characteristic function of A. A set A is K-trivial if its complexity is as low as possible, i.e., $\forall n\ K(A \restriction n) \leq K(n) + O(1)$. See [11,21] for more on C and K-complexity, MLR and trivial sets.

Effective closed sets are captured by Π_1^0-classes. A Π_1^0-class P is a class of sequences such that there is a computable relation R such that $P = \{S \in 2^\omega | \forall n\ R(S \restriction n)\}$.

Definition 2 (Bennett [6]). Let $g(n) \leq n$ be an order. A set S is g-deep$_K$ if for every recursive time bound t and for almost all $n \in \mathbb{N}$, $K^t(S \restriction n) - K(S \restriction n) \geq g(n)$.

A set S is $O(1)$-deep$_K$ (resp. order-deep$_K$) if it is c-deep$_K$ (resp. g-deep$_K$) for every $c \in \mathbb{N}$ (resp. for some recursive order g). A set is said Bennett deep if it is $O(1)$-deep$_K$. We denote by g-deep$_C$ the above notions with K replaced with C. It is easy to see that for every two orders f, g such that $\forall n \in \mathbb{N}\ f(n) \leq g(n)$, every g-deep$_K$ set is also f-deep$_K$.

Bennett's slow growth law (SGL) states that creating depth requires time beyond a "recursive amount", i.e., no shallow set quickly computes a deep one.

Lemma 3 (Bennett [6]; Juedes, Lathrop and Lutz [16]). *Let h be a recursive order, and $A \leq_{tt} B$ be two sets. If A is h-deep$_K$ (resp. $O(1)$-deep$_K$) then B is h'-deep$_K$ (resp. $O(1)$-deep$_K$) for some recursive order h'. Furthermore given indices for the truth-table reduction and for h, one can effectively compute an index for h'.*

The symmetry of information holds in the resource bounded case.

Lemma 4 (Li and Vitányi [18]). *For every time bound t, there is a time bound t' such that for all strings x, y with $|y| \leq t(|x|)$, we have $C^t(x, y) \geq C^{t'}(x) + C^{t'}(y \mid x) - O(\log C^{t'}(x, y))$. Furthermore given an index for t one can effectively compute an index for t'.*

Corollary 5. *Let t be a time bound and x, a be strings. Then there exists a time bound t' such that for every prefix y of x we have $C^t(y \mid a) \geq^+ C^{t'}(x \mid a) - |x| + |y| - O(\log C^t(y \mid a))$. Furthermore given an index for t one can effectively compute an index for t'.*

3 C-Depth

Bennett's original formulation [6] is based on K-complexity. In this section we investigate the depth notion obtained by replacing K with C, which we call plain depth. We study the interactions of plain depth with the notions of Martin-Löf random sets, many-one degrees and the Turing degrees of deep sets.

3.1 MLR is not Order-deep$_C$

The following result is the plain complexity version of Bennett's result that no MLR sets are Bennett deep. Due to lack of space, the proof will appear in the journal version of this paper.

Theorem 6. *For every MLR A and for every recursive order h, A is not h-deep$_C$.*

Sequences that are MLR relative to the halting problem are called 2-random. Equivalently a sequence A is 2-random iff there is a constant c such that $C(A \upharpoonright n) \geq n - c$ for infinitely many n [19,22]. Since there is a constant c' such that $n + c'$ is a trivial upper bound on the plain Kolmogorov complexity of any string of length n, it is clear that no 2-random sequence can be $O(1)$-deep$_C$. Thus most MLR sequences are not $O(1)$-deep$_C$.

3.2   The SGL Fails for C-depth

The following result shows that the Slow Growth Law fails for plain depth. Due to lack of space, the proof will appear in the journal version of this paper.

Theorem 7. *Every many-one degree contains a set which is not $O(1)$-deep$_C$.*

Note that this result shows that order-deep$_K$ does not imply order-deep$_C$: all the sets in the truth-table degree of any order-deep$_K$ set are all order-deep$_K$ (by the SGL), but this degree contains a non order-deep$_C$ set by the previous result.

3.3   Depth Implies Highness or DNR

The following result shows that being constant deep for C implies computational power.

Theorem 8. *Let A be a $O(1)$-deep$_C$ set. Then A is high or DNR.*

Proof. We prove the contrapositive. Suppose that A is neither DNR nor high. Let $f(m)$ be (a coding of) $A \upharpoonright 2^{m+1}$. Because $f \leq_{tt} A$, there are infinitely many m where $\Phi_m(m)$ is defined and equal to $f(m)$. Hence there is an A-recursive increasing function g such that, for almost every m, $g(m)$ is the time to find an $m' \geq m$ with $A(0)A(1)\ldots A(2^{m'+1}) = \Phi_{m'}(m')$ and to evaluate the expression $\Phi_{m'}(m')$ to verify the finding. As A is not high, there is a recursive increasing function h with $h(m) \geq g(m)$ for infinitely many m. Now consider any m where $h(m) \geq g(m)$. Then for the m' found for this m, it holds that $h(m') \geq h(m)$ and $h(m')$ is also larger than the time to evaluate $\Phi_{m'}(m')$. Hence $h(m)$ is larger than the time to evaluate $\Phi_m(m)$ for infinitely many m where $\Phi_m(m)$ codes $A \upharpoonright 2^{m+1}$.

For each such m, let n be a number with $2^m \leq n \leq 2^{m+1}$ and $C(n) \geq m$. Starting with a binary description of such an n, one can compute m from n and run $\Phi_m(m)$ for $h(m)$ steps and, in the case that this terminates with a string σ of length 2^{m+1}, output $\sigma \upharpoonright n$. It follows from this algorithm that there is a resource-bounded approximation to C such that there exist infinitely many n such that, on one hand $C(A \upharpoonright n) \geq \log(n)$ while on the other hand $A \upharpoonright n$ can be described in $\log(n) + c$ bits using this resource bounded description. Hence A is not $O(1)$-deep$_C$.   □

Since there are incomplete r.e. Turing degrees which are high, these are also not DNR and, by Theorem 10, they contain sets which are 0.9-deep$_C$. Thus the preceeding theorem cannot be improved to show that "$O(1)$-deep$_C$ sets are DNR".

Theorem 9. *There exists a set A such that A is $(1-\varepsilon)n$-deep$_C$ (for any $\varepsilon > 0$) but A is not DNR.*

4 K-Depth

Bennett's original depth notion is based on prefix free complexity. He made important connections between depth and truth-table degrees; In particular he proved that the $O(1)$-deep$_K$ sets are closed upward under truth-table reducibility, which he called the slow growth law. In the following section we pursue Bennett's investigation by studying the Turing degrees of deep sets. In the first subsection, we investigate the connections between linear depth and high Turing degrees. We then look at the opposite end by studying the interactions of various lowness notions with logical depth.

4.1 Highness and Depth Coincide

The following result shows that at depth magnitude εn, depth and highness coincide on the Turing degrees. The result holds for both K and C depth.

Theorem 10. *For every set A the following statements are equivalent:*

1. *The degree of A is εn-deep$_C$ for some $\varepsilon > 0$.*
2. *The degree of A is $(1 - \varepsilon)n$-deep$_C$ for every $\varepsilon > 0$.*
3. *A is high.*

Proof. We prove $(1) \Rightarrow (3)$ using the contrapositive: Let $\varepsilon > 0$ and $l \in \mathbb{N}$ such that $\delta < \varepsilon/3$ with $\delta := 1/l$. Let k be the limit inferior of the set $\{0, 1, \ldots, l\}$ such that there are infinitely many n with $C(A \upharpoonright n) \leq n \cdot k \cdot \delta$. Now one can define, relative to A, an A-recursive function g such that for each n there is an m with $n \leq m \leq g(n)$ and $C_{g(n)}(A \upharpoonright m) \leq m \cdot k \cdot \delta$. As A is not high, there is a recursive function h with $h(n) > g(n)$ for infinitely many n; furthermore, $h(n) \leq h(n+1)$ for all n. It follows that there are infinitely many n with $C_{h(n)}(A \upharpoonright n) \leq n \cdot k \cdot \delta$ which is also at most $n \cdot \delta$ away from the optimal value, hence A is not $\varepsilon \cdot n$ deep, which ends this direction's proof.

Let us show $(3) \Rightarrow (2)$. Let $\varepsilon > 0$, A be high, and let $g \leq_T A$ be dominating. We construct $B \equiv_T A$ such that B is $(1 - \varepsilon)n$-deep$_C$.

By definition, if t is a time bound and i an index of t then for every $m \in \mathbb{N}$ $\Phi_i(m)[t(m)] \downarrow = t(m)$. Since g is dominating, we have for almost every $m \in \mathbb{N}$, $t(m) = \Phi_i(m)[g(m)] \downarrow$.

We can thus use g to encode all time bounds that are total on all strings of length less than a certain bound into a set H, where

$$H(\langle i, j \rangle) = 1 \text{ iff } \Phi_i(m)[g(2^j)] \downarrow \text{ for all } m \in \{1, 2, \ldots, 2^j\}.$$

Thus t is a (total) time bound iff for almost every j, $H(\langle i, j \rangle) = 1$ (where i is an index for t).

We have $H \leq_T A$ and we choose the pairing function $\langle \cdot \rangle$ such that $H \upharpoonright n^2 + 1$ encodes the values

$$\{H(\langle i, j \rangle) : i, j \leq n\}.$$

Let $n \in \mathbb{N}$ and suppose $B \upharpoonright 2^n$ is already constructed. Given $A \upharpoonright n + 1$ and $H \upharpoonright n^2 + 1$, we construct $B \upharpoonright 2^{n+1}$. From $H \upharpoonright n^2 + 1$, we can compute the set $L_n = \{i \leq n : H(\langle i, n \rangle) = 1\}$, i.e., a list eventually containing all time bounds that are total on strings of lengths less or equal to 2^n. Let

$$T_n := \max\{\Phi_i(m) : i \in L_n, m \leq 2^n\}.$$

Find the lex first string x_n of length $2^n - 1$ such that

$$C_{T_n}(x_n \mid (B \restriction 2^n)A(n)) \geq 2^n.$$

Let $B \restriction [2^n, 2^{n+1} - 1] := A(n)x_n$. By construction we have $B \equiv_T A$. Also, $C(B \restriction 2^{n+1} \mid H \restriction n^2 + 1, A \restriction n + 1) \leq^+ C(n)$, i.e., $C(B \restriction 2^{n+1}) \leq 2n^2$.

Let us prove B is $\frac{2}{3}n$-deep$_C$; we then extend the argument to show B is $(1-\varepsilon)n$-deep$_C$. Let t be a time bound. Let n be large enough such that $t_1, t_2, t_3 \in L_{n-2}$ and $t'_1, t'_2, t'_3, t'_4 \in L_n$ where the t_i's are derived from t as described below.

Let j be such that $2^n < j \leq 2^{n+1}$ and $j' = j - (2^n - 1)$, i.e., $B \restriction j$ ends with the first $j' - 2$ bits of x_n (One bit is "lost" due to the first bit used to encode $A(n)$).

We consider two cases, first suppose $j' < \log n$. Let t_1 be a time bound (obtained from t) such that $C^t(B \restriction j) \geq^+ C^{t_1}(x_{n-1}, B \restriction 2^{n-1})$, where neither the constant nor t_1 depends on j, n. Let t_2 be derived from t_1 using Lemma 4. We have

$$C^{t_1}(x_{n-1}, B \restriction 2^{n-1})$$
$$\geq C^{t_2}(B \restriction 2^{n-1}) + C^{t_2}(x_{n-1} \mid B \restriction 2^{n-1}) - O(\log 2^n)$$
$$\geq C^{t_2}(B \restriction 2^{n-1}) + C_{T_{n-1}}(x_{n-1} \mid B \restriction 2^{n-1}) - O(n) \qquad \text{because } t_2 \in L_{n-1}$$
$$\geq C^{t_2}(B \restriction 2^{n-1}) + 2^{n-1} - O(n) \qquad \text{by definition of } x_{n-1}$$
$$\geq 2^{n-1} + 2^{n-2} + C^{t_3}(B \restriction 2^{n-2}) - O(n) \qquad \text{reapplying the argument above}$$
$$\geq \frac{3}{4}2^n - O(n) > \frac{2}{3}(2^n + j' + 1) = \frac{2}{3}j.$$

For the second case, suppose $j' > \log n$. We have

$$C^t(B \restriction j) \geq C^{t'_1}(x_n \restriction j', B \restriction 2^n)$$
$$\geq C^{t'_2}(x_n \restriction j' \mid B \restriction 2^n) + C^{t'_2}(B \restriction 2^n) - O(n) \qquad \text{By Lemma 4}$$
$$\geq C^{t'_3}(x_n \mid B \restriction 2^n) - 2^n + j' + C^{t'_2}(B \restriction 2^n) - O(n) \qquad \text{by Corollary 5}$$
$$\geq C_{T_n}(x_n \mid B \restriction 2^n) - 2^n + j' + C^{t'_2}(B \restriction 2^n) - O(n) \qquad \text{because } t'_3 \in L_n$$
$$\geq 2^n - 2^n + j' + C^{t'_2}(B \restriction 2^n) - O(n) \qquad \text{by definition of } x_n$$
$$\geq j' + C^{t'_4}(x_{n-1}, B \restriction 2^{n-1}) - O(n) \qquad \text{same as in the first case}$$
$$\geq j' + \frac{3}{4}2^n - O(n) > \frac{2}{3}j \qquad \text{same as in the first case}$$

Note that each iteration of the argument above yields a 2^{n-k} term ($k = 1, 2, 3, \ldots$), therefore for any $\varepsilon > 0$, there is a number I of iterations, such that B can be shown $(1 - \varepsilon)n$-deep$_C$, for all n large enough such that $t_1, t_2, \ldots, t_{3I} \in L_n$. $\qquad \square$

Corollary 11. *Theorem 10 also holds for K-depth.*

4.2 Depth Implies Highness or DNR

An analogue of Theorem 8 holds for K.

Theorem 12. *Let A be a h-deep$_K$ set for some recursive order h. Then A is high or DNR.*

As a corollary, we show that in the left-r.e. case, depth always implies highness.

Corollary 13. *If A is left-r.e. and h-deep$_K$ (for some recursive order h) then A is high.*

As a second corollary, we prove that every weakly-useful set is either high or DNR. A set A is weakly-useful if there is a time-bound s such that the class of all sets truth-table reducible to A with this time bound s is not small, i.e., does not have measure zero within the class of recursive sets; see [16] for a precise definition. In [16], it was shown that every weakly-useful set is $O(1)$-deep$_K$ (even order-deep$_K$ as observed in [3]) thus generalising the fact that $\emptyset'$ is $O(1)$-deep$_K$, since $\emptyset'$ is weakly-useful.

Theorem 14 (Antunes, Matos, Souto and Vitányi [3]; Juedes, Lathrop and Lutz [16]). *Every weakly-useful set is order-deep$_K$.*

It is shown in [16] that every high degree contains a weakly-useful set. Our results show some type of converse to this fact.

Theorem 15. *Every weakly-useful set is either high or DNR.*

4.3   A Low Deep Set

We showed in Theorem 10 that every εn-deep$_K$ set is high. Also Theorem 12 shows that every order-deep$_K$ set is either high or DNR. Thus one might wonder whether there exists any non-high order-deep$_K$ set. We answer this question affirmatively by showing there exist low order-deep$_K$ sets.

Theorem 16. *If A has PA-complete degree, then there exists a weakly-useful set $B \equiv_T A$.*

Proof. Let $f \leq_T A$ be a Boolean DNR function and let $g(n) := 1 - f(n)$. It follows that if Φ_e is Boolean and total, then $g(e) = \Phi_e(e)$. One can thus encode g into a set $B \leq_T A$ such that for every e such that Φ_e is Boolean and total and for every x, $B(\langle e + 1, x \rangle) = \Phi_e(x)$. One can also encode A into B (for example, $B(\langle 0, x \rangle) = A(x)$) so that $A \equiv_T B$. Thus for every recursive set L there exists e such that for every string x, we have $L(x) = B(r_e(x))$, where $r_e(x) = \langle e, x \rangle$ is computable within $s(n) = n^2$ steps (by using a lookup table on small inputs). It follows that every recursive set is truth-table reducible to B within time $s(n) = n^2$. Because the class of recursive sets does not have measure zero within the class of recursive sets [16], it follows that B is weakly-useful. $\square$

Corollary 17. *If A has PA-complete degree, then there exists an order-deep$_K$ set $B \equiv_T A$. Furthermore, there is a Π_1^0-class only consisting of order-deep$_K$ sets. For each of the properties low, superlow, low for Ω and hyperimmune-free, there exists an order-deep$_K$ set which also has this respective property.*

Here recall that a set A is said low for Ω iff Chaitin's Ω is Martin-Löf random relative to A; a set A has superlow degree if its jump A' is truth-table reducible to the halting problem. This corollary follows from Theorems 14 and 16 as well as from the well-known basis theorems for Π_1^0-classes [10,15]. The reason one uses PA-complete sets instead of merely Martin-Löf random sets (which also satisfy all basis theorems), is that Martin-Löf random sets are not weakly-useful; indeed, it is known that they are not even $O(1)$-deep$_K$. This stands in contrast to the following result.

Corollary 18. *There are two Martin-Löf random sets A and B such that $A \oplus B$ is order-deep$_K$.*

Proof. Barmpalias, Lewis and Ng [5] showed that every PA-complete degree is the join of two Martin-Löf random degrees; hence there are Martin-Löf random sets A, B such that $A \oplus B$ is a hyperimmune-free PA-complete set. Thus, by Theorem 16 there is a weakly-useful set Turing reducible to $A \oplus B$ which, due to the hyperimmune-freeness, is indeed truth-table reducible to $A \oplus B$. It follows that $A \oplus B$ is itself weakly-useful and therefore order-deep$_K$ by Theorem 14. $\square$

4.4 No K-Trivial is $O(1)$-deep$_K$

A key property of depth is that "easy" sets should not be deep. Bennett [6] showed that no recursive set is deep. Here we improve this result by observing that no K-trivial set is deep. As we will see this result is close to optimal.

Theorem 19. *No K-trivial set is $O(1)$-deep$_K$.*

Call a set A ultracompressible if for every recursive order g and all n, $K(A \upharpoonright n) \leq^+ K(n) + g(n)$. The following theorem shows that our result is close to optimal.

Theorem 20 (Lathrop and Lutz [17]). *There is an ultracompressible set A which is $O(1)$-deep$_K$.*

Theorem 21 (Herbert [13]). *There is a set A which is not K-trivial but which satisfies that for every Δ_2^0 order g and all n, $K(A \upharpoonright n) \leq^+ K(n) + g(n)$.*

It would be interesting to know whether such sets as found by Herbert can be $O(1)$-deep$_K$. The result of Herbert is optimal, Csima and Montalbán [8] showed that such sets do not exist when using Δ_4^0 orders and Baartse and Barmpalias [4] improved this non-existence to the level Δ_3^0. We also point to related work of Hirschfeldt and Weber [14].

Theorem 22 (Baartse and Barmpalias [4]). *There is a Δ_3^0 order g such that a set A is K-trivial iff $K(A \upharpoonright n) \leq^+ K(n) + g(n)$ for all n.*

5   Infinitely Often Depth and Conditional Depth

Bennett observed in [6] that being infinitely often Bennett deep is meaningless, because all recursive sets are infinitely often deep. A possibility for a more meaningful notion of infinitely often depth, is to consider a depth notion where the length of the input is given as an advice. We call this notion i.o. depth.

Definition 23. A set A is i.o. $O(1)$-deep$_K$ if for every $c \in \mathbb{N}$ and for every time bound t there are infinitely many n satisfying $K^t(A \upharpoonright n \mid n) - K(A \upharpoonright n \mid n) \geq c$.

If we replace K with C in the above definition, we call the corresponding notion i.o. $O(1)$-deep$_C$.

Lemma 24. *Let A be recursive. Then A is neither i.o. $O(1)$-deep$_C$ nor i.o. $O(1)$-deep$_K$.*

Proof. Let A be recursive and t be a time bound. Wlog A is recursive in time t, i.e., for every $n \in \mathbb{N}$ we have $C^t(A \upharpoonright n \mid n) \leq c$ for some constant c, thus $\forall n\ C^t(A \upharpoonright n \mid n) - C(A \upharpoonright n \mid n) < c$. The K case is similar.   □

The following shows that very little computational power is needed to compute an i.o. deep set.

Theorem 25.   *1. There is a Π^0_1-class such that every member is i.o. εn-deep$_C$ for all $\varepsilon < 1$. In particular there is such a set of hyperimmune-free degree. Furthermore, every hyperimmune Turing degree contains such a set.*

2. Every nonrecursive many-one degree contains an i.o. $O(1)$-deep$_C$ set.

3. If A is not recursive, not DNR and hyperimmune-free, then A is i.o. $O(1)$-deep$_C$.

Proof. This result is obtained by splitting the natural numbers recursively into intervals $I_n = \{a_n, \dots, b_n\}$ such that $b_n = (2 + a_n)^2$. Now one defines the Π^0_1-class such that for each $n = \langle e, k \rangle$ where $t = \Phi_e$ is defined up to b_n, a string $\tau \in \{0,1\}^{b_n - a_n + 1}$ is selected such that for all $\sigma \in \{0,1\}^{a_n}$, $C^t(\sigma\tau) \geq b_n - 2a_n - 2$ and then it is fixed that all members A of the Π^0_1-class have to satisfy $A(x) = \tau(x - a_n)$ for all $x \in I_n$. Since there are $2^{b_n - a_n + 1}$ strings τ and for each program of size below $b_n - 2a_n - 2$ can witness that only 2^{a_n} many τ are violating $C^t(\sigma\tau) \geq |\tau| - |\sigma|$ for some $\sigma \in \{0,1\}^{a_n}$, there will be less than $2^{b_n - a_n + 1} - 2^{b_n - a_n}$ many τ that get disqualified and so the search finds such a τ whenever Φ_e is defined up to b_n. Hence, for every total $t = \Phi_e$, there are infinitely many intervals I_n with n of the form $\langle e, k \rangle$ such that on these I_n, $C^t(A(0)A(1) \dots A(b_n) \mid n) \geq C^t(A(0)A(1) \dots A(b_n)) - \log(n) \geq b_n - 3a_n$ and $C(A(0)A(1) \dots A(b_n) \mid n) \leq a_n + c$ for a constant c, as the program only needs to know how A behaves below a_n and can fill in the values of τ on I_n. So the complexity improves after time $t(b_n)$ from $b_n - 3a_n$ to a_n and, to absorb constants, one can conservatively estimate the improvement by $b_n - 5a_n$. By the choice of a_n, b_n, the ratio $(b_n - 5a_n)/b_n$ tends to 1 and therefore every A in the

Π_1^0-class is εn-deep$_C$ for every $\varepsilon < 1$. Note that there are hyperimmune-free sets inside this Π_1^0-class, as it has only nonrecursive members.

Furthermore, one can see that the proof also can be adjusted to constructing a single set in a hyperimmune Turing degree rather than constructing a full Π_1^0-class. In that case one takes some function f in this degree which is not dominated by any recursive function and then one permits for each $n = \langle e, k \rangle$ the time $\Phi_e(b_n)$ in the case that $\Phi_e(b_n) < f(k)$ and chooses τ accordingly and one takes $\tau = 0^{b_n - a_n + 1}$ in the case that Φ_e does not converge on all values below b_n within time $f(k)$ otherwise. This construction is recursive in the given degree and a slight modification of this construction would permit to code the degree into the set A.

For the second item, consider a set $A \subseteq \{4^n : n \in \mathbb{N}\}$. Every many-one degree contains such a set. For each binary string σ, let

$$S_\sigma = \{\tau \in \{0,1\}^* : 4^{|\sigma|-1} < |\tau| \leq 4^{|\sigma|} \text{ and } \tau(4^n) = \sigma(n) \text{ for all } n < |\sigma|$$
$$\text{and } \tau(n) = 0 \text{ for all } n < |\tau| \text{ which are not a power of } 4\}.$$

In other word, for every $A \subseteq \{4^n : n \in \mathbb{N}\}$, $S_{A(1)A(4)A(16)...A(4^n)}$ contains those τ which are a prefix of A and for which $\tau(4^n)$ is defined but not $\tau(4^{n+1})$. For each e, k, n where Φ_e is a total function t, we now try to find inductively for $m = 4^n + 1, 4^n + 2, \ldots, 4^{n+1}$ strings $\sigma_m \in \{0,1\}^{n+1}$ such that whenever σ_m is found then it is different from all those $\sigma_{m'}$ which have been found for some $m' < m$ and the unique $\tau \in S_{\sigma_m} \cap \{0,1\}^m$ satisfies $C^t(\tau \mid m) \geq e + 3k$. Note that due to the resource-bound on C^t one can for each $m' < m$ check whether $\sigma_{m'}$ exists and take this information into account when trying to find σ_m. Therefore, for those m where σ_m exists, the $\tau \in S_{\sigma_m} \cap \{0,1\}^m$ can be computed from m, e and k and hence $C(\tau \mid m) \leq e + k + c$ for some constant c independent of e, k, n, m.

Now assume that A is not infinitely often $O(1)$-deep$_C$. Then there is a total function $t = \Phi_e$ and a $k > c$ such that $C(\tau \mid |\tau|) \geq C^t(\tau \mid |\tau|) - k$ for all prefixes τ of A. It follows that in particular never a σ_m with S_{σ_m} consisting of prefixes of A is selected in the above algorithm using e, k. This then implies that for almost all n and the majority of the m in the interval from 4^n to 4^{n+1} (which are those for which σ_m does not get defined) it holds that $C^t(\tau \mid m) \leq e + 3k$ for the unique $\tau \in S_{A(1)A(4)A(16)...A(4^n)} \cap \{0,1\}^m$. There are at most 2^{e+3k+2} many strings $\sigma \in \{0,1\}^{n+1}$ such that at least half of the members τ of S_σ satisfy that $C(\tau \mid |\tau|) \leq e + 3k$ and there is a constant c' such that for almost all n the corresponding σ satisfy $C(\sigma|n) \leq e + 3k + c'$. It follows that $C(\tau \mid |\tau|) \leq e + 3k + c''$ for some constant c'' and almost all n and all $\tau \in S_{A(1)A(4)A(16)...A(4^n)}$; in other words, $C(\tau \mid |\tau|) \leq e + 3k + c''$ for some constant c'' and almost all prefixes τ of A. Hence A is recursive [18, Exercise 2.3.4 on page 131].

For the third item, let A be as above, and let t be a time bound. Let $\tilde{h}(n) = \min_{m>n}\{C^t(A \restriction m \mid m) > n\}$. Since $\tilde{h} \leq_T A$ and A is hyperimmune-free, there exists a recursive h such that $\forall n \in \mathbb{N} \; h(n) > \tilde{h}(n)$. Wlog we can choose h such that $\forall n \in \mathbb{N} \; h(n+1) > \tilde{h}(h(n))$. Let $g \leq_T A$ be defined by $g(n) = A \restriction h(n+1)$. Because A is not DNR, we have $\exists^\infty n \; \Phi_n(n) = A \restriction h(n+1)$. Thus, $\forall n \exists m$ $h(n) \leq m < h(n+1)$ and $C^t(A \restriction m \mid m) \geq h(n)$.

Let e be an index of a program such that $\Phi_e(m) = \Phi_n(n) \upharpoonright m$ with n satisfying $h(n) \leq m < h(n+1)$. Thus $\exists^\infty n \exists m\ h(n) \leq m < h(n+1)$ and $C^t(A \upharpoonright m \mid m) \geq h(n)$ and $C(A \upharpoonright m \mid m) \leq O(1)$, i.e., A is i.o. $O(1)$-deep$_C$. □

Franklin and Stephan [12] showed that for every Schnorr trivial set and every order h it holds that A is not i.o. h-deep$_C$. Thus, the second and third points cannot be generalised to order-deep$_C$. Also, there are high truth-table degrees and hyperimmune-free Turing degrees which do not contain any order-deep$_C$ set. Examples for Schnorr trivial sets are all maximal sets and, for every partial $\{0,1\}$-valued function ψ whose domain is a maximal set, all sets A with $\psi(x) \downarrow \Rightarrow A(x) = \psi(x)$.

6   Conclusion

We conclude that the choice of the depth magnitude has consequences on the computational power of the corresponding deep sets, and that larger magnitudes is not necessarily preferable over smaller magnitudes. Therefore choosing the appropriate depth magnitude for one's purpose is delicate, as the corresponding depth notions might be very different. When the depth magnitude is large, we proved that depth and highness coincide. We showed that this is not the case for smaller depth magnitude by constructing a low order deep set, but the set is not r.e. We therefore ask whether there is a low $O(1)$-deep$_K$ r.e. set.

From our results, for magnitudes of order $O(1)$, K-depth behaves better than C-depth. To further strengthen that observation we ask whether there is an MLR $O(1)$-deep$_C$ set.

References

1. Arslanov, M.M.: Degree structures in local degree theory. Complexity Logic, Recursion Theor. **187**, 49–74 (1997)
2. Antunes, L.F.C., Fortnow, L., van Melkebeek, D., Vinodchandran, N.V.: Computational depth. Theor. Comput. Sci. **354**, 391–404 (2006)
3. Antunes, L.F.C., Matos, A., Souto, A., Paul, M.B.P.: Depth as randomness deficiency. Theor. Comput. Sys. **45**(4), 724–739 (2009)
4. Baartse, M., Barmpalias, G.: On the gap between trivial and nontrivial initial segment prefix-free complexity. Theor. Comput. Sys. **52**(1), 28–47 (2013)
5. Barmpalias, G., Lewis, A.E.M.: The importance of Π_1^0-classes in effective randomness. J. Symbolic Logic **75**(1), 387–400 (2010)
6. Bennett, C.H.: Logical Depth and Physical Complexity. The Universal Turing Machine. A Half-Century Survey. Oxford University Press, New York (1988)
7. Chaitin, G.J.: A theory of program size formally identical to information theory. J. Assoc. Comput. Mach. **22**, 329–340 (1975)
8. Csima, B.F., Montalbán, A.: A minimal pair of K-degrees. Proc. Am. Math. soc. **134**, 1499–1502 (2006)
9. Doty, D., Moser, P.: Feasible depth. In: Cooper, S.B., Löwe, B., Sorbi, A. (eds.) CiE 2007. LNCS, vol. 4497, pp. 228–237. Springer, Heidelberg (2007)

10. Downey, R.G., Hirschfeldt, D.R., Miller, J.S., Nies, A.: Relativizing chaitin's halting probability. J. Math. Logic **5**, 167–192 (2005)
11. Downey, R.G., Hirschfeldt, D.R.: Algorithmic Randomness and Complexity. Springer, Heidelberg (2010)
12. Franklin, J., Stephan, F.: Schnorr trivial sets and truth-table reducibility. J. Symbolic Logic **75**, 501–521 (2010)
13. Herbert, I.: Weak Lowness Notions for Kolmogorov Complexity. University of California, Berkeley (2013)
14. Hirschfeldt, D., Weber, R.: Finite Self-Information. Computability **1**(1), 85–98 (2012). Manuscript
15. Jockusch, C.G., Soare, R.I.: Π_1^0 classes and degrees of theories. Trans. Am. Math. Soc. **173**, 33–56 (1972)
16. Juedes, D.W., Lathrop, J.I., Lutz, J.H.: Computational depth and reducibility. Theoret. Comput. Sci. **132**, 37–70 (1994)
17. Lathrop, J.I., Lutz, J.H.: Recursive computational depth. Inf. Comput. **153**(1), 139–172 (1999)
18. Paul, M.L., Vityányi, M.B.: An Introduction to Kolmogorov Complexity and its Applications. Springer Verlag, New York (2008)
19. Miller, J.S.: Every 2-random real is Kolmogorov random. J. Symbolic Logic **69**(3), 907–913 (2004)
20. Moser, P.: On the polynomial depth of various sets of random strings. Theoret. Comput. Sci. **477**, 96–108 (2013)
21. Nies, A.: Computability and Randomness. Oxford University Press, New York (2009)
22. Nies, A., Terwijn, S.A., Stephan, F.: Randomness, relativization and Turing degree. J. Symbolic Logic **70**(2), 515–535 (2005)
23. Odifreddi, P.: Classical Recursion Theory: The Theory of Functions and Sets of Natural Numbers, vol. 1. Elsevier, The Netherlands (1989)

On the Expressive Power of Read-Once Determinants

N.R. Aravind[1]([⊠]) and Pushkar S. Joglekar[2]

[1] Indian Institute of Technology, Hyderabad, India
aravind@iith.ac.in
[2] Vishwakarma Institute of Technology, Pune, India
joglekar.pushkar@gmail.com

Abstract. We introduce and study the notion of read-k projections of the determinant: a polynomial $f \in \mathbb{F}[x_1, \ldots, x_n]$ is called a *read-k projection of determinant* if $f = det(M)$, where entries of matrix M are either field elements or variables such that each variable appears at most k times in M. A monomial set S is said to be expressible as read-k projection of determinant if there is a read-k projection of determinant f such that the monomial set of f is equal to S. We obtain basic results relating read-k determinantal projections to the well-studied notion of determinantal complexity. We show that for sufficiently large n, the $n \times n$ permanent polynomial $Perm_n$ and the elementary symmetric polynomials of degree d on n variables S_n^d for $2 \leq d \leq n-2$ are not expressible as read-once projection of determinant, whereas $mon(Perm_n)$ and $mon(S_n^d)$ are expressible as read-once projections of determinant. We also give examples of monomial sets which are not expressible as read-once projections of determinant.

1 Introduction

In a seminal work [13], Valiant introduced the notion of the determinantal complexity of multivariate polynomials and proved that any polynomial $f \in \mathbb{F}[x_1, \ldots, x_n]$ can be expressed as $f = det(M_{m \times m})$, where the entries of M are affine linear forms in the variables $\{x_1, x_2, \ldots, x_n\}$. The smallest value of m for which $f = det(M_{m \times m})$ holds is called the determinantal complexity of f and denoted by $dc(f)$. Let $Perm_n$ denote the permanent polynomial:

$$Perm_n(x_{11}, \ldots, x_{nn}) = \sum_{\sigma \in S_n} \prod_{i=1}^{n} x_{i,\sigma(i)}$$

Valiant postulated that the determinantal complexity of $Perm_n$ is not polynomially bounded - i.e. $dc(Perm_n) = n^{\omega(1)}$. This is one of the most important conjectures in complexity theory. So far the best known lower bound on $dc(Perm_n)$ is $\frac{n^2}{2}$, known from [1,9].

Another related notion considered in [13] is projections of polynomials: A polynomial $f \in \mathbb{F}[x_1, \ldots, x_n]$ is said to be a projection of $g \in \mathbb{F}[y_1, \ldots, y_m], m \geq$

© Springer International Publishing Switzerland 2015
A. Kosowski and I. Walukiewicz (Eds.): FCT 2015, LNCS 9210, pp. 95–105, 2015.
DOI: 10.1007/978-3-319-22177-9_8

n if f is obtained from g by substituting each variable y_i by some variable in $\{x_1, x_2, \ldots, x_n\}$ or by an element of field $\mathbb{F}$. Valiant's postulate implies that if $Perm_n$ is projection of the Determinant polynomial Det_m then m is $n^{\omega(1)}$. We refer to the expository article by von-zur Gathen on Valiant's result [2].

We define the notion of read-k projection of determinant, which is a natural restriction of the notion of projection of determinant. Let $X = \{x_1, \ldots, x_n\}$ be a set of variables and let $\mathbb{F}$ be a field.

Definition 1. *We say that a matrix $M_{m \times m}$ is a read-k matrix over $X \cup \mathbb{F}$ if the entries of M are from $X \cup \mathbb{F}$ and for every $x \in X$, there are at most k pairs of indices (i, j) such that $M_{i,j} = x$. We say that a polynomial $f \in \mathbb{F}[X]$ is read-k projection of Det_m if there exists a read-k matrix $M_{m \times m}$ over X such that $f = det(M)$.*

Remark: We use the phrase *a polynomial is expressible as read-once determinant* in place of "a polynomial is read-1 projection of determinant" in some places. Note that only a multilinear polynomial can be expressible as a read-once determinant.

The following upper bound on determinantal complexity, proved in Sect. 2, is one of the motivations for studying this model.

Theorem 2. *Let $f \in \mathbb{F}[x_1, \ldots, x_n]$. If f is a read-k projection of determinant, then $dc(f) \leq nk$.*

The above theorem immediately shows that read-k projections of determinant are not universal for any constant k; indeed in the case of finite fields, by simple counting arguments, we can show that most polynomials are not read-k expressible for $k = 2^{o(n)}$.

Ryser's formula for the permanent expresses the permanent polynomial $Perm_n$ as a read-2^{n-1} projection of determinant. In contrast, it follows from Theorem 2 that Valiant's hypothesis implies the following: $Perm_n \neq det(M_{m \times m})$ for a read-$n^{O(1)}$ matrix M of *any size*. So the expressibility question is more relevant in the context of read-k determinant model rather than the size lower bound question. In this paper, we obtain the following results for the simplest case $k = 1$.

Theorem 3. *For $n > 5$, the $n \times n$ permanent polynomial $Perm_n$ is not expressible as a read-once determinant over the field of reals and over finite fields in which -3 is a quadratic non-residue.*

We prove Theorem 3 in Sect. 3 as a consequence of non-expressibility of elementary symmetric polynomials as read-once determinants.

Our interest in this model also stems from the following reason. Most of the existing lower-bound techniques for various models, including monotone circuits, depth-3 circuits, non-commutative ABPs etc., are not sensitive to the coefficients of the monomials of the polynomial for which the lower-bound is proved. For example, the monotone circuit lower-bound for permanent polynomial by Jerrum and Snir [8], carries over to any polynomial with same monomial set as permanent. The same applies to Nisan's rank argument [10] or the various lower bound results based on the partial derivative techniques (see e.g. [4,11]).

On the other hand, for proving lower bounds on the determinantal complexity of the permanent, one must use some properties of the permanent polynomial which are not shared by the determinant polynomial. A natural question is whether there are models more restrictive than determinantal complexity (so that proving lower bounds may be easier) and which are coefficient-sensitive. Read-k determinants appear to be a good choice for such a model.

In light of the above discussion and to formally distinguish the complexity of a polynomial and that of its monomial set, we have the following definition.

Definition 4. *For $f \in F[X]$, we denote by* mon(f) *the set of all monomials with non-zero coefficient in f. We say that a set S of monomials is expressible as read-k determinant if there exists a polynomial $f \in \mathbb{F}[X]$ such that f is a read-k projection of determinant and $S = mon(f)$.*

Let S_n^d denote the elementary symmetric polynomial of degree d:

$$S_n^d(x_1, x_2, \ldots, x_n) = \sum_{A \subseteq \{1,2,\ldots,n\}, |A|=d} \prod_{i \in A} x_i$$

In Sect. 3, we prove the non-expressibility of elementary symmetric polynomials as read-once determinants; a contrasting result also proved in the same section is the following:

Theorem 5. *For all $n \geq d \geq 1$ and $|\mathbb{F}| \geq n$, the monomial set of S_n^d is expressible as projection of read-once determinant.*

The organization of the paper is as follows. In Sect. 2, we prove Theorem 2 and make several basic observations about read-once determinants. In Sect. 3, as our main result, we show the non-expressibility of the elementary symmetric polynomials as read-once determinants and as a consequence deduce non-expressibility of the permanent (Theorem 3). We also prove that the monomial set of any elementary symmetric polynomial is expressible as a read-once determinant (Theorem 5). In Sect. 4, we give examples of monomial sets which are not expressible as read-once determinants.

2 Basic Observations

First we note that read-once determinants are strictly more expressive than occurrence-one algebraic branching programs which in turn are strictly more expressive than read-once formulas. By *occurrence-one ABP* we mean an algebraic branching program in which each variable is allowed to repeat at most once [7]. (We are using the term occurrence-one ABPs rather than read-once ABPs to avoid confusion, as the latter term is sometimes used in the literature to mean an ABP in which any variable can appear at most once on any source to sink path in the ABP).

In the following simple lemma, we compare read-once determinants with read-once formulas and occurrence-one ABPs.

Lemma 6. *Any polynomial computed by a read-once formula can be computed by an occurrence-one ABP, and any polynomial computed by an occurrence-one ABP can be computed by a read-once determinant. Moreover there is a polynomial which can be computed by read-once determinant but can't be computed by occurrence-one ABPs.*

Proof. Let $f \in \mathbb{F}[X]$ be a polynomial computed by a read-once formula or an occurrence-one ABP. Using Valiant's construction [13], we can find a matrix M whose entries are in $X \cup \mathbb{F}$ such that $f = det(M)$. We observe that if we start with a *read-once* formula or an *occurrence-one* ABP then for the matrix M obtained using Valiant's construction, every variable repeats at most once in M. This proves that f can be computed by read-once determinants. To see the other part, consider the elementary symmetric polynomial of degree two over $\{x_1, x_2, x_3\}$:$S_3^2(x_1, x_2, x_3) = x_1x_2 + x_1x_3 + x_2x_3$. It is proved in ([7] Appendix-B) that S_3^2 cannot be computed by an occurrence-one ABP. From the discussion in the beginning of Sect. 3.1 it follows that S_3^2 can be computed by a read-once determinant.

Let $X = \{x_1, x_2, \ldots, x_n\}$. Let $S = \{i_1, i_2, \ldots, i_k\} \subseteq [n]$, let X_S denote the set of variables $\{x_{i_1}, x_{i_2}, \ldots, x_{i_k}\}$. We define $\frac{\partial f}{\partial X_S}$ a partial derivative of f with respect to X_S as $\frac{\partial f}{\partial X_S} = \frac{\partial^k f}{\partial x_{i_1} \partial x_{i_2} \ldots \partial x_{i_k}}$. For a vector $a = (a_1, \ldots, a_k)$ with $a_i \in \mathbb{F}$ let $f|_{S=a}$ denote polynomial g over variables $X \setminus X_S$ which is obtained from f by substituting variable $x_{i_j} = a_j$.

We define the set ROD to be the set of all polynomials in $\mathbb{F}[X]$ that are expressible as read-once projection of determinant. The following simple proposition shows that the set ROD has nice closure properties.

Proposition 7. *Let f be a polynomial over X such that $f \in ROD$ and $S \subseteq X$, $|S| = k$. Let $a \in \mathbb{F}^k$. Then $f|_{S=a}, \frac{\partial f}{\partial X_S} \in ROD$. For any polynomial $g \in ROD$ such that fg is a multilinear polynomial, we have $fg \in ROD$.*

We use the notation $M \sim N$ to mean that $det(M) = det(N)$.

Proof-of Theorem 2. Let $f = det(M)$ for a read-k matrix M. Without loss of generality, we can assume that for some $m \leq kn$, the principal m by m submatrix of M contains all the variables. Let Q denote this submatrix and let R denote the submatrix formed by the remaining columns of the first m rows. Suppose that the number of remaining rows of M is equal to p. Let T denote the submatrix formed by these rows. We note that M has full rank, and hence the row-rank and column-rank of T are both equal to p.

Consider a set of p linearly independent columns in T and let T_1 be the submatrix formed by these columns; let T_2 denote the remaining m columns of T. Further, let Q_1 and R_1 denote the columns of Q and R respectively, corresponding to the columns of T_1 and similarly, let Q_2 and R_2 denote the columns of Q and R, corresponding to the columns of T_2. In other words, the columns of M can be permuted to obtain $M' \sim M$:

$$M' = \begin{pmatrix} Q_1|R_1 & Q_2|R_2 \\ T_1 & T_2 \end{pmatrix}.$$

Let g denote the unique linear transformation such that $[T_2 + g(T_1)] = [0]$.

Applying g to the last m columns of M', we obtain a matrix $N \sim M'$ such that

$$N = \begin{pmatrix} Q_1|R_1 & Q_2|R_2 + g(Q_1|R_1) \\ T_1 & 0 \end{pmatrix}.$$

Let $det(T_1) = c \in \mathbb{F}$; clearly $c \neq 0$. Let N' be a matrix obtained by multiplying some row of $[Q_2|R_2] + g([Q_1|R_1])$ by c. The entries of N' are affine linear forms, $det(N') = f$ and the dimension of N' is $m \leq kn$. This proves Theorem 2. $\qquad\square$

In the next lemma we show that if $f \in \mathbb{F}[x_1, \ldots, x_n]$ such that $f = det(M_{m \times m})$ for a read-once matrix M, then we can without loss of generality assume that $m \leq 3n$. The proof is on similar lines as that of Theorem 2.

Lemma 8. *Let $f \in \mathbb{F}[x_1, \ldots, x_n]$ be expressible as read-once determinant. Then there is a read-once matrix M of size at most $3n$ such that $f = det(M)$:*

3 Elementary Symmetric Polynomials and Permanent

In this section we will prove our main result: the elementary symmetric polynomials S_n^d for $2 \leq n \leq n-2$ and the permanent $Perm_n$ are not expressible as read-once determinants for sufficiently large n. We will first prove that $S_4^2 \notin ROD$ and use it to prove non-expressibility of $Perm_n$ and S_n^d.

We begin with following simple observation based on the closure properties of ROD. We skip the proof due to lack of space.

Lemma 9. *1. If $S_m^k \notin ROD$ then $S_n^d \notin ROD$ for $d \geq k$ and $n \geq m + d - k$.*
2. If $Perm_m \notin ROD$ then $Perm_n \notin ROD$ for $n \geq m$.

From the above lemma, it is clear that, if the polynomials $Perm_n$ or S_n^d are expressible as read-once determinants for some n then these polynomials will be expressible as read-once determinants for some constant value of $n = O(1)$.

3.1 Elementary Symmetric Polynomials

For $d = 1$ or n, the elementary symmetric polynomial S_n^d can be computed by a $O(n)$ size read-once formula so by lemma 6 we can express S_n^d as a read-once determinant. We observe that $S_n^{n-1} \in ROD$ over any field as $S_n^{n-1} = det \begin{pmatrix} D & A \\ C & 0 \end{pmatrix}$ Where D is $n \times n$ diagonal matrix with $(i,i)^{th}$ entry x_i for $i = 1$ to n. C and A are $1 \times n$ and $n \times 1$ matrices such that all entries of C are 1 and all entries of A are -1. So $S_n^d \in ROD$ for $d = 1, n-1, n$. In this section, we show that $S_n^d \notin ROD$ for every other choice of d in the case of field of reals or finite fields in which -3 is quadratic non-residue.

First we consider the case of field of real numbers. Let $S_4^2(x_1, x_2, x_3, x_4) = c' \cdot det(M)$ for a read-once matrix M and a non-zero $c' \in \mathbb{R}$. Rearranging rows and columns of M or taking out a scalar common from either row or column of M will change the value of $det(M)$ only by a scalar, so pertaining to the expressibility question, we can do these operations freely (we will get different scalar than c as a multiplier but that is not a problem). For any $i, j \in \{1, 2, 3, 4\}, i \neq j$, $x_i x_j \in mon(S_4^2)$. So clearly $\frac{\partial S_4^2}{\partial x_i \partial x_j} \neq 0$ which implies that the determinant of the minor obtained by removing the rows and the columns corresponding to the variables x_i, x_j is non-zero. So for any $i, j \in \{1, 2, 3, 4\}, i \neq j$, x_i and x_j appear in different rows and columns in M. By suitably permuting rows and columns of M we can assume that $S_4^2 = c \cdot det(N)$ for a non zero real c and read-once matrix N such that $(i, i)^{th}$ entry of N is variable x_i for $i = 1$ to 4.

So $N = \begin{pmatrix} x_1 & - & - & - & \beta_1 \\ - & x_2 & - & - & \beta_2 \\ - & - & x_3 & - & \beta_3 \\ - & - & - & x_4 & \beta_4 \\ \alpha_1 & \alpha_2 & \alpha_3 & \alpha_4 & L \end{pmatrix}$ Here L is a $m - 4 \times m - 4$ matrix, and α_i, β_i are

column and row vectors of size $m - 4$ for $i = 1$ to 4 and $-$ represents arbitrary scalar entry. Let $p = m - 4$. Index the columns and the rows of the matrix using numbers $1, 2, \ldots, m$. Let S denote a set of column and row indices corresponding to submatrix L. For any set $\{a_1, a_2, \ldots, a_k\} \subset \{1, 2, 3, 4\}$, let $N_{a_1, a_2, \ldots, a_k}$ denote a minor of N obtained by removing rows and columns corresponding to indices $\{1, 2, 3, 4\} \setminus \{a_1, \ldots, a_k\}$ from N.

Definition 10. *Let $X = \{x_1, \ldots, x_n\}$ and M be a matrix with entries from $X \cup \mathbb{F}$. For $a = (a_1, a_2, \ldots, a_n) \in \mathbb{F}^n$ let M_a be the matrix obtained from M by substituting $x_i = a_i$ for $i = 1$ to n. Let $maxrank(M)(respect. minrank(M))$ denote the maximum(respect. minimum) rank of matrix M_a for $a \in \mathbb{F}^n$.*

Now we make some observations regarding ranks of various minors of N.

Lemma 11. *For $i, j \in \{1, 2, 3, 4\}$, $i \neq j$ we have*

1. *$maxrank(N_{i,j}) = minrank(N_{i,j}) = p + 2$*
2. *$maxrank(N_i) = minrank(N_i) = p$*
3. *$rank(L) \in \{p - 1, p - 2\}$*

Proof. Let $\{k, l\} = \{1, 2, 3, 4\} \setminus \{i, j\}$. Monomial $x_k x_l \in mon(S_4^2)$, so the matrix obtained from $N_{i,j}$ by any scalar substitution for x_i and x_j has full rank. So we have $minrank(N_{i,j}) = p + 2$. Since $N_{i,j}$ is a $(p + 2) \times (p + 2)$ matrix with minrank $p + 2$, clearly $maxrank(N_{i,j}) = minrank(N_{i,j}) = p + 2$. To prove the second part, note that the matrix N_i can be obtained by removing a row and a column from the matrix $N_{i,j}$. So clearly $minrank(N_i) \geq minrank(N_{i,j}) - 2 = p$. As S_4^2 doesn't contain any degree 3 monomial we have $maxrank(N_i) \leq p$. Hence $minrank(N_i) = maxrank(N_i) = p$.

The matrix N_i can be obtained from L by adding a row and a column so $rank(L) \geq minrank(N_i) - 2 = p - 2$. Since monomial $x_1 x_2 x_3 x_4 \notin mon(S_4^2)$, L can not be full-rank matrix so $rank(L) \leq p - 1$. Thus proving the lemma.

Suppose $rank(L) = p - 1$. By $cspan(L), rspan(L)$ we denote the space spanned by the columns and the rows of L respectively. Next we argue that for any $i \in \{1, 2, 3, 4\}$, $\alpha_i \in cspan(L)$ iff $\beta_i \notin rspan(L)$. To show that we need to rule out following two possibilities

1. $\alpha_i \notin cspan(L)$ and $\beta_i \notin rspan(L)$. In this case clearly $minrank(N_i) = rank(L) + 2 = p + 1$, a contradiction since $minrank(N_i) = p$ by lemma 11.
2. $\alpha_i \in cspan(L)$ and $\beta_i \in rspan(L)$. As $\beta_i \in rspan(L)$ we can use a suitable scalar value for x_i so that vector $[x_i \ \beta_i]$ is in the row span of the matrix $[\alpha_i \ L]$. Moreover $rank([\alpha_i \ L]) = p - 1$ as $\alpha_i \in cspan(L)$. So we have $minrank(N_i) = rank([\alpha_i \ L]) = rank(L) = p - 1$. But we know that $minrank(N_i)$ is p.

So we have $\alpha_i \in cspan(L)$ iff $\beta_i \notin rspan(L)$. From this it follows immediately that either there exist at least two α_i's $\in cspan(L)$ or there exists atleast two β_i's $\in rspan(L)$. So w.l.o.g. assume that for $i \neq j$, $\alpha_i, \alpha_j \in cspan(L)$, So $rank[\alpha_i \ \alpha_j \ L] = rank(L) = p - 1$. Matrix $N_{i,j}$ can be obtained from $[\alpha_i \ \alpha_j \ L]$ by adding two new rows, so $maxrank(N_{i,j}) \leq rank([\alpha_i \ \alpha_j \ L]) + 2 = p + 1$, a contradiction. This proves that $rank(L)$ can not be $p - 1$.

Now we consider the other case. Let $rank(L) = p - 2$. By applying row and column operation on N we can reduce block L to a diagonal matrix D with all non zero entries 1. Further applying row and column transformations we can drive entries in the vectors α_i's and β_i's corresponding to nonzero part of D to zero. Note that now we can remove all non-zero rows and columns of matrix D still keeping the determinant same. As a result we have $S_4^2 = c_1 \cdot det(N')$ where N' has following structure

$$\begin{pmatrix} x_1 + a_1 & & & & \beta_{1,1} & \beta_{1,2} \\ & x_2 + a_2 & & & \beta_{2,1} & \beta_{2,2} \\ & & x_3 + a_3 & & \beta_{3,1} & \beta_{3,2} \\ & & & x_4 + a_4 & \beta_{4,1} & \beta_{4,2} \\ \alpha_{1,1} & \alpha_{2,1} & \alpha_{3,1} & \alpha_{4,1} & 0 & 0 \\ \alpha_{1,2} & \alpha_{2,2} & \alpha_{3,2} & \alpha_{4,2} & 0 & 0 \end{pmatrix}$$

Note that the coefficient of the monomial $x_i x_j$ in N' is the determinant of the minor obtained by removing rows and columns corresponding to x_i and x_j from N'. It is equal to $(\alpha_{k,1}\alpha_{l,2} - \alpha_{k,2}\alpha_{l,1}) \cdot (\beta_{k,1}\beta_{l,2} - \beta_{k,2}\beta_{l,1})$ where $\{k, l\} = \{1, 2, 3, 4\} \setminus \{i, j\}$. It is easy to see that in fact without loss of generality we can assume that $\alpha_{1,1} = \beta_{1,1} = \alpha_{2,2} = \beta_{2,2} = 1$ and $\alpha_{1,2} = \beta_{1,2} = \alpha_{2,1} = \beta_{2,1} = 0$ (again by doing column and row transformations). So finally we have $S_4^2 = c \cdot det(N)$ where c is a non zero scalar, N is a matrix as shown below, and $a_1, \ldots, a_4$ are real numbers.

$$S_4^2(x_1, x_2, x_3, x_4) = c \cdot det \begin{pmatrix} x_1 + a_1 & & & & 1 & 0 \\ & x_2 + a_2 & & & 0 & 1 \\ & & x_3 + a_3 & & p' & r' \\ & & & x_4 + a_4 & q' & s' \\ 1 & 0 & p & q & 0 & 0 \\ 0 & 1 & r & s & 0 & 0 \end{pmatrix}$$

Comparing coefficients of monomials $x_i x_j$ for $i \neq j$ in S_4^2 and the determinant of corresponding minors of matrix N, we get following system of equations $c = 1, p.p' = q.q' = r.r' = s.s' = 1$ and $(ps - rq)(p's' - r'q') = 1$. Substituting $p' = 1/p, q' = 1/q$ etc. in the equation above, we have $(ps - rq)(1/ps - 1/rq) = 1$ which imply $(ps - rq)^2 = -(ps)(rq)$ i.e. $(ps)^2 + (rq)^2 = (ps)(rq)$ which is clearly false for non-zero real numbers p, q, r, s (as $(ps)^2 + (rq)^2 \geq 2(ps)(rq)$).(Note that we need p, q, r, s to be non zero since we have $pp' = qq' = rr' = ss' = 1$.) This proves that $S_4^2 \notin ROD$ over $\mathbb{R}$.

In the case of finite fields $\mathbb{F}$ in which -3 is a quadratic non-residue the argument is as follows. We have the equation $(ps - rq)^2 = -(ps)(rq)$ as above. Let $x = ps$ and $y = rq$, so we have $y^2 - xy + x^2 = 0$. Considering this as a quadratic equation in variable y, the equation has a solution in the concerned field iff the discriminant $\Delta = -3x^2$ is a perfect square, that happens only when -3 is a quadratic residue. So if -3 is a quadratic non residue, the above equation doesn't have a solution, leading to a contradiction. So we have the following theorem.

Theorem 12. *The polynomial S_4^2 is not expressible as a read-once determinant over the field of reals and over finite fields in which -3 is a quadratic non-residue.*

We note that we can express S_4^2 as a read-once determinant over $\mathbb{C}$ or e.g. over $\mathbb{F}_3$ by solving the quadratic equation in the proof of Theorem 12.

$$S_4^2(x_1, x_2, x_3, x_4) = det \begin{pmatrix} x_1 & 0 & 0 & 0 & 1 & 0 \\ 0 & x_2 & 0 & 0 & 0 & 1 \\ 0 & 0 & x_3 & 0 & 1 & r^{-1} \\ 0 & 0 & 0 & x_4 & 1 & 1 \\ 1 & 0 & 1 & 1 & 0 & 0 \\ 0 & 1 & r & 1 & 0 & 0 \end{pmatrix}$$

In the case $\mathbb{F} = \mathbb{C}$ choose $r = \frac{1+\sqrt{3}i}{2}$ and in case of $\mathbb{F}_3$ choose $r = 2 \pmod 3$.

Remark 13. We speculate that it should be possible to prove $S_6^2 \notin ROD$ over *any field* using similar technique as in the proof of Theorem 12 and that would immediately give us (slightly weaker) non-expressibility results for the general elementary symmetric polynomials and the permanent polynomial as compared to the Theorems 3 and 14. But we haven't worked out the details in the current work.

Theorem 12 together with Lemma 9 proves the desired non-expressibility result for elementary symmetric polynomials.

Theorem 14. *The polynomial $S_n^d \in \mathbb{F}[x_1, x_2, \ldots, x_n]$ is not expressible as a read-once determinant for $n \geq 4$ and $2 \leq d \leq n - 2$ when the field $\mathbb{F}$ is either the field of real numbers or a finite field in which -3 is a quadratic non-residue.*

In contrast, we show that the monomial set of S_n^d is expressible as a read-once determinant.

Proof-of Theorem 5. Let $k = n - d$ and $t = k + 1$.

Let D be a $n \times n$ diagonal matrix with $(i,i)^{th}$ entry x_i for $i = 1$ to n. and $M = \begin{pmatrix} D & A \\ C & B \end{pmatrix}$, where A, B, C are constant block matrices of dimensions $n \times t$, $t \times t$ and $t \times n$, respectively. We shall choose A, B, C such that $mon(det(M)) = mon(S_n^d)$. Let $B = J_{t \times t}$ be the matrix with all 1 entries. Let C be such that $rank(C) = k$, $rank(CB) = t$ and such that any k vectors in $Col(C)$ are linearly independent, where $Col(C)$ denotes set of column vectors of C. For example, we can let the ith column vector of C be $(1, a_i, a_i^2, \ldots, a_i^{k-1}, 0)^T$ for distinct values of a_i. Finally, let $A = C^T$.

It is clear that $det(M)$ is symmetric in the x_i's; thus it suffices to prove that $x_1 x_2 \ldots x_i$ is a monomial of $det(M)$ if and only if $i = d$. For $1 \leq i \leq n$, consider the submatrix M_i of M obtained by removing the first i rows and first i columns. We observe that $det(M_i) = x_{i+1} det(M_{i+1})$. Let r denote the minimum value of i such that $det(M_i) \neq 0$. Then it can be seen that $x_1 x_2 \ldots x_i$ is a monomial of $det(M)$ if and only if $i = r$.

We now prove that $det(M_i) = 0$ if and only if $i > d$. Let N_i denote the matrix formed by the last t rows of M_i. Then $det(M_i) = 0$ if and only if $rank(N_i) < t$. But $rank(N_i) = rank(Col(N_i))$ and by construction, $rank(Col(N_i)) < k$ if and only if $n - i < k$, i.e. if $i > n - k = d$. This completes the proof of Theorem 5. $\square$

3.2   Non-expressibility of Permanent as ROD

Now we prove the non-expressibility result for $Perm_n$ (Theorem 3).

Proof-of Theorem 3. We observe below that the elementary symmetric polynomial S_4^2 is a projection of the read-once 6×6 Permanent over reals.

$$4S_4^2(x_1, x_2, x_3, x_4) = Perm \begin{pmatrix} x_1 & 0 & 0 & 0 & 1 & 1 \\ 0 & x_2 & 0 & 0 & 1 & 1 \\ 0 & 0 & x_3 & 0 & 1 & 1 \\ 0 & 0 & 0 & x_4 & 1 & 1 \\ 1 & 1 & 1 & 1 & 0 & 0 \\ 1 & 1 & 1 & 1 & 0 & 0 \end{pmatrix}$$ So clearly if $Perm_6$ is a read-once

projection of determinant then S_4^2 also is a read-once projection of determinant. But by Theorem 12 we know that $S_4^2 \notin ROD$. So we get $Perm_6 \notin ROD$. From lemma 9 it follows that $Perm_n \notin ROD$ for any $n > 5$. $\square$

4   Non-expressible Monomial Sets

We have seen that the elementary symmetric polynomials and the Permanent polynomial can not be expressed as read-once determinants but their monomial sets are expressible as ROD. In this section we will give examples of monomial sets which can not be expressed as read-once determinant. Let $f \in \mathbb{F}[x_1, \ldots, x_n]$. We say that f is k-full if f contains every monomial of degree k and we say that f is k-empty if f contains no monomial of degree k.

Theorem 15. *Let $f \in \mathbb{F}[x_1, \ldots, x_n]$ and $f \in ROD$ be such that f is n-full, $(n-1)$-empty and $(n-2)$-empty. Then f can not be k-full for any k such that $\lfloor \frac{n-1}{2} \rfloor \leq k < n$.*

Proof. Let $f = det(M_{m \times m})$ for a read-once matrix M. As $x_1 x_2 \ldots x_n \in mon(f)$, without loss of generality assume that the $(i,i)^{th}$ entry of M is x_i for $i = 1$ to n. Since minor corresponding to $x_1 x_2 \ldots x_n$ is invertible we can use elementary row and column operations on M to get a matrix $N_{n \times n}$ such that $f = detN$ and the $(i,i)^{th}$ entry of N is $a_i x_i + b_i$ for $a_i, b_i \in \mathbb{F}$ and $a_i \neq 0$. All the other entries of N are scalars. The assumption that f is $(n-1)$-empty implies that $b_i = 0$ for $i = 1$ to n. Since f is $(n-2)$-empty, we also have $N(i,j)N(j,i) = 0$ for all $i \neq j$, $1 \leq i,j \leq n$. So at least $\binom{n}{2}$ entries of N are zero. So there is a row of N which contains at least $\lceil \frac{n-1}{2} \rceil$ zeros. Let i be the index of that row. For $l \geq \lceil \frac{n-1}{2} \rceil$, let $a_1, a_2, \ldots, a_l$ be the column indices such that $N(i, a_i) = 0$. Note that $i \notin \{a_1, \ldots, a_l\}$. We want to prove that f is not k-full for $\lfloor \frac{n-1}{2} \rfloor \leq k < n$. Let S be a subset of $\{a_1, \ldots, a_l\}$ of size $n - k - 1$. Note that we can pick such a set since $l \geq \lceil \frac{n-1}{2} \rceil$. Let $T = S \cup \{i\}$. Let $m = \prod_{j \notin T} x_j$. Let N' be the minor obtained by removing all the rows and columns in $\{1, 2, \ldots, n\} \setminus T$ from N. Clearly $m \notin mon(f)$ iff the constant term in the determinant of N' is zero. Note that N' contains a row with one entry x_i and the remaining entries in the row are zero. So clearly the constant term in the determinant of N' is zero. This shows that degree k monomial $m \notin mon(f)$. This proves the Theorem.

Let $f = x_1 + x_2 + x_3 + x_4 + x_1 x_2 x_3 x_4$. f is 4-full, 3-empty, 2-empty and 1-full. Applying above theorem for $n = 4$, we deduce that the set $mon(f) = \{x_1 x_2 x_3 x_4, x_1, x_2, x_3, x_4\}$ is not expressible as a read-once determinant.

5   Discussion and Open Problems

Under Valiant's hypothesis we know that $Perm_n$ cannot be expressed as a read-$n^{O(1)}$ determinant. Proving non-expressibility of $Perm_n$ as a read-k determinant for $k > 1$ unconditionally, is an interesting problem. In fact even the simplest case $k = 2$ might be challenging. The corresponding PIT question of checking whether the determinant of a read-2 matrix is identically zero or not, is also open [3].

For the elementary symmetric polynomial of degree d on n variables, Shpilka and Wigderson gave an $O(nd^3 \log d)$ arithmetic formula [12]. Using universality of determinant, we get an $O(nd^3 \log d)$ upper bound on $dc(S_n^d)$, in fact for non constant d this is the best known upper bound on $dc(S_n^d)$ as noted in [6]. Answering the following question in either direction is interesting: Is S_n^d expressible as read-k determinant for $k > 1$? If the answer is NO, it is a nontrivial non-expressibility result and if the answer is YES, for say $k = O(n^2)$, it gives an $O(n^3)$ upper bound on $dc(S_n^d)$, which is asymptotically better than $O(nd^3 \log d)$ for $d = \frac{n}{2}$.

Another possible generalization of read-once determinants is the following. Let $X = \{x_{i,j} | 1 \leq i, j \leq n\}$ and consider the matrix $M_{m \times m}$ whose entries are affine linear forms over X such that the coefficient matrix induced by each variable has rank one. That is if we express M as $B_0 + \sum_{1 \le i,j \le n} x_{i,j} B_{i,j}$ then $rank(B_{i,j}) = 1$ for $1 \le i, j \le n$. B_0 can have arbitrary rank. The question we ask is: can we express $Perm_n$ as the determinant of such a matrix M? This model is clearly a generalization of read-once determinants and has been considered by Ivanyos, Karpinski and Saxena [5], where they give a deterministic polynomial time algorithm to test whether the determinant of such a matrix is identically zero. It would be interesting to address the question of expressibility of permanent in this model.

References

1. Cai, J.-Y., Chen, X., Li, D.: A quadratic lower bound for the permanent and determinant problem over any characteristic $\ne 2$. In: 40th Annual ACM Symposium on Theory of computing, pp. 491–498 (2008)
2. von zur Gathen, J.: Feasible arithmetic computations: valiant's hypothesis. J. Symbolic Comput. **4**, 137–172 (1987)
3. James, F.: Geelen an algebraic matching algorithm. Combinatorica **20**(1), 61–70 (2000)
4. Grigoriev, D., Razborov, A.A.: Exponential lower bounds for depth 3 arithmetic circuits in algebras of functions over finite fields. Appl. Algebra Eng. Commun. Comput. **10**(6), 465–487 (2000)
5. Ivanyos, G., Karpinski, M., Saxena, N.: Deterministic polynomial time algorithms for matrix completion problems. SIAM J. Comput. **39**(8), 3736–3751 (2010)
6. Jansen, M.: Lower bounds for the determinantal complexity of explicit low degree polynomials. In: Frid, A., Morozov, A., Rybalchenko, A., Wagner, K.W. (eds.) CSR 2009. LNCS, vol. 5675, pp. 167–178. Springer, Heidelberg (2009)
7. Jansen, M.J., Qiao, Y., Sarma, J.: Deterministic identity testing of read-once algebraic branching programs, Electron. Colloquium Comput. Complexity (ECCC), 17:84 (2010)
8. Jerrum, M., Snir, M.: Some exact complexity results for straight-line computations over semirings. J. ACM **29**(3), 874–897 (1982)
9. Mignon, T., Ressayre, N.: A quadratic bound for the determinant and permanent problem. Int. Math. Res. Not. **2004**, 4241–4253 (2004)
10. Nisan, N.: Lower bounds for noncommutative computation. In: Proceedings of 23rd ACM Symposium on Theory of Computing, pp. 410–418 (1991)
11. Nisan, N., Wigderson, A.: Lower bounds on arithmetic circuits via partial derivatives. Comput. Complexity **6**(3), 217–234 (1997)
12. Shpilka, A., Wigderson, A.: Depth-3 arithmetic formulae over fiel ds of characteristic zero. J. Comput. Complexity **10**(1), 1–27 (2001)
13. Valiant, L.: Completeness classes in algebra. In: Technical report CSR-40-79, Department of Computer Science, University of Edinburgh, April 1979

Constructive Relationships Between Algebraic Thickness and Normality

Joan Boyar[1] and Magnus Gausdal Find[2]([✉])

[1] Department of Mathematics of Computer Science,
University of Southern Denmark, Odense, Denmark
joan@imada.sdu.dk

[2] Information Technology Laboratory, National Institute of Standards
and Technology, Gaithersburg, USA
magnus.find@nist.gov

Abstract. We study the relationship between two measures of Boolean functions; *algebraic thickness* and *normality*. For a function f, the algebraic thickness is a variant of the *sparsity*, the number of nonzero coefficients in the unique $\mathbb{F}_2$ polynomial representing f, and the normality is the largest dimension of an affine subspace on which f is constant. We show that for $0 < \epsilon < 2$, any function with algebraic thickness $n^{3-\epsilon}$ is constant on some affine subspace of dimension $\Omega\left(n^{\frac{\epsilon}{2}}\right)$. Furthermore, we give an algorithm for finding such a subspace. We show that this is at most a factor of $\Theta(\sqrt{n})$ from the best guaranteed, and when restricted to the technique used, is at most a factor of $\Theta(\sqrt{\log n})$ from the best guaranteed. We also show that a concrete function, majority, has algebraic thickness $\Omega\left(2^{n^{1/6}}\right)$.

1 Introduction and Known Results

Boolean functions play an important role in many areas of computer science. In cryptology, Boolean functions are sometimes classified according to some measure of complexity (also called cryptographic complexity [7], nonlinearity criteria [17] or nonlinearity measures [1]). Examples of such measures are *nonlinearity*, *algebraic degree*, *normality*, *algebraic thickness* and *multiplicative complexity*, and there are a number of results showing that functions that are simple according to a certain measure are vulnerable to a certain attack (see [8] for a good survey).

A significant amount of work in this area presents explicit functions that achieve high (or low) values according to some measure. For the *nonlinearity* measure this was settled by showing the existence of bent functions [21], for *algebraic degree* the problem is trivial, for *multiplicative complexity* this is a well studied problem in circuit complexity [3], for *normality* this is exactly the

Contribution of the National Institute of Standards and Technology. The rights of this work are transferred to the extent transferable according to title 17 § 105 U.S.C.

Partially supported by the Danish Council for Independent Research, Natural Sciences.

M.G. Find—Most of this work was done while at the University of Southern Denmark.

A. Kosowski and I. Walukiewicz (Eds.): FCT 2015, LNCS 9210, pp. 106–117, 2015.
DOI: 10.1007/978-3-319-22177-9_9

problem of finding good *affine dispersers* [22]. The first result in this paper is that the majority function has exponential algebraic thickness.

Another line of work has been to establish relationships between these measures, e.g. considering questions of the form "if a function f is simple (or complex) according to one measure, what does that say about f according to some other measure", see e.g. [1,4,8] and the references therein. In this paper we focus on the relationship between *algebraic thickness* and *normality*. Intuitively, these measures capture, each in their own way, how "far" functions are from being linear [6,7]. In fact, these two measures have been studied together previously (see e.g. [5,6]). The relationship between these measures was considered in the work of Cohen and Tal in [10], where they show that functions with a certain algebraic thickness have a certain normality. For relatively small values of algebraic thickness, we tighten their bounds and present an algorithm to witness this normality. The question of giving a constructive proof of normality is not just a theoretical one. Recently a generic attack on stream ciphers with high normality was successfully mounted in the work [19]. If it is possible to constructively compute a witness of normality given a function with low algebraic thickness, this implies that any function with low algebraic thickness is likely to be vulnerable to the attack in [19], as well as any other attack based on normality. This work suggests that this is indeed possible for functions with small algebraic thickness.

2 Preliminaries and Known Results

Let $\mathbb{F}_2$ be the field of order 2, $\mathbb{F}_2^n$ the n-dimensional vector space over $\mathbb{F}_2$, and $[n] = \{1, \ldots, n\}$. A mapping from $\mathbb{F}_2^n$ to $\mathbb{F}_2$ is called a *Boolean function*. It is a well known fact that any Boolean function f in the variables $x_1, \ldots, x_n$ can be expressed uniquely as a multilinear polynomial over $\mathbb{F}_2$ called the *algebraic normal form* (ANF) or the *Zhegalkin polynomial*. That is, there exist unique constants $c_\emptyset, \ldots, c_{\{1,\ldots,n\}}$ over $\{0, 1\}$, such that

$$f(x_1, \ldots, x_n) = \sum_{S \subseteq [n]} c_S \prod_{j \in S} x_j,$$

where arithmetic is in $\mathbb{F}_2$. In the rest of this paper, most arithmetic will be in $\mathbb{F}_2$, although we still need arithmetic in $\mathbb{R}$. If nothing is mentioned it should be clear from the context what field is referred to. The largest $|S|$ such that $c_S = 1$ is called the *(algebraic) degree* of f, and functions with degree 2 are called *quadratic* functions. We let $\log(\cdot)$ be the logarithm base two, $\ln(\cdot)$ the natural logarithm, and $\exp(\cdot)$ the natural exponential function with base e.

Algebraic Thickness. For a Boolean function, f, let $\|f\| = \sum_{S \subseteq [n]} c_S$, with arithmetic in $\mathbb{R}$. This measure is sometimes called the *sparsity* of f (e.g. [10]). The *algebraic thickness* [4,6] of f, denoted $\mathcal{T}(f)$ is defined as the smallest sparsity after any affine bijection has been applied to the inputs of f. More precisely, letting $\mathcal{A}_n$ denote the set of affine, bijective operators on $\mathbb{F}_2^n$,

$$\mathcal{T}(f) = \min_{A \in \mathcal{A}_n} \|f \circ A\|. \tag{1}$$

Algebraic thickness was introduced and first studied by Carlet in [4–6]. Affine functions have algebraic thickness at most 1, and Carlet showed that for any constant $c > \sqrt{\ln 2}$, for sufficiently large n there exist functions with algebraic thickness

$$2^{n-1} - cn2^{\frac{n-1}{2}},$$

and that a *random* Boolean function will have such high algebraic thickness with high probability. Furthermore *no* function has algebraic thickness larger than $\frac{2}{3}2^n$. Carlet observes that algebraic thickness was also implicitly mentioned in [18, Page 208] and related to the so called "higher order differential attack" due to Knudsen [15] and Lai [16] in that they are dependent on the degree as well as the number of terms in the ANF of the function used.

Normality. A k-dimensional *flat* is an affine (sub)space of $\mathbb{F}_2^n$ with dimension k. A function is k-*normal* if there exists a k-dimensional flat E such that f is constant on E [4,9]. For simplicity define the *normality* of a function f, which we denote $\mathcal{N}(f)$, as the *largest* k such that f is k-normal. We recall that affine functions have normality at least $n-1$ (which is the largest possible for non-constant functions), while for any $c > 1$, a random Boolean function has normality less than $c \log n$ with high probability.

Functions with normality smaller than k are often called *affine dispersers* of dimension k, and a great deal of work has been put into explicit constructions of functions with low normality. Currently the asymptotically best known deterministic function, due to Shaltiel, has normality less than $2^{\log^{0.9} n}$ [22].

Notice the asymmetry in the definitions: linear functions have very low algebraic thickness (0 or 1) but very high normality (n or $n-1$), whereas random functions, with high probability, have very high algebraic thickness (at least $2^{n-1} - 0.92 \cdot n \cdot 2^{\frac{n-1}{2}}$) but low normality (less than $1.01 \log n$) [5].

Remark on Computational Efficiency. In this paper, we say that something is efficiently computable if it is computable in time polynomially bounded in the size of the input. Algorithms in this paper will have a Boolean function with a certain algebraic thickness as input. We assume that the function is represented by the ANF of the function witnessing this small algebraic thickness along with the bijection. That is, if a function f with algebraic thickness $\mathcal{T}(f) = T$ is the input to the algorithm, we assume that it is represented by a function g and an affine bijection A such that $g = f \circ A$ and $\|g\| = T$. In this setting, representing a function f uses $poly(\mathcal{T}(f) + n^2)$ bits.

Quadratic Functions. The normality and algebraic thickness of quadratic functions are well understood due to the following theorem due to Dickson [11] (see also [8] for a proof).

Theorem 1 (Dickson). *Let* $f \colon \mathbb{F}_2^n \mapsto \mathbb{F}_2$ *be quadratic. Then there exist an invertible* $n \times n$ *matrix* A, *a vector* $\mathbf{b} \in \mathbb{F}_2^n$, $t \leq \frac{n}{2}$, *and* $c \in \mathbb{F}_2$ *such that for* $\mathbf{y} = A\mathbf{x} + \mathbf{b}$ *one of the following two equations holds:*

$$f(x) = y_1y_2 + y_3y_4 + \ldots y_{t-1}y_t + c, \ or \ f(x) = y_1y_2 + y_3y_4 + \ldots y_{t-1}y_t + y_{t+1}.$$

Furthermore A, **b** *and c can be found efficiently.*

That is, any quadratic function is affine equivalent to some inner product function. We highlight a simple but useful consequence of Theorem 1. Simply by setting one variable in each of the degree two terms to zero, one gets:

Proposition 1. *Let* $f \colon \mathbb{F}_2^n \to \mathbb{F}_2$ *be quadratic. Then* $\mathcal{N}(f) \geq \lfloor \frac{n}{2} \rfloor$. *Furthermore a flat witnessing the normality of f can be found efficiently.*

Some Relationships. It was shown in [6] that normality and algebraic thickness are logically independent of (that is, not subsumed by) each other. Several other results relating algebraic thickness and normality to other cryptographic measures are given in [6]. We mention a few relations to other measures.

Clearly, functions with degree d have algebraic thickness $O(n^d)$, so having superpolynomial algebraic thickness requires superconstant degree. The fact that there exist functions with low degree and low normality has been established in [4] and [10] independently. In the following, by a *random degree three polynomial*, we mean a function where each term of degree three is included in the ANF independently with probability $\frac{1}{2}$. No other terms are included in the ANF.

Theorem 2 ([4,10][1]). *Let* f *on* n *variables be a random degree three polynomial. Then with high probability,* f *remains nonconstant on any subspace of dimension* $6.12\sqrt{n}$.

In fact, as mentioned in [10] it is not hard to generalize this to the fact that for any constant d, a random degree d polynomial has normality $O\left(n^{1/(d-1)}\right)$. Perhaps surprisingly, this is tight. More precisely the authors give an elegant proof showing that *any* function with degree d has $\mathcal{N}(f) \in \Omega\left(n^{1/(d-1)}\right)$. This result implies the following relation between algebraic thickness and normality.

Theorem 3 (Cohen and Tal [10]). *Let* c *be an integer and let* f *have* $\mathcal{T}(f) \leq n^c$. *Then* $\mathcal{N}(f) \in \Omega\left(n^{1/(4c)}\right)$.

The proof of this has two steps: First they show by probabilistic methods that f has a restriction with a certain number of free variables and a certain degree, and after this they appeal to a relation between degree and normality. Although the authors do study the algorithmic question of finding such a subspace, they do not propose an efficient algorithm for finding a subspace of such dimension. We will pay special attention to the following type of restrictions of Boolean functions.

Definition 1. *Let* $f \colon \mathbb{F}_2^n \to \mathbb{F}_2$. *Setting* $k < n$ *of the bits to* 0 *results in a new function* f' *on* $n - k$ *variables. We say that* f' *is a* 0-restriction *of* f.

---

[1] The constant 6.12 does not appear explicitly in these articles, however it can be derived using similar calculations as in the cited papers. This also follows from Theorem 6 later in this paper. We remark that 6.12 is not optimal.

By inspecting the proof in the next section and the proof of Theorem 3, one can see that *most* of the restrictions performed are in fact setting variables to 0. Furthermore, by inspecting the flat used for the attack performed in [19] (Sect. 5.3), one can see that it is of this form as well. Determining whether a given function represented by its ANF admits a 0-restriction f' on $n-k$ variables with f' constant corresponds exactly to the hitting set problem, and this is well known to be **NP** complete [12]. Furthermore it remains **NP** complete even when restricted to quadratic functions (corresponding to the vertex cover problem).

This stands in contrast to Proposition 1; for quadratic functions and general flats (as opposed to just 0-restrictions) the problem is polynomial time solvable. To the best of our knowledge, the computational complexity of the following problem is open (see also [10]): Given a function, represented by its ANF, find a large(st) flat on which the function is constant.

3   Majority Has High Algebraic Thickness

For many functions, it is trivial to see that the ANF contains many terms, e.g. the function

$$f(\mathbf{x}) = (1 + x_1)(1 + x_2) \cdots (1 + x_n),$$

which is 1 if and only if all the inputs are 0, contains all the possible 2^n terms in its ANF. However, we are not aware of any explicit function along with a proof of a strong (e.g. exponential) lower bound on the algebraic thickness. Using a result from circuit complexity [20], it is straightforward to show that the *majority function*, MAJ_n has exponential algebraic thickness. MAJ_n is 1 if and only if at least half of the n inputs are 1. In the following, an $AC_0[\oplus]$ circuit of depth d is a circuit with inputs $x_1, x_2, \ldots, x_n, (1 \oplus x_1), (1 \oplus x_2), \ldots, (1 \oplus x_n)$. The circuit contains $\wedge, \vee, \oplus$ (AND, OR, XOR) gates of unbounded fan-in, and every directed path contains at most d edges. First we need the following simple proposition:

Proposition 2. *Let* $f \colon \mathbb{F}_2^n \to \mathbb{F}_2$ *have* $T(f) \leq T$. *Then* f *can be computed by an* $AC_0[\oplus]$ *circuit of depth 3 with at most* $n + T + 1$ *gates.*

Proof. Suppose $f = g \circ A$ for some affine bijective mapping A. In the first layer (the layer closest to the inputs) one can compute A using n XOR gates of fan-in at most n. Then by computing all the monomials independently, g can be computed by an $AC_0[\oplus]$ circuit of depth 2 using T AND gates with fan-in at most n and 1 XOR gate of fan-in T.   □

Now we recall a result due to Razborov [20], see also [14, 12.24]

Theorem 4 (Razborov). *Every unbounded fan-in depth-d circuit over* $\{\wedge, \vee, \oplus\}$ *computing* MAJ_n *requires* $2^{\Omega(n^{1/(2d)})}$ *gates.*

Combining these two results, we immediately have the following result that the majority function MAJ_n has high algebraic thickness.

Proposition 3. $T(MAJ_n) \geq 2^{\Omega(n^{1/6})}$.

4   Algebraic Thickness and Normality

This section is devoted to showing that functions with algebraic thickness at most $n^{3-\epsilon}$ are constant on flats of somewhat large dimensions. Furthermore our proof reveals a polynomial time algorithm to find such a subspace. In the following, a term of degree at least 3 will be called a *crucial* term, and for a function f, the number of crucial terms will be denoted $T^{\geq 3}(f)$.

Our approach can be divided into two steps: First it uses 0-restrictions to obtain a quadratic function, and after this we can use Proposition 1. As implied by the relation between 0-restrictions and the hitting set problem, finding the optimal 0-restrictions is indeed a computationally hard task. Nevertheless, as we shall show in this section, the following greedy algorithm gives reasonable guarantees.

The greedy algorithm simply works by continually finding the variable that is contained in the most crucial terms, and sets this variable to 0. It finishes when there are no crucial terms. We show that when the greedy algorithm finishes, the number of variables left, n', is relatively large as a function of n (for a more precise statement, see Theorem 5). Notice that we are only interested in the behavior of n' as a function of n, and that this is not necessarily related to the approximation ratio of the greedy algorithm, which is known to be $\Theta(\log n)$ [13].

We begin with a simple proposition about the greedy algorithm that will be useful throughout the section, and it gives a tight bound.

Proposition 4. *Let $g\colon \mathbb{F}_2^n \to \mathbb{F}_2$ have $T^{\geq 3}(g) \geq m$. Then some variable x_j is contained in at least $\left\lceil 3\frac{m}{n} \right\rceil$ crucial terms.*

Proof. We can assume that no variable occurs twice in the same term. Hence the total number of variable occurrences in crucial terms is at least $3m$. By the pigeon hole principle, some variable is contained in at least $\left\lceil 3\frac{m}{n} \right\rceil$ terms.   □

The following lemma is a special case where a tight result can be obtained. It is included here because the result is tight, and it gives a better constant in Theorem 5 than one would get by simply removing terms one at a time. The result applies to functions with relatively small thickness, and a later lemma reduces functions with somewhat larger thickness to this case.

Lemma 1. *Let $c \leq \frac{2}{3}$ and let $f\colon \mathbb{F}_2^n \to \mathbb{F}_2$ have $T^{\geq 3}(f) \leq cn$. Then f has a 0-restriction f' on $n' = n - \left\lceil \frac{3c-1}{5}n \right\rceil$ variables with $T^{\geq 3}(f') \leq \frac{n'}{3}$.*

Proof. Let the greedy algorithm run until a function f' on n' variables with $T^{\geq 3}(f') \leq \frac{n'}{3}$ is obtained. By Proposition 4 we eliminate at least 2 terms in each step. The number of algorithm iterations is at most $\left\lceil \frac{3c-1}{5}n \right\rceil$. Indeed, let $\left\lceil \frac{3c-1}{5}n \right\rceil = \frac{3c-1}{5}n + \delta$ for some $0 \leq \delta < 1$. After this number of iterations the number of variables left is

$$n' = n - \frac{3c-1}{5}n - \delta = \frac{6-3c}{5}n - \delta$$

and the number of critical terms is at most

$$cn - 2\left(\frac{3c-1}{5}n - \delta\right) = \frac{2-c}{5}n - 2\delta.$$

In particular $\frac{n'}{3} \geq \frac{2-c}{5}n - 2\delta$. □

Lemma 1 is essentially tight.

Proposition 5. *Let $\frac{1}{3} < c \leq \frac{2}{3}$ be arbitrary but rational. Then for infinitely many values of n, there exists a function on n variables with $T^{\geq3}(f) = cn$ such that every 0-restriction f' on $n' > n - \lceil\frac{3c-1}{5}n\rceil$ variables has $T^{\geq3}(f) > \frac{n'}{3}$.*

Proof. Let $\frac{1}{3} < c \leq \frac{2}{3}$ be fixed and consider the function on 6 variables:

$$f(x) = x_1x_2x_3 + x_1x_4x_5 + x_2x_4x_6 + x_3x_5x_6.$$

The greedy algorithm sets this functions to 0 by killing two variables, and this is optimal. Furthermore setting any one variable to 0 kills exactly two terms. Now consider the following function defined on $n = 30m$ variables and having $20m$ terms. For convenience we index the variables by $x_{i,j}$ for $1 \leq i \leq 5m$, $1 \leq j \leq 6$. Let

$$g(x) = \sum_{i=1}^{5m} f(x_{i,1}, x_{i,2}, x_{i,3}, x_{i,4}, x_{i,5}, x_{i,6}).$$

Again here the greedy algorithm is optimal, and setting $6m$ variables to zero leaves $n' = 24m$ variables and $8m$ terms remaining. Thus, the bound from Lemma 1 is met with equality for $c = \frac{2}{3}$.

To see that it is tight for $c < \frac{2}{3}$, consider the function, $\tilde{f}$ on n variables, where n is a multiple of 30 such that $c\frac{4}{3}\frac{n}{2-c}$ is an integer. Run the greedy algorithm until the number of variables is $\tilde{n}$ and $T^{\geq3}(\tilde{f}) = c\tilde{n}$ (assuming $c\tilde{n}$ is an integer). At this point $\tilde{n} = \frac{4}{3}\frac{n}{2-c}$ and the number of terms left is $c\tilde{n}$. Again, by the structure of the function, setting any number, t, of the variables to 0 results in a function with $\tilde{n} - t$ variables and at least $c\tilde{n} - 2t$ terms. When $t < \frac{(3c-1)\tilde{n}}{5}$, we have $c\tilde{n} - 2t > \frac{\tilde{n}-t}{3}$. □

An immediate corollary to Lemma 1 is the following.

Corollary 1. *Let $f\colon \mathbb{F}_2^n \to \mathbb{F}_2$ have $T^{\geq3}(f) \leq \frac{2}{3}n$. Then it is constant on a flat of dimension $n' \geq \left\lfloor\frac{\left\lfloor\frac{2}{3}\lfloor\frac{4}{5}n\rfloor\right\rfloor}{2}\right\rfloor \geq \frac{4}{15}n - 2$. Furthermore, such a flat can be found efficiently.*

Proof. First apply Lemma 1 to obtain a function on $n' = \lfloor\frac{4}{5}n\rfloor$ variables with at most $\frac{n'}{3}$ crucial terms. Now set one variable in each crucial term to 0, so after this we have at least $\frac{2}{3}\lfloor\frac{4}{5}n\rfloor$ variables left and the remaining function is quadratic. Applying Theorem 1 gives the result. □

The following lemma generalizes the lemma above to the case with more terms. The analysis of the greedy algorithm uses ideas similar to those used in certain formula lower bound proofs, see e.g. [23] or [14, Section 6.3].

Lemma 2. *Let* $f\colon \mathbb{F}_2^n \to \mathbb{F}_2$ *with* $\mathcal{T}^{\geq 3}(f) \leq n^{3-\epsilon}$, *for* $0 < \epsilon < 2$. *Then there exists a 0-restriction* f' *on* $n' = \left\lceil \sqrt{\frac{2}{3}n^\epsilon} \right\rceil$ *variables with* $\mathcal{T}^{\geq 3}(f') \leq \frac{2}{3}n'$.

Proof. Let $\mathcal{T}^{\geq 3}(f) = T$. Then, by Proposition 4. Setting the variable contained in the largest number of terms to 0, the number of crucial terms left is at most

$$T - \frac{3T}{n} = T \cdot \left(1 - \frac{3}{n}\right) \leq T \cdot \left(\frac{n-1}{n}\right)^3.$$

Applying this inequality $n - n'$ times yields that after $n - n'$ iterations the number of crucial terms left is at most

$$T \cdot \left(\frac{n-1}{n}\right)^3 \left(\frac{n-2}{n-1}\right)^3 \cdots \left(\frac{n'}{n'+1}\right)^3 = T \cdot \left(\frac{n'}{n}\right)^3.$$

When $n' = \sqrt{\frac{2}{3}n^\epsilon}$ and $T = n^{3-\epsilon}$, this is at most $\frac{2}{3}n'$. $\qquad \square$

Remark: A previous version of this paper [2], contained a version of the lemma with a proof substantially more complicated. We thank anonymous reviewers for suggesting this simpler proof.

It should be noted that Lemma 2 cannot be improved to the case where $\epsilon = 0$, no matter what algorithm is used to choose the 0-restriction. To see this consider the function containing all degree three terms. For this function, *any* 0-restriction (or 1-restriction) leaving n' variables will have at least $\binom{n'}{3}$ crucial terms. On the other hand, restricting with $x_1 + x_2 = 0$ results in all crucial terms with both x_1 and x_2 having lower degree and all crucial terms with just one of them cancelling out. This suggests that for handling functions with larger algebraic thickness, one should use restrictions other than just 0-restrictions.

Combining Lemma 2 with Corollary 1, we get the following theorem.

Theorem 5. *Let* $\mathcal{T}(f) = n^{3-\epsilon}$ *for* $0 < \epsilon < 2$. *Then there exists a flat of dimension at least* $\frac{4}{15}\sqrt{\frac{2}{3}n^\epsilon} - 3$, *such that when restricted to this flat,* f *is constant. Furthermore this flat can be found efficiently.*

This improves on Theorem 3 for functions with algebraic thickness n^s for $1 \leq s \leq 2.82$, and the smaller s, the bigger the improvement, e.g. for $\mathcal{T}(f) \leq n^2$, our bound guarantees $\mathcal{N}(f) \in \Omega(n^{1/2})$, compared to $\Omega(n^{1/8})$.

4.1 Normal Functions with Low Sparsity

How good are the guarantees given in the previous section? The purpose of this section is first to show that the result from Theorem 5 is at most a factor of $\Theta(\sqrt{n})$ from being tight. More precisely, we show that for any $2 < s \leq 3$ there exist functions with thickness at most n^s that are nonconstant on flats of dimension $O(n^{2-\frac{s}{2}})$. Notice that this contains Theorem 2 as a special case where $s = 3$.

Theorem 6. *For any $2 < s \leq 3$, for sufficiently large n, there exist functions with degree 3 and algebraic thickness at most n^s that remain nonconstant on all flats of dimension $6.12n^{2-\frac{s}{2}}$.*

Proof. The proof uses the probabilistic method. We endow the set of all Boolean functions of degree 3 with a probability distribution $\mathcal{D}$, and show that under this distribution a function has the promised normality with high probability.

The proof is divided into the following steps: First we describe the probability distribution $\mathcal{D}$. Then, we fix an arbitrary k-dimensional flat E, and bound the probability that a random f chosen according to $\mathcal{D}$ is constant on E. We show that for $k = Cn^{2-s/2}$, where the constant C is determined later, this probability is sufficiently small that a union bound over all possible choices of E gives the desired result.

We define $\mathcal{D}$ by describing the probability distribution on the ANF. We let each possible degree 3 term be included with probability $\frac{1}{2n^{3-s}}$. The expected number of terms is thus $\frac{1}{2}n^{s-3}\binom{n}{3} \leq n^s/12$, and the probability of having more than n^s terms is less than 0.001 for large n. Now let E be an arbitrary but fixed k-dimensional flat.

One way to think of a function restricted to a k-dimensional flat is that it can be obtained by a sequence of $n - k$ affine variable substitutions of the form $x_i := \sum_{j \in S} x_j + c$. This changes the ANF of the function since x_i is no longer a "free" variable. Assume without loss of generality that we substitute for the variables $x_n, \ldots, x_{k+1}$ in that order. Initially we start with the function f given by

$$f(x) = \sum_{\{a,b,c\} \subseteq [n]} I_{abc} x_a x_b x_c,$$

where I_{abc} is the indicator random variable, indicating whether the $x_a x_b x_c$ is contained in the ANF. Suppose we perform the $n - k$ restrictions and obtain the function $\tilde{f}$. The ANF of $\tilde{f}$ is given by

$$f(x) = \sum_{\{a,b,c\} \subseteq [k]} \left(I_{abc} + \sum_{s \in S_{abc}} I_s \right) x_a x_b x_c,$$

where S_{abc} is some set of indicator random variables depending on the restrictions performed. It is important that I_{abc}, the indicator random variable corresponding to $x_a x_b x_c$, for $\{a, b, c\} \subseteq [k]$ is *only* occurring at $x_a x_b x_c$. Hence we conclude that independently of the outcome of all the indicator random variables $I_{a'b'c'}$ with $\{a', b', c'\} \not\subseteq [k]$, we have that the marginal probability for any I_{abc} with $\{a, b, c\} \subseteq [k]$ occurring remains at least $\frac{1}{2n^{3-s}}$.

Define $t = \binom{k}{3}$ random variables, $Z_1, \ldots, Z_t$, one for each potential term in the ANF of $\tilde{f}$, such that $Z_j = 1$ if and only if the corresponding term is present in the ANF, and 0 otherwise. The obtained function is only constant if there are no degree 3 terms, so the probability of $\tilde{f}$ being constant is thus at most

$$\mathbb{P}\left[Z_1 = \ldots = Z_t = 0\right] \leq \left(1 - \frac{1}{2n^{3-s}}\right)^{\binom{k}{3}}$$

$$\leq \left(1 - \frac{1}{2n^{3-s}}\right)^{\frac{C^3}{27}(n^{6-3s/2})}$$

$$= \left(\left(1 - \frac{1}{2n^{3-s}}\right)^{2n^{3-s}}\right)^{\frac{C^3}{54}(n^{3-s/2})}$$

$$\leq \exp\left(-\frac{C^3}{54}n^{3-s/2}\right).$$

The number of choices for E is at most $2^{n(k+1)}$, so the probability that f becomes constant on *some* affine flat of dimension k is at most

$$\exp\left(-\frac{C^3}{54}n^{3-s/2} + C\ln(2)n^{3-s/2} + n\right).$$

Now if $C > \sqrt{54\ln(2)} \approx 6.11..$, this quantity tends to 0. We conclude that with high probability the function obtained has algebraic thickness at most n^s and normality at most $6.12n^{2-\frac{s}{2}}$.   □

There is factor of $\Theta(\sqrt{n})$ between the existence guaranteed by Theorems 5 and 6 and we leave it as an interesting problem to close this gap.

The algorithm studied in this paper works by setting variables to 0 until all remaining terms have degree at most 2, and after that appealing to Theorem 1. A proof similar to the previous shows that among such algorithms, the bound from Theorem 5 is very close to being asymptotically tight.

Theorem 7. *For any $2 < s < 3$, there exist functions with degree 3 and algebraic thickness at most n^s that have degree 3 on any 0-restriction of dimension $3\sqrt{\ln n}n^{\frac{3-s}{2}}$.*

Proof. We use the same proof strategy as in the proof of Theorem 6. Endow the set of all Boolean functions of degree 3 with the same probability distribution $\mathcal{D}$. For large n, the number of terms is larger than n^s with probability at most 0.001. Now we set all but $C\sqrt{\ln n}n^{\frac{3-s}{2}}$ of the variables to 0, and consider the probability of the function being constant under this fixed 0-restriction. We will show that this probability is so small that a union bound over all such choices gives that with high probability the function is nonconstant under *any* such restriction. We will see that setting $C = 3$ will suffice. There are $\binom{C\sqrt{\ln n}n^{\frac{3-s}{2}}}{3}$ possible degree 3 terms on these remaining variables, and we let each one be included with probability $\frac{1}{n^{3-s}}$. The probability that none of these degree three terms are included is

$$\left(1-\frac{1}{n^{3-s}}\right)^{\left(C\sqrt{\ln n}\,n^{\frac{3-s}{2}}\right)} \le \left(1-\frac{1}{n^{3-s}}\right)^{\frac{C^3}{27}\left(\sqrt{\ln n}\right)^3 n^{\frac{9-3s}{2}}}$$

$$= \left(\left(1-\frac{1}{n^{3-s}}\right)^{n^{3-s}}\right)^{n^{\frac{3-s}{2}}(\ln n)^{3/2}\frac{C^3}{27}}$$

$$\le \exp\left(-\frac{C^3}{27}n^{\frac{3-s}{2}}(\ln n)^{3/2}\right),$$

and the number of 0-restrictions with all but $C\sqrt{\ln n}\,nn^{\frac{3-s}{2}}$ variables fixed is

$$\binom{n}{C\sqrt{\ln n}\,nn^{\frac{3-s}{2}}} \le \frac{n^{C\sqrt{\ln n}\,nn^{\frac{3-s}{2}}}}{(C\sqrt{\ln n}\,nn^{\frac{3-s}{2}})!}$$

$$= \exp(\ln n C\sqrt{\ln n}\,nn^{\frac{3-s}{2}} - \ln((C\sqrt{\ln n}\,nn^{\frac{3-s}{2}})!))$$

$$\le \exp\left(\ln^{3/2}(n)Cn^{\frac{3-s}{2}} - .99C\sqrt{\ln n}\,nn^{\frac{3-s}{2}}\ln\left(C\sqrt{\ln n}\,nn^{\frac{3-s}{2}}\right)\right)$$

$$\le \exp\left((\ln n)^{3/2}Cn^{\frac{3-s}{2}} - \frac{3-s}{2}.98C(\ln n)^{3/2}n^{\frac{3-s}{2}}\right)$$

$$= \exp\left((\ln n)^{3/2}Cn^{\frac{3-s}{2}}\left(1 - 0.98\frac{3-s}{2}\right)\right),$$

where the last two inequalities hold for sufficiently large n. Again, by the union bound, the probability that there exists such a choice on which there are no terms of degree three left is at most

$$\exp\left(-\frac{C^3}{27}n^{\frac{3-s}{2}}(\ln n)^{3/2}\right)\exp\left((\ln n)^{3/2}Cn^{\frac{3-s}{2}}\left(1 - 0.98\frac{3-s}{2}\right)\right).$$

For $C \ge 3$ this probability tends to zero, hence we have that with high probability the function does not have a 0-restriction on $3\sqrt{\ln n}\,nn^{\frac{3-s}{2}}$ variables of degree smaller than 3. $\qquad\square$

References

1. Boyar, J., Find, M., Peralta, R.: Four measures of nonlinearity. In: Spirakis, P.G., Serna, M.J. (eds.) CIAC 2013. LNCS, vol. 7878, pp. 61–72. Springer, Heidelberg (2013)
2. Boyar, J., Find, M.G.: Constructive relationships between algebraic thickness and normality. CoRR abs/1410.1318 (2014)
3. Boyar, J., Peralta, R., Pochuev, D.: On the multiplicative complexity of boolean functions over the basis $(\wedge, \oplus, 1)$. Theor. Comput. Sci. **235**(1), 43–57 (2000)
4. Carlet, C.: On cryptographic complexity of boolean functions. In: Mullen, G., Stichtenoth, H., Tapia-Recillas, H. (eds.) Finite Fields with Applications to Coding Theory, Cryptography and Related Areas, pp. 53–69. Springer, Berlin (2002)
5. Carlet, C.: On the algebraic thickness and non-normality of boolean functions. In: Information Theory Workshop, pp. 147–150. IEEE (2003)

6. Carlet, C.: On the degree, nonlinearity, algebraic thickness, and nonnormality of boolean functions, with developments on symmetric functions. IEEE Trans. Inf. Theor. **50**(9), 2178–2185 (2004)
7. Carlet, C.: The complexity of boolean functions from cryptographic viewpoint. In: Krause, M., Pudlàk, P., Reischuk, R., van Melkebeek, D. (eds.) Complexity of Boolean Functions. Dagstuhl Seminar Proceedings, vol. 06111. Schloss Dagstuhl - Leibniz-Zentrum für Informatik, Germany (2006)
8. Carlet, C.: Boolean functions for cryptography and error correcting codes. In: Crama, Y., Hammer, P.L. (eds.) Boolean Models and Methods in Mathematics, Computer Science, and Engineering, pp. 257–397. Cambridge University Press, Cambridge (2010)
9. Charpin, P.: Normal boolean functions. J. Complexity **20**(2–3), 245–265 (2004)
10. Cohen, G., Tal, A.: Two structural results for low degree polynomials and applications. CoRR abs/1404.0654 (2014)
11. Dickson, L.E.: Linear Groups with an Exposition of the Galois Field Theory. Teubner's Sammlung von Lehrbuchern auf dem Gebiete der matematischen Wissenschaften VL, x+312 (1901)
12. Garey, M.R., Johnson, D.S.: Computers and Intractability: A Guide to the Theory of NP-Completeness. W.H. Freeman, New York (1979)
13. Johnson, D.S.: Approximation algorithms for combinatorial problems. J. Comput. Syst. Sci. **9**(3), 256–278 (1974)
14. Jukna, S.: Boolean Function Complexity - Advances and Frontiers. Algorithms and combinatorics, vol. 27. Springer, Heidelberg (2012)
15. Knudsen, L.R.: Truncated and higher order differentials. In: Preneel, B. (ed.) Fast Software Encryption. LNCS, vol. 1008, pp. 196–211. Springer, Heidelberg (1995)
16. Lai, X.: Higher order derivatives and differential cryptanalysis. In: Blahut, R.E., Costello Jr., D.J., Maurer, U., Mittelholzer, T. (eds.) Communications and Cryptography. The Springer International Series in Engineering and Computer Science, vol. 276, pp. 227–233. Springer, US (1994)
17. Meier, W., Staffelbach, O.: Nonlinearity criteria for cryptographic functions. In: Quisquater, J.-J., Vandewalle, J. (eds.) EUROCRYPT 1989. LNCS, vol. 434, pp. 549–562. Springer, Heidelberg (1990)
18. Menezes, A., van Oorschot, P.C., Vanstone, S.A.: Handbook of Applied Cryptography. CRC Press, Boca Raton (1996)
19. Mihaljevic, M.J., Gangopadhyay, S., Paul, G., Imai, H.: Generic cryptographic weakness of k-normal boolean functions in certain stream ciphers and cryptanalysis of grain-128. Periodica Mathematica Hungarica **65**(2), 205–227 (2012)
20. Razborov, A.A.: Lower bounds on the size of bounded depth circuits over a complete basis with logical addition. Math. Notes **41**(4), 333–338 (1987)
21. Rothaus, O.S.: On "bent" functions. J. Comb. Theory Ser. A **20**(3), 300–305 (1976)
22. Shaltiel, R.: Dispersers for affine sources with sub-polynomial entropy. In: Ostrovsky, R. (ed.) FOCS. pp. 247–256. IEEE (2011)
23. Subbotovskaya, B.A.: Realizations of linear functions by formulas using +,*,-. Math. Dokl. **2**(3), 110–112 (1961)

On the Structure of Solution-Graphs
for Boolean Formulas

Patrick Scharpfenecker[✉]

Institute of Theoretical Computer Science, University of Ulm, Ulm, Germany
patrick.scharpfenecker@uni-ulm.de

Abstract. In this work we extend the study of solution graphs and
prove that for boolean formulas in a class called CPSS, all connected
components are partial cubes of small dimension, a statement which was
proved only for some cases in [16]. In contrast, we show that general
Schaefer formulas are powerful enough to encode graphs of exponential
isometric dimension and graphs which are not even partial cubes.

Our techniques shed light on the detailed structure of st-connectivity
for Schaefer and connectivity for CPSS formulas, problems which were
already known to be solvable in polynomial time. We refine this classification and show that the problems in these cases are equivalent to the
satisfiability problem of related formulas by giving mutual reductions
between (st-)connectivity and satisfiability. An immediate consequence
is that st-connectivity in (undirected) solution graphs of Horn-formulas
is P-complete while for $2SAT$ formulas st-connectivity is NL-complete.

Keywords: Partial cube · Succinct · Embedding · st-Connectivity ·
Connectivity · Satisfiability

1 Introduction

The work of Schaefer [15] first introduced a dichotomy for the complexity of
satisfiability on different classes of boolean formulas. The author proved that
for specific boolean formulas (now called Schaefer formulas), satisfiability is in
P while for all other classes, satisfiability is NP-complete. Surprisingly, there are
no formulas of intermediate complexity. Recently, the work of Gopalan et al.
and Schwerdtfeger [8,16] uncovered a similar behavior for several problems on
solution graphs of boolean formulas. A solution graph is a subgraph of the n-
dimensional hypercube induced by all satisfying assignments, see Definition 1.
Therefore boolean formulas can be seen as a succinct encoding of a solution
graph.

Definition 1. *Let $F(x_1, \ldots, x_n)$ be an arbitrary boolean formula. Then the solution graph G_F is the subgraph of the n-dimensional hypercube H_n induced by all
satisfying solutions x of F.*

P. Scharpfenecker—Supported by DFG grant TO 200/3-1.

© Springer International Publishing Switzerland 2015
A. Kosowski and I. Walukiewicz (Eds.): FCT 2015, LNCS 9210, pp. 118–130, 2015.
DOI: 10.1007/978-3-319-22177-9_10

These works focused on classifying the complexity of the connectivity and *st*-connectivity problem on solution graphs for given classes of formulas. While *st*-connectivity is the problem to determine for a given graph and two nodes if there is a path between these nodes, connectivity asks if a given graph consists only of a single connected component.

Usually, succinct encodings provide a complexity blow-up compared to non-succinct encodings (see for example [3, 5, 14, 17, 18]). Therefore the question arises to what extent the complexity for *st*-connectivity and connectivity change in the case of solution graphs in relation to the power of the encoding formulas.

For this, Gopalan et al. [8] introduced a new class of boolean formulas they call "tight" which lies between Schaefer formulas and general formulas. Their classification shows that for tight formulas, *st*-connectivity is in P while for general formulas it is PSPACE-complete. Similar, for the connectivity problem, they achieve a coNP-algorithm for Schaefer formulas, coNP-completeness for tight formulas and PSPACE-completeness for general formulas and conjecture that this is actually a trichotomy: they suspected connectivity for Schaefer formulas to be in P. More recently, [16] proved a trichotomy by introducing a fourth class of formulas (besides Schaefer, tight and general formulas) the authors call CPSS (constraint-projection separating Schaefer) which is even more restrictive than Schaefer formulas and by modifying the definition of tight formulas to "safely tight" formulas. Figure 1 summarizes the results of [16] and [8].

Function Set R	$Conn(R)$	$stConn(R)$	Diameter
CPSS	P	P	$O(n)$
Schaefer, not CPSS	coNP-complete	P	$O(n)$
safely tight, not Schaefer	coNP-complete	P	$O(n)$
not safely tight	PSPACE-complete	PSPACE-complete	$2^{\Omega(\sqrt{n})}$

Fig. 1. Classification of connectivity problems.

We refine the P-algorithms for *st*-connectivity of tight formulas (which contains all safely tight formulas) and show a close relation to satisfiability of such formulas by an improved analysis of the structure of solution graphs. For all tight formulas, *st*-connectivity reduces to satisfiability of a related formula. So for example, *st*-connectivity on $2SAT$ and $Horn$ formulas can be reduced to satisfiability of the same type. Therefore in the first case, *st*-connectivity is in NL while for the second case, the P-algorithm seems tight. In addition, for $2SAT$ and Horn-formulas, the reverse holds too, that is, satisfiability for these formulas is reducible to *st*-connectivity and connectivity in the solution graph of the same type of formulas. So $stConn(2SAT)$ is NL-complete and $stConn(Horn_3)$ is P-complete.

While [8] proved that for all tight formulas the diameter of connected components is linearly bounded in the number of variables and [16] improved this by showing that bijunctive formulas are even partial cubes, there is still room for improvements. Thereby a partial cube is an induced graph of the hypercube which preserves distances. So if two nodes are connected, their distance in the partial cube has to be the same as in the original hypercube. In our work we study the structure of connected components in solution graphs of Schaefer formulas. For CPSS formulas we show that every connected component is a partial cube of small dimension[1] while general Schaefer formulas are powerful enough to encode partial cubes of exponential dimension or even graphs which are not partial cubes at all. Yet these graphs have still diameter bounded by $O(n)$.

We note that the work of Ekin [6] discusses similar properties like connectedness and geodesy based on the structure of a given DNF formula. The authors discuss recognition of these properties and give a hierarchy of boolean functions which admit these properties. While co-geodetic functions are connected partial cubes, their approach requires the input formula to be a DNF or CNF. In contrast, the work of [8,16] can use arbitrary boolean formulas as clauses.

Another related topic is the so called phase-transition for random kSAT formulas and the clustering of the solution space. The works of [10–12] shed light on the behaviour of random formulas by providing a threshold α_c implying that random kSAT with less than $\alpha_c \cdot n$ clauses on n variables are most likely satisfiable while more than $\alpha_c \cdot n$ clauses imply that the formula is most likely unsatisfiable. Further, the authors of [12] showed that there is another threshold $\alpha_d \leq \alpha_c$ such that formulas with density lower than α_d mainly encode single connected components while formulas with density between α_d and α_c encode many connected components, called clusters. The work of [8,16] and our work can be seen as stepping stones to a better understanding of the structure of solution graphs which may help analyzing the structure of solution graphs of random kSAT formulas.

The rest of this paper is organized as follows. In Sect. 2 we briefly introduce our notation and basic definitions. Section 3 will cover the characterization of CPSS solution graphs as collections of partial cubes. In contrast, Sect. 4 will show that general Schaefer-formulas are powerful enough to encode partial cubes of exponential dimension and even graphs which are not partial cubes at all. Finally, in Sect. 5, we establish reductions from connectivity and st-connectivity problems of solution graphs to the satisfiability problem on related formulas and thereby refine the previous P characterization. We complete the classification of these problems by giving matching lower bounds.

2   Preliminaries

To compare two words $x, y \in \{0,1\}^n$, we use the lexicographic order. For $x, y \in \{0,1\}^n$, $\Delta(x,y)$ denotes the Hamming-distance between x and y and $\Delta(x) := \Delta(x, 0^{|x|})$ denotes the Hamming-weight of x. We associate words in $\{0,1\}^n$ with subsets of $[n] = \{1, \ldots, n\}$ in the standard way. We use graphs $G = (V, E)$ with nodes $V = [n]$ and edge set $E \subseteq V^2$ without self-loops.

---

[1] This was proved for bijunctive formulas in [16], we prove the remaining cases of Horn and dual-Horn formulas.

With P we denote the set of decision problems which can be solved in polynomial-time while L (NL) problems can be solved in (non-deterministic) logarithmic-space. With $\leq_T^L$, $\leq_m^L$ and $\leq_m^P$ we denote logarithmic-space Turing and logarithmic-space as well as polynomial-time many-one reductions.

We recall Definition 1 and note that we talk of solution graphs with the Hamming-distance, implying that two satisfying solutions are connected by an edge iff they differ in exactly one variable. Given a graph G and two nodes u, v, $d(u, v)$ is the length of the shortest path between u and v in G or ∞ if there is no such path.

Definition 2. *An induced subgraph G of H_n is a partial cube iff for all $x, y \in G$, $d(x, y) = \Delta(x, y)$. We call such an induced subgraph "isometric".*

For a $2SAT$ formula $F(x_1, \ldots, x_n)$ we define the implication graph $I(F) = (V, E)$ on nodes $V = \{x_1, \ldots, x_n, \overline{x_1}, \ldots, \overline{x_n}\}$ such that $(k \to l) \in E$ with $k, l \in V$ iff $F \models (k \to l)$.

For all boolean functions $F : \{0,1\}^n \to \{0,1\}$ we can represent F with the subset of all its satisfying assignments in $\{0,1\}^n$. Then a boolean function $F \subseteq \{0,1\}^n$ is closed under a ternary operation $\odot : \{0,1\}^3 \to \{0,1\}$ iff $\forall x, y, z \in F : \odot(x, y, z) := (\odot(x_1, y_1, z_1), \ldots, \odot(x_n, y_n, z_n)) \in F$. Note that we extend the notation of a ternary operation to an operation on three bit-vectors by applying the operation bitwise on the three vectors. We can define a similar closure for binary operations. For R a finite set of boolean functions with arbitrary arities (for example $R = \{(\overline{x} \vee y), (x \oplus y), (x \oplus y \oplus z)\}$, we define $SAT(R)$ to be the satisfiability problem for all boolean formulas which are conjunctions of instantiations of functions in R. For the given example R, $F(x, y, z) = (\overline{z} \vee y) \wedge (x \oplus y)$ is a formula in which every clause is an instantiation of an R-function. With $Conn(R)$ we denote the connectivity problem, given a conjunction F of R-functions, is the solution graph connected? Similarly, $stConn(R)$ is the st-connectivity problem, given a conjunction F of R-functions and s, t, is there a path from s to t in the solution graph? We mostly use F for boolean formulas/functions and R, S for finite sets of functions.

Note that $r \in R$ can be an arbitrary boolean function as for example $r = (x \oplus y)$ or $r = (x \vee \overline{y} \vee \overline{z}) \wedge (\overline{x} \vee z)$. With $2SAT$ we denote the set of all CNF-clauses with two variables and with $Horn_n$ we define the set of all Horn-clauses of size up to n. The ternary majority function $maj : \{0,1\}^3 \to \{0,1\}$ is defined as $maj(a, b, c) = (a \wedge b) \vee (a \wedge c) \vee (b \wedge c)$.

In Definitions 3 to 5 we recall some terms which were partially introduced by [16] and [8].

Definition 3. *A boolean function F is*

- *bijunctive, iff it is closed under $maj(a, b, c)$.*
- *affine, iff it is closed under $a \oplus b \oplus c$.*
- *Horn, iff it is closed under $a \wedge b$.*
- *dual-Horn, iff it is closed under $a \vee b$.*
- *IHSB$-$, iff it is closed under $a \wedge (b \vee c)$.*
- *IHSB$+$, iff it is closed under $a \vee (b \wedge c)$.*

A function has such a property componentwise, iff every connected component in the solution graph is closed under the corresponding operation. A function F has the additional property "safely", iff the property still holds for every function F' obtained by identification of variables².

In the case of Horn-formulas, the usual definition (the conjunction of Horn-clauses, which is a conjunction of literals such that no two literals occur positive) implies that the represented functions are Horn.

Definition 4. *A set of functions R is Schaefer (CPSS) if at least one of the following conditions holds:*

- *every function in R is bijunctive.*
- *every function in R is Horn (and safely componentwise IHSB−).*
- *every function in R is dual-Horn (and safely componentwise IHSB+).*
- *every function in R is affine.*

If we have a boolean formula F which is build from a set of CPSS functions R we say that F is CPSS. Clearly, every CPSS formula is Schaefer. We later use a bigger class of functions which we call *tight*. This class properly contains all Schaefer sets of functions.

Definition 5. *A set R of functions is tight if at least one of the following conditions holds:*

- *every function in R is componentwise bijunctive.*
- *every function in R is OR-free.*
- *every function in R is NAND-free.*

A function is OR-free if we can not derive $(x \lor y)$ by fixing variables. Similar, a function is NAND-free if we can not derive $(\overline{x} \lor \overline{y})$ by fixing variables.

3   Structure of CPSS-Formulas

We now study and refine the properties of connected components in formulas F on n variables which are CPSS. We are going to prove that such connected components are always partial cubes of isometric dimension at most n. Hereby the isometric dimension is the smallest n such that the graph can be isometrically embedded into the hypercube H_n. For this, [8] gives some useful basic properties for bijunctive and affine functions:

Lemma 6 ([8]). *If a boolean function F is bijunctive or affine then it is componentwise bijunctive.*

Lemma 7 ([8]). *Let F be a componentwise bijunctive function. Then the distance of all solutions x and y in the same connected component is exactly $\Delta(x, y)$.*

---

² Identifying two variables corresponds to replacing one of them with the other variable.

Lemma 8 ([8]). *Let R be a set of Horn-functions and let F be built from R-functions. Then every connected component in F has a unique minimal solution x^* and every other solution in this component is connected to x^* with a monotone path with respect to the Hamming distance.*

We can now prove our first statement:

Lemma 9. *Given a CPSS formula F, for two satisfying assignments s and t. Either $d(s,t) = \Delta(s,t)$ or $d(s,t) = \infty$.*

Proof. If F is bijunctive or affine the statements follows by Lemmas 6 and 7.

Now suppose F is Horn and componentwise IHSB− (the last case is dual). Therefore every connected component in F is closed under $x \wedge (y \vee z)$. We only show that for all x, y with $y \leq x$ in the same component there is a path of length $\Delta(x, y)$. Obviously there can not be a shorter path. With this, the statement holds for all a, b: We just use $c = a \wedge b$ as intermediate step and c is in the same connected component: By Lemma 8, we know that every component in a Horn formula has a unique minimal solution x^*. Then $a \wedge (b \vee x^*) = a \wedge b = c$ is in the same component as a and b.

Suppose by contradiction there is no such path from x to y. Then we know that y can not be the unique minimal solution x^*. But then there is a monotone decreasing path from x to x^* which has to bypass y and decrease at least one variable $i \in x \setminus y$. Let a be the first such node below x which decreases exactly one $i \in x \setminus y$. For all other decreased bits we know that $y_j = 1$. Then $x \wedge (a \vee y) = x \setminus \{i\} = x'$ and $d(x', y) = d(x, y) - 1$. An induction over the distance proofs our statement. □

Corollary 10. *Given a CPSS formula $F(x_1, \ldots, x_n)$, every connected component of F is a partial cube of isometric dimension at most n.*

4   Structure of General Schaefer-Formulas

Previously we looked at properties of solution graphs of Schaefer functions which are in addition CPSS. If a given formula F on n variables is CPSS, every connected component is a partial cube of small isometric dimension. If it is Schaefer but not CPSS, the diameter is still linear in n and due to [8], st-connectivity is in P. We now prove that there are Schaefer formulas which encode a partial cube of exponential isometric dimension or even a graph which is not a partial cube at all.

To achieve this, we first create some tools using matrices and their rank. We only use the rank of a matrix with respect to $\mathbb{Z}$. A metric space is a set of elements R equipped with a distance function $d : R \times R \to \mathbb{N}$. We say a matrix $M \in \mathbb{N}^{I \times I}$ with index-set I embeds into a metric space if there is a mapping $\pi : I \to R$ such that for all $i, j \in I$ $M_{i,j} = d(\pi(i), \pi(j))$. An example for such a metric space is the k dimensional integer grid equipped with the L^1-norm (sometimes called Manhattan-norm).

Lemma 11. *The matrix $M = (m_{i,j})_{i,j \in \{0,1\}^n}$ with $m_{i,j} = \Delta(i,j) + 2$ if $i \neq j$ and $m_{i,i} = 0$ has rank at least $2^n - n - 1$.*

Proof. We decompose M as $M = M_1 + M_2$ with $M_1 = (\Delta(i,j))_{i,j \in \{0,1\}^n}$ and $M_2 = M - M_1$. It can be verified that $rank(M_2) = 2^n$ and $rank(M_1) = n+1$. For the latter, a complete basis consists of all the row-vectors of bit-strings w with $\Delta(w) \leq 1$. We denote these vectors as w^0 for the string of weight 0 and w^i the vector for the string setting bit i to 1. Then a row vector of an arbitrary string a can be computed as $(\sum_{i \in [n]} a_i \cdot w^i) - (\Delta(a) - 1) \cdot w^0$. For a given column b, every $a_i \neq 0$ adds $\Delta(b) - 1$ iff $b_i = 1$ and $\Delta(b) + 1$ otherwise. The sum adds up to $\Delta(a)\Delta(b) + \Delta(a \setminus b) - \Delta(a \cap b)$. As $w^0[b] = \Delta(b)$, we subtract $(\Delta(a) - 1)\Delta(b)$. So the result is $\Delta(b) + \Delta(a \setminus b) - \Delta(a \cap b) = \Delta(a \setminus b) + \Delta(b \setminus a) = \Delta(a,b)$.

We know that $rank(-M_1) = rank(M_1)$, and by subadditivity $rank(M_2) \leq rank(M) + rank(-M_1)$. Then $2^n \leq rank(M) + n + 1$ and $rank(M) \geq 2^n - n - 1$. $\qquad\square$

Lemma 12. *If a given point set P with distances $M \in \mathbb{N}^{P^2}$ can be mapped into the metric space $R = \{0,1\}^m$ with L^1 as distance-norm, then $rank(M) \leq 2m$.*

Proof. For a given P we look at the labeling $\pi : P \to R$. Then $d(u,v) = \sum_{i \in [m]} |\pi(u)_i - \pi(v)_i|$. So basically $M = \sum_{i \in [m]} M_i$ with $M_i = (m_i^{j,k})$ and $m_i^{j,k} = |\pi(j)_i - \pi(k)_i|$. So all M_i are 0,1 block matrices. They define sets $A, B \subseteq P$ such that all entries $(a,b) \in A \times B$ are assigned to 1 and everything else is 0. As M_i is symmetric, we can split M_i into two matrices of rank 1: The first one contains all non-negative entries $(a,b) \in A \times B$ and the second one all $(b,a) \in B \times A$. This implies that M is the sum of $2m$ rank 1 matrices and therefore $rank(M) \leq 2m$. $\qquad\square$

Corollary 13. *The matrix $M = (m_{i,j})^{i,j \in \{0,1\}^n}$ with $m_{i,j} = \Delta(i,j) + 2$ if $i \neq j$ and $m_{i,i} = 0$ can not be embedded into the metric space $R = \{0,1\}^m$ for all $m < \frac{2^n - n - 1}{2}$ with L^1 as distance-norm.*

Note that another intuitive argument for this statement is that the second part of the sum basically implies that the embedding contains a part which assigns to all 2^n bit-strings points such that their mutual distance is 2. This is called an equilateral embedding. Moreover, for a given metric space of dimension k the equilateral dimension is the maximal number of points which can be of mutually the same distance. [1] proved that for the integer lattice with L^1 norm, the equilateral dimension is $O(k \log k)$. Therefore the dimension in which this distance matrix can be embedded can not be much smaller than 2^n.

These tools are enough to provide a lower bound for the isometric dimension of Horn-encoded graphs.

Lemma 14. *For every n there is an induced graph G of the hypercube H_{2n+1} of size $2^n + 2^{2n}$ with isometric dimension between $\frac{2^n - n - 1}{2}$ and $2^n + 2n$ which can be encoded in a $Horn_3$ formula of size $poly(n)$.*

Proof. Consider the formula $F(x_1, x_1', \ldots, x_n, x_n') = \bigwedge_{i \in [n]} (x_i \leftrightarrow x_i')$ and $F' = y \to F$ with a new variable y. Obviously $F' \in Horn_3$, F has a solution graph with 2^n isolated vertices and $G = G_{F'}$ is only a single connected component of size $2^n + 2^{2n}$. But by fixing y to 1 we get the original formula with 2^n isolated vertices u. All these vertices u agree on y but for $u \neq u'$, their distance in G is $\Delta(u, u') + 2$. For all vertices u, u' with $y = 0$, their distance is $\Delta(u, u')$.

So an isometric embedding for G implies an embedding for all u with the variable y set to 1. But by Corollary 13 we know that such an embedding needs at least $\frac{2^n - n - 1}{2}$ bits. This proves our lower bound.

For an upper bound we replace y with 2^n bits and every node u of G which is isolated in G_F sets a different bit to 1. Every other node sets all new bits to 0. This is a correct partial cube embedding for F' of dimension $2^n + 2n$.   □

For Horn-encoded graphs which are not even partial cubes, we provide an example:

Lemma 15. *There is a $Horn_4$ formula encoding a single connected component which is not a partial cube.*

Proof. Consider the formula $F(w, x, y, z) = (\overline{y} \to \overline{z}) \wedge ((w \wedge x) \to (y \leftrightarrow z))$. This clearly is a $Horn_4$ formula and the encoded graph is depicted in Fig. 2.

To see this, we note that a classical characterization of partial cubes is that an undirected graph $G = (V, E)$ is a partial cube iff it is bipartite and the relation Θ on the edge set is transitive. Hereby Θ is defined as $\{u, v\} \Theta \{x, y\} \leftrightarrow d(u, x) + d(v, y) \neq d(u, y) + d(v, x)$ for $\{u, v\}, \{x, y\} \in E$ (see for example [13]). It is easy to see that in the given example $e \Theta f$ and $f \Theta g$ but $e \not\Theta g$. Therefore the graph encoded by F is not a partial cube (no matter the isometric dimension).   □

We briefly mention an observation for length-bounded st-connectivity (l-$stConn(S)$). This problem is, given a formula F built from S, $s, t \models F$ and $l \in \mathbb{N}$, is there a path of length at most l from s to t? Clearly, if F is CPSS, then either $d(s, t) = \Delta(s, t)$ or $d(s, t) = \infty$. So l-$stConn(CPSS)$ reduces to counting different bits and checking if there is a path at all. So for CPSS, this problem can be solved in P. In contrast, if F is Schaefer but not CPSS, this problem seems harder although the solution graph still has a small diameter. For l-$stConn(Horn_3)$, [4] proved $W[2]$-hardness.

Theorem 16 ([4]). *l-$stConn(Horn_3)$ is $W[2]$-hard when parameterized by $l' = l - \Delta(s, t)$.*

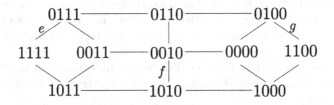

Fig. 2. Induced subgraph of the hypercube which is not a partial cube.

5   Improved Algorithms for Connectivity and *st*-connectivity

In [16], the author gives a polynomial-time algorithm for connectivity on CPSS-formulas. But as this algorithm is basically a logspace-reduction to the satisfiability problem, we can refine this statement for more restricted classes of formulas. We restate this result and prove our corollary.

Theorem 17 *[16]. Given a CPSS set S and a formula $F(x_1, \ldots, x_n)$ over S, the following polynomial-time algorithm decides whether F is connected: For every constraint C_i of F, obtain the projection F_i of F to the variables $\mathbf{x_i}$ occurring in C_i by checking for every assignment a of $\mathbf{x_i}$ whether $F[\mathbf{x_i}/a]$ is satisfiable. Then F is connected iff no F_i is disconnected.*[3]

Corollary 18. *For CPSS sets S, $Conn(S) \leq_T^L Sat(S)$.*

Lemma 19. *$Conn(2SAT) \in NL$.*

Proof. It is easy to see that the solution graph of a satisfiable $2SAT$ formula is disconnected iff the implication graph contains a cycle. The proof of Lemma 22 gives more details on this statement. While checking for a cycle is in $NL = coNL$, checking if a solution graph is disconnected therefore is in $coNL$. It follows that $Conn(2SAT) \in NL$.   □

Now as Corollary 18 is a direct result of [16] in the case of *Conn*, the remaining part of this section will derive a similar statement for *stConn*. In addition, our following work will show that $Conn(2SAT)$ is NL-complete and $Conn(Horn_3)$ is P-complete.

In [8] the authors proved that *st*-connectivity is in P for all tight sets of functions by showing that the diameter in connected components is bounded by a linear function. We now show that even for tight formulas, *st*-connectivity can be reduced to a satisfiability problem.

Theorem 20. *Given a tight formula $F(x_1, \ldots, x_n)$ and $s, t \in \{0,1\}^n$. Then $stConn(F,s,t) \leq_m^L Sat(F \cup \{(x \vee \overline{y})\})$.*

Proof. Given F as well as solutions s, t, we perform a walk on the solution graph starting at s by constructing a formula F' which is satisfiable iff there is a path from s to t in F.

We create a formula F' such that a satisfying assignment describes a walk from s to t in the solution graph. Various copies of the variables $x = (x_1, \ldots, x_n)$ of F simulate steps on the solution graphs. The first copy x^0 gets fixed to s. The additional copies can always only vary from the preceding copy in a specific variable and have to satisfy F. If the distance of consecutive copies x^i and x^{i+1} is at most one and the last copy is equal to t, then there is a path from s to

---
[3] Note that as S is finite, every constraint has finite arity and therefore a solution graph of constant size.

t in F. If we know that s and t have distance at most d, we take dn steps by using d copies of the following construction. The set of variables x^1 is allowed to differ from x^0 only in the variable x_1. The next set of variables, x^2 can only differ from x^1 in x_2. After n such steps and d copies of this construction, we fix the last set of variables to t and know that the formula is satisfiable iff there is a path from s to t.

Note that in each step we only offer the new variable-set to flip a variable. We therefore fix all other variables to the previous value with clauses indicating equivalence and omit such clauses for the variable which is allowed to change. □

Corollary 21. *stConn(2SAT)* $\in$ *NL. stConn(S)* $\in$ *P for S a set of Schaefer functions.*

In addition, we prove the completeness results for st-connectivity problems on solution graphs of $2SAT$- and $Horn$-formulas.

Lemma 22. *stConn(2SAT) is NL-complete.*

Proof. We reduce the complement of the NL-complete problem of acyclic directed connectivity (see for example [2]) to *stConn(2SAT)*. The proof follows as NL = coNL.

Suppose we are given an acyclic directed graph $G = (V, E)$ and two nodes s, t. We now create G' by adding to G the edge (t, s). Clearly, there is a cycle in G' iff there is an s, t path in G. We now interpret G' as implication graph of a $2SAT$ formula F and state that there is a path from 0^n to 1^n in the solution graph G_F iff there is no cycle in G'.

Suppose there is no cycle. We describe a path from 0^n to 1^n. Every satisfying assignment coincides with a $0, 1$ labeling of G'. If a variable x_i is set to 0, then the node $x_i \in G'$ is labeled with 0. An assignment x is satisfying F as long as there is no edge (x_i, x_j) with $x_i = 1$ and $x_j = 0$. We therefore move from 0^n to 1^n by flipping the last bits of all longest path in G'. If all such bits were flipped, we delete the corresponding nodes and repeat until the graph is empty and we reach 1^n.

For the other direction, suppose there is a cycle. Then obviously we can not reach 1^n from 0^n. At one point we have to flip a single variable in this cycle. But then we would have to flip all variables in this cycle at the same time or else there would be an implication $1 \rightarrow 0$. So the graph is disconnected. □

Corollary 23. *Conn(2SAT) is NL-complete.*

Proof. This follows by the observation that in the previous reduction, the constructed solution graph is connected iff there is no cycle in the graph. As $2SAT$ is closed under complementation, our statement follows. □

It would be interesting to give a direct NL algorithm for *stConn(2SAT)* using properties of the solution graph instead of just reducing to the satisfiability problem. This is still open. The main difference to the connectivity for CPSS formulas is that in the case of $2SAT$ formulas, the connected components are median graphs, a subclass of all partial cubes, while formulas which are CPSS consist of partial cubes of dimension n. Interestingly, the st-connectivity on $Horn_3$ formulas is complete for P while finding a shortest path is hard for $W[2]$ as explained in the next section.

Lemma 24. $stConn(Horn_3)$ *is P-complete.*

Proof. We use a similar method as in Lemma 22 and reduce the monotone circuit value problem to $stConn(Horn_3)$. As the monotone circuit value problem is known to be P-complete [7], the hardness result follows.

So given a monotone circuit C with n inputs and bounded in-degree 2, m inner gates and an $x \in \{0,1\}^n$, we first create a hypergraph G such that the marking algorithm[4] on hypergraphs reaches the root node iff $C(x) = 1$. $G = (V, E)$ with V is the set of gates in C and for a gate u with inputs v, w we add the edges

1. $((v, w) \to u)$ iff u is labeled with $\wedge$
2. and $(v \to u)$ and $(w \to u)$ iff u is labeled with $\vee$.

It is easy to see that if we mark all input nodes $x_i = 1$, the marking algorithm reaches the root z iff $C(x) = 1$. Note that the marking algorithm proceeds as follows: if there is an edge $((u_1, \ldots, u_l) \to u)$ and all u_i are marked, we can mark u. We add some additional edges (z, x_i) for every $x_i = 1$. So the marking algorithm can perform a cycle in G iff $C(x) = 1$.

We now interpret this hypergraph as a Horn-formula F with $n + m$ variables and prove that the solution graph of F has a path from $1^{n+m} = x^*$ to 0^{n+m} iff $C(x) = 0$. Suppose first that $C(x) = 0$. We therefore know that the marking algorithm, starting with marked x_i for all $x_i = 1$ never reaches the root node z. Let $A(x)$ be the set of nodes this algorithm marks when starting with all 1 bits in x including exactly the variables which are initially set to 1. Then for all $u \notin A(x)$ without any predecessor in G, it is safe to flip $x^*[u]$ to 0. This corresponds to every input gate of C which is not set to 1. In a second round we flip all nodes with at least one predecessor which was already flipped (and where the premise is therefore false) to 0 and continue this process level by level until we reach the root z. Note that we never violate any clause of F. If we would violate a clause $((u, v) \to w)$ by setting the premise of the clause to 1 but the conclusion to 0, then $w \in A(x)$ which is a contradiction to the way we chose the variables.

We finish this process by reaching the root z. Now, in a second step, we can flip the input variables $x_i \in A(x)$ and perform the same process with all nodes in $A(x)$. This again does not violate any clauses and, in the end, $x^* = 0^{n+m}$.

Now suppose $C(x) = 1$ and $x^* = 1^n$. We just note there is no path in the solution graph of F to reach 0^n. We know that $|A(x)| \geq 2$ and for every single

---

[4] This algorithm starts with a directed hypergraph and an initially marked set of nodes. If there is a hyperedge such that all source-nodes are marked but not all target nodes, we mark all target nodes. The algorithm finishes if there is no hyperedge which would mark a new node.

$u \in A(x)$, flipping u to 0 violates a clause in F. Any $u \in A(x)$ is the conclusion of a clause in F with the premise set to 1. So flipping any single variable in $A(x)$ violates F and there can not be a path from 1^{n+m} to 0^{n+m}. Note that this case did not occur for $C(x) = 0$ because we set z to 0 and then $A(x)$ had elements without predecessors in $A(x)$ (at first the inputs x_i). This finishes our proof. $\square$

Corollary 25. *Conn(Horn$_3$) is P-complete.*

Proof. This follows by the observation that in the previous reduction, the constructed solution graph is connected iff the marking algorithm does not reach the root node. As P is closed under complementation, our statement follows. $\square$

A reduction from satisfiability of tight formulas to st-connectivity of tight formulas is not possible unless P = NP. To see this, we note that the work of [9] implies that satisfiability of tight formulas is NP-complete while st-connectivity is in P, see [16].

6   Conclusions and Open Problems

We have studied solution graphs of different sets of boolean formulas introduced by [8,16]. We showed that all solution graphs of CPSS formulas consist of partial cubes of small isometric dimension and by going to general Schaefer formulas, their dimension may increase exponentially or they may even loose the property of being a partial cube. This gives a sharp separation between solution graphs which behave nicely and solution graphs without any known structure. It would be interesting to further analyze solution graphs of Horn formulas and either show that they behave still nice in another way or if they are already complicated enough for other problems to be much harder for these graphs. One of such problems is the connectivity problem as shown by [8,16] which is coNP-complete. It would be interesting to find more such problems and further understand the origin of this complexity blow-up.

We introduced techniques to reduce connectivity and st-connectivity in CPSS or tight formulas to their satisfiability problem. We even proved the equivalence of these problems and satisfiability for related formulas. These results imply that for solution graphs of $2SAT$ formulas, a collection of undirected partial cubes, the st-connectivity problem is NL-complete while for Horn solution graphs it is P-complete. An explanation for this difference could be the fact that $2SAT$ formulas describe median graphs which are a proper subset of partial cubes. We would like to see an NL-algorithm for $stConn(2SAT)$ which directly exploits this property. A similar statement holds for connectivity.

Simultaneously our results imply that length-bounded st-connectivity is easy for CPSS formulas while a result of Bonsma et al. [4] implied $W[2]$-hardness for general Schaefer-formulas. This implies that there is probably no polynomial-time reduction from $stConn(Horn)$ to $stConn(CPSS)$ which preserves distances.

References

1. Alon, N., Pudlák, P.: Equilateral sets in l_p^n. Geom. Funct. Anal. **13**(3), 467–482 (2003)
2. Arora, S., Barak, B.: Computational Complexity: A Modern Approach, 1st edn. Cambridge University Press, New York (2009)
3. Balcázar, J.L., Lozano, A., Torán, J.: The complexity of algorithmic problems on succinct instances. In: Baeza-Yates, R., Manber, U. (eds.) Computer Science, Research and Applications. Springer, New York (1992)
4. Bonsma, P., Mouawad, A.E., Nishimura, N., Raman, V.: The complexity of bounded length graph recoloring and CSP reconfiguration. In: Cygan, M., Heggernes, P. (eds.) IPEC 2014. LNCS, vol. 8894, pp. 110–121. Springer, Heidelberg (2014)
5. Das, B., Scharpfenecker, P., Torán, J.: Succinct encodings of graph isomorphism. In: Dediu, A.-H., Martín-Vide, C., Sierra-Rodríguez, J.-L., Truthe, B. (eds.) LATA 2014. LNCS, vol. 8370, pp. 285–296. Springer, Heidelberg (2014)
6. Ekin, O., Hammer, P.L., Kogan, A.: On connected Boolean functions. Discrete Appl. Math. **96–97**, 337–362 (1999)
7. Goldschlager, L.M.: The monotone and planar circuit value problems are log space complete for p. SIGACT News **9**(2), 25–29 (1977)
8. Gopalan, P., Kolaitis, P.G., Maneva, E., Papadimitriou, C.H.: The connectivity of boolean satisfiability: computational and structural dichotomies. SIAM J. Comput. **38**(6), 2330–2355 (2009)
9. Juban, L.: Dichotomy theorem for the generalized unique satisfiability problem. In: Ciobanu, G., Păun, G. (eds.) FCT 1999. LNCS, vol. 1684, pp. 327–337. Springer, Heidelberg (1999)
10. Maneva, E., Mossel, E., Wainwright, M.J.: A new look at survey propagation and its generalizations. J. ACM **54**(4), 1–41 (2007)
11. Mézard, M., Zecchina, R.: Random k-satisfiability problem: From an analytic solution to an efficient algorithm. Phys. Rev. E **66**, 056126 (2002)
12. Mézard, M., Ricci-Tersenghi, F., Zecchina, R.: Two solutions to diluted p-spin models and xorsat problems. J. Stat. Phys. **111**(3–4), 505–533 (2003)
13. Ovchinnikov, S.: Graphs and Cubes. Universitext. Springer, New York (2011)
14. Papadimitriou, C.H., Yannakakis, M.: A note on succinct representations of graphs. Inf. Control **71**(3), 181–185 (1986)
15. Schaefer, T.J.: The complexity of satisfiability problems. In: Proceedings of the Tenth Annual ACM Symposium on Theory of Computing - STOC 1978, pp. 216–226. ACM Press, New York, May 1978
16. Schwerdtfeger, K.W.: A Computational Trichotomy for Connectivity of Boolean Satisfiability, p. 24, December 2013. http://arxiv.org/abs/1312.4524
17. Veith, H.: Languages represented by Boolean formulas. Inf. Process. Lett. **63**(5), 251–256 (1997)
18. Veith, H.: How to encode a logical structure by an OBDD. In: Proceedings of the 13th IEEE Conference on Computational Complexity, pp. 122–131. IEEE Computer Society (1998)

Languages

Interprocedural Reachability for Flat Integer Programs

Pierre Ganty[1]([⊠]) and Radu Iosif[2]

[1] IMDEA Software Institute, Madrid, Spain
pierre.ganty@imdea.org
[2] CNRS/VERIMAG, Grenoble, France
radu.iosif@imag.fr

Abstract. We study programs with integer data, procedure calls and arbitrary call graphs. We show that, whenever the guards and updates are given by octagonal relations, the reachability problem along control flow paths within some language $w_1^* \ldots w_d^*$ over program statements is decidable in NEXPTIME. To achieve this upper bound, we combine a program transformation into the same class of programs but without procedures, with an NP-completeness result for the reachability problem of procedure-less programs. Besides the program, the expression $w_1^* \ldots w_d^*$ is also mapped onto an expression of a similar form but this time over the transformed program statements. Several arguments involving context-free grammars and their generative process enable us to give tight bounds on the size of the resulting expression. The currently existing gap between NP-hard and NEXPTIME can be closed to NP-complete when a certain parameter of the analysis is assumed to be constant.

1 Introduction

This paper studies the complexity of the reachability problem for a class of programs featuring procedures and local/global variables ranging over integers. In general, the reachability problem for this class is undecidable [21]. Thus, we focus on a special case of the reachability problem which restricts both the class of input programs and the set of executions considered. The class of input programs is restricted by considering that all updates to the integer variables $\mathbf{x}$ are defined by *octagonal constraints*, that are conjunctions of atoms of the form $\pm x \pm y \leqslant c$, with $x, y \in \mathbf{x} \cup \mathbf{x}'$, where $\mathbf{x}'$ denote the future values of the program variables. The reachability problem is restricted by limiting the search to program executions conforming to a regular expression of the form $w_1^* \ldots w_d^*$ where the w_i's are finite sequences of program statements.

We call this problem *flat-octagonal reachability* (fo-reachability, for short). Concretely, given: (i) a program $\mathcal{P}$ with procedures and local/global variables,

---

P. Ganty—Supported by the EU FP7 2007–2013 program under agreement 610686 POLCA and from the Madrid Regional Government under CM project S2013/ICE-2731 (N-Greens).

© Springer International Publishing Switzerland 2015
A. Kosowski and I. Walukiewicz (Eds.): FCT 2015, LNCS 9210, pp. 133–145, 2015.
DOI: 10.1007/978-3-319-22177-9_11

whose statements are specified by octagonal constraints, and (*ii*) a bounded expression $\mathbf{b} = w_1^* \ldots w_d^*$, where w_i's are sequences of statements of $\mathcal{P}$, the fo-reachability problem $\mathrm{REACH}_{fo}(\mathcal{P}, \mathbf{b})$ asks: can $\mathcal{P}$ run to completion by executing a sequence of program statements $w \in \mathbf{b}$? Studying the complexity of this problem provides the theoretical foundations for implementing efficient decision procedures, of practical interest in areas of software verification, such as bug-finding [9], or counterexample-guided abstraction refinement [14,15].

Our starting point is the decidability of the fo-reachability problem in the absence of procedures. Recently, the precise complexity of this problem was coined to NP-complete [6]. However, this result leaves open the problem of dealing with procedures and local variables, let alone when the graph of procedure calls has cycles, such as in the example of Fig. 1 (a). Pinning down the complexity of the fo-reachability problem in presence of (possibly recursive) procedures, with local variables ranging over integers, is the challenge we address here.

The decision procedure we propose in this paper reduces $\mathrm{REACH}_{fo}(\mathcal{P}, \mathbf{b})$, from a program $\mathcal{P}$ with arbitrary call graphs, to procedure-less programs as follows:

1. we apply a source-to-source transformation returning a procedure-less program $\mathcal{Q}$, with statements also defined by octagonal relations, such that $\mathrm{REACH}_{fo}(\mathcal{P}, \mathbf{b})$ is equivalent to the unrestricted reachability problem for $\mathcal{Q}$, when no particular bounded expression is supplied.
2. we compute a bounded expression $\varGamma_{\mathbf{b}}$ over the statements of $\mathcal{Q}$, such that $\mathrm{REACH}_{fo}(\mathcal{P}, \mathbf{b})$ is equivalent to $\mathrm{REACH}_{fo}(\mathcal{Q}, \varGamma_{\mathbf{b}})$.

The above reduction allows us to conclude that the fo-reachability problem for programs with arbitrary call graphs is decidable and in NEXPTIME. Naturally, the NP-hard lower bound [6] for the fo-reachability problem of procedure-less programs holds in our setting as well. Despite our best efforts, we did not close the complexity gap yet. However we pinned down a natural parameter, called *index*, related to programs with arbitrary call graphs, such that, when setting this parameter to a fixed constant (like 3 in 3-SAT), the complexity of the resulting fo-reachability problem for programs with arbitrary call graphs becomes NP-complete. Indeed, when the index is fixed, the aforementioned reduction computing $\mathrm{REACH}_{fo}(\mathcal{Q}, \varGamma_{\mathbf{b}})$ runs in polynomial time. Then the NP decision procedure for the fo-reachability of procedure-less programs [6] shows the rest.

The index parameter is better understood in the context of formal languages. The control flow of procedural programs is captured precisely by the language of a context-free grammar. A k-index ($k > 0$) underapproximation of this language is obtained by filtering out the derivations containing a sentential form with $k+1$ occurrences of nonterminals. The key to our results is a toolbox of language theoretic constructions of independent interest that enables to reason about the structure of context-free derivations generating words into $\mathbf{b} = w_1^* \ldots w_d^*$, that is, words of the form $w_1^{i_1} \ldots w_d^{i_d}$ for some integers $i_1, \ldots, i_d \geqslant 0$.

To properly introduce the reader to our result, we briefly recall the important features of our source-to-source transformation through an illustrative example. We apply first our program transformation [11] to the program $\mathcal{P}$ shown

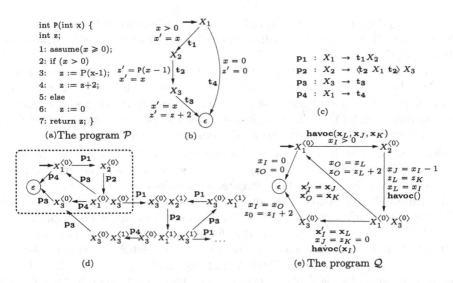

(a)The program $\mathcal{P}$ (b)

$$P_1 : X_1 \rightarrow t_1 X_2$$
$$P_2 : X_2 \rightarrow \langle t_2 \, X_1 \, t_2 \rangle \, X_3$$
$$P_3 : X_3 \rightarrow t_3$$
$$P_4 : X_1 \rightarrow t_4$$

(c)

(d)   (e) The program $\mathcal{Q}$

Fig. 1. $\mathbf{x}_I = \{x_I, z_I\}$ ($\mathbf{x}_O = \{x_O, z_O\}$) are for the input (output) values of x and z, respectively. $\mathbf{x}_{J,K,L}$ provide extra copies. $\mathbf{havoc}(\mathbf{y})$ stands for $\bigwedge_{x \in \mathbf{x}_{I,O,J,K,L} \setminus y} x' = x$, and $\mathbf{x}'_\alpha = \mathbf{x}_\beta$ for $\bigwedge_{x \in \mathbf{x}} x'_\alpha = x_\beta$.

in Fig. 1 (a). The call graph of this program consists of a single state P with a self-loop. The output program $\mathcal{Q}$ given Fig. 1 (e), has no procedures and it can thus be analyzed using any existing intra-procedural tool [4,5]. The relation between the variables x and z of the input program can be inferred from the analysis of the output program. For instance, the input-output relation of the program $\mathcal{P}$ is defined by $z' = 2x$, which matches the precondition $z_O = 2x_I$ of the program $\mathcal{Q}$. Consequently, any assertion such as *"there exists a value $n > 0$ such that $P(n) < n$"* can be phrased as: *"there exist values $n < m$ such that $\mathcal{Q}(n, m)$ reaches its final state"*. While the former can be encoded by a reachability problem on $\mathcal{P}$, by adding an extra conditional statement, the latter is an equivalent reachability problem for $\mathcal{Q}$.

For the sake of clarity, we give several representations of the input program $\mathcal{P}$ that we assume the reader is familiar with including the text of the program in Fig. 1 (a) and the corresponding control flow graph in Fig. 1 (b).

In this paper, the formal model we use for programs is based on context-free grammars. The grammar for $\mathcal{P}$ is given at Fig. 1 (c). The rôle of the grammar is to define the set of *interprocedurally valid* paths in the control-flow graph of the program $\mathcal{P}$. Every edge in the control-flow graph matches one or two symbols from the finite alphabet $\{t_1, \langle t_2, t_2 \rangle, t_3, t_4\}$, where $\langle t_2$ and $t_2 \rangle$ denote the call and return, respectively. The set of nonterminals is $\{X_1, X_2, X_3, X_4\}$. Each edge in the graph translates to a production rule in the grammar, labeled p_1 to p_4. For instance, the call edge $X_2 \xrightarrow{t_2} X_3$ becomes $X_2 \rightarrow \langle t_2 X_1 t_2 \rangle X_3$. The language of the grammar of Fig. 1 (c) (with axiom X_1) is the set

$L = \{(t_1 \langle t_2)^n \, t_4 \, (t_2 \rangle t_3)^n \mid n \in \mathbb{N}\}$ of interprocedurally valid paths in the control-flow graph. Observe that L is included in the language of the regular expression $b = (t_1 \langle t_2)^* \, t_4^* \, (t_2 \rangle t_3)^*$.

Our program transformation is based on the observation that the semantics of $\mathcal{P}$ can be precisely defined on the set of *derivations* of the associated grammar. In principle, one can always represent this set of derivations as a possibly infinite automaton (Fig. 1 (d)), whose states are sequences of nonterminals annotated with priorities (called ranks)[1], and whose transitions are labeled with production rules. Each finite path in this automaton, starting from $X_1^{\langle 0 \rangle}$, defines a valid prefix of a derivation. Since $L \subseteq b$, Luker [18] shows that it is sufficient, to generate L, to restrict derivations to those accepted by a finite sub-automaton. Referring to our example, it consists of the sub-automaton enclosed with a dashed box in Fig. 1 (d), in which each state consists of a at most 2 ranked nonterminals.

Finally, we label the edges of this finite automaton with octagonal constraints that capture the semantics of the relations labeling the control-flow graph from Fig. 1 (b). We give here a brief explanation for the labeling of the finite automaton in Fig. 1 (e), in other words, the output program $\mathcal{Q}$ (see [11] for more details). The idea is to compute, for each production rule p_i, a relation $\rho_i(\mathbf{x}_I, \mathbf{x}_O)$, based on the constraints associated with the symbols occurring in p_i (labels from Fig. 1 (b)). For instance, in the transition $X_2^{\langle 0 \rangle} \xrightarrow{p_2} X_1^{\langle 0 \rangle} X_3^{\langle 0 \rangle}$, the auxiliary variables store intermediate results of the computation of p_2 as follows: $[\mathbf{x}_I] \langle t_2 \, [\mathbf{x}_J] \, X_1 \, [\mathbf{x}_K] \, t_2 \rangle \, [\mathbf{x}_L] \, X_3 \, [\mathbf{x}_O]$. The guard of the transition can be understood by noticing that $\langle t_2$ gives rise to the constraint $x_J = x_I - 1$, $t_2 \rangle$ to $z_L = z_K$, $x_I = x_L$ corresponds to the frame condition of the call, and $\mathbf{havoc}()$ copies all current values of $\mathbf{x}_{I,J,K,L,O}$ to the future ones. It is worth pointing out that the constraints labeling the transitions of the program $\mathcal{Q}$ are necessarily octagonal if the statements of $\mathcal{P}$ are defined by octagonal constraints.

An intra-procedural analysis of the program $\mathcal{Q}$ in Fig. 1 (e) infers the precondition $x_I \geqslant 0 \wedge z_O = 2x_I$ which coincides with the input/output relation of the recursive program $\mathcal{P}$ in Fig. 1 (a), i.e. $x \geqslant 0 \wedge z' = 2x$. The original query $\exists n > 0 : \mathrm{P}(n) < n$ translates thus into the satisfiability of the formula $x_I > 0 \wedge z_O = 2x_I \wedge x_I < z_O$, which is clearly false.

The paper is organised as follows: basic definitions are given Sects. 2 and 3 defines the fo-reachability problem, Sect. 4 presents an alternative program semantics based on derivations and introduces subsets of derivations which are sufficient to decide reachability, Sect. 5 starts with on overview of our decision procedure and our main complexity results and continues with the key steps of our algorithms. A companion technical report [10] contains the missing details.

2   Preliminaries

Let Σ be a finite nonempty set of symbols, called an *alphabet*. We denote by Σ^* the set of finite words over Σ which includes ε, the empty word. The concatenation of two words $u, v \in \Sigma^*$ is denoted by $u \cdot v$ or uv. Given a word $w \in \Sigma^*$,

---

[1] The precise definition and use of ranks will be explained in Sect. 4.

let $|w|$ denote its length and let $(w)_i$ with $1 \leqslant i \leqslant |w|$ be the ith symbol of w. Given $w \in \Sigma^*$ and $\Theta \subseteq \Sigma$, we write $w{\downarrow}_\Theta$ for the word obtained by deleting from w all symbols not in Θ, and sometimes we write $w{\downarrow}_a$ for $w{\downarrow}_{\{a\}}$. A *bounded expression* **b** over alphabet Σ is a regular expression of the form $w_1^* \ldots w_d^*$, where $w_1, \ldots, w_d \in \Sigma^*$ are nonempty words and its size is given by $|\mathbf{b}| = \sum_{i=1}^d |w_i|$. We use **b** to denote both the bounded expression and its language. We call a language L *bounded* when $L \subseteq \mathbf{b}$ for some bounded expression **b**.

A *grammar* is a tuple $G = \langle \Xi, \Sigma, \Delta \rangle$ where Ξ is a finite nonempty set of *nonterminals*, Σ is an alphabet of *terminals*, such that $\Xi \cap \Sigma = \varnothing$, and $\Delta \subseteq \Xi \times (\Sigma \cup \Xi)^*$ is a finite set of *productions*. For a production $(X, w) \in \Delta$, often conveniently noted $X \to w$, we define its *size* as $|(X, w)| = |w| + 1$, and $|G| = \sum_{p \in \Delta} |p|$ defines the size of G.

Given two words $u, v \in (\Sigma \cup \Xi)^*$, a production $(X, w) \in \Delta$ and a position $1 \leqslant j \leqslant |u|$, we define a *step* $u \xrightarrow{(X,w)/j}_G v$ if and only if $(u)_j = X$ and $v = (u)_1 \cdots (u)_{j-1} w (u)_{j+1} \cdots (u)_{|u|}$. We omit (X, w) or j above the arrow when clear from the context. A *control word* is a finite word $\gamma \in \Delta^*$ over the alphabet of productions. A *step sequence* $u \xrightarrow{\gamma}_G v$ is a sequence $u = w_0 \xrightarrow{(\gamma)_1}_G w_1 \ldots w_{n-1} \xrightarrow{(\gamma)_n}_G w_n = v$ where $n = |\gamma|$. If $u \in \Xi$ is a nonterminal and $v \in \Sigma^*$ is a word without nonterminals, we call the step sequence $u \xrightarrow{\gamma}_G v$ a *derivation*. When the control word γ is not important, we write $u \Rightarrow^*_G v$ instead of $u \xrightarrow{\gamma}_G v$, and we chose to omit the grammar G when clear from the context.

Given a nonterminal $X \in \Xi$ and $Y \in \Xi \cup \{\varepsilon\}$, i.e. Y is either a nonterminal or the empty word, we define the set $L_{X,Y}(G) = \{u\,v \in \Sigma^* \mid X \Rightarrow^* uYv\}$. The set $L_{X,\varepsilon}(G)$ is called the *language* of G produced by X, and is denoted $L_X(G)$ in the following. For a set $\Gamma \subseteq \Delta^*$ of control words (also called a *control set*), we denote by $\hat{L}_{X,Y}(\Gamma, G) = \{u\,v \in \Sigma^* \mid \exists \gamma \in \Gamma : X \xrightarrow{\gamma} uYv\}$ the language generated by G using only control words from Γ. We also write $\hat{L}_X(\Gamma, G)$ for $\hat{L}_{X,\varepsilon}(\Gamma, G)$.

Let **x** denote a nonempty finite set of integer variables, and $\mathbf{x}' = \{x' \mid x \in \mathbf{x}\}$. A *valuation* of **x** is a function $\nu : \mathbf{x} \to \mathbb{Z}$. The set of all such valuations is denoted $\mathbb{Z}^{\mathbf{x}}$. A formula $\phi(\mathbf{x}, \mathbf{x}')$ is evaluated with respect to two valuations $\nu, \nu' \in \mathbb{Z}^{\mathbf{x}}$, by replacing each occurrence of $x \in \mathbf{x}$ with $\nu(x)$ and each occurrence of $x' \in \mathbf{x}'$ with $\nu'(x)$. We write $(\nu, \nu') \models \phi$ when the formula obtained from these replacements is valid. A formula $\phi_R(\mathbf{x}, \mathbf{x}')$ *defines* a relation $R \subseteq \mathbb{Z}^{\mathbf{x}} \times \mathbb{Z}^{\mathbf{x}}$ whenever for all $\nu, \nu' \in \mathbb{Z}^{\mathbf{x}}$, we have $(\nu, \nu') \in R$ iff $(\nu, \nu') \models \phi_R$. The composition of two relations $R_1, R_2 \subseteq \mathbb{Z}^{\mathbf{x}} \times \mathbb{Z}^{\mathbf{x}}$ defined by formulae $\varphi_1(\mathbf{x}, \mathbf{x}')$ and $\varphi_2(\mathbf{x}, \mathbf{x}')$, respectively, is the relation $R_1 \circ R_2 \subseteq \mathbb{Z}^{\mathbf{x}} \times \mathbb{Z}^{\mathbf{x}}$, defined by $\exists \mathbf{y} \,.\, \varphi_1(\mathbf{x}, \mathbf{y}) \wedge \varphi_2(\mathbf{y}, \mathbf{x}')$. For a finite set S, we denote its cardinality by $\|S\|$.

3   Interprocedural Flat Octogonal Reachability

In this section we define formally the class of programs and reachability problems considered. An *octagonal relation* $R \subseteq \mathbb{Z}^{\mathbf{x}} \times \mathbb{Z}^{\mathbf{x}}$ is a relation definedby a finite conjunction of constraints of the form $\pm x \pm y \leqslant c$, where $x, y \in \mathbf{x} \cup \mathbf{x}'$ and $c \in \mathbb{Z}$. The set of octagonal relations over the variables in $\mathbf{x}$ and $\mathbf{x}'$ is denoted as $\mathrm{Oct}(\mathbf{x}, \mathbf{x}')$. The *size* of an octagonal relation R, denoted $|R|$ is the size of the binary encoding of the smallest octagonal constraint defining R.

An *octagonal program* is a tuple $\mathcal{P} = \langle G, I, [.] \rangle$, where G is a grammar $G = \langle \Xi, \Sigma, \Delta \rangle$, $I \in \Xi$ is an *initial* location, and $[.] : L_I(G) \to \mathrm{Oct}(\mathbf{x}, \mathbf{x}')$ is a mapping of the words produced by the grammar G, starting with the initial location I, to octagonal relations. The alphabet Σ contains a symbol t for each *internal* program statement (that is not a call to a procedure) and two symbols $\langle\!\!\!\!t, t\!\!\!\rangle$ for each *call* statement t. The grammar G has three kinds of productions: *(i)* (X, t) if t is a statement leading from X to a return location, *(ii)* (X, tY) if t leads from X to Y, and *(iii)* $(X, \langle\!\!\!\!t\, Y\, t\!\!\!\rangle Z)$ if t is a call statement, Y is the initial location of the callee, and Z is the continuation of the call. Each edge t that is not a call has an associated octagonal relation $\rho_t \in \mathrm{Oct}(\mathbf{x}, \mathbf{x}')$ and each matching pair $\langle\!\!\!\!t, t\!\!\!\rangle$ has an associated *frame condition* $\phi_t \in \mathrm{Oct}(\mathbf{x}, \mathbf{x}')$, which equates the values of the local variables, that are not updated by the call, to their future values. The size of an octagonal program $\mathcal{P} = \langle G, I, [.] \rangle$, with $G = \langle \Xi, \Sigma, \Delta \rangle$, is the sum of the sizes of all octagonal relations labeling the productions of G, formally $|\mathcal{P}| = \sum_{(X,t) \in \Delta} |\rho_t| + \sum_{(X,tY) \in \Delta} |\rho_t| + \sum_{(X, \langle\!\!\!t\, Y\, t\!\!\!\rangle Z) \in \Delta} (|\rho_{\langle\!\!\!t}| + |\rho_{t\!\!\!\rangle}| + |\phi_t|)$.

For example, the program in Fig. 1 (a,b) is represented by the grammar in Fig. 1 (c). The terminals are mapped to octagonal relations as: $\rho_{t_1} \equiv x > 0 \wedge x' = x$, $\rho_{\langle\!\!\!t_2} \equiv x' = x - 1$, $\rho_{t_2\!\!\!\rangle} \equiv z' = z$, $\rho_{t_3} \equiv x' = x \wedge z' = z + 2$ and $\rho_{t_4} \equiv x = 0 \wedge z' = 0$. The frame condition is $\phi_{t_2} \equiv x' = x$, as only z is updated by the call $z' = \mathrm{P}(x - 1)$.

Word-Based Semantics. For each word $w \in L_I(G)$, each occurrence of a terminal $\langle\!\!\!t$ in w is matched by an occurrence of $t\!\!\!\rangle$, and the matching positions are nested[2]. The semantics of the word $[w]$ is an octagonal relation defined inductively[3] on the structure of w: (i) $[t] = \rho_t$, (ii) $[t \cdot v] = \rho_t \circ [v]$, and (iii) $[\langle\!\!\!t \cdot u \cdot t\!\!\!\rangle \cdot v] = ((\rho_{\langle\!\!\!t} \circ [u] \circ \rho_{t\!\!\!\rangle}) \cap \phi_t) \circ [v]$, for all $t, \langle\!\!\!t, t\!\!\!\rangle \in \Sigma$ such that $\langle\!\!\!t$ and $t\!\!\!\rangle$ match. For instance, the semantics of the word $w = t_1 \langle\!\!\!t_2 t_4 t_2\!\!\!\rangle t_3 \in L_{X_1}(G)$, for the grammar G given in Fig. 1 (c), is $[w] \equiv x = 1 \wedge z' = 2$. Observe that this word defines the effect of an execution of the program in Fig. 1 (a) where the function P is called twice—the first call is a top-level call, and the second is a recursive call (line 3).

Reachability Problem. The semantics of a program $\mathcal{P} = \langle G, I, [.] \rangle$ is defined as $[\mathcal{P}] = \bigcup_{w \in L_I(G)} [w]$. Consider, in addition, a bounded expression $\mathbf{b}$, we define $[\mathcal{P}]_{\mathbf{b}} = \bigcup_{w \in L_I(G) \cap \mathbf{b}} [w]$. The problem asking whether $[\mathcal{P}]_{\mathbf{b}} \neq \varnothing$ for a pair $\mathcal{P}, \mathbf{b}$ is called the *flat-octagonal reachability problem*. We use $\mathrm{REACH}_{fo}(\mathcal{P}, \mathbf{b})$ to denote a particular instance.

---

[2] A relation $\leadsto \subseteq \{1, \ldots, |w|\} \times \{1, \ldots, |w|\}$ is said to be nested [2] when no two pairs $i \leadsto j$ and $i' \leadsto j'$ cross each other, as in $i < i' \leqslant j < j'$.

[3] Octagonal relations are closed under intersections and compositions [20].

4   Index-Bounded Depth-First Derivations

In this section, we give an alternate but equivalent program semantics based on derivations. Although simple, the word semantics is defined using a nesting relation that pairs the positions of a word labeled with matching symbols $(\!t$ and $t\!)$. In contrast, the derivation-based semantics just needs the control word.

To define our derivation based semantics, we first define structured subsets of derivations namely the depth-first and bounded-index derivations. The reason is two-fold: (a) the correctness proof of our program transformation [11] returning the procedure-less program $\mathcal{Q}$ depends on bounded-index depth-first derivations, and (b) in the reduction of the $\text{REACH}_{fo}(\mathcal{P}, \mathbf{b})$ problem to that of $\text{REACH}_{fo}(\mathcal{Q}, \Gamma_\mathbf{b})$, the computation of $\Gamma_\mathbf{b}$ depends on the fact that the control structure of $\mathcal{Q}$ stems from a finite automaton recognizing bounded-index depth-first derivations. Key results for our decision procedure are those of Luker [18,19] who, intuitively, shows that if $L_X(G) \subseteq \mathbf{b}$ then it is sufficient to consider depth-first derivations in which no step contains more than k simultaneous occurrences of nonterminals, for some $k > 0$ (Theorem 1).

Depth-First Derivations. It is well-known that a derivation can be associated a unique parse tree. A derivation is said to be *depth-first* if it corresponds to a depth-first traversal of the corresponding parse tree. More precisely, given a step sequence $w_0 \xrightarrow{(X_0,v_0)/j_0} w_1 \ldots w_{n-1} \xrightarrow{(X_{n-1},v_{n-1})/j_{n-1}} w_n$, and two integers m and i such that $0 \leqslant m < n$ and $1 \leqslant i \leqslant |w_m|$ define $f_m(i)$ to be the index ℓ of the first word w_ℓ of the step sequence in which the particular occurrence of $(w_m)_i$ appears. A step sequence is *depth-first* [19] iff for all m, $0 \leqslant m < n$:

$$f_m(j_m) = \max\{f_m(i) \mid 1 \leqslant i \leqslant |w_m| \text{ and } (w_m)_i \in \Xi\}.$$

For example, $X \xrightarrow{(X,YY)/1} YY \xrightarrow{(Y,Z)/2} YZ \xrightarrow{(Z,a)/2} Ya$ is depth-first, whereas $X \xrightarrow{(X,YY)/1} YY \xrightarrow{(Y,Z)/2} YZ \xrightarrow{(Y,Z)/1} ZZ$ is not. We have $f_2(1) = 1$ because $(w_2)_1 = Y$ first appeared at w_1, $f_2(2) = 2$ because $(w_2)_2 = Z$ first appeared at w_2, $j_2 = 1$ and $f_2(2) \not\leqslant f_2(j_2)$ since $2 \not\leqslant 1$. We denote by $u \underset{df}{\overset{\gamma}{\Rightarrow}} w$ a depth-first step sequence and call it depth-first derivation when $u \in \Xi$ and $w \in \Sigma^*$.

Depth-First Derivation-Based Semantics. In previous work [11], we defined the semantics of a procedural program based on the control word of the derivation instead of the produced words. We briefly recall this definition here. Given a depth-first derivation $X \underset{df}{\overset{\gamma}{\Rightarrow}} w$, the relation $[\![\gamma]\!] \subseteq \mathbb{Z}^\mathbf{x} \times \mathbb{Z}^\mathbf{x}$ is defined inductively on γ as follows: (i) $[\![(X,t)]\!] = \rho_t$, (ii) $[\![(X,tY) \cdot \gamma']\!] = \rho_t \circ [\![\gamma']\!]$ where $Y \underset{df}{\overset{\gamma'}{\Rightarrow}} w'$, and (iii) $[\![(X, (\!tY t\!)Z) \cdot \gamma' \cdot \gamma'']\!] = [\![(X, (\!tY t\!)Z) \cdot \gamma'' \cdot \gamma']\!] = ((\rho_{(\!t} \circ [\![\gamma']\!] \circ \rho_{t\!)}) \cap \phi_t) \circ [\![\gamma'']\!]$ where $Y \underset{df}{\overset{\gamma'}{\Rightarrow}} w'$ and $Z \underset{df}{\overset{\gamma''}{\Rightarrow}} w''$. We showed [11, Lemma 2] that, whenever $X \underset{df}{\overset{\gamma}{\Rightarrow}} w$, we have $[\![w]\!] \neq \varnothing$ iff $[\![\gamma]\!] \neq \varnothing$.

Index-Bounded Derivations. A step $u \Rightarrow v$ is said to be k-index $(k > 0)$ iff neither u nor v contains $k + 1$ occurrences of nonterminals, i.e. $|u\!\downarrow_\Xi| \leqslant k$ and

$|v\!\downarrow_{\varXi}| \leqslant k$. We denote by $u \overset{\gamma}{\underset{(k)}{\Longrightarrow}} v$ a k-index step sequence and by $u \overset{\gamma}{\underset{\mathbf{df}(k)}{\Longrightarrow}} v$ a step sequence which is both depth-first and k-index. For $X \in \varXi$, $Y \in \varXi \cup \{\varepsilon\}$ and $k > 0$, we define the *k-index language* $L_{X,Y}^{(k)}(G) = \{u\,v \in \varSigma^* \mid \exists \gamma \in \varDelta^* : X \overset{\gamma}{\underset{(k)}{\Longrightarrow}} u\,Y\,v\}$, the *k-index depth-first control set* $\varGamma_{X,Y}^{\mathbf{df}(k)}(G) = \{\gamma \in \varDelta^* \mid \exists u,v \in \varSigma^* :$ $X \overset{\gamma}{\underset{\mathbf{df}(k)}{\Longrightarrow}} u\,Y\,v\}$. We write $L_X^{(k)}(G)$ and $\varGamma_X^{\mathbf{df}(k)}(G)$ when $Y = \varepsilon$, and drop G from the previous notations, when the grammar is clear from the context. For instance, for the grammar in Fig. 1 (c), we have $L_{X_1}^{(2)}(G) = \{(\mathbf{t_1}\langle \mathbf{t_2})^n\,\mathbf{t_4}\,(\mathbf{t_2}\rangle \mathbf{t_3})^n \mid n \in \mathbb{N}\} = L_{X_1}(G)$ and $\varGamma_{X_1}^{\mathbf{df}(2)} = (\mathbf{p_1 p_2 p_3})^* (\mathbf{p_4} \cup \mathbf{p_1 p_2 p_4 p_3})$.

Theorem 1 (Lemma 2 [19], Theorem 1 [18]). *Given a grammar* $G = \langle \varXi, \varSigma, \varDelta \rangle$ *and* $X \in \varXi$:

- *for all* $w \in \varSigma^*$, $X \overset{*}{\underset{(k)}{\Longrightarrow}} w$ *if and only if* $X \overset{*}{\underset{\mathbf{df}(k)}{\Longrightarrow}} w$;
- *if* $L_X(G) \subseteq \mathbf{b}$ *for a bounded expression* $\mathbf{b}$ *over* $\varSigma$ *then* $L_X(G) = L_X^{(K)}(G)$ *where* $K = O(|G|)$.

The introduction of the notion of index naturally calls for an index dependent semantics and an index dependent reachability problem. As we will see later, we have tight complexity results when it comes to the index dependent reachability problem. Given $k > 0$, let $[\mathcal{P}]^{(k)} = \bigcup_{w \in L_I^{(k)}(G)} [w]$ and let $[\mathcal{P}]_{\mathbf{b}}^{(k)} = \bigcup_{w \in L_I^{(k)}(G) \cap \mathbf{b}} [w]$. Thus we define, for a constant k not part of the input, the problem $\mathrm{REACH}_{fo}^{(k)}(\mathcal{P}, \mathbf{b})$, which asks whether $[\mathcal{P}]_{\mathbf{b}}^{(k)} \neq \varnothing$.

Finite Representations of Bounded-Index Depth-First Control Sets. It is known that the set of k-index depth-first derivations of a grammar G is recognizable by a finite automaton [19, Lemma 5]. Below we give a formal definition of this automaton, that will be used to produce bounded control sets for covering (in the sense of generating a superset) the language of G. Moreover, we provide an upper bound on its size, which will be used to prove an upper bound for the time to compute this set (Sect. 5).

Given $k > 0$ and a grammar $G = \langle \varXi, \varSigma, \varDelta \rangle$, we define a labeled graph $A_G^{\mathbf{df}(k)}$ such that its paths defines the set of k-index depth-first step sequences of G. To define the vertices and edges of this graph, we introduce the notion of *ranked words*, where the rank plays the same rôle as the value $f_m(i)$ defined previously. The advantage of ranks is that only k of them are needed for k-index depth-first derivations whereas the set of $f_m(i)$ values grows with the length of derivations. Since we restrict ourselves to k-index depth-first derivations, we thus only need k ranks, from 0 to $k - 1$. The rank based definition of depth-first derivations can be found in the technical report [10].

For a d-dimensional vector $\mathbf{v} \in \mathbb{N}^d$, we write $(\mathbf{v})_i$ for its ith element ($1 \leqslant i \leqslant d$). A vector $\boldsymbol{v} \in \mathbb{N}^d$ is said to be *contiguous* if $\{(\boldsymbol{v})_1, \ldots, (\boldsymbol{v})_d\} = \{0, \ldots, k\}$, for some $k \geqslant 0$. Given an alphabet $\varSigma$ define the ranked alphabet $\varSigma^{\mathbb{N}}$ to be the set $\{\sigma^{\langle i \rangle} \mid \sigma \in \varSigma, i \in \mathbb{N}\}$. A ranked word is a word over a ranked alphabet. Given a word w of length n and an n-dimensional vector $\boldsymbol{\alpha} \in \mathbb{N}^n$, the *ranked word* $w^{\boldsymbol{\alpha}}$ is the sequence $(w)_1^{\langle\langle(\alpha)_1\rangle\rangle}\ldots(w)_n^{\langle\langle(\alpha)_n\rangle\rangle}$, in which the ith element of α annotates the ith symbol of w. We also denote $w^{\langle\langle c\rangle\rangle} = (w)_1^{\langle c\rangle}\ldots(w)_{|w|}^{\langle c\rangle}$ as a shorthand. Let $A_G^{\mathrm{df}(k)} = \langle Q, \Delta, \to\rangle$ be the following labeled graph, where:

$$Q = \{w^\alpha \mid w \in \varXi^*, |w| \leqslant k, \alpha \in \mathbb{N}^{|w|} \text{ is contiguous}, (\alpha)_1 \leqslant \cdots \leqslant (\alpha)_{|w|}\}$$

is the set of vertices, the edges are labeled by the set Δ of productions of G, and the edge relation is defined next. For all vertices $q, q' \in Q$ and labels $(X, w) \in \Delta$, we have $q \xrightarrow{(X,w)} q'$ if and only if

- $q = u\, X^{\langle i\rangle}\, v$ for some u, v, where i is the maximum rank in q, and
- $q' = u\, v\, (w\!\downarrow_\varXi)^{\langle\langle i'\rangle\rangle}$, where $|u\, v\, (w\!\downarrow_\varXi)^{\langle\langle i'\rangle\rangle}| \leqslant k$ and $i' = \begin{cases} 0 & \text{if } u\, v = \varepsilon \\ i & \text{else if } (u\, v)\!\downarrow_{\varXi^{\langle i\rangle}} = \varepsilon \\ i+1 & \text{else} \end{cases}$

We denote by $|A_G^{\mathrm{df}(k)}| = \|Q\|$ the size (number of vertices) of $A_G^{\mathrm{df}(k)}$. In the following, we omit the subscript G from $A_G^{\mathrm{df}(k)}$, when the grammar is clear from the context. For example, the graph $A^{\mathrm{df}(2)}$ for the grammar from Fig. 1 (c), is the subgraph of Fig. 1 (d) enclosed in a dashed line.

Lemma 1. *Given $G = \langle \varXi, \varSigma, \Delta\rangle$, and $k > 0$, for each $X \in \varXi$, $Y \in \varXi \cup \{\varepsilon\}$ and $\gamma \in \Delta^*$, we have $\gamma \in \varGamma_{X,Y}^{\mathrm{df}(k)}(G)$ if and only if $X^{\langle 0\rangle} \xrightarrow{\gamma} Y^{\langle 0\rangle}$ is a path in $A_G^{\mathrm{df}(k)}$. Moreover, we have $|A_G^{\mathrm{df}(k)}| = |G|^{\mathcal{O}(k)}$.*

5 A Decision Procedure for REACH$_{fo}(\mathcal{P}, \mathbf{b})$

In this section we describe a decision procedure for the problem REACH$_{fo}(\mathcal{P}, \mathbf{b})$ where $\mathcal{P} = \langle G, I, [.]\rangle$ is an octagonal program, whose underlying grammar is $G = \langle \varXi, \varSigma, \Delta\rangle$, and $\mathbf{b} = w_1^* \ldots w_d^*$ is a bounded expression over $\varSigma$. The procedure follows the roadmap described next.

First, we compute, in time polynomial in the sizes of $\mathcal{P}$ and $\mathbf{b}$, a set of programs $\{\mathcal{P}_i = \langle G^\cap, X_i, [.]\rangle\}_{i=1}^\ell$, such that $L_I(G) \cap \mathbf{b} = \bigcup_{i=1}^\ell L_{X_i}(G^\cap)$, which implies $[\mathcal{P}]_{\mathbf{b}} = \bigcup_{i=1}^\ell [\mathcal{P}_i]$. The grammar $G^\cap$ is an automata-theoretic product between the grammar G and the bounded expression $\mathbf{b}$. For space reasons, the reader is referred to the technical report [10] for the formal definition of $G^\cap$. We provide Example 1 which exposes the main intuitions of the construction. Deciding REACH$_{fo}(\mathcal{P}, \mathbf{b})$ reduces thus to deciding several instances $\{\text{REACH}_{fo}(\mathcal{P}_i, \mathbf{b})\}_{i=1}^\ell$ of the fo-reachability problem. By the definition of $\mathcal{P}_i$, it suffices to consider the unrestricted reachability in $\mathcal{P}_i$ which, abusing notations, would be written REACH$_{fo}(\mathcal{P}_i)$.

Example 1. Let us consider the bounded expression $\mathbf{b} = (ac)^*\, (ab)^*\, (db)^*$. Consider the grammar $G^{\mathbf{b}}$ with the following productions: $Q_1^{(1)} \to a\, Q_2^{(1)} \mid \varepsilon$, $Q_1^{(2)} \to a$ $Q_2^{(2)} \mid \varepsilon$, $Q_1^{(3)} \to d\, Q_2^{(3)} \mid \varepsilon$, $Q_2^{(1)} \to c\, Q_1^{(1)} \mid c\, Q_1^{(2)} \mid c\, Q_1^{(3)}$, $Q_2^{(2)} \to b\, Q_1^{(2)} \mid b\, Q_1^{(3)}$, $Q_2^{(3)} \to b\, Q_1^{(3)}$. It is easy to check that $\mathbf{b} = \bigcup_{i=1}^3 L_{Q_1^{(i)}}(G^{\mathbf{b}})$. Let $G = \langle\{X, Y, Z, T\}, \{a, b, c, d\}, \Delta\rangle$ where $\Delta =$

$\{X \to aY,\, Y \to Zb,\, Z \to cT,\, Z \to \varepsilon,\, T \to Xd\}$, i.e. we have $L_X(G) = \{(ac)^n\, ab\, (db)^n \mid n \in \mathbb{N}\}$. The following productions define a grammar $G^\cap$:

$$[Q_1^{(j)} X Q_1^{(3)}] \xrightarrow{p_1} a\, [Q_2^{(j)} Y Q_1^{(3)}] \qquad\qquad [Q_2^{(1)} Y Q_1^{(3)}] \xrightarrow{p_2} [Q_2^{(1)} Z Q_2^{(3)}]\, b$$

$$[Q_2^{(1)} Z Q_2^{(3)}] \xrightarrow{p_3} c\, [Q_1^{(j)} T Q_2^{(3)}] \qquad\qquad [Q_2^{(2)} Z Q_2^{(2)}] \xrightarrow{p_4} \varepsilon$$

$$[Q_1^{(j)} T Q_2^{(3)}] \xrightarrow{p_5} [Q_1^{(j)} X Q_1^{(3)}]\, d\,,\ \text{for } j = 1, 2 \qquad [Q_1^{(2)} X Q_1^{(3)}] \xrightarrow{p_6} a\, [Q_2^{(2)} Y Q_1^{(3)}]$$

$$[Q_2^{(2)} Y Q_1^{(3)}] \xrightarrow{p_7} [Q_2^{(2)} Z Q_2^{(2)}]\, b$$

One can check $L_X(G) = L_X(G) \cap \mathbf{b} = L_{[Q_1^{(1)} X Q_1^{(3)}]}(G^\cap) \cup L_{[Q_1^{(2)} X Q_1^{(3)}]}(G^\cap)$. $\blacksquare$

A bounded expression $\mathbf{b} = w_1^* \ldots w_d^*$ over alphabet Σ is said to be *d-letter-bounded* (or simply letter-bounded, when d is not important) when $|w_i| = 1$, for all $i = 1, \ldots, d$. A letter-bounded expression $\widetilde{\mathbf{b}}$ is *strict* if all its symbols are distinct. A language $L \subseteq \Sigma^*$ is (strict, letter-) bounded iff $L \subseteq \mathbf{b}$, for some (strict, letter-) bounded expression $\mathbf{b}$.

Second, we reduce the problem from $\mathbf{b} = w_1^* \ldots w_d^*$ to the strict letter-bounded case $\widetilde{\mathbf{b}} = a_1^* \ldots a_d^*$, by building a grammar $G^\bowtie$, with the same non-terminals as $G^\cap$, such that, for each $i = 1, \ldots, \ell$ (i) $L_{X_i}(G^\bowtie) \subseteq \widetilde{\mathbf{b}}$, (ii) $w_1^{i_1} \ldots w_d^{i_d} \in L_{X_i}^{(k)}(G^\cap)$ iff $a_1^{i_1} \ldots a_d^{i_d} \in L_{X_i}^{(k)}(G^\bowtie)$, for all $k > 0$ (iii) from each control set $\widetilde{\Gamma}$ that covers the language $L_{X_i}^{(k)}(G^\bowtie) \subseteq \hat{L}_{X_i}(\widetilde{\Gamma}, G^\bowtie)$ for some $k > 0$, one can compute, in polynomial time, a control set Γ that covers the language $L_{X_i}^{(k)}(G^\cap) \subseteq \hat{L}_{X_i}(\Gamma, G^\cap)$.

Example 2 (continued from Example 1). Let $\mathcal{A} = \{a_1, a_2, a_3\}$, $\widetilde{\mathbf{b}} = a_1^* a_2^* a_3^*$ and $h : \mathcal{A} \to \Sigma^*$ be the homomorphism given by $h(a_1) = ac, h(a_2) = ab$ and $h(a_3) = db$. The grammar $G^\bowtie$ results from deleting a's and d's in $G^\cap$ and replacing b in p_2 by a_3, b in p_7 by a_2 and c by a_1. Then, it is easy to check that $h^{-1}(L_X(G)) \cap \widetilde{\mathbf{b}} = L_{[Q_1^{(1)} X Q_1^{(3)}]}(G^\bowtie) \cup L_{[Q_1^{(2)} X Q_1^{(3)}]}(G^\bowtie) = \{a_1^n\, a_2\, a_3^n \mid n \in \mathbb{N}\}$. $\blacksquare$

Third, for the strict letter-bounded grammar $G^\bowtie$, we compute a control set $\widetilde{\Gamma} \subseteq (\Delta^\bowtie)^*$ using the result of Theorem 2, which yields a set of bounded expressions $\mathcal{S}_{\widetilde{\mathbf{b}}} = \{\widetilde{\Gamma}_{i,1}, \ldots, \widetilde{\Gamma}_{i,m_i}\}$, such that $L_{X_i}^{(k)}(G^\bowtie) \subseteq \bigcup_{j=1}^{m_i} \hat{L}_{X_i}(\widetilde{\Gamma}_{i,j} \cap \Gamma_{X_i}^{\mathbf{df}(k+1)}, G^\bowtie)$. By applying the aforementioned transformation (*iii*) from $\widetilde{\Gamma}$ to Γ, we obtain that $L_{X_i}^{(k)}(G^\cap) \subseteq \bigcup_{j=1}^{m_i} \hat{L}_{X_i}(\Gamma_{i,j} \cap \Gamma_{X_i}^{\mathbf{df}(k+1)}, G^\cap)$. Theorem 1 allows to effectively compute value $K > 0$ such that $L_{X_i}(G^\cap) = L_{X_i}^{(K)}(G^\cap)$, for all $i = 1, \ldots, \ell$. Thus we obtain[4] $L_{X_i}(G^\cap) = \bigcup_{j=1}^{m_i} \hat{L}_{X_i}(\Gamma_{i,j} \cap \Gamma_{X_i}^{\mathbf{df}(K+1)}, G^\cap)$, for all $i = 1, \ldots, \ell$.

Theorem 2. *Given a grammar $G = \langle \Xi, \mathcal{A}, \Delta \rangle$, and $X \in \Xi$, such that $L_X(G) \subseteq \widetilde{\mathbf{b}}$, where $\widetilde{\mathbf{b}}$ is the minimal strict d-letter bounded expression for $L_X(G)$, for each $k > 0$, there exists a finite set of bounded expressions $\mathcal{S}_{\widetilde{\mathbf{b}}}$ over Δ such that $L_X^{(k)}(G) \subseteq \hat{L}_X(\bigcup \mathcal{S}_{\widetilde{\mathbf{b}}} \cap \Gamma_X^{\mathbf{df}(k+1)}, G)$. Moreover, $\mathcal{S}_{\widetilde{\mathbf{b}}}$ can be constructed in time $|G|^{\mathcal{O}(k)+d}$ and each $\widetilde{\Gamma} \in \mathcal{S}_{\widetilde{\mathbf{b}}}$ can be constructed in time $|G|^{\mathcal{O}(k)}$.*

---

[4] Because $L_{X_i}(G^\cap) \subseteq L_{X_i}^{(K)}(G^\cap) \subseteq \bigcup_{j=1}^{m_i} \hat{L}_{X_i}(\Gamma_{i,j} \cap \Gamma_{X_i}^{\mathbf{df}(K+1)}, G^\cap) \subseteq L_{X_i}(G^\cap)$.

We now sketch the main proof ingredients of Theorem 2:

Constant Case. We solve the constant case computing a bounded control sets for s-letter bounded languages, where $s \geqslant 0$ is a constant (in our case, at most 2). The result is formalized as follows:

Lemma 2. Let $G = \langle \Xi, \mathcal{A}, \Delta \rangle$ be a grammar and $a_1^* \ldots a_s^*$ is a strict s-letter-bounded expression over $\mathcal{A}$, where $s \geqslant 0$ is a constant. Then, for each $k > 0$ there exists a bounded expression Γ over Δ such that, for all $X \in \Xi$ and $Y \in \Xi \cup \{\varepsilon\}$, we have $L_{X,Y}^{(k)}(G) = \hat{L}_{X,Y}(\Gamma \cap \Gamma_{X,Y}^{df(k)}, G)$, provided that $L_{X,Y}(G) \subseteq a_1^* \ldots a_s^*$. Moreover, Γ is computable in time $|G|^{\mathcal{O}(k)}$.

The proof technique relies on the construction of a graph for which we need to compute some cycles of bounded length. This is done by the construction of another graph (basically its unwinding) and the use of Dijkstra's single source shortest path algorithm.

Decomposition Lemma. We lift the constant case to the general case by mean of a decomposition lemma which precisely enables the generalization from s-letter bounded languages where s is a constant to arbitrary letter bounded languages. The lemma decomposes k-index depth-first derivations into a prefix producing a word from the 2-letter bounded expression $a_1^* a_d^*$, and a suffix producing two words included in bounded expressions strictly smaller than $\mathbf{b}$. More precisely, for every k-index depth-first derivation with control word γ, its productions can be rearranged into a $(k+1)$-index depth-first derivation, consisting of (i) a prefix $\gamma^{\sharp}$ producing a word in $a_1^* a_d^*$, then (ii) a *pivot* production (X_i, w) followed by two words γ' and γ'' such that (iii) γ' and γ'' produce words included in two bounded expressions $a_\ell^* \ldots a_m^*$ and $a_m^* \ldots a_r^*$, respectively, where $\max(m - \ell, r - m) < d - 1$. This decomposition is a generalization of a result of Ginsburg [12, Chapter 5.3, Lemma 5.3.3]. Because his decomposition is oblivious to the index or the depth-first policy, it is too weak for our needs. Therefore, we give first a stronger decomposition result for k-index depth-first derivations.

General Case. We compute the set of bounded expressions $\mathcal{S}_{\widetilde{\mathbf{b}}}$ inductively by leveraging the decomposition. We use the result for the constant case applied on the extremities of $\mathbf{b}$ which returns a bounded control set for 2-letter bounded languages. Then we inductively solve the case for the two subexpressions $a_\ell^* \ldots a_m^*$ and $a_m^* \ldots a_r^*$. The main algorithm returns a finite set $\mathcal{S}_{\widetilde{\mathbf{b}}}$ of bounded expressions. The formal statement is given by Theorem 2. The time needed to build each bounded expression $\widetilde{\Gamma} \in \mathcal{S}_{\widetilde{\mathbf{b}}}$ is $|G|^{\mathcal{O}(k)}$ and does not depend of $|\widetilde{\mathbf{b}}| = d$, whereas the time needed to build the entire set $\mathcal{S}_{\widetilde{\mathbf{b}}}$ is $|G|^{\mathcal{O}(k)+d}$. These arguments come in handy when deriving an upper bound on the (non-deterministic) time complexity of the fo-reachability problem for programs with arbitrary call graphs.

The next lemma shows that the exponential dependence on k in the bounds of Theorem 2 is unavoidable.

Lemma 3. *For every $k > 0$ there exists a grammar $G = \langle \Xi, \Sigma, \Delta \rangle$ and $X \in \Xi$ such that $|G| = \mathcal{O}(k)$ and every bounded expression Γ, such that $L_X(G) = \hat{L}_X(\Gamma \cap \Gamma_X^{\mathsf{df}(k+1)}, G)$ has length $|\Gamma| \geq 2^{k-1}$.*

Finally, we turn back to the fo-reachability problem. To solve $\mathrm{REACH}_{fo}(\mathcal{P}_i, \mathbf{b})$, the final step consists in building a finite automaton $A^{\mathsf{df}(K+1)}$ that recognizes the control set $\Gamma_{X_i}^{\mathsf{df}(K+1)}$ (Lemma 1). This yields a procedure-less program $\mathcal{Q}$, whose control structure is given by $A^{\mathsf{df}(K+1)}$, and whose labels are given by the semantics of control words. We recall that, for every word $w \in L_{X_i}(G^{\cap})$ there exists a control word $\gamma \in \Gamma_{X_i}^{\mathsf{df}(K+1)}$ such that $[w] \neq \varnothing$ iff $[\gamma] \neq \varnothing$. We have thus reduced each of the instances $\{\mathrm{REACH}_{fo}(\mathcal{P}_i, \mathbf{b})\}_{i=1}^{\ell}$ of the fo-reachability problem to a set of instances $\{\mathrm{REACH}_{fo}(\mathcal{Q}, \Gamma_{i,j}) \mid 1 \leq i \leq \ell, 1 \leq j \leq m_i\}$. The latter problem, for procedure-less programs, is decidable in NPTIME [6]. A detailed proof of the main result, stated next, is given in the technical report [10].

Theorem 3. *Let $\mathcal{P} = \langle G, I, [.] \rangle$ be an octagonal program, where $G = \langle \Xi, \Sigma, \Delta \rangle$ is a grammar, and $\mathbf{b}$ is a bounded expression over Σ. Then the problem $\mathrm{REACH}_{fo}(\mathcal{P}, \mathbf{b})$ is decidable in NEXPTIME, with a NP-hard lower bound. If, moreover, k is a constant, $\mathrm{REACH}_{fo}^{(k)}(\mathcal{P}, \mathbf{b})$ is NP-complete.*

6   Related Work

The programs we have studied feature unbounded control (the call stack) and unbounded data (the integer variables). The decidability and complexity of the reachability problem for such programs pose challenging research questions. A long standing and still open one is the decidability of the reachability problem for programs where variables behave like Petri net counters and control paths are taken in a context-free language. A lower bound exists [16] but decidability remains open. Atig and Ganty [3] showed decidability when the context-free language is of bounded index. The complexity of reachability was settled for branching VASS by Lazic and Schmitz [17]. When variables updates/guards are given by gap-order constraints, reachability is decidable [1,22]. It is in PSPACE when the set of control paths is regular [7]. More general updates and guard (like octagons) immediately leads to undecidability. This explains the restriction to bounded control sets. Demri *et al.* [8] studied the case of updates/guards of the form $\sum_{i=1}^{n} a_i \cdot x_i + b \leq 0 \wedge \mathbf{x}' = \mathbf{x} + c$. They show that LTL is NP-complete on for bounded regular control sets, hence reachability is in NP. Godoy and Tiwari [13] studied the invariant checking problem for a class of procedural programs where all executions conform to a bounded expression, among other restrictions.

References

1. Abdulla, P.A., Atig, M.F., Delzanno, G., Podelski, A.: Push-down automata with gap-order constraints. In: Arbab, F., Sirjani, M. (eds.) FSEN 2013. LNCS, vol. 8161, pp. 199–216. Springer, Heidelberg (2013)

2. Alur, R., Madhusudan, P.: Adding nesting structure to words. J. ACM **56**(3), 16:1–16:43 (2009)
3. Atig, M.F., Ganty, P.: Approximating petri net reachability along context-free traces. In: FSTTCS 2011, vol. 13. LIPIcs, pp. 152–163. Schloss Dagstuhl (2011)
4. Bardin, S., Finkel, A., Leroux, J., Petrucci, L.: Fast: fast acceleration of symbolic transition systems. In: Hunt Jr., W.A., Somenzi, F. (eds.) CAV 2003. LNCS, vol. 2725, pp. 118–121. Springer, Heidelberg (2003)
5. Bozga, M., Iosif, R., Konečný, F.: Fast acceleration of ultimately periodic relations. In: Touili, T., Cook, B., Jackson, P. (eds.) CAV 2010. LNCS, vol. 6174, pp. 227–242. Springer, Heidelberg (2010)
6. Bozga, M., Iosif, R., Konečný, F.: Safety problems are np-complete for flat integer programs with octagonal loops. In: McMillan, K.L., Rival, X. (eds.) VMCAI 2014. LNCS, vol. 8318, pp. 242–261. Springer, Heidelberg (2014)
7. Bozzelli, L., Pinchinat, S.: Verification of gap-order constraint abstractions of counter systems. Theo. Comput. Sci. **523**, 1–36 (2014)
8. Demri, S., Dhar, A.K., Sangnier, A.: Taming past ltl and flat counter systems. In: Gramlich, B., Miller, D., Sattler, U. (eds.) IJCAR 2012. LNCS, vol. 7364, pp. 179–193. Springer, Heidelberg (2012)
9. Esparza, J., Ganty, P.: Complexity of pattern-based verification for multithreaded programs. In: POPL 2011, pp. 499–510. ACM Press (2011)
10. Ganty, P., Iosif, R.: Interprocedural reachability for flat integer programs. CoRR, abs/1405.3069v3 (2015)
11. Ganty, P., Iosif, R., Konečný, F.: Underapproximation of procedure summaries for integer programs. In: Piterman, N., Smolka, S.A. (eds.) TACAS 2013 (ETAPS 2013). LNCS, vol. 7795, pp. 245–259. Springer, Heidelberg (2013)
12. Ginsburg, S.: The Mathematical Theory of Context-Free Languages. McGraw-Hill Inc., New York (1966)
13. Godoy, G., Tiwari, A.: Invariant checking for programs with procedure calls. In: Palsberg, J., Su, Z. (eds.) SAS 2009. LNCS, vol. 5673, pp. 326–342. Springer, Heidelberg (2009)
14. Hojjat, H., Iosif, R., Konečný, F., Kuncak, V., Rümmer, P.: Accelerating interpolants. In: Chakraborty, S., Mukund, M. (eds.) ATVA 2012. LNCS, vol. 7561, pp. 187–202. Springer, Heidelberg (2012)
15. Kroening, D., Lewis, M., Weissenbacher, G.: Under-approximating loops in c programs for fast counterexample detection. In: Sharygina, N., Veith, H. (eds.) CAV 2013. LNCS, vol. 8044, pp. 381–396. Springer, Heidelberg (2013)
16. Lazic, R.: The reachability problem for vector addition systems with a stack is not elementary. In: RP 2012 (2012)
17. Lazic, R., Schmitz, S.: Non-elementary complexities for branching VASS, MELL, and extensions. In: CSL-LICS 2014. ACM (2014)
18. Luker, M.: A family of languages having only finite-index grammars. Inf. Control **39**(1), 14–18 (1978)
19. Luker, M.: Control sets on grammars using depth-first derivations. Math. Syst. Theo. **13**, 349–359 (1980)
20. Miné, A.: The octagon abstract domain. Higher-Order Symbolic Comput. **19**(1), 31–100 (2006)
21. Minsky, M.: Computation: Finite and Infinite Machines. Prentice-Hall, Upper Saddle River (1967)
22. Revesz, P.Z.: A closed-form evaluation for datalog queries with integer (gap)-order constraints. Theo. Comput. Sci. **116**(1), 117–149 (1993)

Complexity of Suffix-Free Regular Languages

Janusz Brzozowski[1]($\boxtimes$) and Marek Szykuła[2]

[1] David R. Cheriton School of Computer Science,
University of Waterloo, Waterloo, ON N2L 3G1, Canada
brzozo@uwaterloo.ca
[2] Institute of Computer Science, University of Wrocław,
Joliot-Curie 15, 50-383 Wrocław, Poland
msz@cs.uni.wroc.pl

Abstract. A sequence $(L_k, L_{k+1} \ldots)$ of regular languages in some class
$\mathcal{C}$, where n is the state complexity of L_n, is called a *stream*. A stream
is *most complex* in class $\mathcal{C}$ if its languages together with their dialects
(that is, languages that differ only very slightly from the languages in
the stream) meet the state complexity bounds for boolean operations,
product (concatenation), star, and reversal, have the largest syntactic
semigroups, and have the maximal numbers of atoms, each of which has
maximal state complexity. It is known that there exist such most complex streams in the class of regular languages, and also in the classes of
right, left, and two-sided ideals. In contrast to this, we prove that there
does not exist a most complex stream in the class of suffix-free regular
languages. However, we do exhibit one ternary suffix-free stream that
meets the bound for product and whose restrictions to binary alphabets meet the bounds for star and boolean operations. We also exhibit a
quinary stream that meets the bounds for boolean operations, reversal,
size of syntactic semigroup, and atom complexities. Moreover, we solve
an open problem about the bound for the product of two languages of
state complexities m and n in the binary case by showing that it can be
met for infinitely many m and n.

Two transition semigroups play an important role for suffix-free lan-
guages: semigroup $\mathbf{T}^{\leqslant 5}(n)$ is the largest suffix-free semigroup for $n \leqslant 5$,
while semigroup $\mathbf{T}^{\geqslant 6}(n)$ is largest for $n = 2, 3$ and $n \geqslant 6$. We prove that
all witnesses meeting the bounds for the star and the second witness in a
product must have transition semigroups in $\mathbf{T}^{\leqslant 5}(n)$. On the other hand,
witnesses meeting the bounds for reversal, size of syntactic semigroup
and the complexity of atoms must have semigroups in $\mathbf{T}^{\geqslant 6}(n)$.

Keywords: Most complex · Regular language · State complexity · Suffix-
free · Syntactic complexity · Transition semigroup

This work was supported by the Natural Sciences and Engineering Research
Council of Canada grant No. OGP000087, and by Polish NCN grant DEC-
2013/09/N/ST6/01194.

A. Kosowski and I. Walukiewicz (Eds.): FCT 2015, LNCS 9210, pp. 146–159, 2015.
DOI: 10.1007/978-3-319-22177-9_12

1   Introduction

Suffix-Free Languages. A language is *suffix-free* if no word in the language is a suffix of another word in the language. The languages ba^*, $\{a^n b^n \mid n \geq 1\}$, and Σ^n, where Σ is a finite alphabet and n is a positive integer, are all examples of suffix-free languages. Every suffix-free language (except that consisting of the empty word ε) is a suffix code. Suffix codes are an important subclass of general codes, which have numerous applications in cryptography, data compression and error correction. Codes have been studied extensively; see [2] for example. In addition to being codes, suffix-free languages are also a special subclass of suffix-convex languages [1], where a language is *suffix-convex* if, whenever a word w and its suffix u are in the language, then so is every suffix of w that has u as a suffix.

We study complexity properties of suffix-free regular languages.

Quotient Complexity. A basic complexity measure of a regular language L over an alphabet Σ is the number n of distinct left quotients of L, where a *(left) quotient* of L by a word $w \in \Sigma^*$ is $w^{-1}L = \{x \mid wx \in L\}$. We denote the set of quotients of L by $K = \{K_0, \ldots, K_{n-1}\}$, where $K_0 = L = \varepsilon^{-1}L$ by convention. Each quotient K_i can be represented also as $w_i^{-1}L$, where $w_i \in \Sigma^*$ is such that $w_i^{-1}L = K_i$. The number of quotients of L is its *quotient complexity* [3] $\kappa(L)$. A concept equivalent to quotient complexity is *state complexity* [23] of L, which is the number of states in a minimal deterministic finite automaton (DFA) recognizing L.

Let L_n be a regular language of quotient complexity n. The *quotient complexity of a unary operation* $\circ$ on L_n is the maximum value of $\kappa(L_n^\circ)$ as a function of n. To establish the quotient complexity of L_n°, first we need to find an upper bound on this complexity. For example, 2^n is an upper bound on the reversal operation on regular languages [20,22]. Second, we need a sequence $(L_n, n \geq k) = (L_k, L_{k+1}, \ldots)$, called a *stream*, of languages that meet this bound; here k is usually some small integer because the bound may not apply for $n < k$. A language L_n that meets the bound $\kappa(L_n^\circ)$ for the operation $\circ$ is a *witness (language)* for that operation. A stream in which every language meets the bound is called a *witness (stream)*. The languages in a stream are normally defined in the same way, differing only in the parameter n. For example, we might have the stream $(L_n = \{w \in \{a, b\}^* \mid$ the number of a's is 0 modulo $n\}, n \geq 2)$.

Similarly, $\kappa(K_m \circ L_n)$ is the *quotient complexity of a binary operation* $\circ$ on regular languages K_m and L_n of complexities m and n, respectively. Again, we need an upper bound on $\kappa(K_m \circ L_n)$. For example, an upper bound on product (concatenation) is $(m-1)2^n + 2^{n-1}$ [19,22]. Then we have to find two streams $(K_m, m \geq h)$ and $(L_n, n \geq k)$ of languages meeting this bound. In general, the two streams are different, but there are many examples where K_n "differs only slightly" from L_n; such a language K_n has been called a *dialect* [4] of L_n. The notion "differs only slightly" will be made precise below. A pair (K_m, L_n) of languages that meets the bound $\kappa(K_m \circ L_n)$ for the operation $\circ$ is a *witness (pair)* for that operation. A stream in which every pair of languages meets the bound is called a *witness (stream)*.

The quotient/state complexity of an operation gives a worst-case lower bound on the time and space complexity of the operation. For this reason it has been studied extensively; see [3, 23] for additional references. The quotient complexity of suffix-free languages was examined in [14, 15, 17].

We also extend the notions of maximal complexity, stream, and witness to DFAs.

Syntactic Complexity. A second measure of complexity of a regular language is its syntactic complexity. Let Σ^+ be the set of non-empty words of Σ^*. The *syntactic semigroup* of L is the set of equivalence classes of the Myhill congruence $\approx_L$ on Σ^+ defined by $x \approx_L y$ if and only if $uxv \in L \Leftrightarrow uyv \in L$ for all $u, v \in \Sigma^*$. The syntactic semigroup of L is isomorphic to the *transition semigroup* of a minimal DFA $\mathcal{D}$ recognizing L [21], which is the semigroup of transformations of the state set of $\mathcal{D}$ induced by non-empty words. The *syntactic complexity* of L is the cardinality of its syntactic/transition semigroup. It was pointed out in [13] that languages having the same quotient complexity can have vastly different syntactic complexities. Thus syntactic complexity can be a finer measure of complexity. Syntactic complexity of suffix-free languages was studied in [9, 11].

Complexities of Atoms. A possible third measure of complexity of a regular language L is the number and quotient complexities, which we call simply complexities, of certain languages, called atoms, uniquely defined by L. Atoms arise from an equivalence on Σ^* which is a left congruence refined by the Myhill congruence, where two words x and y are equivalent if $ux \in L$ if and only if $uy \in L$ for all $u \in \Sigma^*$ [16]. Thus x and y are equivalent if $x \in u^{-1}L \Leftrightarrow y \in u^{-1}L$. An equivalence class of this relation is called an *atom* [12] of L. It follows that an atom is a non-empty intersection of complemented and uncomplemented quotients of L. The quotients of a language are unions of its atoms.

Most Complex (Streams of) Languages. The concept of *most complex* languages in a class C of languages was introduced in [4]. Such languages, with some of their dialects, must meet the bounds in the class C on the quotient complexities of the unary operations reversal and (Kleene) star, and on the product and the binary boolean operations. Moreover, they must also have the largest possible syntactic semigroups and the most complex atoms. It is surprising that there exists a single stream of languages that meets all these conditions for maximal complexity [4]. Moreover, most complex right, left, and two-sided ideals also exist [7]. We show in this paper, however, that this is not the case for suffix-free languages.

Most complex languages are useful for testing the efficiency of systems. The complexity of operations on languages gives a measure of time and space requirements for these operations. Hence to check the maximal size of the objects that a system can handle, we can use most complex languages. It is certainly simpler to have just one or two universal worst-case examples.

Terminology and Notation. A *deterministic finite automaton (DFA)* is a quintuple $\mathcal{D} = (Q, \Sigma, \delta, q_0, F)$, where Q is a finite non-empty set of *states*, Σ is a finite non-empty *alphabet*, $\delta: Q \times \Sigma \to Q$ is the *transition function*, $q_0 \in Q$ is the *initial* state, and $F \subseteq Q$ is the set of *final* states. We extend δ to a function $\delta \colon Q \times \Sigma^* \to Q$ as usual. A DFA $\mathcal{D}$ *accepts* a word $w \in \Sigma^*$ if $\delta(q_0, w) \in F$. The language accepted by $\mathcal{D}$ is denoted by $L(\mathcal{D})$. If q is a state of $\mathcal{D}$, then the language L^q of q is the language accepted by the DFA $(Q, \Sigma, \delta, q, F)$. A state is *empty* if its language is empty. Two states p and q of $\mathcal{D}$ are *equivalent* if $L^p = L^q$. A state q is *reachable* if there exists $w \in \Sigma^*$ such that $\delta(q_0, w) = q$. A DFA is *minimal* if all of its states are reachable and no two states are equivalent. Usually DFAs are used to establish upper bounds on the quotient complexity of operations and also as witnesses that meet these bounds.

A *nondeterministic finite automaton (NFA)* is a quintuple $\mathcal{D} = (Q, \Sigma, \delta, I, F)$, where Q, Σ and F are defined as in a DFA, $\delta \colon Q \times \Sigma \to 2^Q$ is the *transition function*, and $I \subseteq Q$ is the *set of initial states*. An *ε-NFA* is an NFA in which transitions under the empty word ε are also permitted.

The *quotient DFA* of a regular language L with n quotients is defined by $\mathcal{D} = (K, \Sigma, \delta_\mathcal{D}, K_0, F_\mathcal{D})$, where $\delta_\mathcal{D}(K_i, w) = K_j$ if and only if $w^{-1} K_i = K_j$, and $F_\mathcal{D} = \{K_i \mid \varepsilon \in K_i\}$. To simplify the notation, without loss of generality we use the set $Q_n = \{0, \ldots, n-1\}$ of subscripts of quotients as the set of states of $\mathcal{D}$; then $\mathcal{D}$ is denoted by $\mathcal{D} = (Q_n, \Sigma, \delta, 0, F)$, where $\delta(p, w) = q$ if $\delta_\mathcal{D}(K_p, w) = K_q$, and F is the set of subscripts of quotients in $F_\mathcal{D}$. The quotient DFA of L is unique and it is isomorphic to each complete minimal DFA of L.

A *transformation* of Q_n is a mapping $t \colon Q_n \to Q_n$. The *image* of $q \in Q_n$ under t is denoted by qt. The *range* of t is $\mathrm{rng}(t) = \{q \in Q_n \mid pt = q \text{ for some } p \in Q_n\}$. In any DFA, each letter $a \in \Sigma$ induces a transformation δ_a of the set Q_n defined by $q\delta_a = \delta(q, a)$; we denote this by $a \colon \delta_a$. By a slight abuse of notation we use the letter a to denote the transformation it induces; thus we write qa instead of $q\delta_a$. We also extend the notation to sets of states: if $P \subseteq Q_n$, then $Pa = \{pa \mid p \in P\}$. If s, t are transformations of Q, their composition is denoted by $s \circ t$ and defined by $q(s \circ t) = (qs)t$; the $\circ$ is usually omitted. Let $\mathcal{T}_{Q_n}$ be the set of all n^n transformations of Q_n; then $\mathcal{T}_{Q_n}$ is a monoid under composition.

For $k \geqslant 2$, a transformation (permutation) t of a set $P = \{q_0, q_1, \ldots, q_{k-1}\} \subseteq Q$ is a *k-cycle* if $q_0 t = q_1, q_1 t = q_2, \ldots, q_{k-2} t = q_{k-1}, q_{k-1} t = q_0$. A k-cycle is denoted by $(q_0, q_1, \ldots, q_{k-1})$. A 2-cycle (q_0, q_1) is called a *transposition*. A transformation that changes only one state p to a state $q \neq p$ is denoted by $(p \to q)$. A transformation mapping a subset P of Q to a single state and acting as the identity on $Q \setminus P$ is denoted by $(P \to q)$. We also denote by $[q_0, \ldots, q_{n-1}]$ the transformation that maps $p \in \{0, \ldots, n-1\}$ to q_p.

In this paper we consider only three types of simple dialects of DFAs and languages. Let $\Sigma = \{a_1, \ldots, a_k\}$, and let π be a permutation of Σ:

1. DFA $\mathcal{D}_n(a_1, \ldots, a_k)$ is *permutationally equivalent* to DFA $\mathcal{D}'_n(a_1, \ldots, a_k)$ if $\mathcal{D}'_n(a_1, \ldots, a_k) = \mathcal{D}_n(\pi(a_1), \ldots, \pi(a_k))$. In other words, the transformation induced by a_i in $\mathcal{D}'_n$ is the transformation induced by $\pi(a_i)$ in $\mathcal{D}_n$.
2. Let $\Gamma \subseteq \Sigma$. DFA $\mathcal{D}'_n$ is the *restriction* of a DFA $\mathcal{D}_n$ to Γ if all the transitions induced by letters in $\Sigma \setminus \Gamma$ are deleted from $\mathcal{D}_n$. For example, $\mathcal{D}_n(-, b, -, d)$ is the DFA $\mathcal{D}_n(a, b, c, d)$ restricted to the alphabet $\{b, d\}$. However, if the dashes appear only at the end, they may be omitted.

3. DFA $\mathcal{D}'_n$ is the *permutational restriction* of a DFA $\mathcal{D}_n$ to Γ if the letters of Σ are first permuted and then the letters in $\Sigma \setminus \Gamma$ are deleted.

The same notational conventions are used for languages.

Contributions.

1. We prove that a most complex stream of suffix-free languages does not exist. This is in contrast with the existence of streams of most complex regular languages [4], right ideals [5,6], and left and two-sided ideals [6,7].
2. We exhibit a single ternary witness that meets the bounds for star, product, and boolean operations.
3. We exhibit a single quinary witness that meets the bounds for boolean operations, reversal, number of atoms, syntactic complexity, and quotient complexities of atoms.
4. We show that when $m, n \geqslant 6$ and $m - 2$ and $n - 2$ are relatively prime, there are binary witnesses that meet the bound $(m - 1)2^{n-2} + 1$ for product.
5. We prove that any witness DFA for star and any second witness DFA for product must have transition semigroups that are subsemigroups of the suffix-free semigroup of transformations $\mathbf{T}^{\leqslant 5}(n)$ which is largest for $n \leqslant 5$; that the witness DFAs for reversal, syntactic complexity and quotient complexities of atoms must have transition semigroups that are subsemigroups of the suffix-free semigroup of transformations $\mathbf{T}^{\geqslant 6}(n)$ which is largest for $n = 2, 3$ and $n \geqslant 6$; and that the witness DFAs for boolean operations can have transition semigroups that are subsemigroups of $\mathbf{T}^{\leqslant 5} \cap \mathbf{T}^{\geqslant 6}$.

The full version of this paper is available at [10].

2   Suffix-Free Transformations

In this section we discuss some properties of suffix-free languages with emphasis on their syntactic semigroups as represented by the transition semigroups of their quotient DFAs. We assume that our basic set is always $Q_n = \{0, \ldots, n - 1\}$.

Let $\mathcal{D}_n = (Q_n, \Sigma, \delta, 0, F)$ be the quotient DFA of a suffix-free language L, and let T_n be its transition semigroup. For any transformation t of Q_n, the sequence $(0, 0t, 0t^2, \ldots)$ is called the 0-*path* of t. Since Q_n is finite, there exist i, j such that $0, 0t, \ldots, 0t^i, 0t^{i+1}, \ldots, 0t^{j-1}$ are distinct but $0t^j = 0t^i$. The integer $j - i$ is the *period* of t and if $j - i = 1$, t is *initially aperiodic*. The following properties of suffix-free languages are known [9,15]:

Lemma 1. *If L is a suffix-free language, then*

1. *There exists $w \in \Sigma^*$ such that $w^{-1}L = \emptyset$; hence $\mathcal{D}_n$ has an empty state, which is state $n - 1$ by convention.*
2. *For $w, x \in \Sigma^+$, if $w^{-1}L \neq \emptyset$, then $w^{-1}L \neq (xw)^{-1}L$.*
3. *If $L \neq \emptyset$ and $w^{-1}L = L$, then $w = \varepsilon$. This is known as the* non-returning *property [15] and also as* unique reachability *[8].*
4. *For any $t \in T_n$, the 0-path of t in $\mathcal{D}_n$ is aperiodic and ends in $n - 1$.*

An (unordered) pair $\{i, j\}$ of distinct states in $Q_n \setminus \{0, n-1\}$ is *colliding* (or p *collides* with q) in T_n if there is a transformation $t \in T_n$ such that $0t = p$ and $rt = q$ for some $r \in Q_n \setminus \{0, n-1\}$. A pair of states is *focused* by a transformation u of Q_n if u maps both states of the pair to a single state $r \notin \{0, n-1\}$. We then say that $\{p, q\}$ is *focused to state r*. If L is a suffix-free language, then from Lemma 1 (2) it follows that if $\{p, q\}$ is colliding in T_n, there is no transformation $t' \in T_n$ that focuses $\{p, q\}$. So colliding states can be mapped to a single state by a transformation in T_n only if that state is the empty state $n-1$.

Following [9], for $n \geq 2$, we let

$$\mathbf{B}(n) = \{t \in T_Q \mid 0 \notin \mathrm{rng}(t),\ (n-1)t = n-1,\ \text{and for all } j \geq 1,$$

$$0t^j = n-1 \text{ or } 0t^j \neq qt^j,\ \forall q \text{ such that } 0 < q < n-1\}.$$

Proposition 1 ([9]). *If L is a regular language having quotient DFA $\mathcal{D}_n = (Q_n, \Sigma, \delta, 0, F)$ and syntactic semigroup T_L, then the following hold:*

1. *If L is suffix-free, then T_L is a subset of $\mathbf{B}(n)$.*
2. *If L has the empty quotient, only one final quotient, and $T_L \subseteq \mathbf{B}(n)$, then L is suffix-free.*

Since the transition semigroup of a minimal DFA of a suffix-free language must be a subsemigroup of $\mathbf{B}(n)$, the cardinality of $\mathbf{B}(n)$ is an upper bound on the syntactic complexity of suffix-free regular languages with quotient complexity n. This upper bound, however, cannot be reached since $\mathbf{B}$ is not a semigroup for $n \geq 4$: We have $s = [1, 2, n-1, \ldots, n-1]$ and $t = [n-1, 2, 2, \ldots, 2, n-1]$ in $\mathbf{B}(n)$, but $st = [2, 2, n-1, \ldots, n-1]$ is not in $\mathbf{B}(n)$.

We now consider semigroups that are largest for $n \leq 5$. For $n \geq 2$, let $\mathbf{T}^{\leq 5}(n) = \{t \in \mathbf{B}(n) \mid$ for all $p, q \in Q_n$ where $p \neq q$, either $pt = qt = n-1$ or $pt \neq qt\}$.

Proposition 2. *For $n \geq 4$, semigroup $\mathbf{T}^{\leq 5}(n)$ is generated by the following set $\mathbf{H}^{\leq 5}(n)$ of transformations of Q:*

- *a: $(0 \rightarrow n-1)(1, \ldots, n-2)$,*
- *b: $(0 \rightarrow n-1)(1, 2)$,*
- *for $1 \leq p \leq n-2$, c_p: $(p \rightarrow n-1)(0 \rightarrow p)$.*

For $n = 4$, a and b coincide, and so $\mathbf{H}^{\leq 5}(4) = \{a, c_1, c_2\}$. Also, $\mathbf{H}^{\leq 5}(3) = \{a, c_1\} = \{[2, 1, 2], [1, 2, 2]\}$ and $\mathbf{H}^{\leq 5}(2) = \{c_1\} = \{[1, 1]\}$.

A DFA using these transformations is illustrated in Fig. 1 for $n = 5$.

Proposition 3. *For $n \geq 2$, $\mathbf{T}^{\leq 5}(n)$ is the unique maximal semigroup of a suffix-free language in which all possible pairs of states are colliding.*

Proposition 4. *For $n \geq 5$, the number n of generators of $\mathbf{T}^{\leq 5}(n)$ cannot be reduced.*

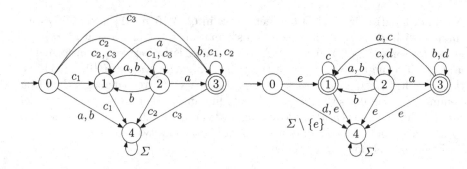

Fig. 1. DFAs with $\mathbf{T}^{\leqslant 5}(5)$ (left) and $\mathbf{T}^{\geqslant 6}(5)$ (right) as their transition semigroups

Next, we present semigroups that are largest for $n \geqslant 6$. For $n \geqslant 2$, let

$$\mathbf{T}^{\geqslant 6}(n) = \{t \in \mathbf{B}(n) \mid 0t = n - 1 \text{ or } qt = n - 1, \forall q \text{ such that } 1 \leqslant q \leqslant n - 2\}.$$

Proposition 5 ([11]). *For $n \geqslant 4$, $\mathbf{T}^{\geqslant 6}(n)$ is a semigroup contained in $\mathbf{B}(n)$, its cardinality is $(n-1)^{n-2} + (n-2)$, and it is generated by the set $\mathbf{G}^{\geqslant 6}(n)$ of the following transformations:*

- a: $(0 \to n-1)(1, \ldots, n-2)$;
- b: $(0 \to n-1)(1, 2)$;
- c: $(0 \to n-1)(n-2 \to 1)$;
- d: $(\{0, 1\} \to n-1)$;
- e: $(Q \setminus \{0\} \to n-1)(0 \to 1)$.

For $n = 4$, a and b coincide, and so $\mathbf{G}^{\geqslant 6}(4) = \{a, c, d, e\}$. Also $\mathbf{G}^{\geqslant 6}(3) = \{a, e\} = \{[2, 1, 2]\}, \{[1, 2, 2]\}$ and $\mathbf{G}^{\geqslant 6}(2) = \{e\} = \{[1, 1]\}$.

A DFA using the transformations of Proposition 5 is shown in Fig. 1 for $n = 5$. Semigroups $\mathbf{T}^{\geqslant 6}(n)$ are the largest suffix-free semigroups for $n \geqslant 6$ [11].

3   Witnesses with Transition Semigroups in $\mathbf{T}^{\leqslant 5}(n)$

In this section we consider DFA witnesses whose transition semigroups are sub-semigroups of $\mathbf{T}^{\leqslant 5}(n)$. We show that there is one witness that satisfies the bounds for star, product and boolean operations.

Definition 1. *For $n \geqslant 6$, we define the DFA $\mathcal{D}_n = (Q_n, \Sigma, \delta, 0, \{1\})$, where $Q_n = \{0, \ldots, n-1\}$, $\Sigma = \{a, b, c\}$, and δ is defined by the transformations*

- a: $(0 \to n-1)(1, 2, 3)(4, \ldots, n-2)$,
- b: $(2 \to n-1)(1 \to 2)(0 \to 1)(3, 4)$,
- c: $(0 \to n-1)(1, \ldots, n-2)$.

Theorem 1 (Star, Product, Boolean Operations). *Let* $\mathcal{D}_n(a,b,c)$ *be the DFA of Definition 1, and let the language it accepts be* $L_n(a,b,c)$. *For* $n \geqslant 6$, L_n *and its dialects meet the bounds for star, product and boolean operations as follows:*

1. $L_n^*(a,b,-)$ *meets the bound* $2^{n-2}+1$. *[Cmorik and Jirásková [14]]*
2. $L_m(a,b,c) \cdot L_n(b,c,a)$ *meets the bound* $(m-1)2^{n-2}+1$.
3. $L_m(a,b,-)$ *and* $L_n(c,b,-)$ *meet the bounds* $mn-(m+n-2)$ *for union and symmetric difference,* $mn-2(m+n-3)$ *for intersection, and* $mn-(m+2n-4)$ *for difference.*

The claim about the star operation was proved in [14]. We add a result about the transition semigroup of the star witness and prove the remaining two claims.

In 2009 Han and Salomaa [15] showed that the language of a DFA over a four-letter alphabet meets the bound $2^{n-2}+1$ for the star operation for $n \geqslant 4$. The transition semigroup of this DFA is a subsemigroup of $\mathbf{T}^{\leqslant 5}(n)$. In 2012 Cmorik and Jir'asková [14] showed that for $n \geqslant 6$ a binary alphabet $\{a,b\}$ suffices. The transition semigroup of this DFA is again a subsemigroup of $\mathbf{T}^{\leqslant 5}(n)$. We prove that these are special cases of the following general result:

Theorem 2. *For* $n \geqslant 4$, *the transition semigroup of the quotient DFA* $\mathcal{D}$ *of a suffix-free language* L *that meets the bound* $2^{n-2}+1$ *for the star operation is a subsemigroup of* $\mathbf{T}^{\leqslant 5}(n)$ *and is not a subsemigroup of* $\mathbf{T}^{\geqslant 6}(n)$.

For the product, to avoid confusing the states of the two DFAs, we label the states of the first DFA differently. Let $\mathcal{D}'_m = \mathcal{D}'_m(a,b,c) = (Q'_m, \Sigma, \delta', 0', \{1'\})$, where $Q'_m = \{0', \ldots, (m-1)'\}$, and $\delta'(q', x) = p'$ if $\delta(q, x) = p$, and let $\mathcal{D}_n = \mathcal{D}_n(b,c,a)$. We use the standard construction of the ε-NFA $\mathcal{N}$ for the product: the final state of $\mathcal{D}'_m$ becomes non-final, and an ε-transition is added from that state to the initial state of $\mathcal{D}_n$. This is illustrated in Fig. 2 for $m = 9, n = 8$.

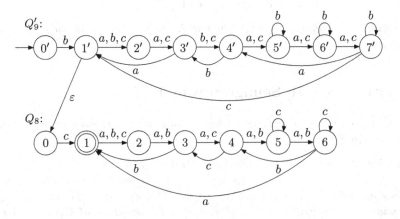

Fig. 2. The NFA $\mathcal{N}$ for product $L'_9(a,b,c) \cdot L_8(b,c,a)$. The empty states $8'$ and 7 and the transitions to them are omitted

We use the subset construction to determinize $\mathcal{N}$ to get a DFA $\mathcal{P}$ for the product. The states of $\mathcal{P}$ are subsets of $Q'_m \cup Q_n$ and have one of three forms: $\{0'\}$, $\{1', 0\} \cup S$ or $\{p'\} \cup S$, where $p' = 2', \ldots, (m-1)'$ and $S \subseteq \{1, \ldots, n-1\}$.

Note that for each $x \in \Sigma$ every state $q \in Q_n \setminus \{0, n-1\}$ has a unique predecessor state $p \in Q_n \setminus \{n-1\}$ such that $px = q$. For $w \in \Sigma^*$, the w-predecessor of $S \subseteq Q_n \setminus \{0, n-1\}$ is denoted by Sw^{-1}.

Lemma 2. *For each $n \geqslant 6$ and each $q \in Q_n$ there exists a word $w_q \in c\{a, b\}^*$ such that $1'w_q = 3'$, $0w_q = q$, and each state of $Q_n \setminus \{0, q, n-1\}$ has a unique w_q-predecessor in $Q_n \setminus \{0, n-1\}$.*

Theorem 3 (Product: Ternary Case). *For $n \geqslant 6$, the product $L'_m(a, b, c) \cdot L_n(b, c, a)$ meets the bound $(m-1)2^{n-2} + 1$.*

Cmorik and Jirásková [14, Theorem 5] also found binary witnesses that meet the bound $(m-1)2^{n-2}$ in the case where $m-2$ and $n-2$ are relatively prime. It remained unknown whether the bound $(m-1)2^{n-2} + 1$ is reachable with a binary alphabet. We show that a slightly modified first witness of [14] meets the upper bound exactly. For $m \geqslant 6, n \geqslant 3$, let the first DFA be that of [14], except that the set of final states is changed to $\{2', 4'\}$; thus let $\Sigma = \{a, b\}$, $\mathcal{D}'_m(a, b) = (Q'_m, \Sigma, \delta', 0', \{2', 4'\})$, and let $\mathcal{D}_n(a, b) = (Q_n, \Sigma, \delta, 0, \{1\})$ be the second DFA [14]. Let $L'_m(a, b)$ and $L_n(a, b)$ be the corresponding languages.

Theorem 4 (Product: Binary Case). *For $m, n \geqslant 6$, $L'_m(a, b)$ is suffix-free and $L'_m(a, b) \cdot L_n(a, b)$ meets the bound $(m-1)2^{n-2} + 1$ when $m-2$ and $n-2$ are relatively prime.*

Theorem 5. *Suppose $m, n \geqslant 4$ and $L'_m L_n$ meets the bound $2^{n-2} + 1$. Then the transition semigroup T_n of a minimal DFA $\mathcal{D}_n$ of L_n is a subsemigroup of $\mathbf{T}^{\leqslant 5}(n)$ and is not a subsemigroup of $\mathbf{T}^{\geqslant 6}$.*

For boolean operations we use the witnesses $L'_m(a, b, -)$ and $L_n(c, b, -)$ and relabel them as $L'_m(a, b)$ and $L_n(a, b)$. See Fig. 3.

Theorem 6. *For $m, n \geqslant 6$, $L'_m(a, b)$ and $L_n(a, b)$ meet the bounds for boolean operations.*

4   Witnesses with Semigroups in $\mathbf{T}^{\geqslant 6}(n)$

We now turn to the operations which cannot have witnesses with transition semigroups in $\mathbf{T}^{\leqslant 5}$.

Definition 2. *For $n \geqslant 4$, we define the DFA $\mathcal{D}_n(a, b, c, d, e) = (Q_n, \Sigma, \delta, 0, F)$, where $Q_n = \{0, \ldots, n-1\}$, $\Sigma = \{a, b, c, d, e\}$, δ is defined by the transformations of Proposition 5, and $F = \{q \in Q_n \setminus \{0, n-1\} \mid q \text{ is odd}\}$. For $n = 4$, a and b coincide, and we can use $\Sigma = \{b, c, d, e\}$. The structure of $\mathcal{D}_5(a, b, c, d, e)$ is illustrated in Fig. 1.*

Our main result in this section is the following theorem:

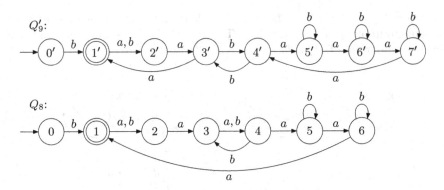

Fig. 3. Witnesses $\mathcal{D}'_9(a,b)$ and $D_8(a,b)$ for boolean operations. The empty states $8'$ and 7 and the transitions to them are omitted figure moved

Theorem 7 (Boolean Operations, Reversal, Number and Complexities of Atoms, Syntactic Complexity). *Let $\mathcal{D}_n(a,b,c,d,e)$ be the DFA of Definition 2, and let the language it accepts be $L_n(a,b,c,d,e)$. Then $L_n(a,b,c,d,e)$ meets the following bounds:*

1. *For $n,m \geqslant 4$, $L_m(a,b,-,d,e)$ and $L_n(b,a,-,d,e)$ (Fig. 4) meet the bounds $mn - (m+n-2)$ for union and symmetric difference, $mn - 2(m+n-3)$ for intersection, and $mn - (m+2n-4)$ for difference.*
2. *For $n \geqslant 4$, $L_n(a,-,c,-,e)$ meets the bound $2^{n-2}+1$ for reversal and number of atoms.*
3. *For $n \geqslant 6$, $L_m(a,b,c,d,e)$ meets the bound $(n-1)^{n-2} + n - 2$ for syntactic complexity, and the bounds on the quotient complexities of atoms.*

The claim about syntactic complexity is known from [11]. It was shown in [12] that the number of atoms of a regular language L is equal to the quotient complexity of L^R. In the next subsections we prove the claim about boolean operations, reversal, and atom complexities. First we state some properties of $\mathcal{D}_n$.

Proposition 6. *For $n \geqslant 4$ the DFA of Definition 2 is minimal, accepts a suffix-free language, and its transition semigroup T_n has cardinality $(n-1)^{n-2}+n-2$. In particular, T_n contains (a) all $(n-1)^{n-2}$ transformations that send 0 and $n-1$ to $n-1$ and map $Q \setminus \{0,n-1\}$ to $Q \setminus \{0\}$, and (b) all $n-2$ transformations that send 0 to a state in $Q \setminus \{0,n-1\}$ and map all the other states to $n-1$. Also, T_n is generated by $\{a,b,c,d,e\}$ and cannot be generated by a smaller set of transformations.*

We now show that witness DFAs for boolean operations may have transition semigroups in $\mathbf{T}^{\geqslant 6}$.

Theorem 8. *For $n,m \geqslant 4$, $L_m(a,b,-,d,e)$ and $L_n(b,a,-,d,e)$ meet the bounds for boolean operations.*

Since $t_d = t_{c_1} t_{c_1}$ and $t_e = t_{c_1} t_{c_2} \cdots t_{c_{n-1}}$, where the c_i are from Proposition 2, the semigroup of $\mathcal{D}_n(a, b, -, d, e)$ is in $\mathbf{T}^{\leqslant 5}(n) \cap \mathbf{T}^{\geqslant 6}(n)$. In fact, one can verify that the semigroup of $\mathcal{D}_n(a, b, -, d, e)$ is $\mathbf{T}^{\leqslant 5}(n) \cap \mathbf{T}^{\geqslant 6}(n)$.

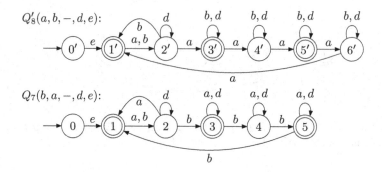

Fig. 4. The DFAs $\mathcal{D}'_8$ and $\mathcal{D}_7$ for boolean operations; empty states omitted

Han and Salomaa [15] showed that to meet the bound for reversal one can use the binary DFA of Leiss [18] and add a third input to get a suffix-free DFA. Cmorik and Jirásková [14] showed that a binary alphabet will not suffice. We show a different ternary witness below, and prove that any witness must have its transition semigroup in $\mathbf{T}^{\geqslant 6}$.

Theorem 9 (Semigroup of Reversal Witness). *For $n \geqslant 4$, the transition semigroup of a minimal DFA of a suffix-free language L_n that meets the bound $2^{n-2} + 1$ for reversal is a subsemigroup of $\mathbf{T}^{\geqslant 6}(n)$ but not of $\mathbf{T}^{\leqslant 5}(n)$.*

Theorem 10 (Reversal Complexity). *If $n \geqslant 4$, then $L_n(a, -, c, -, e)$ of Definition 2 meets the bound $2^{n-2} + 1$ for reversal.*

Although $L_n(a, -, c, -, e)$ meets the bound for the number of atoms, it does not meet the bounds on the quotient complexities of atoms; we now show that $L_n(a, b, c, d, e)$ does.

Let $Q_n = \{0, \ldots, n-1\}$ and let L be a non-empty regular language with quotients $K = \{K_0, \ldots, K_{n-1}\}$. Let $\mathcal{D} = (Q_n, \Sigma, \delta, 0, F)$ be the minimal DFA of L in which the language of state q is K_q.

Denote the complement of a language L by $\overline{L} = \Sigma^* \setminus L$. Each subset S of Q_n defines an *atomic intersection* $A_S = \bigcap_{i \in S} K_i \cap \bigcap_{i \in \overline{S}} \overline{K_i}$, where $\overline{S} = Q_n \setminus S$. An *atom* of L is a non-empty atomic intersection. Since atoms are pairwise disjoint, every atom A_S has a unique atomic intersection associated with it, and this atomic intersection has a unique subset S of K associated with it.

Let $A_S = \bigcap_{i \in S} K_i \cap \bigcap_{i \in \overline{S}} \overline{K_i}$ be an atom. For any $w \in \Sigma^*$ we have

$$w^{-1} A_S = \bigcap_{i \in S} w^{-1} K_i \cap \bigcap_{i \in \overline{S}} \overline{w^{-1} K_i}.$$

Since a quotient of a quotient of L is also a quotient of L, $w^{-1}A_S$ has the form:

$$w^{-1}A_S = \bigcap_{i \in X} K_i \cap \bigcap_{i \in Y} \overline{K_i},$$

where $|X| \leqslant |S|$ and $|Y| \leqslant n - |S|$, $X, Y \subseteq Q_n$.

Proposition 7. *Suppose L is a suffix-free language with $n \geqslant 4$ quotients. Then L has at most $2^{n-2} + 1$ atoms. Also, the complexity $\kappa(A_S)$ of atom A_S satisfies*

$$\kappa(A_S) \begin{cases} \leqslant 2^{n-2} + 1, & \text{if } S = \emptyset; \\ = n, & \text{if } S = \{0\}; \\ \leqslant 1 + \sum_{x=1}^{|S|} \sum_{y=0}^{n-2-|S|} \binom{n-2}{x} \binom{n-2-x}{y}, & \emptyset \neq S \subseteq \{1, \ldots, n-2\}. \end{cases} \quad (1)$$

Following Iván [16] we define a DFA for each atom:

Definition 3. *Suppose $\mathcal{D} = (Q, \Sigma, \delta, q_0, F)$ is a DFA and let $S \subseteq Q$. Define the DFA $\mathcal{D}_S = (Q_S, \Sigma, \Delta, (S, \overline{S}), F_S)$, where*

- $Q_S = \{(X, Y) \mid X, Y \subseteq Q, X \cap Y = \emptyset\} \cup \{\bot\}$.
- *For all $a \in \Sigma$, $(X, Y)a = (Xa, Ya,)$ if $Xa \cap Ya = \emptyset$, and $(X, Y)a = \bot$ otherwise; and $\bot a = \bot$.*
- $F_S = \{(X, Y) \mid X \subseteq F, Y \subseteq \overline{F}\}$.

DFA $\mathcal{D}_S$ recognizes the atomic intersection A_S of L. If $\mathcal{D}_S$ recognizes a non-empty language, then A_S is an atom.

Theorem 11. *For $n \geqslant 4$, the language $L_n(\mathcal{D}(a, b, c, d, e))$ of Definition 2 meets the bounds of Proposition 7 for the atoms.*

Remark 1. The complexity of atoms in left ideals [6] is

$$\kappa(A_S) \begin{cases} = n, & \text{if } S = Q_n; \\ \leqslant 2^{n-1}, & \text{if } S = \emptyset; \\ \leqslant 1 + \sum_{x=1}^{|S|} \sum_{y=1}^{n-|S|} \binom{n-1}{x} \binom{n-1-x}{y-1}, & \text{otherwise.} \end{cases} \quad (2)$$

The formula for $S \notin \{\emptyset, Q_n\}$ evaluated for $n-1$ and $S \subseteq \{1, \ldots, n-2\}$ becomes $1 + \sum_{x=1}^{|S|} \sum_{y=1}^{n-2-|S|} \binom{n-2}{x} \binom{n-2-x}{y-1}$, which is precisely the formula for suffix-free languages. ∎

5   Conclusions

It may appear that semigroup $\mathbf{T}^{\leqslant 5}(n)$ should not be of great importance, since it exceeds $\mathbf{T}^{\geqslant 6}(n)$ only for $n = 4$ and $n = 5$, and therefore should not matter when n is large. However, our results show that this is not the case. We conclude with our result about the non-existence of single universal suffix-free witness.

Theorem 12. *There does not exist a most complex stream in the class of suffix-free languages.*

The first four studies of most complex languages were done for the classes of regular languages [4], right ideals [5,6], left ideals [6,7], and two-sided ideals [6,7]. In those cases there exists a single witness stream of languages over a minimal alphabet which, together with their dialects, cover all the complexity measures. In the case of suffix-free languages such a stream does not exist. Our study is an example of a general problem: Given a class of regular languages, find the smallest number of streams over minimal alphabets that together cover all the measures. The witness of Definition 1 is conjectured to be over a minimal alphabet, unless the bound for product can be met by binary DFAs for every $n, m > c$, for some c; this is an open problem. The witness of Definition 2 is over a minimal alphabet, since five letters are required to meet then bound for syntactic complexity.

References

1. Ang, T., Brzozowski, J.: Languages convex with respect to binary relations, and their closure properties. Acta Cybernet. **19**(2), 445–464 (2009)
2. Berstel, J., Perrin, D., Reutenauer, C.: Codes and Automata. Cambridge University Press, Cambridge (2009)
3. Brzozowski, J.: Quotient complexity of regular languages. J. Autom. Lang. Comb. **15**(1/2), 71–89 (2010)
4. Brzozowski, J.: In search of the most complex regular languages. Int. J. Found. Comput. Sc. **24**(6), 691–708 (2013)
5. Brzozowski, J., Davies, G.: Most complex regular right-ideal languages. In: Jürgensen, H., Karhumäki, J., Okhotin, A. (eds.) DCFS 2014. LNCS, vol. 8614, pp. 90–101. Springer, Heidelberg (2014)
6. Brzozowski, J., Davies, S.: Quotient complexities of atoms in regular ideal languages (2015). http://arxiv.org/abs/1503.02208
7. Brzozowski, J., Davies, S., Liu, B.Y.V.: Most complex regular ideals (2015). (in preparation)
8. Brzozowski, J., Jirásková, G., Li, B., Smith, J.: Quotient complexity of bifix-, factor-, and subword-free regular languages. Acta Cybernet. **21**, 507–527 (2014)
9. Brzozowski, J., Li, B., Ye, Y.: Syntactic complexity of prefix-, suffix-, bifix-, and factor-free regular languages. Theoret. Comput. Sci. **449**, 37–53 (2012)
10. Brzozowski, J., Szykuła, M.: Complexity of suffix-free regular languages (2015). http://arxiv.org/abs/1504.05159
11. Brzozowski, J., Szykuła, M.: Upper bound for syntactic complexity of suffix-free languages. In: Okhotin, A., Shallit, J. (eds.) DCFS 2015. LNCS, vol. 9118, pp. 33–45. Springer, Heidelberg (2015). http://arxiv.org/abs/1412.2281
12. Brzozowski, J., Tamm, H.: Theory of átomata. Theoret. Comput. Sci. **539**, 13–27 (2014)
13. Brzozowski, J., Ye, Y.: Syntactic complexity of ideal and closed languages. In: Mauri, G., Leporati, A. (eds.) DLT 2011. LNCS, vol. 6795, pp. 117–128. Springer, Heidelberg (2011)

14. Cmorik, R., Jirásková, G.: Basic operations on binary suffix-free languages. In: Kotásek, Z., Bouda, J., Černá, I., Sekanina, L., Vojnar, T., Antoš, D. (eds.) MEMICS 2011. LNCS, vol. 7119, pp. 94–102. Springer, Heidelberg (2012)

15. Han, Y.S., Salomaa, K.: State complexity of basic operations on suffix-free regular languages. Theoret. Comput. Sci. **410**(27–29), 2537–2548 (2009)

16. Iván, S.: Complexity of atoms, combinatorially (2015). http://arxiv.org/abs/1404.6632

17. Jirásková, G., Olejár, P.: State complexity of union and intersection of binary suffix-free languages. In: Bordihn, H., et al. (eds.) NMCA, pp. 151–166. Austrian Computer Society, Wien (2009)

18. Leiss, E.: Succinct representation of regular languages by boolean automata. Theoret. Comput. Sci. **13**, 323–330 (2009)

19. Maslov, A.N.: Estimates of the number of states of finite automata. Dokl. Akad. Nauk SSSR **194**, 1266–1268 (1970). (Russian): English translation: Soviet Math. Dokl. **11**, 1373–1375 (1970)

20. Mirkin, B.G.: On dual automata. Kibernetika (Kiev) **2**, 7–10 (1966). (Russian): English translation: Cybernetics **2**, 6–9 (1966)

21. Pin, J.-E.: Syntactic semigroups. In: Rozenberg, G., Salomaa, A. (eds.) Handbook of Formal Languages, pp. 679–746. Springer, New York (1997)

22. Yu, S., Zhuang, Q., Salomaa, K.: The state complexities of some basic operations on regular languages. Theoret. Comput. Sci. **125**, 315–328 (1994)

23. Yu, S.: State complexity of regular languages. J. Autom. Lang. Comb. **6**, 221–234 (2001)

Alternation Hierarchies of First Order Logic with Regular Predicates

Luc Dartois[2,3] and Charles Paperman[1(✉)]

[1] University of Warsaw, Warsaw, Poland
charles.paperman@gmail.com
[2] École Centrale Marseille, Marseille, France
luc.dartois@lif.univ-mrs.fr
[3] LIF, UMR7279, Aix-Marseille Université and CNRS,
Marseille, France

Abstract. We investigate the decidability of the definability problem for fragments of first order logic over finite words enriched with regular numerical predicates. In this paper, we focus on the quantifier alternation hierarchies of first order logic. We obtain that deciding this problem for each level of the alternation hierarchy of both first order logic and its two-variable fragment when equipped with all regular numerical predicates is not harder than deciding it for the corresponding level equipped with only the linear order.

Relying on some recent results, this proves the decidability for each level of the alternation hierarchy of the two-variable first order fragment while in the case of the first order logic the question remains open for levels greater than two.

The main ingredients of the proofs are syntactic transformations of first-order formulas as well as the infinitely testable property, a new algebraic notion on varieties that we define.

1 Introduction

The equivalence between regular languages and automata as well as monadic second order logic [3] and finite monoids [14] was the start of a domain of research that is still active today. In this article, we are interested in the logic on finite words, and more precisely the question we address is the *definability problem* for fragments of logic. Fragments of logic are defined as sets of monadic second order formulas satisfying some restrictions, and are equipped with a set of predicates called a *signature*. Then the definability problem of a fragment of logic **F** consists in deciding if a regular language can be defined by a formula of **F**.

This question has already been considered and solved in many cases where the signature contains only the predicate <, which denotes the linear order over the positions of the word. For instance, a celebrated result by Schützenberger [19] and McNaughton and Papert [13] gave an effective algebraic characterization of languages definable by first order formulas. The decidability has often been achieved through algebraic means, showing a deep connection between algebraic

© Springer International Publishing Switzerland 2015
A. Kosowski and I. Walukiewicz (Eds.): FCT 2015, LNCS 9210, pp. 160–172, 2015.
DOI: 10.1007/978-3-319-22177-9_13

and logical properties of a given regular language. In this article, we follow this approach.

We investigate the question of the behaviour of the decidability of some fragments when their signature is enriched with *regular numerical predicates*. These predicates are exactly the formulas of monadic second order logic without letter predicates. Intuitively they correspond to the maximal class of *numerical predicates* that can enrich the signature of a fragment of **MSO**, while keeping the definable languages regular. This question was already considered in the case of first order logic (**FO**) in [2] and one of its fragments: the formulas without quantifier alternation in [15].

The enrichment by regular numerical predicates arose in the context of the *Straubing's conjectures* [23]. Roughly speaking, these conjectures state that deciding the definability of a regular language to a fragment of enriched logic corresponds to deciding its circuit complexity. It is known [15,23] that an enrichment of the classical fragments by regular numerical predicates is equivalent to an enrichment by the signature [<, +1, MOD], where +1 denotes the *local predicates* and MOD the *modular predicates*. A first step toward the study of fragments of logic with these predicates was initiated by Straubing [22]. He obtained that adding the local predicates preserves the decidability for a large number of fragments. As a corollary of this work, Straubing obtained that the decidability of the alternation hierarchy of first order logic ($\mathcal{B}\Sigma_k$) equipped with [<, +1] reduces to the decidability of the simpler one [<]. More recently, Kufleitner and Lauser [11] proved the decidability of the alternation hierarchy of the two-variable first order fragment ($\mathbf{FO}_k^2$) equipped with [<, +1] by using the recent results [10,12] on the decidability of this hierarchy with [<].

In this context, the case of modular predicates is poorly understood. The study of this enrichment was first considered for first order logic in [2], and had been extended to the first level of its alternation hierarchy with the successor predicate in [15], and later without it in [4]. The enrichment by a finite set of modular predicate was considered in [8]. Finally, the authors provided a characterization of the two-variable first order logic over the signature [<, MOD] in [6].

In this paper, we focus on the enrichment by all regular predicates and let aside the question of the signature [<, MOD], which surprisingly turns out to be more intricate. The fragments we consider here are the *quantifier alternations* hierarchy of the first order logic and its two-variable counterpart. Our main result states that for both of these hierarchies, the decidability of each level equipped with regular numerical predicates reduces to decidability of the same level with the signature [<, +1]. Then by using the recent decidability result of Kufleitner and Lauser [11], as well as the decidability of $\mathcal{B}\Sigma_2[<]$ by Place and Zeitoun [18], we deduce that the fragments $\mathbf{FO}_k^2[\text{Reg}]$, for any positive k, and $\mathcal{B}\Sigma_2[\text{Reg}]$ are decidable. Our main contributions are summarized in the next table.

Proofs Methods. The proofs of the main results can be decomposed in two major steps. The first part is rather classical and shows that in the cases we consider, adding a finite number of modular predicates does not affect the decidability. The second part is dedicated to finding a systematic way to select, for a given

	$\mathcal{B}\Sigma_1 = \mathbf{FO}_1^2$	$\mathbf{FO}_k^2$	$\mathbf{FO}^2$	$\mathcal{B}\Sigma_2$	$\mathcal{B}\Sigma_k$	$\mathbf{FO}$
[<]	Decidable [20, 26]	Decidable [10, 12]	Decidable [25]	Decidable [18]	Open	Decidable [13, 19]
[<, +1]	Decidable [9]	Decidable [11]	Decidable [1, 25]	Decidable [18, 22]	Reduces to [<] [22]	Decidable [13, 19]
[Reg]	Decidable [16]	Decidable **New result**	Decidable **New result**	Decidable **New result**	Reduces to [<, +1] **New result**	Decidable [2]

regular language and a fragment, a finite number of modular predicates that can serve as witness of its definability. This is done through a heavy use of the algebraic framework of varieties of semigroups. We introduce a new notion for varieties of semigroups that we call the infinitely testable property and show that this property is satisfied by the considered fragments. We then conclude by proving that this property allows us to find such a witness set for modular predicates that only depends on the input language.

Generalizations. While we are focused in this article on the levels of the quantifier alternation hierarchies, our approach can be generalized to other fragments under certain conditions. The generality of our results are discussed in Remarks 4, 6, 10 and 12.

Organization of the Paper. Section 2 defines the logical and algebraic notions that will be used in the paper. The main results of the paper are presented in Sect. 3. The Sects. 4 and 5 are then dedicated to the proofs. Section 4 first discusses adding a finite number of predicates and reduces our decidability problems to a *delay question*, which can be summarized as being able to choose the proper finite set of modular predicates. Then Sect. 5 defines a new notion, the *infinitely testable* property, which is satisfied by the fragments that we consider and whereby gives a delay. Finally, we discuss in Sect. 5 some other results that can be directly obtained from our approach, as well as a related algebraic characterization of the two-variable first order logic with the regular numerical signature.

2   Preliminaries

Logic. We consider the monadic second order logic on finite words $\mathbf{MSO}[<]$ as usual (see [23] for example). We denote by A an *alphabet* and by a a *letter* of A. A word u over an alphabet A is a set of labelled positions ordered from 0 to $|u| - 1$. The set of words over A is denoted A^* and a subset L of A^* is called a *language*. We also denote by A^+ the set of non-empty words. A language is said to be *defined* by a formula if it corresponds exactly to the set of words that satisfy this formula. It is said to be *regular* if it is defined by a $\mathbf{MSO}[<]$ formula. When syntactic restrictions are applied to $\mathbf{MSO}[<]$, one defines fragments of logic that characterize subclasses of regular languages. The most well-known fragment is probably the first order logic, whose expressive power was characterized thanks to the results of [13, 19]. The first order logic itself gave birth to its own zoo of fragments. These were defined using syntactical restrictions such as limiting the number of variables, or by enrichment of its signature. A fragment **F** with signature σ will be denoted $\mathbf{F}[\sigma]$ and will refer to the formulas as well as the class of languages it defines.

We first define the different signatures that will appear through this paper, and then formally define the fragments that are considered here: the quantifier alternation hierarchies.

Signatures. We are interested in regular numerical predicates, which are numerical predicates that can only define regular languages. Simultaneously, Straubing [23] and Péladeau [15] defined three sets of regular numerical predicates that can be used as a base for all the regular numerical predicates. The first set is the singleton order $\{<\}$ which is a binary predicate corresponding to the natural order on the positions of the input word. The second set is $\{\mathbf{min}, \mathbf{max}, \mathbf{S}\}$ and is called the *local predicates*. It is usually denoted $+1$. The predicates **min** and **max** are unary predicates that are satisfied respectively on the first and last positions. The predicate **S**, the *successor*, is a binary predicate satisfied if the second variable quantifies the successor of the first one.

Finally, we define, for each positive integer d, the *modular predicates on d*, denoted MOD^d, as the set, for $i < d$, of predicates $\mathbf{MOD_i^d}(x)$ which are unary predicates satisfied if the position quantified by x is congruent to i modulo d, and the predicates $\mathbf{D_i^d}$ which are constants holding if the length of the input word is congruent to $i \bmod d$. We denote by MOD the union of the classes MOD^d, for any positive d.

Example 1. The language $(A^2)^* a A^*$ is defined by the formula: $\exists x \; \mathbf{a}(x) \wedge \mathbf{MOD_0^2}(x)$.

The signatures that we will consider for our fragments are unions of these three sets of regular numerical predicate, and will always contain the letter predicates. Abusing notations, we will also write $\mathrm{Reg} = \{<\} \cup +1 \cup \mathrm{MOD}$.

Fragments and Alternation Hierarchies. While $\mathbf{MSO}[\mathrm{Reg}] = \mathbf{MSO}[<]$, the equality does not hold for subclasses of **MSO**. For a signature σ, we denote by $\mathbf{FO}[\sigma]$ the class of first order formulas whose predicates belong to σ. Since the local predicates can be expressed in $\mathbf{FO}[<]$, the fragments $\mathbf{FO}[<]$ and $\mathbf{FO}[<, +1]$ define the same classes of languages, called the Star-Free languages [13]. On the other hand the fragment $\mathbf{FO}[<, \mathrm{MOD}]$ is strictly more expressive [2].

The fragment $\mathbf{FO}^2$ is the subclass of formulas of **FO** using only two symbols of variables which can be reused (see Example 2). Here, the class of languages defined by $\mathbf{FO}^2[<]$ is strictly contained in $\mathbf{FO}^2[<, +1]$ and $\mathbf{FO}^2[<, \mathrm{MOD}]$ (see [6,25]).

Example 2. The language $A^* a A^* b A^* a A^*$ can be described by the first order formula $\exists x \exists y \exists z \; x < y < z \wedge \mathbf{a}(x) \wedge \mathbf{b}(y) \wedge \mathbf{a}(z)$. This formula uses three variables x, y and z. However, by reusing x we get an equivalent formula that uses only two variables:

$$\exists x \; \mathbf{a}(x) \wedge \Big(\exists y \; x < y \wedge \mathbf{b}(y) \wedge \big(\exists x \; y < x \wedge \mathbf{a}(x)\big)\Big). \tag{a}$$

Now given a first order formula, one can compute a prenex normal form using the De Morgan's laws. We define the *quantifier alternation* of a formula as the number of blocks of quantifiers $\forall$ and $\exists$ in its prenex normal form. For example, the formula $\exists x \exists y \forall z \; x < z < y \wedge \mathbf{a}(x) \wedge \mathbf{a}(y) \wedge \mathbf{c}(z)$ has a quantifier alternation of 2. It describes the language $A^* ac^* aA^*$. Then given a signature σ and a positive integer k, we denote by $\mathcal{B}\Sigma_k[\sigma]$ the set of prenex normal formulas of $\mathbf{FO}[\sigma]$ whose quantifier alternation is smaller or equal to k. They form the levels of the *quantifier alternation hierarchy* over $\mathbf{FO}[\sigma]$.

When σ is reduced to $\{<\}$, this hierarchy is called the Straubing-Thérien hierarchy [21,24]. Only the first [20] and second [18] levels are known to be decidable. For $\sigma = \{<\} \cup +1$, this hierarchy is called the Dot-Depth hierarchy [5]. The decidability of each level reduces to the decidability of the corresponding level of the Straubing-Thérien hierarchy [22]. In both cases, the hierarchies are known to be strict, and cover all Star-Free languages. In this article, we also consider the alternation hierarchy of $\mathbf{FO}^2$. To define formally the number of alternations of a formula, we cannot rely on the prenex normal form since the construction increases the number of variables. In particular, remark that $\mathbf{FO}^2[<]$ is equivalent to $\Sigma_2[<] \cap \Pi_2[<]$ which is a subclass of $\mathcal{B}\Sigma_2[<]$ [7]. That said, the number of alternations is still a relevant parameter that could be defined as follows: Consider the parse tree naturally associated to a formula. For instance, (a) has $\exists$ as a root and the atomic formulas as the leaves. In a two-variable first order formula we count the maximal number of alternations appearing on a branch, i.e. between the root and a leaf, once the negations have been pushed on to the leaves. A more precise definition can be found in [28]. We denote by $\mathbf{FO}^2_k[\sigma]$ the formulas of $\mathbf{FO}^2[\sigma]$ that have at most $k - 1$ quantifier alternations. The hierarchy induced by $\mathbf{FO}^2_k[<]$ is known to be strict [28] and its definability problem is decidable [10,12]. Note that the hierarchy $\mathbf{FO}^2_k[<, +1]$ is also known to be decidable [11].

Algebra. We quickly present here the fundamental notions used by the proofs of the article (mainly Sect. 5) and refer the reader to [17] for a detailed approach. A (finite) *semigroup* is a finite set equipped with an associative internal law. A semigroup with a neutral element for this law is called a *monoid*. Recall that a semigroup S *divides* another semigroup T if S is a quotient of a subsemigroup of T. This defines a partial order on finite semigroups. Given a finite semigroup S, an element e of S is *idempotent* if $ee = e$. We denote by $E(S)$ the set of idempotents of S. For any element x of S, there exists a positive integer n such that x^n is idempotent. We call this element the *idempotent power* of x and denote it by x^ω. One can check that the application $x \to x^\omega$ is well defined.

A semigroup S recognizes a language L over an alphabet A via a *morphism* $\eta : A^+ \to S$. Given a regular language L, we can compute its *syntactic semigroup* as the smallest semigroup that recognizes L, in the sense of division. For a morphism $\eta : A^+ \to S$, the set $\eta(A)$ is an element of the powerset semigroup of S. As such it has an idempotent power. The *stability index* of a morphism η is then defined as the smallest positive integer s such that $\eta(A^s) = \eta(A^{2s})$.

Remark that $\eta(A^s)$ forms a subsemigroup of S, that we call the *stable semigroup*. A subset T of S is an *ideal* if the sets TS and ST are both included in T. A (pseudo-)*variety* of semigroups is a non empty class of finite semigroups closed under division and finite product.

A fragment of logic is *characterized* by a variety if they recognize the same languages. By extension, a variety $\mathbf{V}$ will also refers to the class of languages it recognizes. The most famous example is the equality $\mathbf{FO}[<] = \mathbf{A}$ [13,19], where $\mathbf{A}$ denotes the class of aperiodic semigroups, which are finite semigroups that are not divided by any group. As for $\mathbf{FO}[<]$, the definability problem for a fragment of logic has often been solved thanks to an algebraic characterization ([20,24,25] for example). This decidability is sometimes obtained through *profinite equations*. For example, the variety of aperiodic semigroups $\mathbf{A}$ is defined by the equation $x^{\omega+1} = x^{\omega}$.

3 Main Results

We present here the main results of this paper, which are reductions of decidability from any level of the first order hierarchies equipped with the regular complete signature to the corresponding level whose signature is reduced to the order. As the decidability of each level of the two-variable hierarchy is known, we get a decidability result. But as the decidability of both the Straubing-Thérien hierarchy, and consequently the Dot-Depth hierarchy as well as their decidability are equivalent, is still open for any level greater than 2, we only get a transfer result.

Theorem 3. *Let k be a positive integer.*

1. *The fragment $\mathcal{B}\Sigma_k[\mathrm{Reg}]$ is decidable if $\mathcal{B}\Sigma_k[<]$ is decidable.*
2. *The fragment $\mathbf{FO}_k^2[\mathrm{Reg}]$ is decidable.*

Let us remark first that this theorem implies that both hierarchies are strict, which is a new result. The recent result of Place and Zeitoun [18] allows us to state as a direct corollary that $\mathcal{B}\Sigma_2[\mathrm{Reg}]$ is decidable.

Remark 4. This approach could be applied to any *abstract fragment* characterized by a variety and expressive enough to contain the languages $(ab)^+$ and A^*a. At this level of abstraction, the operation of adding modular predicates corresponds to a *wreath product* by *modular morphisms*. However, for the sake of concise presentation, we focus on what we assume to be the most interesting corollaries of this approach: the alternation hierarchies with successor. This method can also be generalized to varieties that do not contain $(ab)^+$ and is therefore not dependant on the presence of the *successor* relation. However, this requires to introduce the more involved framework of *finite categories* [27]. In this context, the infinitely testable property of a variety of semigroups, which is the key ingredient of the proof, lifts to the associated *variety of semigroupoids*.

Proof Scheme. First, we reduce the decidability of $\mathcal{B}\Sigma_k[\mathrm{Reg}]$ and $\mathbf{FO}_k^2[\mathrm{Reg}]$ to the decidability of $\mathcal{B}\Sigma_k[<, +1]$ and $\mathbf{FO}_k^2[<, +1]$, respectively. Then we conclude by using the result of Straubing [22], that reduces the decidability of $\mathcal{B}\Sigma_k[<, +1]$ to the decidability of $\mathcal{B}\Sigma_k[<]$, and the result of Kufleitner and Lauser [11] that prove the decidability of $\mathbf{FO}_k^2[<, +1]$. The main issue is therefore to prove the first reduction. In order to obtain it, we decompose the proof in two important steps. The first one proves that adding a finite number of modular predicates is decidable, while the second one allows us to compute such a finite set that serves as a witness for a language to belong to the fragment. If the first step is quite standard, the second introduces a new notion, the infinitely testable property, which allows us to solve the delay question for the fragments we consider.

4   The Delay Question

The objective of this section is to reduce the decidability question to another question, the delay. Informally, the delay question is: which modular predicates would be used by a formula of the fragment to describe the input language. Firstly, we deal with adding the modular predicates ranging over one specific congruence. The idea is to reduce the decidability of a partially enriched fragment to the one of the input fragment. As in [6], this is done by transferring the modular information to an enriched alphabet. For any positive integer d, we denote by $A_d = A \times \mathbb{Z}_d$ the *enriched alphabet* of A and by $\pi_d : A_d^+ \to A^+$ the projection on the first component. To link this enrichment to the modular information, we also define the *well-formed words* K_d as the language of words $(a_0, i_0) \ldots (a_n, i_n)$ such that for any $0 \leqslant j \leqslant n$ $i_j = j \mod d$. Finally, given a language L, we denote by $L_d = \pi_d^{-1}(L) \cap \mathrm{K}_d$. The following theorem proves the reduction from the partially enriched fragment to the initial one by deriving formulas for one language to a formula to the other.

Proposition 5. *Let $\mathbf{F}[\sigma]$ be one of the fragments $\mathcal{B}\Sigma_k[<, +1]$ or $\mathbf{FO}_k^2[<, +1]$ for $k \geqslant 1$. Then, for any regular language L and any $d > 0$, $L \in \mathbf{F}[\sigma, \mathrm{MOD}^d]$ if, and only if, $L_d \in \mathbf{F}[\sigma]$.*

Remark that since one can compute the well-formed enrichment of given a regular language, we obtain as a direct consequence that if $\mathbf{F}[\sigma]$ is one of the fragments $\mathcal{B}\Sigma_k[<, +1]$ or $\mathbf{FO}_k^2[<, +1]$ for $k \geqslant 1$, the fragment $\mathbf{F}[\sigma]$ is decidable if, and only if, $\mathbf{F}[\sigma, \mathrm{MOD}^d]$ is decidable.

Remark 6. Even if the previous proposition is only stated for the fragments considered in this article, its applications range over many more fragments. Indeed, it would hold for any expressive enough fragment, i.e. any fragment that can define the set of well-formed words over the enriched alphabet and that satisfies some closure properties.

Now that we proved that adding predicates according to one congruence we make the following easy remark. Let $\mathbf{F}[\sigma]$ be one of the fragments $\mathcal{B}\Sigma_k[<, +1]$

or $\mathbf{FO}_k^2[<,+1]$ and d, p be two positive integers. Then $\mathbf{F}[\sigma, \mathrm{MOD}^d, \mathrm{MOD}^p] \subseteq \mathbf{F}[\sigma, \mathrm{MOD}^{dp}]$. Then as a formula can only use a finite number of modular predicates, for any language of $\mathcal{F}[\sigma, \mathrm{MOD}]$, there exists an integer d such that it belongs to $\mathcal{F}[\sigma, \mathrm{MOD}^d]$. In fact, there exists an infinite number of such witnesses. Thanks to Proposition 5, the decidability of the fragments we study reduces to the following question:

The Delay Question: Given a regular language L, is it possible to compute an integer d_L such that L belongs to $\mathcal{F}[\sigma, \mathrm{MOD}]$ *if, and only if,* it belongs to $\mathcal{F}[\sigma, \mathrm{MOD}^{d_L}]$?

The denomination stems from the Delay Theorem of [22] that solves a similar question for the enrichment by the successor predicate.

5   The Infinitely Testable Property

In this Section, we conclude the proof of the main theorem by solving the delay question for the fragments considered. We actually solve the delay question for the fragments we consider via an algebraic property on varieties satisfied by their characterization. This property, which we call the *infinitely testable* property, is a new notion that we introduce and which is defined below. Informally, a variety is infinitely testable if the membership of a language to the variety only depends on words *long enough*.

Definition. Given a semigroup S, the *idempotents' ideal* of S, denoted $\mathcal{I}_E(S)$, is the ideal of S generated by its idempotents. We have then $\mathcal{I}_E(S) = SE(S)S$, where $E(S)$ denotes the set of idempotents of S. Note also that given a morphism $\eta : A^+ \to S$, it is the semigroup of all elements of S having an infinite number of preimages by η. An aware reader could notice that $\mathcal{I}_E(S)$ is the set of all elements of S that are $\mathcal{J}$-below an idempotent. A variety of semigroups $\mathbf{V}$ is said to be *infinitely testable* if the membership of a semigroup to $\mathbf{V}$ is equivalent to the membership of its idempotents' ideal. Informally, a variety is infinitely testable if its membership can be reduced to an algebraic condition on the idempotents' ideal. By extension, we say that a fragment of logic is infinitely testable if it is characterized by an infinitely testable variety.

Example 7. The fragment $\mathbf{FO}[=]$ is equivalent to the aperiodic and commutative variety $\mathbf{ACom}$. This fragment is also described by the equations $xy = yx$ and $x^{\omega+1} = x^\omega$. This fragment is not infinitely testable. For instance the language equal to the singleton $\{ab\}$ has a trivial idempotents' ideal while it is not definable in $\mathbf{FO}[=]$.

Example 8. The fragment $\mathbf{FO}[+1]$ is equivalent to the languages whose syntactic semigroup belongs to the variety: $\mathbf{ACom} * \mathbf{LI}$ [23, Theorem VI.3.1]. This fragment is also described by the profinite equation

$$x^\omega u y^\omega v x^\omega w y^\omega = x^\omega w y^\omega v x^\omega u y^\omega. \tag{b}$$

We now show that it is an infinitely testable fragment. Let L be a regular language and S its syntactic semigroup. We simply prove that if the Eq. (b) is not satisfied by S, then it is not satisfied by $\mathcal{I}_E(S)$. Suppose that there exists $x, y, u, v, w \in S$ such that the Eq. (b) is not satisfied. Then by setting: $x' = x^\omega$, $y' = y^\omega$, $u' = x^\omega u y^\omega$, $v' = y^\omega v x^\omega$, $w' = x^\omega w y^\omega$. All new variables belong to $\mathcal{I}_E(S)$ and they also fail to satisfy (b).

Infinitely Testable Fragments. The infinitely testable property of levels of the $\mathbf{FO}^2[<, +1]$ hierarchy is proved using the equational characterization obtained in [11], following Example 8. Because of the lack of equational description for $\mathcal{B}\Sigma_k[<, +1]$, we use a more involved algebraic argument for this latter case.

Proposition 9. *Let k be a positive integer. The fragments $\mathbf{FO}_k^2[<, +1]$ and $\mathcal{B}\Sigma_k[<, +1]$ are infinitely testable.*

Remark 10. The infinitely testable property of $\mathcal{B}\Sigma_k[<, +1]$ can be stated in a more general framework. Indeed, in the article of Tilson [27], a version of the delay theorem states that a semigroup belongs to $\mathbf{V} * \mathbf{LI}$ if, and only if, the *idempotents' category* belongs to the *variety of finite categories* generated by $\mathbf{V}$. In this framework of finite categories, the idempotents categories is defined as the semigroup S_E by removing the absorbing element 0. Therefore, one could argue that all varieties of semigroups of the form $\mathbf{V} * \mathbf{LI}$ have the property to be infinitely testable.

Delay Theorem for Quantifier Hierarchies. We reach the key theorem of our presentation. It proves a delay for each levels of the quantifier hierarchies over the first order logic and its two-variable counterpart. The delay we obtain here is the stability index.

Theorem 11. *Let $\mathbf{F}[\sigma]$ be one of the fragments $\mathcal{B}\Sigma_k[<, +1]$ or $\mathbf{FO}_k^2[<, +1]$ and L a regular language of stability index s. Then L belongs to $\mathcal{F}[\sigma, \mathrm{MOD}]$ if, and only if, L belongs to $\mathcal{F}[\sigma, \mathrm{MOD}^s]$.*

Proof. We denote by $\mathbf{V}$ the infinitely testable variety of semigroups equivalent to $\mathcal{F}[\sigma]$. Consider a regular language L that belongs to $\mathcal{F}[\sigma, \mathrm{MOD}]$. Then as there exists an integer d such that L belongs to $\mathcal{F}[\sigma, \mathrm{MOD}^d]$, it is sufficient to show that if there exists $d > 0$ such that if L belongs to $\mathcal{F}[\sigma, \mathrm{MOD}^{ds}]$, then it belongs to $\mathcal{F}[\sigma, \mathrm{MOD}^s]$. Thus, by Proposition 5, it suffices to prove that if L_{ds} is definable in $\mathcal{F}[\sigma]$, then L_s is in $\mathcal{F}[\sigma]$ as well. We recall that $L_d = \pi_d^{-1}(L) \cap K_d$ for any $d > 0$. We set $\eta_s : A_s^+ \to S_s$ and $\eta_{ds} : A_{ds}^+ \to S_{ds}$ the syntactic morphisms of L_s and L_{ds} respectively.

Claim. The semigroup $\mathcal{I}_E(S_s)$ divides $\mathcal{I}_E(S_{ds})$.

Before proving this claim, let us remark that since a variety of semigroups is closed by division, this claim ends the proof. Since if L belongs to $\mathcal{F}[\sigma, \mathrm{MOD}^{ds}]$

then S_{ds} belongs to $\mathbf{V}$ and therefore $\mathcal{I}_E(S_{ds})$ belongs to $\mathbf{V}$ as well. By division, $\mathcal{I}_E(S_s)$ belongs to $\mathbf{V}$, and thanks to the infinitely testable hypothesis, we have that S_s belongs to $\mathbf{V}$. Finally, we deduce that L_s belongs to $\mathbf{F}[\sigma]$. We now aim to construct a division from $\mathcal{I}_E(S_s)$ to $\mathcal{I}_E(S_{ds})$. This is done through the enriched alphabet. We introduce the following projection

$$h\colon \begin{cases} A_{ds}^+ & \to & A_s^+ \\ (a,i) & \mapsto & (a,i \bmod s) \end{cases}$$

and F_d the language of well-formed factors, which is the set of well-formed words that do not necessarily start by a letter of the form $(a,0)$. Note that $L_{ds} = h^{-1}(L_s) \cap K_s$. Let us remark also that the image a word not in F_s (resp. F_{ds}) by η_s (resp. η_{ds}) has an absorbing zero as image by η_s (resp. η_{ds}). This zero being idempotent, it belongs to $\mathcal{I}_E(S_s)$ (resp. $\mathcal{I}_E(S_{ds})$). Finally, if two words of F_s have the same image by η_s, then they have the same length modulo s and their first (and consequently last) letters have the same enrichment.

Consider then x a non-zero element of $\mathcal{I}_E(S_s)$. We show that

$$h^{-1}(\eta_s^{-1}(x)) \cap \eta_{ds}^{-1}(\mathcal{I}_E(S_{ds})) \neq \emptyset.$$

Since x belongs to $\mathcal{I}_E(S_s)$, there exists a word u of A_s^+ of length greater than s in the preimage of x. And since $\eta_s(A_s^s) = \eta_s(A_s^{2s})$ by definition of the stability index, for any $k > 0$ there exists a word v_k of A_s^+ of length greater than ks such that $u \equiv_L v_k$ and $|u| = |v_k| \bmod s$, since $\eta_s(u) = \eta_s(v_k)$. Then for k sufficiently large, there exists a word w in $h^{-1}(v_k)$, such that $\eta_{ds}(w)$ belongs to $\mathcal{I}_E(S_{ds})$. Note that by taking k as a multiple of d, we obtain a word w such that $|u| \bmod s = |w| \bmod ds$. Thus for each element $x \in \mathcal{I}_E(S_s)$, we can choose such an element, that we denote w_x. This justifies the definition of the following function:

$$f\colon \begin{cases} \mathcal{I}_E(S_s) \to \mathcal{I}_E(S_{ds}) \\ x \quad \mapsto \eta_{ds}(w_x) \text{ if } x \neq 0 \\ 0 \quad \mapsto \quad 0 \quad \text{otherwise.} \end{cases}$$

We conclude by proving that f is an injective morphism, and thus $\mathcal{I}_E(S_s)$ is a subsemigroup of $\mathcal{I}_E(S_{ds})$.

The Application f Is a Morphism. Let $x, y \in \mathcal{I}_E(S_s)$. We show that $f(xy) = f(x)f(y)$. First, we can assume without loss of generality that $x \neq 0$ and $y \neq 0$. We remark that since $|w_x| \bmod ds = |h(w_x)| \bmod s$, the concatenated word $w_x w_y$ is well-formed if, and only if, $h(w_x)h(w_y)$ is well-formed too. If $xy \neq 0$.Then, xy have a well-formed preimage and $w_x w_y$ is well-formed. Then as w_{xy} and $w_x w_y$ are syntactically equivalent with respect to both F_{ds} and $h^{-1}(L_s)$, $\eta_{ds}(w_{xy}) = \eta_{ds}(w_x w_y) = \eta_{ds}(w_x)\eta_{ds}(w_y)$, meaning that $f(xy) = f(x)f(y)$.

Now if $xy = 0$, then either xy has no well-formed preimage or xy is a zero for $\pi_s^{-1}(L)$. In the latter case, then $f(x)f(y) = 0$ according to the previous point. If xy has no well-formed preimage, then $w_x w_y$ is not well-formed and consequently $f(x)f(y) = 0$.

The Application f Is Injective. Let $x, y \in \mathcal{I}_E(S_s)$ be such that $x \neq y$. Without loss of generality, we assume that $x \neq 0$. Necessarily, there exist $p, q \in S_s$ such that $pxq \in \eta_s(L_s)$ if, and only if, $pyq \notin \eta_s(L_s)$. Let u and v be words from the preimage of p and q respectively. Then there exists two words $u' \in h^{-1}(u) \cap F_{ds}$ and $v' \in h^{-1}(v) \cap F_{ds}$ such that $u'w_x v' \in L_{ds}$ if, and only if, $u'w_y v' \notin L_{ds}$. Therefore, we have $f(x) \neq f(y)$ and f is injective.

Remark 12. Theorem 11 is only stated for the levels of the quantifier alternation hierarchies that we consider. The main reason for that is that it makes use of Proposition 5 which was also stated for these fragments. Actually, the theorem would hold for any infinitely testable fragment for which we can obtain a result similar to Proposition 5 (see Remark 6).

Discussion. The main result gives the decidability of the alternation hierarchy of $\mathbf{FO}^2[\mathrm{Reg}]$. However, the decidability of this fragment is still an open problem. But one can notice that Proposition 9 proves that $\mathbf{FO}^2[<, +1]$ is infinitely testable, and that Proposition 5 holds. Therefore, Theorem 11 gives the decidability of $\mathbf{FO}^2[\mathrm{Reg}]$ as well. However, we prefer to give an elegant algebraic characterization of this fragment that one could transfer into an equational description. This characterization draws a parallel with the characterization $\mathbf{FO}[\mathrm{Reg}] = \mathbf{QA}$ obtained in [2] and extends the characterization $\mathbf{FO}^2[<, \mathrm{MOD}] = \mathbf{QDA}$ obtained by the authors in [6]. A language L belongs to $\mathbf{LDA}$ if for any idempotent e of S_L, the monoid $eS_L e$ belongs to $\mathbf{DA}$. It belongs to $\mathbf{QLDA}$ if its stable semigroup belongs to $\mathbf{LDA}$.

Theorem 13. $\mathbf{FO}^2[\mathrm{Reg}] = \mathbf{QLDA}$.

Conclusion. In this paper, we proved that regarding the quantifier alternation hierarchy of the first order and its two-variable counterpart, dealing with all the regular numerical predicates is as difficult as dealing with the order predicate only. We chose a generic algebraic approach which introduced a new notion, the infinitely testable property, and proved that for fragments that are expressive enough, the decidability with enriched signature reduces to the simpler one.

While mainly applied to the levels of the quantifier alternation hierarchies, this approach can be used on other fragments that satisfy the same hypotheses, as the fragment $\mathbf{FO}[+1]$. This approach appears in fact to be a part of some more generic results that could also be applied to less expressive fragments. These results stem from the more intricate framework of *varieties of finite categories*, as considered in [4]. In this case, if the delay question is solved, then the decidability of the modular enriched fragment reduces to the decidability of the *global* of the initial variety. It is possible to adapt the definition of the infinitely testable property for varieties of categories, and extend the equational proofs like the one proposed in Example 8 to prove that this property holds. This generalized approach might provide the decidability of the hierarchy $\mathbf{FO}_k^2[<, \mathrm{MOD}]$, which is not covered by our results.

An interesting fact is that despite the different methods used to obtain a delay when adding modular predicates, it was always revealed that the stability index is a delay, even in cases not covered by the approach mentioned above. The question of solving the adding of modular predicate in a general setting seems then achievable, but one has first to solve many questions, like for example what is a good notion of fragment of logic. Surprisingly, a good case of study would be the quite simple fragment **FO**[=]. Indeed, the global of this fragment is not infinitely testable, and it is unknown if it accepts the stability index as a delay.

References

1. Almeida, J.: A syntactical proof of locality of DA. Internat. J. Algebra Comput. **6**(2), 165–177 (1996)
2. Barrington, D.A.M., Compton, K., Straubing, H., Thérien, D.: Regular languages in NC^1. J. Comput. System Sci. **44**(3), 478–499 (1992)
3. Büchi, J.R.: Weak second-order arithmetic and finite automata. Z. Math. Logik Grundlagen Math. **6**, 66–92 (1960)
4. Chaubard, L., Pin, J.-É., Straubing, H.: First order formulas with modular predicates. In: LICS, pp. 211–220. IEEE (2006)
5. Cohen, R.S., Brzozowski, J.A.: Dot-depth of star-free events. J. Comput. Syst. Sci. **5**(1), 1–16 (1971)
6. Dartois, L., Paperman, C.: Two-variable first order logic with modular predicates over words. In: Portier, N., Wilke, T. (eds.) STACS, pp. 329–340. Schloss Dagstuhl - Leibniz-Zentrum fuer Informatik, Dagstuhl (2013)
7. Diekert, V., Gastin, P., Kufleitner, M.: A survey on small fragments of first-order logic over finite words. Internat. J. Found. Comput. Sci. **19**(3), 513–548 (2008)
8. Ésik, Z., Ito, M.: Temporal logic with cyclic counting and the degree of aperiodicity of finite automata. Acta Cybernet. **16**(1), 1–28 (2003)
9. Knast, R.: A semigroup characterization of dot-depth one languages. RAIRO Inform. Théor. **17**(4), 321–330 (1983)
10. Krebs, A., Straubing, H.: An effective characterization of the alternation hierarchy in two-variable logic. In: FSTTCS, pp. 86–98 (2012)
11. Kufleitner, M., Lauser, A.: Quantifier alternation in two-variable first-order logic with successor is decidable. In: STACS, pp. 305–316 (2013)
12. Kufleitner, M., Weil, P.: The FO^2 alternation hierarchy is decidable. In: Computer Science Logic 2012, Volume 16 of LIPIcs. Leibniz Int. Proc. Inform., pp. 426–439. Schloss Dagstuhl. Leibniz-Zent. Inform., Wadern (2012)
13. McNaughton, R., Papert, S.: Counter-free Automata. The M.I.T. Press, Cambridge (1971)
14. Nerode, A.: Linear automaton transformation. Proc. AMS **9**, 541–544 (1958)
15. Péladeau, P.: Logically defined subsets of N^k. Theoret. Comput. Sci. **93**(2), 169–183 (1992)
16. Perrin, D., Pin, J.É.: First-order logic and star-free sets. J. Comput. System Sci. **32**(3), 393–406 (1986)
17. Pin, J.-É.: Syntactic semigroups. In: Rozenberg, G., Salomaa, A. (eds.) Handbook of Formal Languages, vol. 1, pp. 679–746. Springer, Heidelberg (1997)
18. Place, T., Zeitoun, M.: Going higher in the first-order quantifier alternation hierarchy on words. In: Esparza, J., Fraigniaud, P., Husfeldt, T., Koutsoupias, E. (eds.) ICALP 2014, Part II. LNCS, vol. 8573, pp. 342–353. Springer, Heidelberg (2014)

19. Schützenberger, M.P.: On finite monoids having only trivial subgroups. Inf. Control **8**, 190–194 (1965)

20. Simon, I.: Piecewise testable events. In: Brakhage, H. (ed.) Automata Theory and Formal Languages (Second GI Conference Kaiserslautern, 1975). LNCS, vol. 33, pp. 214–222. Springer, Heidelberg (1975)

21. Straubing, H.: A generalization of the Schützenberger product of finite monoids. Theoret. Comput. Sci. **13**(2), 137–150 (1981)

22. Straubing, H.: Finite semigroup varieties of the form $V * D$. J. Pure Appl. Algebra **36**(1), 53–94 (1985)

23. Straubing, H.: Finite Automata, Formal Logic, and Circuit Complexity. Birkhäuser Boston Inc., Boston (1994)

24. Thérien, D.: Classification of finite monoids: the language approach. Theoret. Comput. Sci. **14**(2), 195–208 (1981)

25. Thérien, D., Wilke, T.: Over words, two variables are as powerful as one quantifier alternation. In: STOC 1998 (Dallas. TX), pp. 234–240. ACM, New York (1999)

26. Thomas, W.: Classifying regular events in symbolic logic. J. Comput. System Sci. **25**(3), 360–376 (1982)

27. Tilson, B.: Categories as algebra: an essential ingredient in the theory of monoids. J. Pure Appl. Algebra **48**(1–2), 83–198 (1987)

28. Weis, P., Immerman, N.: Structure theorem and strict alternation hierarchy for FO^2 on words. Log. Methods Comput. Sci. **5**(3:3:4), 23 (2009)

A Note on Decidable Separability by Piecewise Testable Languages

Wojciech Czerwiński[1], Wim Martens[2]([✉]), Lorijn van Rooijen[3],
and Marc Zeitoun[3]

[1] University of Warsaw, Warsaw, Poland
[2] University of Bayreuth, Bayreuth, Germany
wim.martens@uni-bayreuth.de
[3] Bordeaux University, Talence, France

Abstract. The separability problem for word languages of a class C by languages of a class S asks, for two given languages I and E from C, whether there exists a language S from S that includes I and excludes E, that is, $I \subseteq S$ and $S \cap E = \emptyset$. It is known that separability for context-free languages by any class containing all definite languages (such as regular languages) is undecidable. We show that separability of context-free languages by piecewise testable languages is decidable. This contrasts with the fact that testing if a context-free language is piecewise testable is undecidable. We generalize this decidability result by showing that, for every full trio (a class of languages that is closed under rather weak operations) which has decidable diagonal problem, separability with respect to piecewise testable languages is decidable. Examples of such classes are the languages defined by labeled vector addition systems and the languages accepted by higher order pushdown automata of order two. The proof goes through a result which is of independent interest and shows that, for *any* kind of languages I and E, separability can be decided by testing the existence of common patterns in I and E.

1 Introduction

We say that language I can be *separated* from E by language S if S includes I and excludes E, that is, $I \subseteq S$ and $S \cap E = \emptyset$. In this case, we call S a *separator*. We study the *separability problem* of classes C by classes S:

Given: Two languages I and E from a class C
Question: Can I and E be separated by some language from S?

Separability is a classical problem in mathematics and computer science that recently found much new interest. For example, recent work investigated the separability problem of regular languages by piecewise testable languages [10, 26], locally testable and locally threshold testable languages [25] or by first order

This work was supported by DFG grant MA 4938/2-1, by Poland's National Science Centre grant no. UMO-2013/11/D/ST6/03075, and Agence Nationale de la Recherche ANR 2010 BLAN 0202 01 FREC.

© Springer International Publishing Switzerland 2015
A. Kosowski and I. Walukiewicz (Eds.): FCT 2015, LNCS 9210, pp. 173–185, 2015.
DOI: 10.1007/978-3-319-22177-9_14

definable languages [28]. Another recent example, which uses separation and goes beyond regularity, is the proof of Leroux [19] for the decidability of reachability for vector addition systems or Petri nets. It greatly simplifies earlier proofs by Mayr [21] and Kosaraju [18].

In this paper we focus on the theoretical underpinnings of separation by piecewise testable languages. Our interest in piecewise testable languages is mainly because of the following two reasons. First, it was shown recently [10,26] that separability of regular languages (given by their non-deterministic automata) by piecewise testable languages is in PTIME. We found the tractability of this problem to be rather surprising.

Second, piecewise testable languages form a very natural class in the sense that they only reason about the *order* of symbols. More precisely, they are finite Boolean combinations of regular languages of the form $A^* a_1 A^* a_2 A^* \cdots A^* a_n A^*$ in which $a_i \in A$ for every $i = 1, \ldots, n$ [30]. We are investigating to which extent piecewise testable languages and fragments thereof can be used for computing simple explanations for the behavior of complex systems [15].

Separation and Characterization. For classes C effectively closed under complement, separation of C by S is a natural generalization of the *characterization problem* of C by S, which is defined as follows. For a given language $L \in C$ decide whether L is in S. Indeed, L is in S if and only if L can be separated from its complement by a language from S. The characterization problem is well studied. The starting points were famous works of Schützenberger [29] and Simon [30], who solved it for the regular languages by the first-order definable languages and piecewise testable languages, respectively. There were many more results showing that, for regular languages and a subclass thereof (often characterized by a given logic), the problem is decidable, see for example [4,22,24,27,32,34]. Similar problems have been considered for trees [3,5,7,8].

Decidability. To the best of our knowledge, all the above work and in general all the decidable characterizations were obtained in cases where C is the class of regular languages, or a subclass of it. This could be due to several negative results which may seem to form a barrier for any nontrivial decidability beyond regular languages. For a context-free language (given by a grammar or a pushdown automaton) it is undecidable to determine whether it is a regular language, by Greibach's theorem [14]. Furthermore, it is also undecidable to determine whether a given context-free languIage is piecewise testable.

Concerning context-free languages, there is a strong connection between the intersection emptiness problem and separability. Trivially, testing intersection emptiness of two given context-free languages is the same as deciding if they can be separated by some context-free language. However, in general, the negative result is even more overwhelming. Szymanski and Williams [33] proved that separability of context-free languages by regular languages is undecidable. This was then generalized by Hunt [16], who proved that separability of context-free languages by any class containing all the *definite languages* is undecidable. A language L is *definite* if it can be written as $L = F_1 A^* \cup F_2$, where F_1 and F_2 are finite languages over alphabet A. As such, for definite languages, it can be decided whether a given word w belongs to L by looking at the prefix of w of a given fixed length. (The same statement holds for *reverse definite* languages, in which we are looking at suffixes.) Containing all the definite, or reverse definite, languages is a very weak condition. Note that if a logic can test what is the i-th letter of a word and is closed under boolean combinations, it already can define all the definite languages. In his paper, Hunt makes an explicit link between intersection emptiness and separability. Hunt writes: *"We show that separability is undecidable in general for the same reason that the emptiness-of-intersection problem is undecidable. Therefore, it is unlikely that separability can be used to circumvent the undecidability of the emptiness-of-intersection problem."*

Our Contribution. In this paper, we show that the above mentioned quote does not apply for separability by piecewise testable languages (PTLs): we show that it can be decided whether two given context-free languages are separable by a PTL. This may come as a surprise in the light of the undecidability results we already discussed.

In fact, we prove a stronger result that implies that separability by PTLs is also decidable for some rather expressive classes such as Petri net languages (also known as labeled vector addition system languages). This result is an equivalence between decidability of separability by PTLs and decidability of a problem that we call *diagonal problem*. One direction of the equivalence is proved here: First we show that (arbitrary) languages I and E are not separable by PTL if and only if they possess a certain *common pattern*. Then, we use this fact to reduce to the diagonal problem. The other direction of the equivalence is due to Georg Zetzsche [35].

A curiosity of this work is perhaps the absence of algebraic methods. Most decidability results we are aware of have considered syntactic monoids of regular languages and investigated properties thereof. The exceptions are the recent studies of separability of regular languages by piecewise testable languages (e.g., [10,26]). However, since the algebraic framework for regular languages is so rich, some may not find it clear whether [10,26] do or do not rely on algebraic methods; perhaps simply in a rephrased way. Here, the situation is different in the sense that for context-free languages the syntactic monoid is infinite and it is difficult to design any algebraic framework for them. So the work shows that it is not always necessary to use algebraic techniques to prove separability questions.

2   Preliminaries

The set of all integers and nonnegative integers are denoted by $\mathbb{Z}$ and $\mathbb{N}$ respectively. A *word* is a concatenation $w = a_1 \cdots a_n$ of symbols a_i that come from a finite alphabet A. The *length* of w is n, the number of its symbols. The *alphabet of w* is the set $\{a_1, \ldots, a_n\}$ and is denoted $\mathsf{alph}(w)$. For a subalphabet $B \subseteq A$, a word $v \in A^*$ is a *B-subsequence* of w, denoted $v \preceq_B w$, if $v = b_1 \cdots b_m$ and $w \in B^* b_1 B^* \cdots B^* b_m B^*$. (We do not require that $\{b_1, \ldots, b_m\} \subseteq B$ or $B \subseteq \{b_1, \ldots, b_m\}$.) We refer to the relation $\preceq_A$ as the *subsequence* relation and denote it by $\preceq$. A regular word language over alphabet A is a *piece language* if it is of the form $A^* a_1 A^* \cdots A^* a_n A^*$ for some $a_1, \dots, a_n \in A$, that is, it is the set of words having $a_1 \cdots a_n$ as a subsequence. A regular language is a *piecewise testable language* if it is a (finite) boolean combination of piece languages. The class of all piecewise testable languages is denoted PTL.

Separability and Common Patterns. The first main result of the paper proves that two (not necessarily regular) languages are not separable by PTL if and only if they have a common subpattern. We now make this more precise.

A *factorization pattern* is an element of $(A^*)^{p+1} \times (2^A \setminus \emptyset)^p$ for some $p \geq 0$. In other terms, if $(\overrightarrow{u}, \overrightarrow{B})$ is such a factorization pattern, there exist words $u_0, \dots, u_p \in A^*$ and nonempty alphabets $B_1, \dots, B_p \subseteq A$ such that $\overrightarrow{u} = (u_0, \dots, u_p)$ and $\overrightarrow{B} = (B_1, \dots, B_p)$. For $B \subseteq A$, we denote by $B^\circledast$ the set of words with alphabet exactly B, that is, $B^\circledast = \{w \in B^* \mid \mathsf{alph}(w) = B\}$. Given a factorization pattern $(\overrightarrow{u}, \overrightarrow{B})$, with $\overrightarrow{u} = (u_0, \dots, u_p)$ and $\overrightarrow{B} = (B_1, \dots, B_p)$, let

$$\mathcal{L}(\overrightarrow{u}, \overrightarrow{B}, n) = u_0 (B_1^\circledast)^n u_1 \cdots u_{p-1} (B_p^\circledast)^n u_p.$$

In other terms, in a word of $\mathcal{L}(\overrightarrow{u}, \overrightarrow{B}, n)$, the infix between u_{k-1} and u_k is required to be the concatenation of n words, each containing *all* letters of B_k (for each $1 \leq k \leq p$). A sequence $(w_n)_n$ is said $(\overrightarrow{u}, \overrightarrow{B})$-*adequate* if

$$\forall n \in \mathbb{N}, \ w_n \in \mathcal{L}(\overrightarrow{u}, \overrightarrow{B}, n).$$

Finally, language L *contains the pattern* $(\overrightarrow{u}, \overrightarrow{B})$ if there exists an infinite sequence of words $(w_n)_n$ in L that is $(\overrightarrow{u}, \overrightarrow{B})$-adequate. We prove the following Theorem in Sect. 3:

Theorem 1. *Two word languages I and E are not separable by PTL if and only if they contain a common pattern $(\overrightarrow{u}, \overrightarrow{B})$.*

A Characterization for Decidable Separability. The second main result is an algorithm that decides separability for *full trios* that have a decidable *diagonal* problem. Full trios, also called *cones*, are language classes that are closed under rather weak operations [6,12].

Fix a language L over alphabet A. For an alphabet B, the *B-projection* of a word is its longest subsequence consisting of symbols from B. The *B-projection* of a language L is the set of all B-projections of words belonging to L. Therefore, the B-projection of L is a language over alphabet $A \cap B$. The *B-upward closure* of a language L is the set of all words that have a B-subsequence in L, i.e.,

$$\{w \in (A \cup B)^* \mid \exists v \in L \text{ such that } v \preceq_B w\}.$$

In other words, the B-upward closure of L consists of all words that can be obtained from taking a word in L and padding it with symbols from B.

A class of languages C is *closed* under an operation OP if $L \in C$ implies that $\text{OP}(L) \in C$. We use term *effectively closed* if, furthermore, the representation of $\text{OP}(L)$ can be effectively computed from the representation of L.

A nonempty class C of languages is a *full trio* if it is effectively closed under:

1. B-projection for every finite alphabet B,
2. B-upward closure for every finite alphabet B, and
3. intersection with regular languages.

We note that full trios are usually defined differently (through closures under homomorphisms or rational transductions [6,12]) but we use the above mentioned properties in the proofs and they are easily seen to be equivalent.

The problem that we will require to be decidable is the *diagonal problem*, which we explain next. Let $A = \{a_1, \ldots, a_n\}$. For a symbol $a \in A$ and a word $w \in A^*$, let $\#_a(w)$ denote the number of occurrences of a in w. The *Parikh image* of a word w is the n-tuple $(\#_{a_1}(w), \ldots, \#_{a_n}(w))$. The *Parikh image* of a language L is the set of all Parikh images of words from L. A tuple $(m_1, \ldots, m_n) \in \mathbb{N}^n$ is *dominated* by a tuple $(d_1, \ldots, d_n) \in \mathbb{N}^n$ if $d_i \geq m_i$ for every $i = 1, \ldots, n$. The *diagonal problem* for language L asks whether there exist infinitely many $m \in \mathbb{N}$ such that the tuple $(m, \ldots, m)$ is dominated by some tuple in the Parikh image of L. We are now ready to state the second main result:

Theorem 2. *For each full trio C, the diagonal problem for C is decidable if and only if separability of C by PTL is decidable.*

In Sect. 4 we present an algorithm to decide separability for full trios that have a decidable diagonal problem, showing one direction of the equivalence. The algorithm does not rely on semilinearity of Parikh images. For example, in Sect. 5 we apply the lemma to Vector Addition System languages, which do not have a semilinear Parikh image. Very recently, we were informed by Georg Zetzsche [35] that the other implication also holds and that, therefore, there actually is an equivalence.

3   Common Patterns

In this section we prove Theorem 1. We say that a sequence is *adequate* if it is $(\overrightarrow{u}, \overrightarrow{B})$-adequate for some factorization pattern. The following statement can be shown using Simon's Factorization Forest Theorem [31].

Lemma 3. *Every sequence $(w_n)_n$ of words admits an adequate subsequence.*

For a word w, denote its first (resp., last) letter by $\text{first}(w)$, (resp., $\text{last}(w)$). We call a factorization pattern $(\overrightarrow{u}, \overrightarrow{B}) = ((u_0, \ldots, u_p), (B_1, \ldots, B_p))$ *proper* if (i) for all i, $\text{last}(u_i) \notin B_{i+1}$ and $\text{first}(u_i) \notin B_i$, and (ii) for all i, $u_i = \varepsilon \Rightarrow (B_i \not\subseteq B_{i+1}$ and $B_{i+1} \not\subseteq B_i)$.

Note that if a sequence $(w_n)_n$ is adequate, then there exists a proper factorization pattern $(\overrightarrow{u}, \overrightarrow{B})$ such that $(w_n)_n$ is $(\overrightarrow{u}, \overrightarrow{B})$-adequate. This is easily seen from the following observations and their symmetric counterparts:

$$u = a_1 \cdots a_k \text{ and } a_k \in B \Rightarrow a_1 \cdots a_k (B^{\circledast})^n \subseteq a_1 \cdots a_{k-1} (B^{\circledast})^n,$$
$$B_{i-1} \subseteq B_i \Rightarrow (B_{i-1}{}^{\circledast})^n (B_i{}^{\circledast})^n \subseteq (B_i{}^{\circledast})^n.$$

The following lemma gives a condition under which two sequences share a factorization pattern and is very similar to [2, Theorem 8.2.6]. In its statement, we write $v \sim_n w$ for two words v and w if they have the same subsequences up to length n, that is, for every word u of length at most n, $u \preceq v$ iff $u \preceq w$.

Lemma 4. *Let* $(\overrightarrow{u}, \overrightarrow{B})$ *and* $(\overrightarrow{t}, \overrightarrow{C})$ *be proper factorization patterns. Let* $(v_n)_n$ *and* $(w_n)_n$ *be two sequences of words such that*

- $(v_n)_n$ *is* $(\overrightarrow{u}, \overrightarrow{B})$-*adequate*
- $(w_n)_n$ *is* $(\overrightarrow{t}, \overrightarrow{C})$-*adequate*
- $v_n \sim_n w_n$ *for every* $n \geq 0$.

Then, $\overrightarrow{u} = \overrightarrow{t}$ *and* $\overrightarrow{B} = \overrightarrow{C}$.

Now we are equipped to prove Theorem 1. We only show the "only if" direction here, due to space restrictions.

Proof (of Theorem 1, "only-if"). It is not difficult to see that I and E are *not* PTL-separable iff for every $n \in \mathbb{N}$, there exist $v_n \in I$ and $w_n \in E$ such that $v_n \sim_n w_n$. This defines an infinite sequence of pairs $(v_n, w_n)_n$, from which we will iteratively extract infinite subsequences to obtain additional properties, while keeping $\sim_n$-equivalence.

By Lemma 3, one can extract from $(v_n, w_n)_n$ a subsequence whose first component forms an adequate sequence. From this subsequence of pairs, using Lemma 3 again, we extract a subsequence whose second component is also adequate (note that the first component remains adequate). Therefore, one can assume that both $(v_n)_n$ and $(w_n)_n$ are themselves adequate. This means there exist proper factorization patterns for which $(v_n)_n$ resp. $(w_n)_n$ are adequate. Lemma 4 shows that one can choose the *same* proper factorization pattern $(\overrightarrow{u}, \overrightarrow{B})$ such that both $(v_n)_n$ and $(w_n)_n$ are $(\overrightarrow{u}, \overrightarrow{B})$-adequate. This means that I and E contain a common pattern $(\overrightarrow{u}, \overrightarrow{B})$.   □

4 The Algorithm for Separability

We prove one direction of Theorem 2 by showing that, for full trios with decidable diagonal problem, we can decide separability by PTL. Fix two languages I and E from a full trio $\mathcal{C}$ which has decidable diagonal problem.

To test whether I is separable from E by a piecewise testable language S, we run two semi-procedures in parallel. The *positive* one looks for a witness that

I and E are separable by PTL, whereas the *negative* one looks for a witness that they are *not* separable by a PTL. Since one of the semi-procedures always terminates, we have an effective algorithm that decides separability. It remains to describe the two semi-procedures.

Positive Semi-procedure. We first note that, when a full trio has decidable diagonal problem, it also has decidable emptiness[1]. The positive semi-procedure enumerates all PTLs over the union of the alphabets of I and E. For every PTL S it checks whether S is a separator, so if $I \subseteq S$ and $E \cap S = \emptyset$. The first test is equivalent to $I \cap (A^* \setminus S) = \emptyset$. Thus both tests boil down to checking whether the intersection of a language from the class $\mathcal{C}$ (I or E, respectively) and a regular language (S and $A^* \setminus S$, respectively) is empty. This is decidable, as $\mathcal{C}$ is effectively closed under taking intersections with regular languages and has decidable emptiness problem.

Negative Semi-procedure. Theorem 1 shows that there is always a finite witness for inseparability: a pattern $(\overrightarrow{u}, \overrightarrow{B})$. The negative semi-procedure enumerates all possible patterns and for each one, checks the condition of Theorem 1. We now show how to test this condition, i.e., for a pattern $(\overrightarrow{u}, \overrightarrow{B})$ test whether for all $n \in \mathbb{N}$ the intersection of $\mathcal{L}(\overrightarrow{u}, \overrightarrow{B}, n)$ with both I and E is nonempty. *Checking the condition.* Here we show for an arbitrary language from $\mathcal{C}$ how to check whether for all $n \in \mathbb{N}$ its intersection with the language $\mathcal{L}(\overrightarrow{u}, \overrightarrow{B}, n)$ is nonempty. Fix $L \in \mathcal{C}$ over an alphabet A and a pattern $(\overrightarrow{u}, \overrightarrow{B})$, where $\overrightarrow{u} = (u_0, \ldots, u_k)$ and $\overrightarrow{B} = (B_1, \ldots, B_k)$. Intuitively, we just consider a diagonal problem with some artifacts: we are counting the number of occurrences of alphabets B_i and checking whether those numbers can simultaneously become arbitrarily big.

We show decidability of the non-separability problem by a formal reduction to the diagonal problem. We perform a sequence of steps. In every step we will slightly modify the considered language L and appropriately customize the condition to be checked. Using the closure properties of the class $\mathcal{C}$ we will assure that the investigated language still belongs to $\mathcal{C}$.

First we add special symbols $\$_i$, for $i \in \{1, \ldots, k\}$, which do not occur in A. These symbols are meant to count how many times alphabet B_i is "fully occurring" in the word. Then we will assure that words are of the form

$$u_0 (B_1 \cup \{\$_1\})^* u_1 \cdots u_{k-1} (B_k \cup \{\$_k\})^* u_k,$$

which already is close to what we need for the pattern. Then we will check that between every two symbols $\$_i$ (with the same i), every symbol from B_i occurs, so that the $\$_i$ are indeed counting the number of iterations through the entire alphabet B_i. Finally we will remove all the symbols except those

---

[1] Emptiness of L over alphabet A can be decided by taking the $\{x\}$-upward closure of L, where $x \notin A$, intersecting the resulting language with the regular language $(A \cup \{x\})^* A (A \cup \{x\})^*$, and then taking the $\{x\}$-projection. In the resulting language, the diagonal problem returns true iff L is nonempty [36].

from $\{\$_1, \ldots, \$_k\}$. The resulting language will contain only words of the form $\$_1^* \$_2^* \cdots \$_k^*$ and the condition to be checked will be exactly the diagonal problem.

More formally, let $L_0 := L$. We modify iteratively L_0, resulting in $L_1, L_2, L_3,$ and L_4. Each of them will be in $\mathcal{C}$ and we describe them next.

Language L_1 is the $\{\$_1, \ldots, \$_k\}$-upward closure of L_0. Thus, L_1 contains, in particular, all words where the $\$_i$ are placed "correctly", i.e., in between two $\$_i$-symbols the whole alphabet B_i should occur. However at this moment we do not check it. By closure under B-upward closures, language L_1 belongs to $\mathcal{C}$.

Note that L_1 also contains words in which the $\$_i$-symbols are placed totally arbitrary. In particular, they can occur in the wrong order. The idea behind L_2 is to consider only those words in which the $\$_i$-symbols were guessed at least in the good areas. Concretely, L_2 is an intersection of L_1 with the language

$$u_0 \left(B_1 \cup \{\$_1\}\right)^* u_1 \cdots u_{k-1} \left(B_k \cup \{\$_k\}\right)^* u_k.$$

By the closure under intersection with regular languages, L_2 belongs to $\mathcal{C}$.

Language L_2 still may contain words, such that in between two $\$_i$-symbols not *all* the symbols from B_i occur. We get rid of these by intersecting L_2 with

$$u_0(\$_1 B_1^{\circledast})^* \$_1 u_1 \cdots u_{k-1}(\$_k B_k^{\circledast})^* \$_k u_k.$$

As such, we obtain L_3 which, again by closure under intersection with regular languages, belongs to $\mathcal{C}$.[2]

Note that intersection of $L = L_0$ with the language $\mathcal{L}(\overrightarrow{u}, \overrightarrow{B}, n)$ is nonempty if and only if L_3 contains a word with precisely $n + 1$ symbols $\$_i$ for every $i \in \{1, \ldots, k\}$. Indeed, L_3 just contains the (slightly modified versions of) words from L_0 which fit into the pattern and in which the symbols $\$_i$ "count" occurrences of $B_i^{\circledast}$. Furthermore, for every word in L_3, the word obtained by removing some occurrences of some $\$_i$ is in L_3 as well. It is thus enough to focus on the $\$_i$-symbols. Language L_4 is therefore the $\{\$_1, \ldots, \$_k\}$-projection of L_3. By the closure under projections, language L_4 belongs to $\mathcal{C}$. The words contained in L_4 are therefore of the form

$$\$_1^{a_1} \cdots \$_k^{a_k},$$

such that there exists $w \in L$ with at least $a_i - 1$ occurrences of $B_i^{\circledast}$. Therefore intersection of L with $\mathcal{L}(\overrightarrow{u}, \overrightarrow{B}, n)$ is nonempty for all $n \geq 0$ if and only if the tuple $(n, \ldots, n)$ belongs to the Parikh image of L_4 for infinitely many $n \geq 0$. This is precisely the diagonal problem, which we know to be decidable for $\mathcal{C}$.

5   Decidable Classes

In this section we show that separability by piecewise testable languages is decidable for a wide range of classes, by proving that they meet the conditions of Theorem 2. In particular, we show this for context-free languages, languages of

---

[2] Of course, one could also immediately obtain L_3 from L_1 by performing a single intersection with a regular language.

labeled vector addition systems (which are the same as languages of labeled Petri nets). We comment also on other natural classes of languages containing all the regular languages.

Theorem 5. *Separability by piecewise testable languages is decidable for*

1. *context-free languages; and for*
2. *languages of labeled vector addition systems.*

Our approach also allows to mix the above scenarios. That is, separability of a context-free language from the language of a labeled vector addition system is also decidable. In the remainder of this section, we prove the theorem.

Context-Free Languages. Context-free languages are well-known to be a full trio. The only nontrivial condition is deciding the diagonal problem. A set $S \subseteq \mathbb{N}^k$ is *linear* if it is of the form

$$S = \{v + n_1 v_1 + \ldots + n_m v_m \mid n_1, \ldots, n_m \in \mathbb{N}\}$$

for some *base* vector $v \in \mathbb{N}^k$ and *period* vectors $v_1, \ldots, v_m \in \mathbb{N}^k$. A *semilinear* set is a finite union of linear sets. Parikh's theorem [23] states that the Parikh image of a context-free language is semilinear, moreover the computation of its description as a (finite) union of linear sets if effective[3]. It is enough to check whether for infinitely many $n \geq 0$ the mentioned semilinear set contains a tuple that dominates $(n, \ldots, n)$.

Semilinear sets are exactly these, which can be defined by Presburger logic. Moreover, the translation can be done effectively. Assume that $|A| = k$, so the Parikh image P of the considered language is a subset of $\mathbb{N}^k$ and ϕ is a Presburger formula describing P having exactly k free variables. Then

$$\psi = \forall_{n \in \mathbb{N}} \exists_{x_1, x_2, \ldots, x_k} \left(\bigwedge_{i \in \{1, \ldots, k\}} (x_i \geq n) \right) \wedge \phi(x_1, x_2, \ldots, x_k)$$

is true if and only if the diagonal problem for the considered language is answered positively. Decidability of the Presburger logic finishes the proof of decidability of the diagonal problem for context-free languages. We refer for the details of semilinear sets and Presburger logic to [13]. Finally, by Theorem 2, separability for context-free languages by piecewise testable languages is decidable.

Languages of Labeled Vector Addition Systems and Petri Nets. A k-dimensional *labeled vector addition system*, or *labeled VAS* $M = (A, T, \ell, s, t)$ over alphabet A consists of a set of *transitions* $T \subseteq \mathbb{Z}^k$, a labeling $\ell : T \to A \cup \{\varepsilon\}$, where ε stands for the empty word and *source* and *target* vectors $s, t \in \mathbb{N}^k$. A labeled VAS defines a transition relation on the set $\mathbb{N}^k$ of *markings*. For two markings $u, v \in \mathbb{N}^k$ we write $u \xrightarrow{a} v$ if there is $r \in T$ such that $u + r = v$ and $\ell(r) = a$, where the addition of vectors is defined as an addition on every

---

[3] A simple proof of this fact can be found in [11].

coordinate. For two markings $u, v \in \mathbb{N}^k$ we say that u *reaches* v *via a word* w if there is a sequence of markings $u_0 = u, u_1, \ldots, u_{n-1}, u_n = v$ such that $u_i \xrightarrow{a_i} u_{i+1}$ for all $i \in \{0, \ldots, n-1\}$ and $w = a_0 \cdots a_{n-1}$. For a given labeled VAS M the *language* of M, denoted $L(M)$, is the set of all words $w \in A^*$ such that source reaches target via w. We note that languages of labeled VASs are the same as languages of labeled Petri nets.

Since labeled VAS languages are known to be a full trio [17], we only need to prove decidability of the diagonal problem. First we will show that it is enough to consider VASs in which the target marking equals $(0, \ldots, 0)$. To this end, let M be a k-dimensional labeled VAS with source vector $s = (s_1, \ldots, s_k)$ and target vector $t = (t_1, \ldots, t_k)$. We transform M to a new VAS M' in which we add two auxiliary coordinates, called *life* coordinates. The source coordinate is enriched by 0 on one life coordinate and by 1 on the other one, so it is $s' = (s_1, \ldots, s_k, 0, 1) \in \mathbb{N}^{k+2}$. Every original transition has two copies. One of these transitions subtracts one from the first life coordinate and adds one to the second life coordinate, the second transition does the opposite. Note that nonemptiness of life coordinates serve just as a necessary condition for firing any transition, as every transition subtracts one from one of these coordinates. Therefore, the original source marking s reaches the original target marking t via the same set of words by which the new source marking s' reaches either $(t_1, \ldots, t_k, 0, 1)$ or $(t_1, \ldots, t_k, 1, 0)$. We add also two *final* transitions, which subtract the original target vector, subtract one from one of the life coordinates and are labeled ε. Therefore, s can reach t by a word w in M if and only if s' can reach 0^{k+2} by w in M'. Indeed, the implication from left to right is immediate. On the other hand, in order to reach the marking 0^{k+2} in M', the last transition has to be the final transition, so implication from right to left also holds. Thus it is enough to solve the diagonal problem for VASs in which the target marking is $(0, \ldots, 0)$.

We will show that this diagonal problem is decidable by a reduction to the place-boundedness problem for VASs with one zero test, which is decidable due to Bonnet et al. [9]. We modify the considered VAS in the following way. For every letter $a \in A$ we add a new *letter-coordinate*, which is counting how many times we read the letter a, that is, for every transition which is labeled by $a \in A$, we add 1 in the letter-coordinate corresponding to a and 0 in the letter-coordinates corresponding to other letters. The set of letter-coordinates computes the Parikh image of a word. We also add one new *minimum-coordinate* and a new transition, labeled by ε, which subtracts one from all the letter-coordinates and adds one to the minimum-coordinate. It is easy to see that minimum-coordinate can maximally reach the minimum number from the Parikh image tuple. Additionally, for every letter-coordinate we add a transition, labeled by ε, which can decrease this coordinate by one. The diagonal problem for the original VAS is equivalent to the question whether for infinitely many $n \geq 0$ the source marking, enriched by zeros in the new coordinates, reaches a marking $(0, \ldots, 0, n)$, with zeros everywhere beside the minimum-coordinate with number n. This can be easily reduced to the place-boundedness for a VAS with one zero test. We do not show the details. Intuitively, the zero test checks whether there are zeros everywhere else than the minimum-coordinate and we check whether under this condition the minimum-coordinate can get unbounded. This finishes the proof of decidability of the diagonal problem for labeled VASs.

Other Classes. Among another natural language classes extending regular languages one can think about context-sensitive languages. Unfortunately context-sensitive languages do not meet the conditions of Theorem 2, as they are no full trio, nor is their emptiness problem decidable.

Very recently, Zetzsche [36] showed that *indexed languages* [1] or, equivalently, languages accepted by higher-order pushdown automata of order two [20] fulfill the conditions of Theorem 2 and therefore have decidable separability by PTL. His proof showing that indexed languages have a decidable diagonal problem is much more involved than the one for context-free languages we presented here. This shows that separability of languages definable by pushdown automata of order two by PTL is decidable as well. It would be interesting to know if it is decidable for pushdown automata for even higher order as well.

6   Concluding Remarks

Since the decidability results we presented seem to be in strong contrast with the remark of Hunt in the introduction, we briefly comment on this. What we essentially do is show that undecidable emptiness-of-intersection for a class C does not always imply undecidability for separability of C with respect to some nontrivial class of languages. In the case of separability with respect to piecewise testable languages, the main reason is basically that we only need to construct intersections of languages from C with languages that are regular (or even piecewise testable). Here, the fact that such intersections can be effectively constructed, together with decidable emptiness and diagonal problems seem to be sufficient for decidability.

Regarding future work, we see many interesting directions and new questions. Which language classes have a decidable diagonal problem? Which other characterizations are there for decidable separability by PTL? Can Theorem 2 be extended to also give complexity guarantees? Can we find similar characterizations for separability by subclasses of PTL, as considered in [15]?

Acknowledgments. We would like to thank Tomáš Masopust for pointing us to [16] and Thomas Place for pointing out to us that determining if a given context-free language is piecewise testable is undecidable. We are also grateful to the anonymous reviewers for many helpful remarks that simplified proofs. We are much indebted to Georg Zetzsche for many useful remarks and most of all for sending us a simple proof that showed that, for full trios, separability by PTL implies decidability of the diagonal problem, thereby turning Theorem 2 into an equivalence. We plan to incorporate his proof in the full version of this paper.

References

1. Aho, A.V.: Indexed grammars – an extension of context-free grammars. J. ACM **15**(4), 647–671 (1968)

2. Almeida, J.: Finite Semigroups and Universal Algebra, Volume 3 of Series in Algebra. World Scientific Publishing Company, Singapore (1994)
3. Antonopoulos, T., Hovland, D., Martens, W., Neven, F.: Deciding twig-definability of node selecting tree automata. In: ICDT, pp. 61–73 (2012)
4. Arfi, M.: Polynomial operations on rational languages. In: STACS, pp. 198–206 (1987)
5. Benedikt, M., Segoufin, L.: Regular tree languages definable in FO and in FO_{mod}. ACM Trans. Comput. Logi $11(1)$, 4:1–4:32 (2009)
6. Berstel, J.: Transductions and context-free languages. Teubner, Stuttgart (1979)
7. Bojańczyk, M., Idziaszek, T.: Algebra for infinite forests with an application to the temporal logic EF. In: Bravetti, M., Zavattaro, G. (eds.) CONCUR 2009. LNCS, vol. 5710, pp. 131–145. Springer, Heidelberg (2009)
8. Bojanczyk, M., Segoufin, L., Straubing, H.: Piecewise testable tree languages. LMCS $8(3)$, 1–20 (2012)
9. Bonnet, R., Finkel, A., Leroux, J., Zeitoun, M.: Model checking vector addition systems with one zero-test. LMCS $8(2)$, 1–25 (2012)
10. Czerwiński, W., Martens, W., Masopust, T.: Efficient separability of regular languages by subsequences and suffixes. In: Fomin, F.V., Freivalds, R., Kwiatkowska, M., Peleg, D. (eds.) ICALP 2013, Part II. LNCS, vol. 7966, pp. 150–161. Springer, Heidelberg (2013)
11. Esparza, J., Ganty, P., Kiefer, S., Luttenberger, M.: Parikh's theorem: a simple and direct automaton construction. Inf. Process. Lett. $111(12)$, 614–619 (2011)
12. Ginsburg, S., Greibach, S.A.: Abstract families of languages. In: SWAT / FOCS, pp. 128–139 (1967)
13. Ginsburg, S., Spanier, E.H.: Semigroups, Presburger formulas, and languages. Pacific J. Math. 16, 285–296 (1966)
14. Greibach, S.: A note on undecidable properties of formal languages. Math. Sys. Theor. $2(1)$, 1–6 (1968)
15. Hofman, P., Martens, W.: Separability by short subsequences and subwords. In: ICDT, pp. 230–246 (2015)
16. Hunt III, H.B.: On the decidability of grammar problems. J. ACM $29(2)$, 429–447 (1982)
17. Jantzen, M.: On the hierarchy of Petri net languages. RAIRO Informatique Théorique $13(1)$, 19–30 (1979)
18. Kosaraju, S.R.: Decidability of reachability in vector addition systems (preliminary version). In: STOC, pp. 267–281 (1982)
19. Leroux, J.: The general vector addition system reachability problem by Presburger inductive invariants. LMCS $6(3)$, 1–25 (2010)
20. Maslov, A.N.: Multilevel stack automata. Probl. Inf. Transm. $12(1)$, 38–42 (1976)
21. Mayr, E.W.: An algorithm for the general Petri net reachability problem. SIAM J. Comput. $13(3)$, 441–460 (1984)
22. McNaughton, R.: Algebraic decision procedures for local testability. Math. Syst. Theor. $8(1)$, 60–76 (1974)
23. Parikh, R.: On context-free languages. J. ACM $13(4)$, 570–581 (1966)
24. Pin, J.-E., Weil, P.: Polynomial closure and unambiguous product. Theory Comput. Syst. $30(4)$, 383–422 (1997)
25. Place, T., van Rooijen, L., Zeitoun, M.: Separating regular languages by locally testable and locally threshold testable languages. In: FSTTCS, pp. 363–375 (2013)
26. Place, T., van Rooijen, L., Zeitoun, M.: Separating regular languages by piecewise testable and unambiguous languages. In: Chatterjee, K., Sgall, J. (eds.) MFCS 2013. LNCS, vol. 8087, pp. 729–740. Springer, Heidelberg (2013)

27. Place, T., Zeitoun, M.: Going higher in the first-order quantifier alternation hierarchy on words. In: Esparza, J., Fraigniaud, P., Husfeldt, T., Koutsoupias, E. (eds.) ICALP 2014, Part II. LNCS, vol. 8573, pp. 342–353. Springer, Heidelberg (2014)

28. Place, T., Zeitoun, M.: Separating regular languages with first-order logic. In: CSL-LICS, pp. 75:1–75:10. ACM (2014)

29. Schützenberger, M.P.: On finite monoids having only trivial subgroups. Inf. Control **8**(2), 190–194 (1965)

30. Simon, I.: Piecewise testable events. In: Simon, I. (ed.) Automata Theory and Formal Languages. LNCS, pp. 214–222. springer, Heidelberg (1975)

31. Simon, I.: Factorization forests of finite height. Theor. Comput. Sci. **72**(1), 65–94 (1990)

32. Straubing, H.: Semigroups and languages of dot-depth two. Theor. Comput. Sci. **58**, 361–378 (1988)

33. Szymanski, T., Williams, J.: Noncanonical extensions of bottom-up parsing techniques. SIAM J. Comput. **5**(2), 231–250 (1976)

34. Zalcstein, Y.: Locally testable languages. J. Comput. Syst. Sci. **6**(2), 151–167 (1972)

35. Zetzsche, G.: Personal communication

36. Zetzsche, G.: An approach to computing downward closures. In: ICALP (2015). To appear, Accessed on http://arxiv.org/abs/1503.01068

Set Algorithms, Covering, and Traversal

Multidimensional Binary Vector Assignment Problem: Standard, Structural and Above Guarantee Parameterizations

Marin Bougeret[1], Guillerme Duvillié[1], Rodolphe Giroudeau[1], and Rémi Watrigant[2(✉)]

[1] LIRMM, Université Montpellier 2, Montpellier, France
{bougeret,duvillie,rgirou}@lirmm.fr
[2] Computing Department, Hong Kong Polytechnic University,
Hong Kong, China
csrwatrigant@comp.polyu.edu.hk

Abstract. In this article we focus on the parameterized complexity of the Multidimensional Binary Vector Assignment problem (called bMVA). An input of this problem is defined by m disjoint sets $V^1, V^2, \ldots, V^m$, each composed of n binary vectors of size p. An output is a set of n disjoint m-tuples of vectors, where each m-tuple is obtained by picking one vector from each set V^i. To each m-tuple we associate a p dimensional vector by applying the bit-wise AND operation on the m vectors of the tuple. The objective is to minimize the total number of zeros in these n vectors. bMVA can be seen as a variant of multidimensional matching where hyperedges are implicitly locally encoded via labels attached to vertices, but was originally introduced in the context of integrated circuit manufacturing.

We provide for this problem FPT algorithms and negative results (ETH-based results, $W[2]$-hardness and a kernel lower bound) according to several parameters: the standard parameter k (*i.e.* the total number of zeros), as well as two parameters above some guaranteed values.

1 Introduction

1.1 Definition of the Problem

In this paper, we consider the parameterized version of the MULTIDIMENSIONAL BINARY VECTOR ASSIGNMENT problem (bMVA). An input of this problem is described by m sets $V^1, V^2, \ldots, V^m$, each of these sets containing n p-dimensional binary vectors. We note $V^i = \{v_1^i, \ldots, v_n^i\}$ for all $i \in [m]$[1], and for all $j \in [n]$ and $r \in [p]$, we denote by $v_j^i[r] \in \{0, 1\}$ the rth component of v_j^i.

In order to define the output of the problem, we need to introduce the notion of stack. A stack $s = (v_1^s, v_2^s, \ldots, v_m^s)$ is an m-tuple of vectors such that $\forall i \in [m], v_i^s \in V^i$. The output of bMVA is a set S of n stacks such that for all

---

[1] $[m]$ stands for $\{1, \ldots, m\}$.

© Springer International Publishing Switzerland 2015
A. Kosowski and I. Walukiewicz (Eds.): FCT 2015, LNCS 9210, pp. 189–201, 2015.
DOI: 10.1007/978-3-319-22177-9_15

$i, j \in [m] \times [n]$, v_j^i belongs to only one stack (in that case, the stacks are said *disjoint*). An example of an instance together with a solution is depicted in Fig. 1.

We are now ready to define the objective function. We define the operator $\wedge$ that, given two p-dimensional vectors u and v, computes the vector $w = (u[1] \wedge v[1], u[2] \wedge v[2], \ldots, u[p] \wedge v[p])$. We associate to any stack s a unique vector $v_s = \bigwedge_{i \in [m]} v_i^s$.

We define the cost of a binary vector v as the number of zeros in it. More formally, if v is p-dimensional, $c(v) = p - \sum_{r \in [p]} v[r]$. We extend this definition to a set of stacks $S = \{s_1, \ldots, s_n\}$ as follows: $c(S) = \sum_{j \in [n]} c(v_{s_j})$. Finally, the objective of bMVA is to obtain a set S of n disjoint stacks while minimizing $c(S)$. In the decision version of the problem, we are given an integer k, and we ask whether there exists a solution S of cost at most k. The problem is thus defined formally as follows:

Problem 1. MULTIDIMENSIONAL BINARY VECTOR ASSIGNMENT (bMVA)

Input: m sets of n binary p-dimensional vectors, an integer k
Question: Is there a set S of n disjoint stacks such that $c(S) \leq k$?

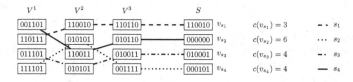

Fig. 1. Example of bMVA instance with $m = 3, n = 4, p = 6$ and of a feasible solution S of cost $c(S) = 17$.

In order to avoid heavy notations throughout the paper, we will denote an instance of bMVA only by $\mathcal{I}[m, n, p, k]$, the notations of the sets and vectors being implicitly given as previously.

1.2   Application and Related Work

bMVA can be seen as a variant of multidimensional matching where hyperedges are implicitly locally encoded via labels attached to vertices. However, this kind of problem was originally introduced in [17] in the context of semiconductor industry as the "yield maximization problem in wafer-to-wafer 3-D integration technology". In this context, each vector v_j^i represents a wafer, which is seen as a string of bad dies (0) and good dies (1). Integrating two wafers corresponds to superimposing the two corresponding strings. In this operation, a position in the merged string is "good" when the two corresponding dies are good, and is "bad" otherwise. The objective of Wafer-to-Wafer Integration is to form n

stacks, while maximizing their overall quality, or equivalently, minimizing the number of errors (depending on the objective function). In the following, we will denote by $max \sum 1$-bMVA the dual version of bMVA where given the same input and ouput, the objective is to maximize $np - c(S)$, the total number of ones.

The results obtained so far concerning these problems mainly concern their approximability. The **NP**-hardness of bMVA is provided in [4] even when $m = 3$, as well as a $\frac{4}{3}$-approximation (still for $m = 3$). We can also mention [5] which provides a $f(m)$-approximation for general m, and an **APX**-hardness for $m = 3$. The main related article is [8] where it is proved that $max \sum 1$-bMVA has no $O(p^{1-\epsilon})$ nor $O(m^{1-\epsilon})$-approximation for any $\epsilon > 0$ unless $\mathbf{P} = \mathbf{NP}$ (even when $n = 2$), but admits a $\frac{p}{c}$-approximation algorithm for any constant $c \in \mathbb{N}$, and is **FPT** when parameterized by p (which also holds for bMVA). Notice that one of the reductions provided in [8] is a parameter-preserving reduction from the CLIQUE problem to $max \sum 1$-bMVA, immediately proving **W**[1]-hardness for $max \sum 1$-bMVA when parameterized by the objective function. This is why our motivation in this paper is to consider the parameterized complexity of bMVA. As we will see in the next section, we provide an analysis for several parameters related to this problem.

For formal definitions and detailed concepts on Fixed-Parameter Tractability, we refer to the monograph of Downey and Fellows [6]. Moreover, in order to define lower bounds on the running time of parameterized algorithms, we will rely on the *Exponential Time Hypothesis (ETH)* of Impagliazzo, Paturi and Zane [11], stating that 3-SAT cannot be solved in $O^*(2^{o(n)})$ where n is the number of variables ($\mathcal{O}^*(.)$ hides polynomial terms). For more results about lower bounds obtained under ETH, we refer the reader to the survey of [12].

1.3 Parameterizations

One of the main purposes of Fixed-Parameter Tractability is to obtain efficient algorithms when the considered parameter is small in practice. When dealing with the decision version of an optimization problem, the most natural parameter is perhaps the value of the desired solution (*e.g.* k for bMVA). Such a parameter is often referred to in the literature as the "standard parameter" of the problem. In some cases, this parameter might not be very interesting, either because it usually takes high values in practice, or because **FPT** algorithms with respect to this parameterization are trivial to find. When this happens, it is possible to obtain more interesting results by subtracting to the objective function a known lower bound of it. For instance, if one can prove that any solution of a given minimization problem is of cost at least $\mathcal{B}$, then one can ask for a solution of cost $\mathcal{B} + c$ and parameterize by c. This idea, called "above guarantee parameterization" was introduced by [15] and first applied to MAX SAT and MAX CUT problems. It then became a fruitful line of research with similar results obtained for many other problems (among others, see [3,9,10,16]).

In this paper, we analyze the parameterized complexity of bMVA using three types of parameters. Given an instance $\mathcal{I}[m, n, p, k]$ of our problem, we will consider the following parameters:

- the standard parameter, k, the number of zeros to minimize in the optimization version of the problem.
- three natural structural parameters: m, the number of sets of the input, n, the number of vectors in each set, and p, the size of each vector.

As we said previously, it was already proved in [8] that bMVA is **FPT** parameterized by p. As we will notice in Lemma 2 that we can obtain an equivalent instance with $p \leq k$ after a polynomial pre-processing step, this implies that bMVA is also **FPT** with its standard parameter. Our idea here is to use this previous inequality in order to obtain smaller parameters. Thus, we define our first above guarantee parameter $\zeta_p = k - p$.

Finally, in order to define our last parameter, we first need to describe the corresponding lower bound $\mathcal{B}$, that will represent the maximum, over all sets of vectors, of the total number of zeros for each set. More formally, we define $\mathcal{B} = \max_{i \in [m]} c(V^i)$ where $c(V^i) = \sum_{j=1}^{n} c(v_j^i)$. Since we perform a bit-wise AND over each m-tuple, it is easily seen that any solution will be of cost at least $\mathcal{B}$. Thus, we define our last parameter $\zeta_\mathcal{B} = k - \mathcal{B}$.

1.4 Our Results

In the next section, we present some pre-processing rules leading to a kernel of size $O(k^2 m)$, and prove that even when $m = 3$ we cannot improve it to $p^{O(1)}$ unless **NP** $\subseteq$ **coNP/poly** (remember that bMVA is known to be **FPT** when parameterized by p). Section 3 is mainly focused on results associated with parameter $\zeta_\mathcal{B}$: we prove that bMVA can be solved in $O^*(4^{\zeta_\mathcal{B} \log(n)})$, while it is **W**[2]-hard when parameterized by $\zeta_\mathcal{B}$ only, and cannot be solved in $O^*(2^{o(\zeta_\mathcal{B}) \log(n)})$ nor in $O^*(2^{\zeta_\mathcal{B} o(\log(n))})$ assuming ETH. In Sect. 4, we focus on the parameterization by ζ_p: we show that when $n = 2$, the problem can be solved in single exponential time with this parameter, but is not in **XP** for any fixed $n \geq 3$ (unless **P** = **NP**). The reduction we use also shows that for fixed $n \in \mathbb{N}$, the problem cannot be solved in $2^{o(k)}$ (and thus in $2^{o(\zeta_\mathcal{B})}$) unless ETH fails, which matches the upper bound obtained in Sect. 3. Due to space constraints, results marked with a $(\star)$ are proven in the full version of the paper [2]. A summary of our results is depicted in the following table.

2 First Remarks and Kernels

Let us start with two simple lemmas allowing us to bound the size of the input. Notice first that it is not always safe to create a 1-stack (*i.e.* a stack with ones on every component) when possible. Indeed, in instance $V^1 = \{\langle 111 \rangle, \langle 101 \rangle, \langle 011 \rangle\}$,

Positive results	Negative results
$O(k^2m)$ kernel (Theorem 1)	no $p^{O(1)}$ kernel unless $\mathbf{NP} \subseteq \mathbf{coNP/poly}$ (Theorem 2)
$\mathcal{O}^*(4^{\zeta_B \log(n)})$ algorithm (Theorem 3)	$\mathbf{W}[2]$-hard for ζ_B only (Theorem 4)
	no $2^{o(\zeta_B)\log(n)}$ nor $2^{\zeta_B o(\log(n))}$ under ETH (Theorem 5)
	no $2^{o(k)}$ for fixed n under ETH (Theorem 7)
$O^*(d^{\zeta_p})$ algorithm for $n = 2$ (Theorem 6)	$\mathbf{NP}$-hard for $\zeta_p = 0$ and fixed $n \geq 3$ (Theorem 7)

$V^2 = \{\langle 111 \rangle, \langle 101 \rangle, \langle 110 \rangle\}$, $V^3 = \{\langle 111 \rangle, \langle 011 \rangle, \langle 110 \rangle\}$, no optimal solution creates a 1-stack. However, as we will see in Lemma 1, creating 1-stacks becomes safe if $n > k$.

Lemma 1 ($\star$). *There exists a polynomial algorithm which, given any instance $\mathcal{I}[m, n, p, k]$ of bMVA, either detects that $\mathcal{I}$ is a negative instance, or outputs an equivalent instance $\mathcal{I}'[m, n', p, k]$ such that $n' \leq k$.*

Lemma 2 ($\star$). *There exists a polynomial algorithm which, given any instance $\mathcal{I}[m, n, p, k]$ of bMVA, either detects that $\mathcal{I}$ is a negative instance, or outputs an equivalent instance $\mathcal{I}'[m, n, p', k]$ such that $p' \leq k$.*

Given the two previous lemmas, we can suppose from now on that for any instance of bMVA we have $n \leq k$ and $p \leq k$. This immediately implies a polynomial kernel parameterized by k and m.

Theorem 1. *bMVA admits a kernel with $O(k^2m)$ bits.*

Let us now turn to the main result of this section. To complement Theorem 1, we show that even when $m = 3$, we cannot obtain a polynomial kernel with the smaller parameter p under some classical complexity assumptions (notice however that the existence of a polynomial kernel in k is still open). Notice also that as bMVA was known to be **FPT** when parameterized by p, it was a natural question to ask for a polynomial kernel.

In order to establish kernel lower bounds, we use the concept of AND-cross-composition of Bodlaender et al. [1], together with the recently proved AND-conjecture of Drucker [7].

Theorem 2. *Even for $m = 3$, bMVA parameterized by p does not admit a polynomial kernel unless* **NP** $\subseteq$ **coNP/poly**.

Proof. The proof is an AND-cross-composition from a sequence of instances of 3-DIMENSIONAL PERFECT MATCHING, inspired by the **NP**-hardness reduction for bMVA provided in [4]. 3-DIMENSIONAL PERFECT MATCHING is formally defined as follows:

Problem 2. 3-Dimensional Perfect Matching

Input: Three sets X, Y and Z of size n, a set of hyperedges $S \subseteq X \times Y \times Z$
Question: Does there exist a subset $S' \subseteq S$ such that:
                 – for all $e, e' \in S'$ with $e = (x, y, z)$ and $e' = (x', y', z')$, we have $x \neq x'$,
                 $y \neq y'$ and $z \neq z'$ (that is, S' is a matching)
                 – $|S'| = n$ (that is, S' is perfect)

Let $(X_1, Y_1, Z_1, S_1), \cdots, (X_t, Y_t, Z_t, S_t)$ be a sequence of t equivalent instances of 3-DPM, with respect to the following polynomial equivalence relation: (X, Y, Z, S) and (X', Y', Z', S') are equivalent if $|X| = |X'|$ (and thus $|Y| = |Y'| = |Z| = |Z'| = |X|$), and $|S| = |S'|$. In the following we denote by n the cardinality of the sets X_i (and equivalently the sets Y_i and Z_i), and by m the cardinality of the sets S_i. Moreover, for all $i \in \{1, \cdots, t\}$ we define $X_i = \{x_{i,1}, \cdots, x_{i,n}\}, Y_i = \{y_{i,1}, \cdots, y_{i,n}\}, Z_i = \{z_{i,1}, \cdots, z_{i,n}\}$, and $S_i = \{s_{i,1}, \cdots, s_{i,m}\}$. We also assume that $t = 2^q$ for some $q \in \mathbb{N}$ (if it is not the case, we add a sufficiently number of dummy yes-instances).

In the following we construct three sets (X^*, Y^*, Z^*) of nt vectors each: $X^* = \{x_{i,j}^*\}_{i=1,\cdots,t}^{j=1,\cdots,n}$, $Y^* = \{y_{i,j}^*\}_{i=1,\cdots,t}^{j=1,\cdots,n}$ and $Z^* = \{z_{i,j}^*\}_{i=1,\cdots,t}^{j=1,\cdots,n}$, where each vector is composed of $p^* = m + 2mq$ components. Let us first describe the first m components of each vector. For all $i \in \{1, \cdots, t\}, j \in \{1, \cdot, n\}$ and $k \in \{1, \cdots, m\}$ we set:

$$x_{i,j}^*[k] = \begin{cases} 1 \text{ if the hyperedge } s_{i,k} \text{ contains } x_{i,j} \\ 0 \text{ otherwise} \end{cases}$$

$$y_{i,j}^*[k] = \begin{cases} 1 \text{ if the hyperedge } s_{i,k} \text{ contains } y_{i,j} \\ 0 \text{ otherwise} \end{cases}$$

$$z_{i,j}^*[k] = \begin{cases} 1 \text{ if the hyperedge } s_{i,k} \text{ contains } z_{i,j} \\ 0 \text{ otherwise} \end{cases}$$

Then, for all $i \in \{1, \cdots, t\}$, we append two vectors b_i and $\bar{b}_i$ to all vectors $\{x_{i,j}^*\}_{j=1,\cdots,n}, \{y_{i,j}^*\}_{j=1,\cdots,n}$ and $\{z_{i,j}^*\}_{j=1,\cdots,n}$. The vector b_i is composed of mq coordinates, and is defined as the binary representation of the integer i, where each bit is duplicated m times. Finally, $\bar{b}_i$ is obtained by taking the complement of b_i (i.e. replacing all zeros by ones, and conversely). It is now clear that each vector $x_{i,j}^*$ (resp. $y_{i,j}^*, z_{i,j}^*$) is composed of $p^* = m + 2mq$ coordinates. Thus, the parameter of the input instance is a polynomial in n, m and $\log t$ whereas the total size of the instance is a polynomial in the size of the sequence of inputs, as required in cross-compositions. It now remains to prove that (X^*, Y^*, Z^*) contains an assignment of cost $k^* = nt(mq + m - 1)$ if and only if for all $i \in \{1, \cdots, t\}, S_i$ contains a perfect matching S_i'.

- Suppose that for all $i \in \{1, \cdots, t\}$ we have a perfect matching $S_i' \subseteq S_i$. W.l.o.g. suppose that $S_i' = \{s_{i,1}, \cdots, s_{i,n}\}$. Then, for each $j \in \{1, \cdots, n\}$, we have $s_{i,j} = (x_{i,j_1}, y_{i,j_2}, z_{i,j_3})$ for some $j_1, j_2, j_3 \in \{1, \cdots, n\}$. We assign x_{i,j_1}^* with y_{i,j_2}^* and z_{i,j_3}^*. It is easy to see that the cost of this triple is $m - 1 + mq$. Indeed, they all have a one at the jth coordinate, corresponding to the jth hyperedge of S_i (and this is the only shared one, since we can suppose that all hyperedges are pairwise distinct), and they all contain the same vectors b_i and $\bar{b}_i$. Summing up for all instances, we get the desired solution value.

- Conversely, first remark that in any assignment, the cost of every triple $(x_{i_1,j_1}^*, y_{i_2,j_2}^*, z_{i_3,j_3}^*)$ is at least $m - 1 + mq$, and let us prove that this bound is tight when (1) all elements are chosen within the same instance, i.e. $i_1 = i_2 = i_3 = i$, and (2) this triple corresponds to an element of S_i, i.e. $(x_{i,j_1}, y_{i,j_2}, z_{i,j_3}) \in S_i$. Indeed, suppose first that $i_1 \neq i_2$. Then, since the binary representation of i_1 and i_2 differs on at least one bit, it is clear that the resulting vector is of cost at least $m(q + 1) > mq + m - 1$. Now if $i_1 = i_2 = i_3 = i$, then the result is straightforward, since at most one hyperedge of S_i can contain x_{i_1,j_1}^*, y_{i_2,j_2}^* and z_{i_3,j_3}^*. Finally, using the same arguments as previously, we can easily deduce a perfect matching $S_i' \subseteq S_i$ for each $i \in \{1, \cdots, t\}$, and the result follows. $\square$

3   Parameterizing According to $\zeta_{\mathcal{B}}$

In this section, we present an **FPT** algorithm when parameterized by $\zeta_{\mathcal{B}}$ and n (recall that both $\zeta_{\mathcal{B}}$ and n are smaller parameters than the standard one k, since $k = \mathcal{B} + \zeta_{\mathcal{B}}$ and $n \leq k$ in any reduced instance). Notice first that it is easy to get a $\mathcal{O}^*(2^{\zeta_{\mathcal{B}}(log(n)+log(p))})$ algorithm. Indeed, by considering a set $i \in [m]$ where $c(V^i) = \mathcal{B}$, and guessing the positions of the $\zeta_{\mathcal{B}}$ new zeros (among np possible positions) that will appear in an optimal solution, we can actually guess in $\mathcal{O}^*((np)^{\zeta_{\mathcal{B}}})$ the vectors $\{v_{s_j^*}\}$ of an optimal solution, and it remains to check in polynomial time that every V^j can be "matched" to $\{v_{s_j^*}\}$. Now we show how to get rid of the $log(p)$ term in the exponent.

Theorem 3. *bMVA can be solved in $\mathcal{O}^*(4^{\zeta_{\mathcal{B}} \log(n)})$.*

Proof. Let $\mathcal{I}[m, n, p, k]$ be an instance of our problem and, w.l.o.g., suppose that V^1 is a set whose number of zeros reaches the upper bound $\mathcal{B}$, i.e. $c(V^1) = \mathcal{B}$. The algorithm consists in constructing a solution by finding an optimal assignment between V^1 and $V^2, \ldots, V^m$, successively.

We first claim that we can decide in polynomial time whether there is an assignment between V^1 and V^2 which does not create any additional zero. To that end, we create a bipartite graph G with bipartization (A, B), $A = \{a_1, \ldots, a_n\}$, $B = \{b_1, \ldots, b_n\}$, and link a_{j_1} and b_{j_2} for all $(j_1, j_2) \in [n] \times [n]$ iff assigning vector $v_{j_1}^1$ from V^1 and vector $v_{j_2}^2$ from V^2 does not create any additional zero in V^1 ($v_{j_1}^1 \wedge v_{j_2}^2 = v_{j_1}^1$). If a perfect maching can be found in G, then we can safely delete the set V^2 and continue. In order to avoid heavy notations, we consider this first step as a polynomial pre-processing, and we re-label V^i into V^{i-1} for all $i \in \{3, \ldots, m\}$ (and m is implicitly decreased by one).

In the following, we suppose that the previous pre-processing step cannot apply (*i.e.* there is no perfect matching in G). Intuitively, in this case any assignment (including an optimal one) between V^1 and V^2 must lead to at least one additional zero in V^1. In this case, we perform a branching to guess one couple of vectors from $V^1 \times V^2$ which will induce such an additional zero. More formally, we branch on every couple $(j_1, j_2) \in [n] \times [n]$, and create a new instance as a copy of $\mathcal{I}$ in which $v_{j_1}^1$ is replaced by $v_{j_1}^1 \wedge v_{j_2}^2$. This operation increases $c(V^1)$ by at least one, and thus $\mathcal{B}$ by at least one as well. If we denote by $\mathcal{I}'$ this new instance, we can check that a solution of cost at most k for $\mathcal{I}'$ will immediately imply a solution of cost at most k for $\mathcal{I}$, as $\mathcal{I}'$ is constructed from $\mathcal{I}$ by adding some zeros. The converse is also true as one assignment we enumerate corresponds to one from an optimal solution.

As the value of $\mathcal{B}$ in this branching increases by at least one while we still look for a solution of cost k, this implies that this branching will be applied at most $\zeta_\mathcal{B}$ times. Summing up, we have one polynomial pre-processing and one branching of size n^2 which will be applied at most $\zeta_\mathcal{B}$ times. The total running time of this algorithm is thus bounded by $\mathcal{O}^*(4^{\zeta_\mathcal{B} \log(n)})$. □

Despite its simplicity, we now show that, when considering each parameter (n and $\zeta_\mathcal{B}$) separately, this algorithm is the best we can hope for (whereas the existence of an $\mathcal{O}^*(2^k)$ algorithm is still open). Indeed, we first show in Theorem 5 that the linear dependence in $\zeta_\mathcal{B}$ and $log(n)$ in the exponent is necessary (unless ETH fails), and also that we cannot hope for an **FPT** algorithm parameterized by $\zeta_\mathcal{B}$ only unless **FPT** = **W**[2] (Theorem 4). Finally, as we will see in the next section (Theorem 7), this result is matched by a $2^{o(\zeta_\mathcal{B})}$ lower bound when $n \in \mathbb{N}$ is fixed. We now present a reduction from the HITTING SET problem which produces an instance of bMVA in which parameters $\zeta_\mathcal{B}$ and n are preserved.

Problem 3. Hitting Set

Input: m subsets $R_1, \ldots, R_m$ of $[n]$, and an integer k

Question: Is there a set R of k elements of $[n]$ such that $R \cap R_i \neq \emptyset$ for any $i \in [m]$?

Lemma 3. *There is a polynomial reduction from* HITTING SET *to bMVA that given an instance composed of m subsets of $[n]$ and an integer k, constructs an instance of bMVA $\mathcal{I}[m', n', p', k']$ such that $n' = n$ and $\zeta_\mathcal{B} = k$.*

Proof. Let $R_1, \ldots, R_m$ be subsets of $[n]$, and $k \in \mathbb{N}$. We construct m sets $V^1, \ldots, V^m$ of n vectors each, where, for all $i \in [m]$ we have $V^i = \{v_1^i, \ldots, v_n^i\}$, each vector being composed of n components. For all $i \in [m]$ and all $j \in [n]$, if $j \in R_i$, then the vector v_j^i is composed of ones everywhere except at the jth component. If $j \notin R_i$, then v_j^i is a 0-vector (*i.e.* a vector with zero in every component). We also add a set V^* composed of $(n-1)$ 0-vectors and one 1-vector.

For this constructed instance, it is clear that $\mathcal{B} = n(n-1)$ because of the set V^*. In other words, any assignment will lead to a solution with $(n-1)$ 0-vectors, and thus with at least $n(n-1)$ zeros. We will actually show that this instance has a solution with $n(n-1) + k$ zeros if and only if $R_1, ..., R_m$ has a hitting set of size k. By the foregoing, we only need to focus on the only vector of each set which is assigned to the 1-vector of V^*.

$\Rightarrow$ Let $J \subseteq [n]$ be a hitting set of size k. By the definition of a hitting set, for all $i \in [m]$, there exists $j_i \in J \cap R_i$. Thus, for all $i \in [m]$, we select the vector $v_{j_i}^i$ from the set V^i to be assigned to the 1-vector of V^*. By construction, this vector has only one zero at the j_i^{th} component, which implies that the conjunction of all such vectors $\bigwedge_{i=1}^m v_{j_i}^j$ will have a 1 everywhere except at the components corresponding to J. We thus have the desired number of zeros in our solution.

$\Leftarrow$ Conversely, for each $i \in [m]$, let $j_i \in [n]$ be the vector from V^i which is assigned to the 1-vector of V^*. Since the resulting conjunction of all these vectors has only k zeros, $v_{j_i}^i$ cannot be a 0-vector, and we thus have $j_i \in R_i$. Using the same arguments as previously, $\{u_{j_i}\}_{i \in [m]}$ corresponds to a hitting set of $R_1, ..., R_m$ of size k.

As we seen previously, $\mathcal{B} = n(n-1)$ for the obtained instance, $k' = n(n-1) + k$, (which implies $\zeta_\mathcal{B} = k$), and the size of all sets is n, as desired. $\square$

As we can see, the reduction is parameter-preserving for $\zeta_\mathcal{B}$. From the **W**[2]-hardness of HITTING SET [6], we have the following:

Theorem 4. *bMVA is* **W**[2]*-hard when parameterized by* $\zeta_\mathcal{B}$.

As said previously, we also use this reduction to show the following result:

Theorem 5. *bMVA cannot be solved in* $O^*(2^{o(\zeta_\mathcal{B}) \log(n)})$ *nor* $O^*(2^{\zeta_\mathcal{B} o(\log(n))})$, *unless ETH fails.*

Proof. To show this, we use our previous hardness result, but using a constrained version of the HITTING SET problem obtained by [13], where the element set is $[k] \times [k]$ and can thus be seen as a table with k rows and k columns:

Problem 4. $k \times k$ Hitting Set

Input:	An integer k, and $R_1, ..., R_t \subseteq [k] \times [k]$
Question:	Is there a set R containing exactly one element from each row such that $R \cap R_i \neq \emptyset$ for any $i \in [t]$?

Authors of [13] show that assuming ETH this problem cannot be solved in $2^{o(k \log(k))} n^{O(1)}$ (whereas a simple brute force solves it in $O^*(2^{k \log(k)})$). Notice that we can modify the question of this problem by dropping the constraint that S contains at least one element from each row. Indeed, let us add to the instance a set of k sets $\{R_1', \ldots, R_k'\}$, where R_i' contains all elements of row i for $i \in [k]$. Now, finding a (classical) hitting set of size k on this modified instance is equivalent to finding a solution of size k for the original instance of $k \times k$-HITTING SET. Moreover, it is easy to check that a $2^{o(k \log(k))} n^{O(1)}$ algorithm for this relaxed problem would also contradict ETH. To summarize, we know that unless ETH fails, there is no $2^{o(k \log(k))} n^{O(1)}$ algorithm for the classical HITTING SET problem, even when the ground set has size k^2. This allows us to perform the reduction of Lemma 3 on these special instances, leading to an instance $\mathcal{I}[m', n', p', k']$ with associated parameter ζ_B such that $\zeta_B = k$ and $n' = k^2$. Suppose now that there exists an algorithm for bMVA running in $2^{o(\zeta_B) \log(n)} (k + m + n + p)^{O(1)}$. Using the reduction above, we would be able to solve the instance of $k \times k$-HITTING SET in $2^{o(k) \log(k^2)} n^{O(1)}$, and thus in $2^{o(k \log(k))} n^{O(1)}$, which would violate ETH. A similar idea also rules out any algorithm running in $2^{\zeta_B o(\log(n))}$ under ETH. $\qquad\square$

4   Parameterizing According to ζ_p

We now consider the problem parameterized by $\zeta_p = k - p$ (recall that $p \leq k$). Notice that one motivation of this parameterization is the previous reduction of Lemma 3 from HITTING SET. Indeed, when applied for $n = 2$, it reduces an instance of VERTEX COVER to an instance of bMVA with $k = p + \zeta_p$ where ζ_p is equal to the size of the vertex cover. Our intuition is confirmed by the following result: we show that when parameterized by ζ_p, the problem is indeed **FPT** when $n = 2$ (Theorem 6). We complement this by showing that for any $n \geq 3$, it becomes **NP**-hard even when $\zeta_p = 0$ (Theorem 7), and is thus not in **XP** (unless **P** = **NP**). The reduction we use even proves that for any fixed $n \geq 3$, the problem cannot be solved in $2^{o(k)}$ (and thus in $2^{o(\zeta_B)}$) unless ETH fails, while the algorithm of Theorem 3 runs in $O^*(2^{O(\zeta_B)})$. In the following, n-bMVA denotes the problem bMVA where the size of all sets is fixed to some constant $n \in \mathbb{N}$.

4.1   Positive Result for $n = 2$

In this subsection, we prove that 2-bMVA is **FPT** parameterized by ζ_p. To do so, we reduce to the ODD CYCLE TRANSVERSAL problem (OCT for short), which consists, given a graph $G = (V, E)$ and an integer $c \in \mathbb{N}$, to decide whether there exists a partition (X, S_1, S_2) of V with $|X| \leq c$ such that S_1 and S_2 are independent sets.

We first introduce a generalized version of OCT, called BIP-OCT. In this problem, we are given a set of vertices V, an integer c, and a set of m pairs $(A_1, B_1), ..., (A_m, B_m)$ with $A_i, B_i \subseteq V$ for all $i \in [m]$ and $A_i \cap B_i = \emptyset$. Informally, each pair (A_i, B_i) can be seen as a complete bipartite subgraph. The output of BIP-OCT is described by a partition (X, S_1, S_2) of V such that for any $i \in [m]$, either $(A_i \setminus X \subseteq S_1$ and $B_i \setminus X \subseteq S_2)$ or $(A_i \setminus X \subseteq S_2$ and $B_i \setminus X \subseteq S_1)$. The question is whether there exists such a partition with $|X| \leq c$. As we can see, if all A_i and B_i are singletons (and thus form edges), then BIP-OCT corresponds to OCT. Notice that in the following, the considered parameter of OCT and

BIP-OCT will always be the standard parameter, *i.e. c*. We first show that there is a linear parameter-preserving reduction from 2-bMVA parameterized by ζ_p to BIP-OCT, and then that there is also a linear parameter-preserving transformation from BIP-OCT to OCT.

Lemma 4 ($\star$). *There is a linear parameter-preserving reduction from 2-bMVA parameterized by ζ_p to* BIP-OCT.

Lemma 5 ($\star$). *There is a linear parameterized reduction from* BIP-OCT *to* OCT.

As ODD CYCLE TRANSVERSAL can be solved in $O^*(2.3146^c)$ [14], and since our parameters are exactly preserved in our two reductions, we obtain the following result:

Theorem 6. *2-bMVA can be solved in $O^*(d^{\zeta_p})$ where $d \leq 2.3146$ is such that* OCT *can be solved in $O^*(d^c)$.*

4.2   Negative Results for $n \geq 3$

We now complement the previous result by proving that the problem is intractable with respect to the parameter ζ_p for larger values of n.

Theorem 7. *For any fixed $n \geq 3$, n-bMVA is not in* **XP** *when parameterized by ζ_p (unless* **P**=**NP***), and cannot be solved in $2^{o(k)}$ (unless ETH fails).*

Proof. Let $\chi \geq 3$. We present a reduction from χ-COLORING, which consists in, given a graph $G = (V, E)$, to ask for a mapping $f : V \longrightarrow [\chi]$ such that for all $\{u, v\} \in E$ we have $f(u) \neq f(v)$. Let $E = \{e_1, ..., e_{m_G}\}$ and $V = [n_G]$. Let us construct an instance $\mathcal{I}$ of n-bMVA with $n = \chi$, $p = n_G$, $m = m_G$ and such that G admits a χ-coloring iff I has a solution of cost p (*i.e.* $\zeta_p = 0$). To each edge $e_i = \{u, v\} \in E$, $i \in [m_G]$, we associate a set V^i with $|V^i| = \chi$, where:

- v_1^i represents the vertex u, that is $v_1^i[u] = 0$ and $v_1^i[r] = 1$ for any $r \in [n_G]$, $r \neq u$,
- v_1^i represents the vertex v, that is $v_2^i[v] = 0$ and $v_2^i[r] = 1$ for any $r \in [n_G]$, $r \neq v$,
- for all $j \in \{3, ..., \chi\}$, v_j^i is a 1-vector, *i.e.* it has a 1 at every component.

Let us now prove that G admits a χ-coloring iff $\mathcal{I}$ has a solution of cost $p = n_G$.

$\Rightarrow$ Let $S_j \subseteq V$, $j \in [\chi]$ be the χ color classes (notice that the S_j are pairwise disjoint, some of them may be empty, and $\bigcup_{j \in [\chi]} S_j = V$). To each S_j we associate a stack s_j such that $v_{s_j}[r] = 0$ iff $r \in S_j$. It remains to prove that the solution $S = \{s_1, ..., s_\chi\}$ is feasible, as its cost is exactly p by construction. Let us consider a set V^i where v_1^i (resp. v_2^i) represents a vertex u (resp. v). As $\{u, v\}$ is an edge of G, we know that u and v have two different colors, *i.e.* that $u \in S_j$ and $v \in S_{j'}$, for some $j, j' \in [\chi]$ with $j \neq j'$. Thus, we can add v_1^i to stack s_j, v_2^i to stack $s_{j'}$, and the $\chi - 2$ other v_j^i ($j \geq 3$) in an arbitrary way. Since the only 0 in v_1^i (resp. v_2^i) is at the uth (resp. v^{th}) component, we have $v_1^i \wedge v_{s_j} = v_{s_j}$ (resp. $v_2^i \wedge v_{s_{j'}} = v_{s_{j'}}$), which proves that S is feasible.

$\Leftarrow$. Let $S = \{s_1, \ldots, s_\chi\}$ be the stacks of an optimal solution. For $j \in [\chi]$, let $S_j = \{r \in [p] | v_{s_j}[r] = 0\}$. Notice that $\bigcup_{j=1}^{\chi} S_j = V$, and as I is of cost p, all the S_j are pairwise disjoints and form a partition of V. Moreover, as for any $i \in [m]$, v_1^i and v_2^i have been assigned to different stacks, the corresponding vertices have been assigned to different colors, and thus each S_j induces an independent set, which completes the reduction.

It is known [11] that there is no $2^{o(|V|)}$ algorithm for deciding whether a graph $G = (V, E)$ admits a χ-COLORING, for any $\chi \geq 3$ (under ETH). As we can see, the value of the optimal solution for n-bMVA in the previous reduction equals the number of vertices in the instance of χ-COLORING, which proves that n-bMVA cannot be solved in $2^{o(k)}$ for any $n \geq 3$. $\square$

Finally, remark that as for the parameterization by p, one could ask if bMVA is **FPT** when parameterized by the first lower bound $\mathcal{B}$. However, we can see in the previous reduction that we obtain a graph with $\mathcal{B} = 2$, and thus the problem is even not in **XP** unless **P** = **NP**.

5 Conclusion

In this article, we presented some negative and positive results for a multidimensional binary vector assignment problem in the framework of parameterized complexity. Notice that neither lower bounds of Theorem 5 nor Theorem 7 are able to rule out an algorithm running in $O^*(2^k)$ (when n is part of the input), hence the existence of such an algorithm seems a challenging open problem. Another interesting question concerns the improvement of the $O(k^2m)$ kernel of Theorem 1 by getting rid of the parameter m: does bMVA admit a polynomial kernel when parameterized by k only?

References

1. Bodlaender, H.L., Jansen, B.M.P., Kratsch, S.: Kernelization lower bounds by cross-composition. SIAM J. Discrete Math. **28**(1), 277–305 (2014)
2. Bougeret, M., Duvillie, G., Giroudeau, R., Watrigant, R.: Multidimensional binary vector assignment problem: standard, structural and above guarantee parameterizations. CoRR, abs/1506.03282 (2015)
3. Cygan, M., Pilipczuk, M., Pilipczuk, M., Wojtaszczyk, J.O.: On multiway cut parameterized above lower bounds. In: Proceedings of the 6th International Conference on Parameterized and Exact Computation (IPEC2011), pp. 1–12, Springer-Verlag, Heidelberg (2012)
4. Dokka, T., Bougeret, M., Boudet, V., Giroudeau, R., Spieksma, F.C.R.: Approximation algorithms for the wafer to wafer integration problem. In: Erlebach, T., Persiano, G. (eds.) WAOA 2012. LNCS, vol. 7846, pp. 286–297. Springer, Heidelberg (2013)
5. Dokka, T., Crama, T.Y., Spieksma, F.C.R.: Multi-dimensional vector assignment. Discrete Optim. **14**, 111–125 (2014)

6. Downey, R.G., Fellows, M.R.: Fundamentals of Parameterized Complexity. Texts in Computer Science. Springer, Hiedelberg (2013)
7. Drucker, A.: New limits to classical and quantum instance compression. In: FOCS, pp. 609–618, (2012)
8. Duvillié, G., Bougeret, M., Boudet, V., Dokka, T., Giroudeau, R.: On the complexity of wafer-to-wafer integration. In: Paschos, V.T., Widmayer, P. (eds.) CIAC 2015. LNCS, vol. 9079, pp. 208–220. Springer, Heidelberg (2015)
9. Gutin, G., Rafiey, A., Szeider, S., Yeo, A.: The linear arrangement problem parameterized above guaranteed value. In: Calamoneri, T., Finocchi, I., Italiano, G.F. (eds.) CIAC 2006. LNCS, vol. 3998, pp. 356–367. Springer, Heidelberg (2006)
10. Gutin, G., Yeo, A.: Constraint satisfaction problems parameterized above or below tight bounds: a survey. In: Bodlaender, H.L., Downey, R., Fomin, F.V., Marx, D. (eds.) Fellows Festschrift 2012. LNCS, vol. 7370, pp. 257–286. Springer, Heidelberg (2012)
11. Impagliazzo, R., Paturi, R., Zane, F.: Which problems have strongly exponential complexity? J. Comput. Sys. Sci. **63**(4), 512–530 (2001)
12. Lokshtanov, D., Marx, D., Saurabh, S.: Lower bounds based on the exponential. Bull. EATCS **105**, 41–72 (2011)
13. Lokshtanov, D., Marx, D., Saurabh, S.: Slightly superexponential parameterized problems. In: Proceedings of the Twenty-second Annual ACM-SIAM Symposium on Discrete Algorithms (SODA2011), pp. 760–776. SIAM (2011)
14. Lokshtanov, D., Narayanaswamy, N.S., Raman, V., Ramanujan, M.S., Saurabh, S.: Faster parameterized algorithms using linear programming. ACM Trans. Algorithms **11**(2), 151–153 (2014)
15. Mahajan, M., Raman, V.: Parameterizing above guaranteed values: Maxsat and maxcut. J. Algorithms **31**(2), 335–354 (1999)
16. Mahajan, M., Raman, V., Sikdar, S.: Parameterizing above or below guaranteed values. J. Comput. Sys. Sci. **75**(2), 137–153 (2009)
17. Reda, S., Smith, G., Smith, L.: Maximizing the functional yield of wafer-to-wafer 3-d integration. IEEE Trans. Very Large Scale Integr. (VLSI) Sys. **17**(9), 1357–1362 (2009)

Incremental Complexity of a Bi-objective Hypergraph Transversal Problem

Ricardo Andrade[2,3,4,5], Etienne Birmelé[1]([✉]), Arnaud Mary[2,3,4,5],
Thomas Picchetti[1], and Marie-France Sagot[2,3,4,5]

[1] MAP5, UMR CNRS 8145, Université Paris Descartes, Paris, France
etienne.birmele@parisdescartes.fr
[2] Université de Lyon, 69000 Lyon, France
[3] Université Lyon 1, Villeurbanne, France
[4] CNRS, UMR5558, Laboratoire de Biométrie et Biologie Evolutive,
69622 Villeurbanne, France
[5] INRIA Grenoble Rhône-Alpes - ERABLE, Lyon, France

Abstract. The hypergraph transversal problem has been intensively studied, both from a theoretical and a practical point of view. In particular, its incremental complexity is known to be quasi-polynomial in general and polynomial for bounded hypergraphs. Recent applications in computational biology however require to solve a generalization of this problem, that we call bi-objective transversal problem. The instance is in this case composed of a pair of hypergraphs $(\mathcal{A}, \mathcal{B})$, and the aim is to enumerate minimal sets which hit all the hyperedges of $\mathcal{A}$ while intersecting a minimal set of hyperedges of $\mathcal{B}$. In this paper, we formalize this problem and relate it to the enumeration of minimal hitting sets of bundles. We show cases when under degree or dimension contraints these problems remain NP-hard, and give a polynomial algorithm for the case when $\mathcal{A}$ has bounded dimension, by building a hypergraph whose transversals are exactly the hitting sets of bundles.

1 Introduction

Let $\mathcal{A} \subseteq 2^V$ be a hypergraph on a finite set V. A *transversal* of $\mathcal{A}$ is any set $S \subseteq V$ intersecting all hyperedges of $\mathcal{A}$. It is straightforward to see that being a transversal is a monotone property on the subsets of V, so that the collection of minimal transversals characterizes all of them. This collection is called the *dual* or *transversal hypergraph* of $\mathcal{A}$, and is denoted by $tr(\mathcal{A})$.

The problem of computing the transversal hypergraph of any $\mathcal{A}$ is equivalent to enumerating maximal independent sets in hypergraphs [4] or to solving the Boolean function dualization problem [10]. Furthermore, it has many applications, for instance in artificial intelligence [8]. This problem thus received much attention in the last decades, both from a theoretical and a practical point of view (see [10] for a review).

R. Andrade—Financially supported by CNPq – Brazil.

A. Kosowski and I. Walukiewicz (Eds.): FCT 2015, LNCS 9210, pp. 202–213, 2015.
DOI: 10.1007/978-3-319-22177-9_16

The first method was proposed by Berge [2], who considered the hyperedges iteratively, updating the partial solutions obtained at each step. This algorithm may however have to store a high number of partial solutions, and no full solution will be available until the algorithm stops. More recent work thus focused on methods that compute the minimal transversals incrementally, studying the following problem [16]:

Problem 1 **DUAL**$(\mathcal{A}, \mathcal{X})$. Given a hypergraph $\mathcal{A}$ and a set $\mathcal{X}$ of minimal transversals of $\mathcal{A}$, decide whether $tr(\mathcal{A}) = \mathcal{X}$ or find a new minimal transversal in $tr(\mathcal{A}) \setminus \mathcal{X}$.

The complexity of this problem remains an open question. However, Fredman and Khachiyan [11] showed that it is quasi-polynomial by proposing two algorithms of respective complexities $N^{\mathcal{O}(\log^2 N)}$ and $N^{o(\log N)}$, where $N = |\mathcal{A}| + |\mathcal{X}|$ is the size of the input. We define the dimension $dim(\mathcal{A})$ of a hypergraph $\mathcal{A}$ as the size of its largest hyperedge, and the degree of a vertex as the number of hyperedges it belongs to, $deg(\mathcal{A})$ being the maximum degree in $\mathcal{A}$. For hypergraphs of bounded dimension, the problem is polynomial [7]. It is also polynomial for hypergraphs of bounded degree [9]. Moreover, the complexity class of the problem does not change if multiple minimal transversals or partial minimal transversals are required [5]. There are parallel algorithms for some classes of hypergraphs. For bounded edge size hypergraphs the problem is known to be in RNC [4] and a global parallel algorithm exists [17]. Also, there is a parallel polylog algorithm [18] for uniformly sparse hypergraphs.

The performance of the algorithms in practice was also studied in several publications. Khachiyan et al. [16] gave an algorithm with the same worst-case complexity than the one of Fredman and Khachiyan but with a better performance in practice. More recently, Toda [21] and Murakami and Uno [19] compared the existing algorithms and proposed new ones which can deal with large scale hypergraphs.

Our extension of the problem is motivated by several already studied or potential applications in computational biology. It was for example proposed for elaborating knock-out strategies in metabolic networks [14], the hyperedges representing metabolic pathways whose activity should be suppressed. One may also consider the vertices as genes and a hyperedge as the set of mutated genes in a tumoral tissue. The transversal hypergraph then lists the collection of minimal mutation sets hitting all the tumors. The mutation scenarios would be described by sets of genes, rather than by a single ranking of the genes based on the p-value of a statistical over-representation test.

However, due to the complexity of cellular mechanisms, in both previous cases it appears there actually are two types of hyperedges, some of them having to be intersected while others should be avoided. Indeed, if one wants to knock-out a given set of metabolic pathways, one needs to maintain the biomass production of the cell in order to avoid cellular death. Hädicke and Klamt [13] introduced thus the notion of *constrained minimal cut sets* corresponding to vertex sets hitting all target pathways while avoiding at least n pathways among a prescribed set.

An adaptation of the Berge algorithm was proposed and was compared to binary integer programming on real data sets [15].

Coming back to the tumoral mutation example, a similar bi-objective problem appears. Mutations may indeed not be related to cancer, and the goal is to discriminate driver mutations from so-called back-seat mutations. Bertrand et al. [3] showed that this is equivalent to the *minimal set cover* and used a greedy approximation algorithm to solve it. An alternative way to deal with the problem would be to use other mutation data on similar but non tumoral tissues, and to look for mutation collections covering all tumors while avoiding healthy samples as much as possible.

We therefore propose to consider a bi-objective generalization of the hypergraph transversal problem, in which two distinct hypergraphs represent, respectively, the sets of nodes to hit and those to avoid, and to search for the minimal sets of vertices fulfilling both criteria.

In Sect. 2 we formalize the problem and relate it to the enumeration of hitting sets of bundles. In Sect. 3 we show that both associated decision problems are NP-hard, even under some degree or dimension restrictions. Finally in Sect. 4 we give a polynomial algorithm for the case where $dim(\mathcal{A})$ is bounded by some constant C.

2   The Bi-objective Transversal Problem

2.1   The Problem

We consider two hypergraphs on the same set of vertices V. The first hypergraph $\mathcal{A}$ will be denoted as the red hypergraph and represents the sets of vertices that have to be intersected. The second hypergraph $\mathcal{B}$ will be denoted as the blue hypergraph and represents the sets of vertices which should not be intersected if possible. We will represent such an instance as a tripartite graph, as shown in Fig. 1.

Notation: for $S \subseteq V$, we define $B_S = \{b \in \mathcal{B}; S \cap b \neq \emptyset\}$, the blue neighborhood of S.

Definition 1. *A bi-objective minimal transversal of the couple $(\mathcal{A}, \mathcal{B})$ is a set $S \subseteq V$ such that:*

1. S is a minimal transversal of $\mathcal{A}$
2. there exists no S' verifying condition 1 and such that $B_{S'} \subsetneq B_S$.

The collection of bi-objective minimal transversals of $(\mathcal{A}, \mathcal{B})$ is denoted by $btr(\mathcal{A}, \mathcal{B})$.

Example: in Fig. 1, consider the sets $S = \{u, v\}$ and $T = \{u, w\}$. Both are minimal transversals of $\mathcal{A}$. However S is not a bi-objective minimal transversal as $B_T = \{b_1\}$ is a strict subset of $B_S = \{b_1, b_2\}$.

If a_1, a_2, a_3 are tumoral samples, b_1, b_2, b_3 healthy samples and u, v, w, x mutations, then the set T is more likely linked to cancer than S because, while both sets hit all tumors, T hits only a subset of the healthy samples hit by S.

The problem we introduce is:

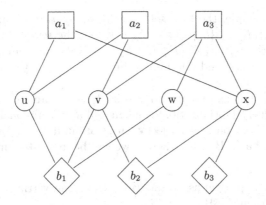

Fig. 1. Tripartite representation of an instance of the problem. The circled vertices are the vertices of the hypergraphs. The squared (resp. diamond) vertices represent the hyperedges of $\mathcal{A}$ (resp. $\mathcal{B}$).

Problem 2 **Bi-objective Hypergraph Transversal Problem.** Given hypergraphs $\mathcal{A}$ and $\mathcal{B}$ on the same vertices, enumerate $btr(\mathcal{A}, \mathcal{B})$.

A first approach for this problem would be to enumerate all minimal transversals of $\mathcal{A}$ and then to check for the minimality condition with respect to $\mathcal{B}$. However, such a procedure may spend an exponential time on enumerating minimal transversals of $\mathcal{A}$ which will be ruled out in the second step. Indeed, consider a hypergraph $\mathcal{A}$ on a vertex set V such that $\mathcal{A}$ has an exponential number of minimal transversals. Let S be one of them. Consider the hypergraph $\mathcal{B}$ having $V \setminus S$ as unique hyperedge. S is then the unique bi-objective minimal transversal of $(\mathcal{A}, \mathcal{B})$.

As the dual hypergraph problem corresponds to the special case $\mathcal{B} = \emptyset$, we propose to adopt the same strategy of an incremental search of the solutions. We therefore introduce the following problem:

Problem 3 **BIDUAL**$(\mathcal{A}, \mathcal{B}, \mathcal{X})$. Given hypergraphs $\mathcal{A}$ and $\mathcal{B}$ on the same vertices and a set $\mathcal{X}$ of bi-objective minimal transversals of $(\mathcal{A}, \mathcal{B})$, decide whether $btr(\mathcal{A}, \mathcal{B}) = \mathcal{X}$ or find a new minimal transversal in $btr(\mathcal{A}, \mathcal{B}) \setminus \mathcal{X}$.

2.2   Another Enumeration Problem

We now link Problem 2 with another, pre-existing problem, which involves a slight shift of perspective on the objects.

Definition 2. *Let $\mathcal{B}$ be a set of nodes, and $\mathcal{A}$ a set of sets of* bundles, *each bundle being a subset of $\mathcal{B}$. A Hitting Set of Bundles (HSB) of $(\mathcal{A}, \mathcal{B})$ is a subset $B \subseteq \mathcal{B}$ such that for each $a \in \mathcal{A}$ at least one bundle $x \in a$ is contained in B.*

If $\mathcal{A}$ and $\mathcal{B}$ are hypergraphs as before, we can talk about HSBs of $(\mathcal{A}, \mathcal{B})$ by identifying every vertex $x \in V$ with the bundle B_x.

In Fig. 1 for example, the base set contains b_1, b_2, b_3, the bundles are $B_u = \{b_1\}, B_v = \{b_1, b_2\}, B_w = \{b_1\}, B_x = \{b_2, b_3\}$, and the sets of bundles are $a_1 = \{B_u, B_x\}, a_2 = \{B_u, B_v\}, a_3 = \{B_v, B_w, B_x\}$. A HSB must contain one bundle in a_1, one bundle in a_2, and one bundle in a_3, e.g. $\{b_1, b_2\}$ contains $B_u \in a_1$, $B_v \in a_2$, and $B_w \in a_3$.

Hitting sets of bundles were introduced in [1] as a more abstract formulation of the Multiple Query Optimisation problem [20] and can model other interesting problems [6]. Work on this subject focuses on finding or approximating the minimum weight of a HSB [1,6], whereas we will be interested in minimality for inclusion:

Problem 4. Given hypergraphs $\mathcal{A}$ and $\mathcal{B}$ on the same vertices, enumerate the minimal (for inclusion) HSBs of $(\mathcal{A}, \mathcal{B})$.

Our interest in this problem comes from the following lemma and theorem:

Lemma 1. *Let $\mathcal{A}$ and $\mathcal{B}$ be two hypergraphs on the same vertex set V. A minimal transversal S of $\mathcal{A}$ is a bi-objective minimal transversal of $(\mathcal{A}, \mathcal{B})$ if and only if B_S is a minimal HSB of $(\mathcal{A}, \mathcal{B})$. Also, for every minimal HSB of $(\mathcal{A}, \mathcal{B})$ there exists such a transversal.*

Proof. The proof is a combination of the following simple facts:

- If S is a transversal of $\mathcal{A}$, then B_S is a HSB of $(\mathcal{A}, \mathcal{B})$.
- Conversely if B is a HSB of $(\mathcal{A}, \mathcal{B})$ there is a transversal S of $\mathcal{A}$ for which $B_S \subseteq B$.
- These two facts imply that if B is a HSB of $(\mathcal{A}, \mathcal{B})$, it is minimal if and only if there is no transversal S such that $B_S \subsetneq B$.
- Using the fact that any transversal S of $\mathcal{A}$ contains a minimal transversal S', and that $B_{S'} \subseteq B_S$, the claims are proved. □

Theorem 1. *If Problem 4 can be solved in quasi-polynomial time in the combined input and output size, then the same is true of Problem 2.*

Proof. Thanks to the Lemma 1, this can be done by enumerating the minimal HSBs and, for each B among them, enumerating the minimal transversals of $\mathcal{A}$ such that $B_S = B$. Notice that it can be done by computing the minimal transversals of the hypergraph $\mathcal{A}$ restricted to the vertices whose corresponding bundles are included in B.

The first part is done in quasi-polynomial time in the input and its own output (the minimal HSBs), which, by the second claim of our lemma, means quasi-polynomial time in the input and the final output (the bi-objective minimal transversals). The second part consists in computing as many transversal hypergraphs as there are minimal HSBs: this is done in quasi-polynomial time for each one, and again, by the second claim of the lemma, they are fewer than the final output size.

Hence the whole process runs in quasi-polynomial time in the combined input and output size. □

To study Problem 4, we consider again its incremental version:

Problem 5 **mHSB**$(\mathcal{A}, \mathcal{B}, \mathcal{X})$. Given hypergraphs $\mathcal{A}$ and $\mathcal{B}$ on the same vertices and a set $\mathcal{X}$ of minimal HSBs of $(\mathcal{A}, \mathcal{B})$, decide whether all the minimal HSBs are in $\mathcal{X}$ or find a new one.

The traditional transversal hypergraph problem becomes polynomially solvable when the degree is bounded in the hypergraph, or its dimension is bounded. We will now study these situations for the problems *BIDUAL* and *mHSB*.

3   Situations in Which the Problems Remain Hard

3.1   Bounded Degree

Here we show that unlike the classical transversal hypergraph problem, the enumeration of minimal HSBs and of bi-objective minimal transversals remain hard when both hypergraphs $\mathcal{A}$ and $\mathcal{B}$ have bounded degree.

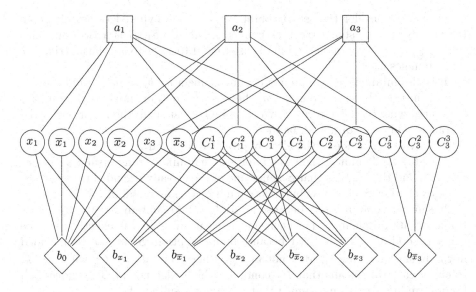

Fig. 2. Instance for the reduction of Theorem 2. The considered clauses are $C_1 = \overline{x}_1 \vee x_2 \vee \overline{x}_3$, $C_2 = x_1 \vee \overline{x}_2$ and $C_3 = x_3$.

Theorem 2. *Problems BIDUAL and mHSB are NP-complete, even restricted to $deg(\mathcal{A}) = 1$ and $deg(\mathcal{B}) \leq 3$.*

Proof. This reduction is inspired by Theorem 4 in [12], although new arguments were needed to prove the present theorem.

Let $f(x_1, \ldots, x_n) = C_1 \wedge \ldots \wedge C_m$ be a 3-CNF formula (without loss of generality, we assume that all clauses are different). We polynomially construct two hypergraphs $\mathcal{A}$ and $\mathcal{B}$ as follows (see Fig. 2):

- The vertex set V is composed of the $2n$ literals, and kn "clause" vertices C_k^i for all $1 \leq i \leq n$, $1 \leq k \leq m$.
- For all $1 \leq i \leq n$, a red hyperedge a_i contains $x_i, \overline{x}_i, C_1^i, \ldots, C_m^i$.
- For every literal l, a blue hyperedge b_l contains l and all the clauses in which its *negation* appears.
- Finally, a blue hyperedge b_0 contains all the literals.

Note that every vertex belongs to exactly one hyperedge in $\mathcal{A}$ (red), and at most 3 hyperedges in $\mathcal{B}$ (blue).

For all $1 \leq k \leq m$, the set $S_k = \{C_k^i, 1 \leq i \leq n\}$ is a bi-objective minimal transversal of $(\mathcal{A}, \mathcal{B})$ and its blue neighborhood B_{S_k} is a minimal HSB (note: B_{S_k} consists in the negated literals of clause C_k. This is exactly the bundle associated to C_k^i, for all i).

The result follows from the equivalence of these three conditions:

1. 3-SAT(f) has a positive answer.
2. $mHSB(\mathcal{A}, \mathcal{B}, \{B_{S_k}, 1 \leq k \leq m\})$ does not have a positive answer.
3. $BIDUAL(\mathcal{A}, \mathcal{B}, \{S_k, 1 \leq k \leq m\})$ does not have a positive answer.

$(1) \Rightarrow (2)$: Consider the set B containing the blue hyperedges associated to the literals of a satisfying truth assignment, and also b_0. B is a HSB since for every $1 \leq i \leq n$ it contains the bundle associated to x_i or to $\overline{x}_i$, thus hitting the set of bundles a_i.

It is also minimal. Indeed, removing any element b_0, b_{x_i} or $b_{\overline{x}_i}$ would prevent B from hitting the set of bundles a_i, because B contains none of the bundles associated with the C_k^i (since every clause in f is satisfied, the negations of the literals in a clause can not all be in B).

Finally B is not one of the already provided minimal HSBs.

$(2) \Rightarrow (3)$: By Lemma 1 there is a bi-objective minimal transversal S such that $B_S = B$, and this implies that S is not one of the already provided transversals.

$(3) \Rightarrow (1)$: Let S be a new bi-objective minimal transversal of $(\mathcal{A}, \mathcal{B})$.

It can not contain a C_k^i, because then B_S would contain strictly B_{S_k}, contradicting its minimality. The inclusion is clear since B_{S_k} is exactly the blue neighborhood of C_k^i. It is strict because by assumption, S can not be included in S_k: it must contain a vertex not in S_k, either a literal or a C_h^j with $h \neq k$. In either case this means that B_S contains a blue edge not in B_{S_k} (indeed: all literals are in b_0, and we assumed that C_h is not a subset of C_k).

Since S contains no C_k^i, it contains only literals. Being a minimal transversal of $\mathcal{A}$ it must contain exactly one literal for each variable, i.e. represent a truth assignment. Suppose this assignment does not satisfy some clause C_k: this means that S contains the negations of all the literals in C_k, so B_S contains B_{S_k}. Again, since B_S also contains b_0, the inclusion is strict, contradicting the minimality of B_S. $\square$

3.2 $\mathcal{B}$ of Bounded Dimension

We show a similar result for the case where $dim(\mathcal{B})$ is restricted.

Theorem 3. *Problems BIDUAL and mHSB are NP-complete, even restricted to $dim(\mathcal{B}) \leq 2$.*

Proof. Let us consider a 3-SAT instance $f(x_1, \ldots, x_n) = C_1 \wedge \ldots \wedge C_m$. We can consider, without loss of generality, that there exists no i such that all clauses contain either x_i or $\overline{x}_i$ (in that case the instance would easily be solved as two instances of 2-SAT).

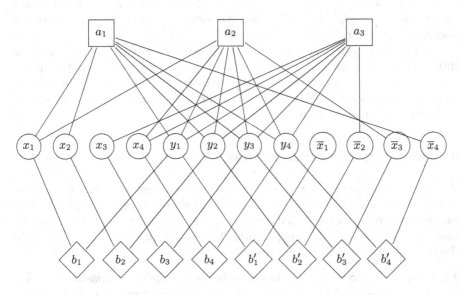

Fig. 3. Instance for the reduction of Theorem 3. The considered clauses are $C_1 = x_1 \vee x_2 \vee \overline{x}_4$, $C_2 = x_1 \vee \overline{x}_3 \vee x_4$ and $C_3 = \overline{x}_2 \vee x_3 \vee x_4$.

Construct the following hypergraphs (see Fig. 3):

1. Consider $3n$ vertices $V = \{x_1, \overline{x}_1, \ldots, x_n, \overline{x}_n, y_1, \ldots, y_n\}$.
2. For every $1 \leq j \leq m$, define a red hyperedge a_j including the x_i's and $\overline{x}_i$'s defining C_j as well as $\{y_1, \ldots, y_n\}$.
3. For every $1 \leq i \leq n$, define a blue hyperedge $b_i = \{x_i, y_i\}$ and a blue hyper-edge $b'_i = \{\overline{x}_i, y_i\}$. Observe that $\mathcal{B}$ is of dimension 2.

For every $1 \leq i \leq n$, consider $S_i = \{y_i\}$. It covers all the red hyperedges as well as b_i and b'_i. As neither x_i nor $\overline{x}_i$ is contained in all the clauses, it is a minimal solution to the bi-objective problem and its blue neighborhood $B_i = \{b_i, b'_i\}$ is a minimal HSB.

The theorem results from the equivalence of these conditions:

1. 3-SAT(f) has a positive answer.
2. $mHSB(\mathcal{A}, \mathcal{B}, \{B_i, \ldots, B_n\})$ does not have a positive answer.
3. $BIDUAL(\mathcal{A}, \mathcal{B}, \{S_1, \ldots, S_n\})$ does not have a positive answer.

(1)$\Rightarrow$(2): Consider a satisfying truth assignment and let B contain b_i when x_i is true in the assignment, and b'_i when x_i is false in the assignment. B is visibly a HSB and contains none of the B_i. It therefore contains a new minimal HSB.

$(2) \Rightarrow (1)$: Let B be a new minimal HSB. For any $1 \le i \le n$, it can not contain $\{b_i, b_i'\}$, hence B can be extended into a truth assignment. The fact that B hits all hyperedges of $\mathcal{A}$ means that such a truth assignment must satisfy all the clauses in f.

$(2) \Rightarrow (3)$: Let B be a new minimal HSB. There exists a bi-objective minimal transversal S such that $B_S = B$. Since B_S does not contain $\{b_i, b_i'\}$ for any i, S does not contain any of the y_i.

$(3) \Rightarrow (2)$: Let S be a new minimal bi-objective transversal, B_s is a minimal HSB. Let us prove that it does not belong to $\{\{b_i, b_i'\}, 1 \le i \le n\}$. Suppose $B_s = \{b_i, b_i'\}$. Since S can not be a superset of $\{y_i\}$, it is a subset of $\{x_i, \overline{x}_i\}$, and the assumption we made earlier contradicts the fact that S is a transversal of $\mathcal{A}$ (not all clauses contain either x_i or $\overline{x}_i$).   □

4   Bounding the Dimension of $\mathcal{A}$

We show in this section that when $dim(A) \le C$ for some constant C, one can reduce both enumeration Problems 2 and 4 to the transversal hypergraph problem in a hypergraph of bounded dimension, and thus solve them in polynomial time. To do this we build a hypergraph $\mathcal{H}$ whose transversals are exactly the HSBs of $(\mathcal{A}, \mathcal{B})$ (this can be of interest in itself).

Theorem 4. *mHSB can be solved in polynomial time if the dimension of $\mathcal{A}$ is bounded by a constant C. As a consequence, Problem 2 can be solved in incremental polynomial time, where the polynomial bounds depend on C.*

When the dimension of $\mathcal{A}$ is bounded, given a set $B \in \mathcal{B}$, the enumeration of all minimal transversals whose blue neighborhood is B corresponds to the enumeration of minimal transversals of a hypergraph of bounded dimension, and therefore can be done in incremental polynomial time. To show that Problem 2 is polynomial, it is sufficient to show that Problem 4 is polynomial. Actually we will show that Problem 4 can also be reduced to the enumeration of all minimal transversals of a hypergraph of bounded dimension.

Definition 3. *For $a \in \mathcal{A}$, let us denote by $\mathcal{H}_a$ the hypergraph induced on $\mathcal{B}$ by the bundles of a:*

- $V(\mathcal{H}_a) = \bigcup\limits_{x \in a} B_x$
- $\mathcal{E}(\mathcal{H}_a) = \{B_x \mid x \in a\}$

The following proposition gives a characterisation of the subsets B of $\mathcal{B}$ that contain a bundle of a given hyperedge $a \in \mathcal{A}$. Given $a \in \mathcal{A}$, by construction of the hypergraph $\mathcal{H}_a$, B contains a bundle of a if and only if B contains a hyperedge of $\mathcal{H}_a$.

For our purpose, we need to reformulate this simple fact. The formulation given in Lemma 2 can be seen as a direct consequence of the following observation. A subset of vertices X of a hypergraph $\mathcal{H}$ contains a hyperedge if and only if X is a transversal of the hypergraph $tr(\mathcal{H})$. Indeed, by the duality property between a hypergraph and its transversal hypergraph (see [2]), the minimal transversals of $tr(\mathcal{H})$ are exactly the minimal hyperedges of $\mathcal{H}$. Thus, a subset of vertices contains a hyperedge of $\mathcal{H}$ if and only if it contains a minimal transversal of $tr(\mathcal{H})$, i.e. if it is a transversal of $tr(\mathcal{H})$.

Lemma 2. *Let $B \subseteq \mathcal{B}$ and $a \in \mathcal{A}$. B contains a bundle of a if and only if B is a transversal of $tr(\mathcal{H}_a)$.*

Proof. ($\Rightarrow$) Assume that B contains the bundle B_x for some $x \in a$. Let $t \in tr(\mathcal{H}_a)$. Since B_x is a hyperedge of $\mathcal{H}_a$, t must intersect B_x and then $t \cap B \neq \emptyset$. We conclude that B is a transversal of $tr(\mathcal{H}_a)$.

($\Leftarrow$) Assume now that B is a transversal of $tr(\mathcal{H}_a)$ and contains no bundle in a. This means that for all $x \in a$, there exists $b_x \in B_x$ such that $b_x \notin B$. Let $t \subseteq \mathcal{B}$ be the set formed by all b_x for all $x \in a$ i.e. $t := \bigcup_{x \in a} B_x \setminus B$. Since for all $x \in a$, $B_x \setminus B \neq \emptyset$, t is a transversal of $\mathcal{H}_a$ and then contains a minimal transversal t' of $\mathcal{H}_a$. However by construction of t, we have $t' \cap B = \emptyset$, contradicting the fact that B is a transversal of $tr(\mathcal{H}_a)$. $\qquad\qquad\square$

Now since we require that B contains a bundle in every hyperedge of $\mathcal{A}$, B must be a transversal of $\mathcal{H}_a$ for every $a \in \mathcal{A}$.

Proposition 1. *The set of minimal HSBs of $(\mathcal{A}, \mathcal{B})$ is $tr(\bigcup_{a \in \mathcal{A}} tr(\mathcal{H}_a))$.*

Proof. Let $\mathcal{H} = \bigcup_{a \in \mathcal{A}} tr(\mathcal{H}_a)$.

$$B \in HSB(\mathcal{A}, \mathcal{B}) \Longleftrightarrow \forall a \in \mathcal{A}, B \text{ contains a bundle of } a \quad \text{by definition.}$$
$$\Longleftrightarrow \forall a \in \mathcal{A}, B \text{ is a transversal of } tr(\mathcal{H}_a) \quad \text{by Lemma 2.}$$
$$\Longleftrightarrow B \text{ is a transversal of } \mathcal{H}.$$

Thus, the HSBs of $(\mathcal{A}, \mathcal{B})$ are exactly the transversals of $\mathcal{H}$. Therefore, the set of minimal HSBs of $(\mathcal{A}, \mathcal{B})$ is $tr(\bigcup_{a \in \mathcal{A}} tr(\mathcal{H}_a))$. $\qquad\square$

If $dim(\mathcal{A})$ is bounded by C, for all $a \in \mathcal{A}$, $\mathcal{H}_a$ has at most C hyperedges and then each minimal transversal t of $\mathcal{H}_a$ is of size at most C. Then $\bigcup_{a \in \mathcal{A}} tr(\mathcal{H}_a))$ is a hypergraph of dimension at most C having at most $|\mathcal{A}||\mathcal{B}|^C$ hyperedges. We can then construct it in polynomial time and enumerate its minimal transversals in incremental polynomial time as it is of bounded dimension [7].

This implies that if $\mathcal{A}$ is of bounded dimension, the minimal HSBs of $(\mathcal{A}, \mathcal{B})$ can be enumerated in incremental polynomial time, thus proving Theorem 4.

Acknowledgements. We would like to thank the reviewers for their remarks which helped us to make the paper clearer, and particularly for pointing out the pre-existing notion of hitting sets of bundles.

References

1. Angel, E., Bampis, E., Gourvès, L.: On the minimum hitting set of bundles problem. Theoret. Comput. Sci. **410**(45), 4534–4542 (2009)
2. Berge, C.: Hypergraphs: Combinatorics of Finite Sets. North-Holland, Amsterdam (1989)
3. Bertrand, D., Chng, K.R., Sherbaf, F.G., Kiesel, A., Chia, B.K.H., Sia, Y.Y., Huang, S.K., Hoon, D.S.B., Liu, T., Hillmer, A., Hillmer, A., Nagarajan, N.: Patient-specific driver gene prediction and risk assessment through integrated network analysis of cancer omics profiles. Nucleic Acids Res. **43**(3), 1332–1344 (2015)
4. Boros, E., Elbassioni, K., Gurvich, V., Khachiyan, L.: An efficient incremental algorithm for generating all maximal independent sets in hypergraphs of bounded dimension. Parallel Process. Lett. **10**, 253–266 (2000)
5. Boros, E., Gurvich, V., Khachiyan, L., Makino, K.: Generating partial and multiple transversals of a hypergraph. In: Welzl, E., Montanari, U., Rolim, J.D.P. (eds.) ICALP 2000. LNCS, vol. 1853, pp. 588–599. Springer, Heidelberg (2000)
6. Damaschke, P.: Parameterizations of hitting set of bundles and inverse scope. J. Comb. Optim. **36**(2012), 1–12 (2013)
7. Eiter, T., Gottlob, G.: Identifying the minimal transversals of a hypergraph and related problems. SIAM J. Comput. **24**(6), 1278–1304 (1995)
8. Eiter, T., Gottlob, G.: Hypergraph transversal computation and related problems in logic and AI. In: Flesca, S., Greco, S., Leone, N., Ianni, G. (eds.) JELIA 2002. LNCS (LNAI), vol. 2424, pp. 549–564. Springer, Heidelberg (2002)
9. Eiter, T., Gottlob, G., Makino, K.: New results on monotone dualization and generating hypergraph transversals. SIAM J. Comput. **32**(2), 514–537 (2003)
10. Eiter, T., Makino, K., Gottlob, G.: Computational aspects of monotone dualization: a brief survey. Discrete Appl. Math. **156**(11), 2035–2049 (2008)
11. Fredman, M.L., Khachiyan, L.: On the complexity of dualization of monotone disjunctive normal forms. J. Algorithms **21**(3), 618–628 (1996)
12. Gurvich, V., Khachiyan, L.: On generating the irredundant conjunctive and disjunctive normal forms of monotone boolean functions. Discrete Appl. Math. **96–97**, 363–373 (1999)
13. Hädicke, O., Klamt, S.: Computing complex metabolic intervention strategies using constrained minimal cut sets. Metab. Eng. **13**(2), 204–213 (2011)
14. Haus, U.-U., Klamt, S., Stephen, T.: Computing knock-out strategies in metabolic networks. J. Comput. Biol. **15**(3), 259–268 (2008)
15. Jungreuthmayer, C., Nair, G., Klamt, S., Zanghellini, J.: Comparison and improvement of algorithms for computing minimal cut sets. BMC Bioinf. **14**(1), 318 (2013)
16. Khachiyan, L., Boros, E., Elbassioni, K., Gurvich, V.: An efficient implementation of a quasi-polynomial algorithm for generating hypergraph transversals and its application in joint generation. Discrete Appl. Math. **154**(16), 2350–2372 (2006)
17. Khachiyan, L., Boros, E., Elbassioni, K., Gurvich, V.: A global parallel algorithm for the hypergraph transversal problem. Inf. Process. Lett. **101**(4), 148–155 (2007)
18. Khachiyan, L., Boros, E., Gurvich, V., Elbassioni, K.: Computing many independent sets for hypergraphs in parallel. Parallel Process. Lett. **17**(02), 141–152 (2007)
19. Murakami, K., Uno, T.: Efficient algorithms for dualizing large-scale hypergraphs. Discrete Appl. Math. **170**, 83–94 (2014)

20. Sellis, T.K.: Multiple-query optimization. ACM Trans. Database Sys. **13**(1), 23–52 (1988)
21. Toda, T.: Hypergraph transversal computation with binary decision diagrams. In: Demetrescu, C., Marchetti-Spaccamela, A., Bonifaci, V. (eds.) SEA 2013. LNCS, vol. 7933, pp. 91–102. Springer, Heidelberg (2013)

Pairs Covered by a Sequence of Sets

Peter Damaschke[(✉)]

Department of Computer Science and Engineering,
Chalmers University, 41296 Göteborg, Sweden
ptr@chalmers.se

Abstract. Enumerating minimal new combinations of elements in a sequence of sets is interesting, e.g., for novelty detection in a stream of texts. The sets are the bags of words occuring in the texts. We focus on new pairs of elements as they are abundant. By simple data structures we can enumerate them in quadratic time, in the size of the sets, but large intersections with earlier sets rule out all pairs therein in linear time. The challenge is to use this observation efficiently. We give a greedy heuristic based on the twin graph, a succinct description of the pairs covered by a set family, and on finding good candidate sets by random sampling. The heuristic is motivated and supported by several related complexity results: sample size estimates, hardness of maximal coverage of pairs, and approximation guarantees when a few sets cover almost all pairs.

1 Introduction

1.1 Motivation and Aim

In a chronological sequence of texts about some topic, such as a news stream, posts in social media, a timeline, etc., we may want to quickly understand what is novel in each entry, or what caused peaks in the volume of news about a topic. A simple approach is to determine new combinations of words. Ignoring the order of words, grammar, etc., let us consider the data as a sequence of sets ("bags of words"). The given bags may already be preprocessed: ignoring stop words, stemming, identifying synonyms, etc.

Definition 1. *Let $B_0, B_1, B_2, \ldots, B_{m-1}$ be a sequence of sets that we call* bags. *For another bag $B := B_m$ we call a subset $X \subseteq B$* new *at m, if X was not already subset of an earlier bag: $\forall i < m : X \setminus B_i \neq \emptyset$. Otherwise $X \subseteq B$ is said to be* old *at m. We call $X \subseteq B$* minimal new *at m, if X is new and also minimal (with respect to inclusion) with this property.*

For knowing *all* new sets it suffices to know the *minimal* new sets. In the case of word sets, X can be a single new name or term, or a new pair, triple, etc., of old words, indicating new connections. They often give a good intuitive description of what is novel. Others can be understood only in context, or they are unrelated and meet only by chance. But before judging the new sets semantically one

© Springer International Publishing Switzerland 2015
A. Kosowski and I. Walukiewicz (Eds.): FCT 2015, LNCS 9210, pp. 214–226, 2015.
DOI: 10.1007/978-3-319-22177-9_17

needs to find them first. Examples as below suggest that minimal new pairs are abundant, and minimal new sets of $h > 2$ words are rare. This is expected since X can be minimal new only if all its subsets appeared earlier.

Example. In a timeline of major discoveries in physics we should find an article about the first formulation of the law of conservation of energy. Then {conservation, energy} is minimal new, as this combination is new, but the term "energy" was coined earlier, and other conservation laws were formulated earlier, e.g., conservation of matter. Other articles will deal with the prediction and confirmation of electromagnetic waves. The pair {electromagentic, wave} is minimal new: They were totally unknown before, but physics had already dealt with other waves (such as light waves, being unaware that they are electromagnetic, too), and with other electromagnetic phenomena like induction.

Novelty mining and novelty detection in streams is extensively studied [2,11,12]. The subject is also related to minimal infrequent itemsets and mining emerging patterns [1,7,8]. However, in the present work we do not apply any language processing or machine learning to extract news or to produce online summaries, rather we explore the complexity of a combinatorial approach that relies on a very simple idea but gives meaningful hints to novelty. We consider the following problem, applied to sequences of texts like short articles.

Problem. Given a sequence of bags $B_0, B_1, B_2, \ldots$, enumerate the minimal new subsets X in each bag.

X is minimal new at m if and only if X is a minimal hitting set in the family of sets $\{B \setminus B_i \mid i < m\}$, that we call *hyperedges*. A set family is also called a *hypergraph*, and a *hitting set* (or *transversal*) intersects all hyperedges.

Define $n := \max |B_i|$, let m be the number of hyperedges (bags), and let c denote the number of minimal hitting sets (new sets at index m). Note that c is known only in hindsight, nevertheless one can express time bounds in terms of c. We may also fix a small number h and enumerate only hitting sets of size at most h. In particular, we focus on $h = 2$ due to the above motivation.

Algorithms for several parameterizations of minimal transversal enumeration are given in [4]: One can enumerate them with $O(n^2 m^2 e^{m/e})$ delay, hence in $O(n^2 m^2 e^{m/e} c)$ time, or alternatively in $O(n^2 c^3 e^{c/e} m)$ time. Actually one can enumerate any number $j \leq c$ of minimal hitting sets in $O(n^2 j^3 e^{j/e} m)$ time.)

A time bound is also provided in [4] for the case when the elements have *complementary degree* at most $q \ll m$, that is, every element appears in all but at most q hyperedges. (However, their result addresses only the verification of a given enumeration of the minimal transversals, not the construction.) In our application, the assumption means that words appear in at most q bags B_i, with $q \ll m$. This is sensible, because more frequent words are common words or stop words that are not informative and may be ignored.

1.2 Overview of Contributions

First we fill the mentioned gap in the parameterized complexity of transversal enumeration, but then we focus on small transversal sizes h, due to our motivation. The minimal new subsets of size h in each bag of size n in a sequence can be enumerated in $O(n^h)$ time. The rest of the work deals with the most relevant case $h = 2$, that is, new pairs. Note that we do not improve upon the $O(n^2)$ worst-case time per bag, but we study heuristics to save time for special structures that are however likely to appear in streams of topic-wise related texts: Large overlaps with earlier bags allow to recognize old pairs in (ideally) linear time. Saving a factor up to n in the processing time is worthwhile. Twin graphs give a succinct description of the pairs covered by a family of bags. We use them within a simple greedy heuristic to cover many old pairs. The greedy approach has known performance guarantees. We combine it with random sampling to find good candidate bags fast enough. Several complexity results justify this approach: We derive a logarithmic upper bound on the number of elements to be sampled in order to find bags with nearly largest overlaps with a given bag. We show that covering the exact maximum number of pairs by a prescribed number r of bags is $W[1]$-hard. We also propose to build larger bags from the given ones and show that some ternary set operation is sufficient for that. A final technical section is devoted to approximation guarantees for the number of covered pairs, if a few bags cover almost all pairs. We give a combinatorial approach to prove such results, based on special minimal set families and symmetries in convex optimization problems. Much of the latter parts is work in progress. We point to several open problems, highlighted as "Research question". The practical scenario might also be turned suitably into online problems. Besides the complexity aspects it would be interesting to test the approach on extensive data sets. – Due to lack of space, several proofs and proof details are omitted in this version.

2   Finding Minimal New Sets

2.1   Transversals for Small Complementary Degree

Theorem 1 below might result from [4] by reductions, however we present a self-contained algorithm description. Given a hitting set H and an element $s \in H$, we call E a *private hyperedge* if $s \in E$ and s is the only element of $H \cap E$. A hitting set H is minimal if and only if every $s \in H$ has at least one private hyperedge. We can assume that no hyperedge is subset of another one.

Theorem 1. *In hypergraphs $\mathcal{H}$ with complementary degree at most q we can enumerate all minimal hitting sets in $O(n^3mq^2e^{q/e}c)$ time.*

Proof. Loop through the $O(nm)$ pairs (s, E) of any element s and any hyperedge $E \ni s$ and do the following from scratch. Delete all hyperedges $F \ni s$ (they are hit by s) and all elements in E (in order to keep E private for s). At most q hyperedges remain, from which the elements of E are deleted. No hyperedge becomes empty, since no other hyperedge is a subset of E. In this "small" instance enumerate all minimal hitting sets H. Every $H \cup \{s\}$ is a minimal hitting set of $\mathcal{H}$: It intersects all hyperedges, every element has a private hyperedge since H was minimal, and s has the private hyperedge E. Conversely, all c minimal hitting sets of $\mathcal{H}$ are obtained in this way. To avoid duplicates, first sort the elements arbitrarily, and demand the above s to be the first element in the hitting sets, that is, delete also elements before s. Now some pairs (s, E) yield no solution since some hyperedges lose all their elements, but we recognize these cases instantly. The "small" instances are solved by the $O(n^2m^2e^{m/e})$-delay algorithm for hitting set enumeration from [4]; replace m with q. Some original edges may become identical due to deletions, which does not affect the time bound. Finally note that every "small" instance is prepared in $O(nm)$ time. $\qquad\square$

2.2   Enumerating the Minimal New Sets in a Sequence

Unlike the previous parameterizations, for the text stream applications we need small transversal size h and time bounds that are polynomial in q and m and also have a better dependency on n. Naive exhaustive search takes every $X \subseteq B_m$, $|X| \le h$, and checks in $O(hq)$ time whether X intersects all hyperedges $B_m \setminus B_i$. (For each element s we maintain the at most q indices with $s \in B_i$. An $s \in X$ misses at most q hyperedges, and we see whether the other elements of X hit them.) This way we would need $O(hqn^h/h!) = O(hq(en/h)^h) = O(n^h \cdot q \cdot h(e/h)^h)$ time to find all minimal new sets of size h in B_m. However we can avoid checking all $X \subseteq B_m$ against all previous $B_m \setminus B_i$ ($i < m$): We generate the candidate sets X with increasing sizes and, due to minimality, stop as soon as all further X would be supersets of already detected minimal new sets at m. More importantly, we can use information about the minimal new sets at earlier $i < m$, as detailed below.

Let $f(X) := \min\{i \mid X \subseteq B_i\}$, and let $f(X)$ be undefined if no such i exists. Note that $f(X) = i$ if and only if X is new at i (but not necessarily minimal). Furthermore, X is then old at any further index $j > i$. In the following we assume for simplicity that a dictionary operation costs constant time per element. We store some sets X along with $f(X)$ in a dictionary. In particular, whenever a set X is minimal new at i, we store "$f(X) = i$". This is no extra work, since our aim is to enumerate all minimal new sets. So suppose that we have already determined all minimal new sets with at most h elements at all $i < m$. In particular, every single element is new as soon as it appears for the first time. Hence we can check every single element for being new, in $O(nm)$ time in total. After these precautions we get for $B = B_m$:

Theorem 2. *The minimal new subsets of size at most h in a bag of n elements can be enumerated in $O(n^h \cdot 2^h/h!)$ time.*

Proof. Suppose that we have already determined all minimal new sets with fewer than j elements. Consider any set $X \subseteq B_m$ of j old elements, such that no $Y \subset X$ is new at m. In order to check whether X is new at m, thus minimal new, we proceed as follows. If a value $f(X) < m$ is stored, then X is old. Suppose that no $f(X)$ value is stored. Then we must figure out whether $X \subseteq B_i$ for some $i < m$. Assume that such an index exists, and let i be the smallest one. Since no $f(X)$ value is in place, X is not minimal new at i, and this is possible only if some nonempty $Y \subset X$ is minimal new at i, in particular, $f(Y) = i$ has been stored. Thus we must only look up the $2^j - 2$ values $f(Y)$ and, if $f(Y)$ is stored, check whether $X \subseteq B_{f(Y)}$. If we find such Y, then we conclude that X is old at m, and we can also store $f(X)$ which equals the smallest such $f(Y)$. Otherwise we conclude that X is minimal new at m, and we also store $f(X) := m$. Altogether we can decide in $O(2^j)$ time whether a set X of size j is minimal new. The number of sets X to check is at most $\binom{n}{j} < n^j/j!$. The procedure is repeated for increasing j up to h, and $j = h$ dominates the time bound. $\qquad\square$

3   A Heuristic for Minimal New Pairs

3.1   Below the Worst-Case Quadratic Time

The remainder of the paper deals with the case $h = 2$ only. As argued earlier, minimal new pairs, i.e., new pairs of old elements, seem to be very good indicators of novelty in text streams. (Also note that the case $h = 1$ is trivial.) Let $C_j := B_j \cap \bigcup_{i<j} B_i$ denote the set of old elements in B_j. We define the *index set* of an element $s \in C_j$ as $\{i|\ i < j,\ s \in B_i\}$. While watching the sequence we can collect the old elements and thus obtain the sets C_j in $O(\sum_j |B_j|)$ overall time, and maintain the index sets in $O(\sum_j |C_j|)$ overall time. After this trivial auxiliary processing, the problem for each bag can be stated as follows:

Problem. In a sequence of bags $B_0, B_1, B_2, \ldots$ enumerate the new pairs in $C := C_m = B_m \cap \bigcup_{j<m} B_j$, that is, pairs that are not covered by earlier B_j, $j < m$. Besides the B_j, the index sets of all elements in C are already given. Let $n := |C|$. We also refer to the $B_j \cap C$ as *bags*, without risk of confusion.

By Theorem 2 we can recognize the new pairs in C in $O(n^2)$ time in the worst case. The interesting matter is to enumerate them faster if only a minority of pairs in C is new: Note that, once we detect a large bag $B_j \cap C$, we can immediately exclude the pairs therein as candidates for new pairs, in $O(|B_j \cap C|)$ time rather than $O(|B_j \cap C|^2)$. Large intersections are likely in streams of related texts, as they tend to form topics clusters with similar word content.

Driving the idea one step further, we may select a number r of bags, exclude all pairs covered by them, and test only the uncovered pairs for being new, each in $O(1)$ time. However the total time for finding such bags and listing the uncovered pairs must be $o(n^2)$, therefore we will have to use, in general, some $r < m$ bags. In the following we elaborate on this idea.

3.2   The Twin Graph

Suppose that we have already selected r bags. The set of pairs (not) covered by them is completely described by the following structure.

Definition 2. *With respect to a set of r bags, any two elements of C with the same index set (i.e., elements being in exactly the same bags) are called* twins. *This yields an equivalence relation on C whose t equivalence classes are called* twin classes. *The* twin graph *is the graph whose vertices are the twin classes, and where any two vertices with disjoint index sets are joined by an edge.*

Some properties of the twin graph are obvious: It has a self-loop only at the vertex representing elements that are in no bag (if existing). The uncovered pairs of elements are exactly those in any two adjacent twin classes. The twin graph has $t \leq \min(n, 2^r)$ vertices and at most $\frac{1}{2}\min(t^2, 3^r)$ edges, since the r indices can be partitioned in $\frac{1}{2} \cdot 3^r$ ways in two disjoint index sets and the rest.

Proposition 1. *The twin graph of r bags can be constructed iteratively, that is, by inserting the bags one by one, in $O(r \cdot (\min(t^2, 3^r) + n))$ time.*

Proof. Every bag may split some twin classes in two smaller ones. These splittings are done in $O(n)$ time per bag, hence $O(rn)$ time overall. Index sets are stored as a tree in an obvious way. To obtain the edges we either check the $O(t^2)$ pairs of twin classes for disjointness of their index sets in $O(rt^2)$ time, or we take all $O(3^r)$ possible pairs of disjoint index sets and check their existence, which can be done in $O(r)$ time for each pair.   □

3.3   Greedy Partial Set Cover of Pairs

We need to determine the bags to be inserted and to update their twin graph. We can stop as soon as inserting another bag, that covers p new pairs, requires $O(p)$ time, i.e., the time needed to simply test these pairs individually for being new, as in Theorem 2. As Proposition 1 indicates that the time to update the twin graph can grow exponentially in the number r of bags, let us fix a number r of bags (which may however depend on n, say, some $r = o(\log n)$), and aim at solving the following problem.

PARTIAL SET COVER OF PAIRS. Given a family of bags and a number r, identify r bags that together cover the maximum number of pairs.

Proposition 2. PARTIAL SET COVER OF PAIRS *is NP-complete, and also $W[1]$-complete in the parameter r.*

Proof. Reduction from INDEPENDENT SET. Omitted due to space limits.   □

Due to this negative observation we resort to a greedy approach: The PARTIAL SET COVER problem asks to cover a maximum number of elements by a prescribed number r of bags. The greedy algorithm for (PARTIAL) SET COVER iteratively adds to the solution a bag with the largest number of yet uncovered elements. The number of elements covered in r greedy steps is at least a $1 - (1 - 1/b)^r$ fraction of the optimum that could be covered by b bags. This was shown in [5], generalizing an earlier result in [9,10] for $r = b$. This bound is also tight for $r = b$ [10], and the worst-case example for $r = b$ also works in general.

Now let P be the set of the $\binom{n}{2}$ pairs of elements in C, and let P_j be similarly defined for each $B_j \cap C$. We refer to the P_j also as bags, without risk of confusion. Since the P_j and P are made of pairs of other elements, PARTIAL SET COVER

OF PAIRS is a special case of PARTIAL SET COVER. Thus, the greedy algorithm gives at least the same approximation guarantee.

Research Question. Figure out the approximation ratio for greedy PARTIAL SET COVER OF PAIRS. We conjecture that it is significantly better than for PARTIAL SET COVER. However, the case of pairs appears to be intrinsically more difficult. While the proof in [5] is merely based on the pigeonhole principle, we must also deal with the twin graph structure. (Section 5 will give a method to obtain some results for the case when r bags cover almost all pairs.)

4   Supplementary Results

4.1   Sampling Large Intersections

Since $o(n^2)$ time is mandatory, implementation of the greedy heuristic needs some care. Besides updating the twin graph we have to count in the time for finding the next bag that covers as many further pairs as possible. As long as m is small compared to n we can afford computing all intersections $B_j \cap C$ in $O(mn)$ time. For larger m we may sample some random elements or pairs from C, use their index sets (of size at most q, typically much smaller than m) to count their occurences in the given bags, and take the bags with most hits.

Theorem 3. *Suppose that the largest bag has $(1-x)n$ elements, and we sample s random elements and return the bag with the largest number of hits. Then we fail to find a bag with at least $(1-y)n$ elements with probability at most $2m \cdot \exp(-(y-x)^2 s/16y)$. In particular, we get a failure probability below any prescribed constant by choosing $s = \Theta(\log m \cdot y/(y-x)^2)$.*

Proof. By Chernoff bounds. Omitted due to space limits.   □

In particular, we need $O((q \log m)/y)$ time to find a bag whose size is at least $1 - y$ of the maximum size of a bag, where q denotes the maximum size of the index sets. (Fix x in Theorem 3 and note that we must traverse the index set of every sample.) While Theorem 3 was formulated for elements, it applies literally to sampling of pairs, too, and can be used in each step of the greedy heuristic: Some bags are already selected, and they form a twin graph with t vertices. Then the uncovered pairs form the edge sets of $O(t^2)$ cliques and bicliques of known sizes, thus one can easily sample random uncovered pairs from them.

4.2   Building Larger Bags

For a sequence of bags $B_0, B_1, B_2, \ldots, B_{m-1}$ consider the graph G whose vertices are the elements, where two vertices u, v are adjacent if and only if $u, v \in B_i$ for some i. A *clique edge cover* in a graph is a set of cliques that cover all edges. Hence the bags form a clique edge cover of G. However, G may contain further, larger cliques, and using them besides the given bags within our heuristic is beneficial, since they cover more of the old pairs. If we can quickly find and build some of these larger cliques, we can use them later in the sequence and make later steps more efficient. Bags with large intersections inside the current bag B_m (that are anyway used to cover many old pairs in B_m) are likely to have large intersections also outside B_m. We may apply the following ternary set operation Δ on them: $\Delta(X, Y, Z) := (X \cap Y) \cup (X \cap Z) \cup (Y \cap Z)$. Note that each pair of elements in $\Delta(X, Y, Z)$ is also contained in some of X, Y, Z. In particular, if X, Y, Z are any three cliques, then $\Delta(X, Y, Z)$ is a clique, too. A neat fact is that all maximal cliques can be generated from any clique edge cover using only the Δ operation. This supports the idea to apply Δ to bags that are anyhow considered in a step of the heuristic.

Proposition 3. *Given a graph along with a clique edge cover $\mathcal{K}$, we can obtain every maximal clique solely by repeated Δ operations applied to cliques of $\mathcal{K}$.*

Proof. By induction on the size. Omitted due to space limits. $\qquad \square$

The sizes of bags produced by Δ can grow quickly, by a factor up to $\frac{3}{2}$ (if $X = Y' \cup Z'$, $Y = X' \cup Z'$, $Z = X' \cup Y'$ for three disjoint sets X', Y', Z' of equal size. On the combinatorial side it would be interesting to know what cliques size are guaranteed to exist in a graph with few non-edges compared to $\binom{n}{2}$. Turán's theorem [13] states, informally, that a graph with few non-edges always has a large clique. We would be interested in a generalization to unions of r cliques:

Research Question. Given r, n, u, what is the guaranteed number of edges that can be covered by r cliques, in any graph of n vertices and $\binom{n}{2} - u$ edges? While some ideas from the proof of Turán's theorem generalize to a union of r cliques, the extremal problem apparently becomes harder.

We remark that another conceivable approach for obtaining larger bags would be to see if some r bags can be replaced with $k < r$ new bags that cover the same edges. This amounts to the CLIQUE EDGE COVER problem parameterized by k. This NP-hard problem is fixed-parameter tractable in parameter k [6], however with doubly exponential time bound, and non-existence of a polynomial kernel [3] leaves little hope to reduce this asymptotic worst-case bound.

5   Approximations if a Few Bags Cover All Pairs

5.1   Setup and Preparations

Proposition 2 raises the question: What fraction of pairs in C can we cover within some $O(m^{O(1)}n)$ time bound? Due to the context, we are interested in the case that r bags exist that cover all pairs in C, subject to some small fraction. Note that m may here denote the number of sampled candidate bags (having large intersections with C) rather than the length of the entire sequence of sets.

More specifically, suppose that some r bags cover $(1-\delta)\binom{n}{2}$ pairs, where $\delta > 0$ is a small number. Consider their twin graph (Definition 2), with t vertices. If two adjacent twin classes have $\sqrt{\delta}n$ elements each, then already δn^2 pairs of elements are uncovered, contradicting the assumption. Hence the twin graph has a vertex cover of twin classes with fewer than $\sqrt{\delta}n$ elements each. The complement of this vertex cover is an independent set in the twin graph, thus representing a subset C' of C with at least $(1 - t\sqrt{\delta})n$ elements in which all pairs are covered. (Actually the fraction is closer to 1; the given bound is coarse only due to the general argument.) Thus we may clean up our question as follows:

Problem. In a family of m bags, suppose that r of them cover all pairs in C', for some $C' \subset C$ of size $n' := |C'| > (1 - \epsilon)|C|$. (But note that C' is not specified in the input.) What number $\gamma\binom{n'}{2}$ of pairs can we at least cover within some $O(m^{o(r)}n)$ time bound? We call the fraction γ the *coverage*.

In the following we work with the complement of the twin graph restricted to C', which is a clique of some $t' \leq t$ vertices whose edges are covered by r smaller cliques (bags). We suppose that r is minimal, that is, $r - 1$ of the bags would not cover all pairs. Let A be the *incidence matrix* of these bags: A has a row for every bag, a column for every twin class, and entries $a_{ij} = 1$ if bag i contains the twin class j, and $a_{ij} = 0$ otherwise. Hence every column is the characteristic vector of an index set. A pair of twin classes is *private* for a bag if that bag covers that pair but none of the other $r - 1$ bags does. We establish some properties of A. Since r is minimal, no row is contained in another row, and since the bags cover all pairs, the columns pairwise intersect. In other words:

(1) For any two rows i and i' there exists some columns j and j' such that $a_{ij} = a_{i'j'} = 1$ and $a_{ij'} = a_{i'j} = 0$.

(2) For any two columns j and j' there exists some row i with $a_{ij} = a_{ij'} = 1$.

Since we are only interested in worst-case approximation ratios, we can assume further restrictions: Deletion of any twin class from any bag must destroy some private pair, since otherwise there would exist a worse instance with smaller bags, such that the coverage can only decrease. In other words, every twin class in a bag belongs to some private pair of that bag. More formally:

(3) For each entry $a_{ij} = 1$ there exists a column j' such that $a_{ij'} = 1$, and $a_{i'j} = 0$ or $a_{i'j'} = 0$ holds for each row $i' \neq i$.

We remark that (3) implies (1). Another conclusion is that any two index sets are incomparable, that is, not in subset relation:

(4) For any two columns j and j' there exists some rows i and i' such that $a_{ij} = a_{i'j'} = 1$ and $a_{ij'} = a_{i'j} = 0$.

To show (4), let i'' be some row according to (2): $a_{i''j} = a_{i''j'} = 1$. There must be some row i where $a_{ij} \neq a_{ij'}$, since equal columns would represent the same twin class. Suppose $a_{ij} = 1$ and $a_{ij'} = 0$. Due to (3), there exists a column j'' such that $a_{ij''} = 1$. Since the pair represented by $a_{ij} = a_{ij''} = 1$ is private and $a_{i''j} = 1$, it also follows $a_{i''j''} = 0$. Condition (2) applied to j' and j'' yields the existence of another row i' with $a_{i'j'} = a_{i'j''} = 1$. Finally, since the pair represented by $a_{ij} = a_{ij''} = 1$ is private and $a_{i'j''} = 1$, it follows $a_{i'j} = 0$.

The following consideration of symmetries will help reduce case distinctions. Automorphisms of an optimization problem are permutations of the variables that leave the set of constraints invariant. An orbit of the automorphism group is a set of variables mapped onto each other by automorphisms; clearly they form equivalence classes. For convex minimization problems it is known that the convex combination of any two minimal solutions is a minimal solution, too. From this it follows: If we take any minimal solution, apply all automorphisms to the variables, and take component-wise the average of all these solutions, we obtain a minimal solution where all variables in each orbit have equal values.

Consider an optimization problem with variables $x_1, \ldots, x_t$ and y, with the objective $\min y$, and constraints $g_j(x_1, \ldots, x_t) \leq y$ and $h_j(x_1, \ldots, x_t) \leq 0$, where all g_j and h_j are convex functions. Such a problem is convex and can be rephrased as $\min \max_j g_j(x_1, \ldots, x_t)$ under the constraints $h_j(x_1, \ldots, x_t) \leq 0$.

5.2 Illustration: Some Approximation Guarantees

Now we apply these tools to determine set families that minimize certain guaranteed approximation ratios. Since only $o(n^2)$ time bounds matter, the smallest r are most relevant for us. To avoid technicalities we assume large enough n such that we can neglect lower-order terms and the effects of rounding fractional numbers to integers. The case of $r \leq 2$ bags is trivial. Let $O(m^\omega)$ be a time bound for multiplying $m \times m$ matrices.

Theorem 4. *If r of $m < n$ bags cover all pairs in C, we find r such bags in $O(m^{\omega-1}n) = o(m^2 n)$ time if $r \leq 4$, and in $O(m^{r-2}n + (2m)^r)$ time if $r > 4$.*

Proof. Let A again denote the $0, 1$-incidence matrix whose rows represent bags. First we exclude in $O(mn)$ time the trivial case that some row has only 1s. We call a set R of rows unsuitable if R has some all-0 column, and suitable otherwise. Let J be the graph whose m vertices are the rows, with an edge in every suitable pair. We compute J by switching 0s and 1s in A and multiplying this matrix with its transpose in $O((n/m)m^\omega = O(m^{\omega-1}n)$ time. Let R denote a set of r rows (bags) or the corresponding $r \times n$ submatrix of A. For $r = 3$ we have: R covers all pairs if and only if any pair of rows in R is suitable. For $r \geq 4$ we have: R covers all pairs if and only if R is not the union of two unsuitable sets S, T with $|S| + |T| = r$ and $2 \leq |S| \leq |T| \leq r - 2$. (Recall that the trivial case was ruled out.) For $r = 3$ we need to find a triangle in J, which is well known to work in time $O(m^\omega) = O(m^{\omega-1}n)$, as $m < n$ was assumed. For $r = 4$ we also need to find a triangle plus any fourth vertex, or a star of three edges in J. The latter is trivially done in $O(m^2) = O(mn)$ time. For $r > 4$ we determine all unsuitable sets of at most $r - 2$ rows in $O(m^{r-2}n)$ time. The condition for each R is then checked in $O(2^r)$ time. □

A slight relaxation of this algorithm handles the case when r bags cover all pairs in C': Then we call a pair of rows suitable when at most ϵn columns lack a 1, apply the arguments to C', and find r bags with coverage $1 - \Theta(\epsilon)$ in the same time. – The following proofs show the existence of bags with certain minimum numbers of elements in C', implying coverage values if we would choose these bags. Actually we select bags of that size in C rather than in the unknown $C' \subset C$, which can only increase the coverage due to smaller overlaps.

Proposition 4. *If three out of m bags cover all pairs in C', then we can find one bag with coverage 4/9 in $O(mn)$ time, and two bags with coverage 7/9 exist.*

Proof. We remark that the pigeonhole principle would only yield coverage $1/3$ and $2/3$, respectively. Any three bags satisfying conditions (1)–(4) have the form $T_1 \cup T_2$, $T_1 \cup T_3$, $T_2 \cup T_3$, with twin classes T_1, T_2, T_3. From the m given bags we take one bag, and a pair of bags, respectively, with the largest coverage. Let $x_i := |T_i|$. For a lower bound on the coverage we minimize the maximum number of pairs covered by one or two of the considered bags. The numbers of covered pairs are convex functions of the x_i, the only orbit is $\{x_1, x_2, x_3\}$, and the constraints $x_i \geq 0$ and $x_1 + x_2 + x_3 = n'$ are linear. Now the above symmetry consideration yields that $x_1 = x_2 = x_3 = \frac{1}{3}$ is the minimal solution in both cases, and finally, simple calculations yield the ratios. □

Two bags with coverage 7/9 could be found in $O(m^2 n)$ time, but Theorem 4 achieves already more. We mention the coverage 7/9 only as a benchmark for the next result that sacrifices some coverage for speed.

Proposition 5. *If three out of m bags cover all pairs in C', then we can find two bags with coverage 56/81 > 0.69 in $O(mn)$ time.*

Proof. A greedy strategy first takes a largest bag G and then adds a second bag that covers a maximum number of further pairs. Three bags that minimally cover all pairs have a structure as in Proposition 4, with each twin class further split in two by G. This results in six twin classes $T_1, \ldots, T_6$ such that $G = T_1 \cup T_2 \cup T_3$, and $T_1 \cup T_2 \cup T_4 \cup T_5$, $T_2 \cup T_3 \cup T_5 \cup T_6$, $T_1 \cup T_3 \cup T_4 \cup T_6$ are mentioned three bags. Note that $x_1 + x_2 + x_3 \geq \frac{2}{3}n'$, and the second greedy step can take some of these three bags (or a better one). For a lower bound on the coverage we minimize the maximum number of pairs covered by these three choices. The numbers of covered pairs are convex functions of the x_i, the orbits are $\{x_1, x_2, x_3\}$ and $\{x_4, x_5, x_6\}$, and the constraints $x_i \geq 0$, $x_1 + x_2 + x_3 \geq \frac{2}{3}n'$ and $x_1 + \cdots + x_6 = n'$ are linear. Symmetry consideration yields the existence of a minimal solution where $x_1 = x_2 = x_3$ and $x_4 = x_5 = x_6$. Moreover, $x_1 + x_2 + x_3 = \frac{2}{3}n'$ is the worst case, and simple calculations yield the ratios. □

Proposition 6. *If four out of m bags cover all pairs in C', then we can find one bag with coverage 9/25 in $O(mn)$ time.*

Proof. By case inspections, any four bags satisfying the conditions (1)–(4) have the form $T_1 \cup T_2 \cup T_3$, $T_1 \cup T_4$, $T_2 \cup T_4$, $T_3 \cup T_4$, with twin classes T_1, T_2, T_3, T_4, and then similar arguments as above apply, with orbits $\{x_1, x_2, x_3\}$ and $\{x_4\}$. □

Research Question. Systematically explore the trade-off between time and coverage for any r and number s of greedy bags. The numbers of pairs covered by bags are sums of squares and products of the induced twin class sizes, thus always convex. But with growing r and s the task becomes more challenging, as we need to understand the combinatorics of the "minimal" coverings and their binary incidence matrices that obey conditions (1)–(4). It might be possible to combine the pigeonhole principle with our stronger method. Moreover, does the relationship to matrix multiplication also yield lower time bounds (see [14])?

Acknowledgment. This work has been supported by the Swedish Foundation for Strategic Research (SSF) through Grant IIS11-0089 for a data mining project entitled "Data-driven secure business intelligence". The author also wishes to thank the referees for careful reading.

References

1. Boros, E., Gurvich, V., Khachiyan, L., Makino, K.: On Maximal Frequent and Minimal Infrequent Sets in Binary Matrices. Ann. Math. Artif. Intell. **39**, 211–221 (2003)
2. Ceci, M., Appice, A., Loglisci, C., Caruso, C., Fumarola, F., Valente, C., Malerba, D.: Relational frequent patterns mining for novelty detection from data streams. In: Perner, P. (ed.) MLDM 2009. LNCS, vol. 5632, pp. 427–439. Springer, Heidelberg (2009)
3. Cygan, M., Kratsch, S., Pilipczuk, M., Pilipczuk, M., Wahlström, W.: Clique cover and graph separation: new incompressibility results. ACM Trans. Comput. Theory **6**, Article 6 (2014),
4. Elbassioni, K.M., Hagen, M., Rauf, I.: Some fixed-parameter tractable classes of hypergraph duality and related problems. In: Grohe, M., Niedermeier, R. (eds.) IWPEC 2008. LNCS, vol. 5018, pp. 91–102. Springer, Heidelberg (2008)
5. Elomaa, T., Kujala, J.: Covering analysis of the greedy algorithm for partial cover. In: Elomaa, T., Mannila, H., Orponen, P. (eds.) Ukkonen Festschrift 2010. LNCS, vol. 6060, pp. 102–113. Springer, Heidelberg (2010)
6. Gramm, J., Guo, J., Hüffner, F., Niedermeier, R.: Data Reduction and Exact Algorithms for Clique Cover. ACM J. Exper. Algor. 13, Article No. 2 (2008)
7. Gupta, A., Mittal, A., Bhattacharya, A.: Minimally infrequent itemset mining using pattern-growth paradigm and residual trees. In: Haritsa, J.R., Dayal, U., Deshpande, P.M., Sadaphal, V.P. (eds.) 17th International Conference on Management of Data, pp. 57–68. Allied Publishers, Bangalore (2011)
8. Haglin, D.J., Manning, A.M.: On minimal infrequent itemset mining. In: Stahlbock, R., Crone, S.F., Lessmann, S. (eds.) DMIN 2007, pp. 141–147, CSREA Press (2007)
9. Hochbaum, D.S.: Approximating covering and packing problems: set cover, vertex cover, independent set, and related problems. In: Hochbaum, D.S. (ed.) Approximation Algorithms for NP-hard Problems, pp. 94–143. PSW Publishing, Boston (1997)
10. Hochbaum, D.S., Pathria, A.: Analysis of the greedy approach of maximum k-coverage. Naval Res. Q. **45**, 615–627 (1998)
11. Karkali, M., Rousseau, F., Ntoulas, A., Vazirgiannis, M.: Efficient online novelty detection in news streams. In: Lin, X., Manolopoulos, Y., Srivastava, D., Huang, G. (eds.) WISE 2013, Part I. LNCS, vol. 8180, pp. 57–71. Springer, Heidelberg (2013)

12. Karkali, M., Rousseau, F., Ntoulas, A., Vazirgiannis, M.: Using temporal IDF for efficient novelty detection in text streams. CoRR abs/1401.1456 (2014)
13. Turán, P.: On an extremal problem in graph theory. Matematikai és Fizikai Lapok **48**, 436–452 (1941)
14. Williams, V.V., Williams, R.: Subcubic equivalences between path, matrix and triangle problems. In: FOCS 2010, pp. 645–654. IEEE Computer Society (2010)

Recurring Comparison Faults: Sorting and Finding the Minimum

Barbara Geissmann[1]([✉]), Matúš Mihalák[1,2], and Peter Widmayer[1]

[1] Department of Computer Science, ETH Zurich,
Zurich, Switzerland
barbara.geissmann@inf.ethz.ch
[2] Department of Knowledge Engineering, Maastricht University,
Maastricht, The Netherlands

Abstract. In a faulty environment, comparisons between two elements with respect to an underlying linear order can come out right or go wrong. A wrong comparison is a *recurring* comparison fault if comparing the same two elements yields the very same result each time we compare the elements. We examine the impact of such faults on the elementary problems of sorting a set of distinct elements and finding a minimum element in such a set. The more faults occur, the worse the approaches to solve these problems can become and we parametrize our analysis by an upper bound k on the number of faults.

We first explain that reconstructing the sorted order of the elements is impossible in the presence of even one fault. Then, we focus on the maximum information content we get by performing all possible comparisons. We consider two natural approaches for sorting the elements that involve knowledge of the outcomes of all comparisons: the first approach finds a permutation (compatible solution) that contradicts at most k times the outcomes of comparisons, and the second approach sorts the elements by the number of times an element is returned to be larger in the outcomes of its comparisons with all other elements (score solution). In such permutations the elements can be dislocated from their positions in the linear order. We measure the quality of such permutations by three measures: the maximum dislocation of an element, the sum of dislocations of all elements, and the Kemeny distance compared to the linear order. We show for compatible solutions that the Kemeny distance is at most $2k$, the sum of dislocations at most $4k$, and the maximum dislocation at most $2k$. In score solutions the Kemeny distance is smaller than $4k$, the sum of dislocations smaller than $8k$, and the maximum dislocation at most $k + 1$. Our upper bounds are tight for compatible solutions, but possibly not tight for score solutions. It turns out that none of the two approaches is better than the other in all measures.

For the problem of finding a minimum element, we first observe that there is no deterministic algorithm that guarantees to return one of the smallest $k+1$ elements. This implies that computing the first element of a score solution is optimum and we derive an algorithm that guarantees to

We gratefully acknowledge the support of the Swiss National Science Foundation in the project *Algorithm Design for Microrobots with Energy Constraints*.

© Springer International Publishing Switzerland 2015
A. Kosowski and I. Walukiewicz (Eds.): FCT 2015, LNCS 9210, pp. 227–239, 2015.
DOI: 10.1007/978-3-319-22177-9_18

find one of the $k + 2$ smallest elements in time $O(\sqrt{k}n)$ making $O(\sqrt{k}n)$ comparisons, where n is the number of elements, and we generalize this algorithm to find all elements of score at most a given target t.

1   Introduction

In standard algorithm design, one assumes that elementary operations on a set of elements, such as the comparison of two elements, always work correctly. New emerging alternative ways to CPU-design trade the correct, exact computation for low power consumption. In such settings, elementary operations, such as comparisons, may occasionally fail. Naturally, one asks for the consequences of a failing comparison. Consider, for example, the problem of finding a minimum element in an unsorted array. The standard algorithm for this problem iterates linearly over all elements of the array, compares the currently considered element with the so-far computed minimum, and adjusts this minimum if the considered element is smaller. This algorithm can fail miserably, returning the maximum instead of the targeted minimum, even if only one comparison fails: just make the maximum appear last in the array, and let the very last comparison fail. Thus, if comparisons can go wrong, standard algorithms may fail, and one needs to develop new algorithms even maybe for the most elementary tasks.

Several computational models reflect comparison faults. They can be classified in two main groups: one group of models assumes independent faults, in the sense that two comparisons of the same two elements can return different results; the second group assumes recurring faults, in the sense that a comparison between the same two elements will always lead to the same result, regardless of when and how many times it is performed.

We study recurring comparison faults on a set of n distinct elements underlying a strict linear order. Our comparison operation takes two elements as input and outputs a claimed order of the two elements. We assume that the comparison operation fails on at most k inputs, where k is a parameter of the model. We concentrate on the following two elementary algorithmic questions: *finding a minimum element*, and *sorting the elements*. We are interested in knowing what guarantees can be achieved *at all*: Even with infinite computational power, one cannot find a minimum, if a single fault can appear. Hence, we look for worst-case differences between a computed solution and the linear order.

Related Work. There is a probabilistic model where every comparison can only be made once and fails with a probability p. It is possible to find a permutation of the elements in which each element is placed with high probability not more than $O(\log n)$ positions away from its true position. This needs $O(n \log n)$ comparisons and $O(n^{3+24c})$ time, where $c > 0$ is a constant depending on p [5]. Also with high probability, one can compute the minimum element with $O(n^2)$ comparisons and $O(n)$ time [12]. Searching for any element in a presorted array of elements with unreliable comparisons or corrupted memory is studied in [6].

A different probabilistic model assumes independent comparison faults with probability p, but allows the repetition of comparisons in order to boost the probability of obtaining the correct result between two elements at the cost of additional work [8,10]. If at most a fraction $p < 1/2$ of all comparisons made so far might be wrong, a maximum (or minimum) element can be determined making $O((\frac{1}{1-p})^n)$ comparisons [1]. In the same model but with a fixed maximum number k of faults an extremum is found with $(k+1)n-1$ comparisons [14].

Our Model Refined. We consider a set S of n distinct elements that possess a strict linear order. We call the permutation π of the elements of S in which the elements appear in their linear order the *sorted order*. We denote the position of an element e in π by $\pi(e)$, and call it the *rank* of e. The elements can be compared by a comparison operation that takes two distinct elements a and b as input and outputs a claim for their order, i.e., for the input $\{a, b\}$ the output is either (a, b) or (b, a). For at most k pairs of elements, the comparison operation may return a result that differs from the sorted order π. We say that a is *supposedly* smaller than b and b is *supposedly* larger than a if their comparison returns (a, b).

A claimed solution to the sorting problem is represented as a permutation of the elements, where each element might be dislocated by several positions when compared to its position in the sorted order. We measure the quality of a permutation in three ways: the *maximum dislocation* is the maximum number of positions any element is dislocated; the *dislocation distance* is the sum of dislocations of all elements; the *Kemeny distance* is the number of inversions between the computed permutation and the sorted order.

Our Results. We first focus on the sorting problem in Sect. 2. We are primarily interested in what is possible, at all, using as much information as we can get, and thus let the algorithm compare all $\binom{n}{2}$ pairs of elements. We study two approaches for sorting. The first one computes a permutation that disagrees with at most k comparisons among the observed $\binom{n}{2}$ comparisons. Such a permutation exists since the sorted order has this property. We call such a permutation a compatible *solution*. The second approach scores for every element the number of comparisons in which the element was supposedly larger. We call a permutation of the elements sorted by these scores a *score solution*. We study the quality of the two sorting approaches. The Kemeny distance in any compatible solution is at most $2k$ and in any score solution strictly smaller than $4k$. The dislocation distance is at most $4k$ in a compatible solution and smaller than $8k$ in a score solution. Finally, the maximum dislocation of an element is at most $2k$ in a compatible solution, whereas it is only at most $k+1$ in a score solution. The last result surprises at first sight, since a score solution uses less information of the comparison outcomes than a compatible solution. In fact, there is no deterministic way to sort the elements achieving a smaller maximum dislocation.

We study the problem of finding a minimum element in Sect. 3. Using the bounds on the maximum dislocation we can conclude that the maximum rank of an element placed first in a compatible solution is at most $2k+1$ and in a score solution at most $k+2$. We show that computing an element with minimum score is optimum for the problem of minimum finding, but performing all comparisons and sorting all elements is an unnecessary overhead. Finally, we present an algorithm that computes an element with minimum score in time $O(\sqrt{k}n)$ making $O(\sqrt{k}n)$ comparisons.

2   Sorting

We have demonstrated in the introduction that no algorithm can compute the sorted order π, even if at most one comparison can fail. We are interested to know the limits of deterministic algorithms with respect to the quality of computed solutions. We therefore allow the algorithms to make all $\binom{n}{2}$ comparisons, and focus on what we can deduce from the obtained information.

We can think of the $\binom{n}{2}$ outcomes of all comparisons between any two elements in S as a complete directed graph (a tournament) with the vertex set S such that there is an arc from u to v if and only if u is supposedly smaller than v. We call this graph the *comparison graph* of S. Without any fault, the comparison graph features a unique structure: For every $i \in \{0, \ldots, n-1\}$, there is exactly one vertex with i incoming and $n-1-i$ outgoing arcs. Such a comparison graph contains no directed cycles and its vertices can be arranged on a line such that all arcs point from left to right. We say that such an arrangement contains only *forward arcs* and no *backward arcs*, and that the comparison graph is *acyclic*.

Proposition 1. *A cyclic comparison graph implies that a fault happened.*

2.1   Compatible Solutions

If we flip the direction of the arcs corresponding to failed comparisons, we get the acyclic comparison graph that represents the sorted order π. This motivates the following natural approach: After making all $\binom{n}{2}$ comparisons and constructing the comparison graph, flip some arcs so that the result is an acyclic graph.

If we know k, a natural aim is to flip at most k arcs to obtain an acyclic graph. We call such an acyclic graph, and the underlying linear order, a *compatible solution*. If we do not know k, a natural approach is to flip a minimum number of arcs that result in an acyclic graph. We call such a solution a *minimum compatible solution*.

Definition 1. *A permutation σ for which $|\{\{a,b\} \subset S : \sigma(a) < \sigma(b) \wedge a \succ b\}| \leq k$ is called a* compatible *solution. A permutation σ that minimizes $|\{\{a,b\} \subset S : \sigma(a) < \sigma(b) \wedge a \succ b\}|$ is called a* minimum compatible solution.

Proposition 2. *A compatible solution always exists.*

A minimum compatible solution is not necessarily better than any other compatible solution. For example, there are instances where exactly k comparisons fail, but a minimum compatible solution for the resulting comparison graph flips strictly less than k arcs. Such an example is shown in Fig. 1.

Fig. 1. Let $k = 3$. There are three permutations with one inconsistency: (a, b, e, c, d), (b, e, a, c, d), and (e, a, b, c, d). There are permutations with three inconsistencies, e.g., among others, (a, b, c, d, e), (e, c, a, b, d), and (e, b, c, a, d).

An equivalent way to flip arcs is to find a linear arrangement of the vertices on a line such that there are at most k (or a minimum number of) backward arcs. The MINIMUMFEEDBACKARCSET problem asks to *delete* a minimum number of arcs of a directed graph such that the resulting graph is acyclic. Obviously, the backward arcs of a compatible solution are a feedback arc set of size at most k, since deleting the arcs also makes the original comparison graph acyclic. Note that the other way round is not true, except for feedback arc sets that are minimal, as reversing an arc instead of deleting it might lead to a directed cycle (deleting any arc in an acyclic tournament will result in an acyclic graph, but flipping some arc might create a cycle). Hence, finding a minimum feedback arc set in a tournament is equivalent to finding a minimum compatible solution.

Note on the Complexity. MINIMUMFEEDBACKARCSET is NP-hard [9], even for tournaments [2,3]. For tournaments, it was shown that there exists a PTAS [11], and that it is fixed parameter tractable [4,7,13].

2.2 Score Solutions

The rank of an element e is equal to one plus the number of elements that are smaller than e. In the comparison graph, each incoming arc to e comes from an element that is supposedly smaller than e. Hence, one plus the number of these arcs can be viewed as a claim for the rank of e, which we call the *score* of e, and denote it by score(e). Instead of constructing the comparison graph and counting arcs, we can directly compute the scores of the elements from all the $\binom{n}{2}$ comparisons, which is more space-effective.

A natural attempt to sorting is to simply order the elements in a non-decreasing order of their scores. We call every such solution a *score solution*.

Definition 2. *A permutation σ such that $\forall a, b \in S \colon \sigma(a) < \sigma(b) \Rightarrow score(a) \leq score(b)$ is called is a* score solution.

Observe that knowing k is not needed for computing a score solution. Recall that a cycle in the comparison graph gives evidence to the presence of a fault. We observe that scores of the elements have the same revealing power.

Proposition 3. *Elements have unique scores if and only if the comparison graph is acyclic.*

2.3   Quality of Compatible Solutions and Score Solutions

We will compare a delivered solution permutation σ with the sorted order π using the following three measures.

Definition 3. *The* Kemeny distance *between two permutations σ and π is* $d_K(\pi, \sigma) := |\{(i, j) \colon \pi(i) < \pi(j) \wedge \sigma(i) > \sigma(j)\}|$.

Definition 4. *The* dislocation distance *between two permutations σ and π is* $d_L(\pi, \sigma) := \sum_{i=1}^{n} |\pi(i) - \sigma(i)|$.

Definition 5. *The* maximum dislocation *of any element between two permutations σ and π is* $d_{\max}(\pi, \sigma) := \max_{i=1}^{n} |\pi(i) - \sigma(i)|$.

The scores of the elements reveal less information than the entire comparison graph. As a result, there exist situations where a score solution is different from any compatible solution. Figure 2 shows an example in which every permutation is a score solution but only few permutations are compatible solutions. It may seem that compatible solutions shall perform better than score solutions, which is indeed the case for the Kemeny distance. However, as we will see, score solutions have a better worst-case guarantee for the maximum dislocation of an element.

Fig. 2. Let $k = 3$. There are five compatible solutions: (a, b, c, d, e), (b, c, d, e, a), (c, d, e, a, b), (d, e, a, b, c), and (e, a, b, c, d). And each permutation is a score solution.

Quality of Compatible Solutions. Recall that each permutation featuring at most k inconsistencies with the outcomes of all comparisons is a compatible solution. An *inverse pair* is a pair of elements whose order in the computed permutation disagrees with the sorted order. We derive an upper bound on their number in any compatible solution.

Theorem 1. *For any compatible solution σ and the sorted order π it holds that* $d_K(\pi, \sigma) \leq 2k$. *This bound is tight.*

Proof. The comparison graph contains at most k wrong arcs, which are potential inverse pairs. A compatible solution takes a comparison graph and flips at most k arcs, possibly different ones. That is, in total, there are at most $2k$ wrong arcs, and therefore at most $2k$ inverse pairs. *Tightness:* Assume faults between elements of ranks i and $i + 2$, for $i = 3, 6, \ldots, 3k$. Consider a permutation σ' where elements from i to $i + 2$ appear in the order $i + 1, i + 2, i$, and all other elements appear in their sorted order. Then, σ' is a compatible solution and there are two inverse pairs per fault, namely, $(i + 1, i)$ and $(i + 2, i)$.   □

We continue with upper bounds for the sum of dislocation of all elements and the maximum dislocation of a single element.

Theorem 2. *For any compatible solution σ and the sorted order π it holds that $d_L(\pi,\sigma) \leq 4k$. This is tight.*

Proof. Consider a sorting algorithm that only swaps adjacent elements, e.g., the bubble sort. If applied to σ (and with no faults), the algorithm swaps exactly the inverse pairs, and each exactly once. By Theorem 1, there are at most $2k$ swaps. Observe that in a swap, the two elements are shifted by one position each. Hence, the total dislocation is at most $4k$. *Tightness:* Consider the example in the proof for the Kemeny distance (Theorem 1). The elements with ranks $i + 1$ and $i + 2$ are dislocated each by one position, and elements with ranks i are shifted by two positions.   □

Theorem 3. *For any compatible solution σ and the sorted order π it holds that $d_{\max}(\pi,\sigma) \leq 2k$. This is tight.*

Proof. By Theorem 1, an element e forms an inverse pair with at most $2k$ elements. Each of these elements causes e to shift by one position in either direction. Hence, e's dislocation is at most $2k$. *Tightness:* Consider k faults between the element of rank 1, and the elements of ranks $k + 2, \ldots 2k + 1$. Consider the compatible solution σ' described by the ranks $(2, \ldots, 2k + 1, 1, 2k + 2, \ldots, n)$. The dislocation of the element of rank 1 is exactly $2k$.   □

Quality of Score Solutions. We now present upper bounds for the Kemeny distance, dislocation distance, and maximum dislocation in score solutions. The score of an element can differ from its rank. We call this difference the *score change* of the element. The total number of score changes, $\sum_e |\pi(e) - \text{score}(e)|$, is at most $2k$, since every fault decreases the score of one and increases the score of another element by one.

Theorem 4. *Let C denote the number of elements in a score solution σ with different scores and ranks. For σ and the sorted order π it is $d_K(\pi,\sigma) \leq 4k - \lceil \frac{C}{2} \rceil$. This is a tight upper bound.*

Proof. If no fault happens, the score of an element is its rank. Let us consider for the moment that we can change the scores of the elements arbitrarily (not only by faults) within a budget $i \geq 0$: set $\text{score}(e) = \pi(e) + i(e)$, $-i \leq i(e) \leq i$, such that $\sum_e |i(e)| \leq i$. Let $I(i, m)$ denote the maximum number of inverse pairs in any score sorted permutation ρ with a total of i score changes and with m elements having scores different to their ranks. Obviously, the Kemeny distance between π and σ is at most $\max_m I(2k, m)$. We now show by induction on i that $I(i, m) \leq 2i - \lceil \frac{m}{2} \rceil$, which proves the theorem. *Base cases:* For $i = 0$ we necessarily have $m = 0$. Then, $\sigma = \pi$, and thus $I(0, 0) = 0$. For $i = 1$, we always have $m = 1$, and therefore $I(1, 1) = 1$, since the score of a single element is increased or decreased by one. Hence, there are exactly two elements with the same score, which can appear inversely in ρ. *Step case:* Assume that for all $j < i$ and for any score sorted permutation ρ' with j score changes affecting m' elements, $S(j, m') \leq 2j - \lceil \frac{m'}{2} \rceil$. Consider a permutation ρ with j score changes where m elements changed their score. Take an element e in ρ with maximum score change $l = |\pi(e) - \text{score}(e)|$. Assume $\pi(e) < \text{score}(e)$ (the other case is analogue). We decrease the score of e to fit its rank and upper bound the number of inverse pairs that disappear. By the maximality of l, only inverse pairs formed by e with elements of larger rank than $\pi(e)$ can disappear. Such elements e' can form an inverse pair only if $\pi(e') - \pi(e) \leq 2l$. We distinguish two cases. *Case 1:* element e' with rank $\pi(e) + 2l$ is not inverse with e. Hence, the number of inverse pairs that disappear is at most $2l - 1$. The total number of score changes decreased from i to $i - l$. Therefore, $I(i, m) \leq 2l - 1 + I(i - l, m - 1)$. *Case 2:* element e' with rank $\pi(e) + 2l$ is inverse with e. Hence, it has a decreasing score change of l. We now increase the score of e' such that it is equal to its rank. The elements that are possibly inverse with e' have ranks in $[\pi(e), \dots, \pi(e) + 2l - 1]$. Hence, the maximum number of inverse pairs that disappear if we modify the score of both e and e is $4l - 1$ (we can count the pair (e, e') only once). The total number of score changes decreased to $i - 2l$. Therefore, we get $I(i, m) \leq 4l - 1 + I(i - 2l, m - 2)$. Obviously, the bound of case 2 is the larger of the two. Hence, we get $I(i, m) \leq 4l - 1 + I(i - 2l, m - 2) \leq 4l - 1 + 2(i - 2l) - \lceil \frac{m-2}{2} \rceil = 2i - \lceil \frac{m}{2} \rceil$. *Tightness:* Let $k = \binom{j+1}{2}$. Consider $2j + 1$ elements with consecutive ranks. Assume the smallest has faults with the j largest elements, the second smallest with the $j - 1$ largest elements, and so on. Finally the j-th smallest with the largest. Hence, all $2j + 1$ elements have the same score, the number of score changes is exactly $2k$, and the number of elements with score changes is $C = 2j$. Since all elements have the same score, they can appear in reverse rank order. Hence, the number of inverse pairs is $\binom{2j+1}{2} = 2j^2 + j = 4\binom{j+1}{2} - j = 4k - \frac{C}{2} = 4k - \lceil \frac{C}{2} \rceil$. $\qquad\square$

Theorem 5. *The dislocation distance between any score solution σ and the sorted order π is at most two times the Kemeny distance.*

Proof. For each element e, let X_e be the set of elements e' with $\pi(e) < \pi(e')$ and $\sigma(e) > \sigma(e')$ and Y_e the number of elements with $\pi(e) > \pi(e')$ and $\sigma(e) < \sigma(e')$. Clearly, the Kemeny distance is $\frac{1}{2} \sum_e |X_e| + |Y_e|$ and the dislocation distance is $\sum_e ||X_e| - |Y_e|| \leq \sum_e |X_e| + |Y_e|$. $\qquad\square$

A lower bound on the worst-case dislocation distance in score solutions is $4k$: Consider k sets of three rank adjacent elements with a fault between the largest and the smallest element. The three elements can appear in reversed rank order, where the dislocation is zero for one element and two for the other elements.

Theorem 6. *For any score solution σ and the sorted order π it is $d_{\max}(\pi, \sigma) \leq k + 1$. This is tight.*

Proof. Consider an element e with $\text{score}(e) \geq \sigma(e)$ (the other case is analogous). At least $\text{score}(e) - \sigma(e)$ elements must increase their scores to at least $\text{score}(e)$.

This requires at least $\binom{\text{score}(e)-\sigma(e)+1}{2}$ faults. Element e has at most one fault with each element, thus, at most $\text{score}(e) - \sigma(e) + k - \binom{\text{score}(e)-\sigma(e)+1}{2}$ faults. An element with zero faults has the same rank and score. For every fault an element can have we can increase this rank value by one. Hence, $\pi(e) \leq \text{score}(e) + \text{score}(e) - \sigma(e) + k - \binom{\text{score}(e)-\sigma+1}{2}$. The dislocation is at most $\pi(e) - \sigma(e) \leq 2(\text{score}(e)-\sigma(e))+k-\binom{\text{score}(e)-\sigma(e)+1}{2}$, which is maximum for $\text{score}(e) = \sigma(e)+1$ and $\text{score}(e) = \sigma(e) + 2$, where it equals $k + 1$. $\square$

3   Finding a Minimum Element

We have learned that we cannot reconstruct the sorted order of the elements in the presence of faults. Obviously, finding a minimum element is impossible, as well. We are thus interested in computing an element whose guaranteed worst-case rank is as small as possible.

Sorting all elements gives an immediate way to find a minimum element: Return the element that comes first in the solution. In Sect. 2 we have presented two approaches to sorting: the compatible solution and the score solution. As a corollary of Theorems 3 and 6, we obtain the following upper bounds on ranks of the computed elements in the respective solutions.

Corollary 1. *The rank of the first element in any score solution is at most $k+2$. The rank of the first element in any compatible solution is at most $2k+1$.*

It follows that computing an element of minimum score (i.e., an element that comes first in a score solution) gives the best guarantee on the rank.

Theorem 7. *There is no deterministic algorithm that guarantees to find an element of rank at most $k + 1$, even if an upper bound $k > 0$ on the number of faults is known to the algorithm.*

Proof. Consider the structure of the comparison graph G obtained if the comparison fails between the smallest and the third smallest element. The graph has exactly one cycle between three vertices with in-degree one. All elements in the graph can be arranged on a line such that these three elements are the first elements, and there is only one backward arc. Let us call the vertices of G by their index j in the linear arrangement. We show that for every vertex j there exists a set of k arc flips (i.e., faults) that transform the acyclic graph G' corresponding to the sorted order into graph G and where an element of rank at least $k + 2$ becomes vertex j. Thus, if the algorithm returns vertex j, its rank is at least $k + 2$ in the presented set of k faults. We distinguish three cases. *Case 1: $j \in \{1, 2, 3\}$.* Observe that k faults between the element of rank $k + 2$ with elements of rank $1, 3, 4, \ldots, k, k+1$ result in comparison graph G where element of rank $k+2$ is exactly the vertex j. *Case 2: $j \in \{4, \ldots, k+1\}$.* One fault between the elements of ranks 1 and 3, and $k - 1$ faults of the element of rank $k + j - 1$ with elements of rank $j, j + 1, \ldots, k + j - 2$ result in comparison graph G where vertex j is the element of rank $k + j - 1$. *Case 3: $j \in \{k + 2, \ldots, n\}$.* Observe that a single fault between elements of rank 1 and 3 results in graph G where vertex j is the element of rank j. □

Making all $\binom{n}{2}$ comparisons seems an unnecessary overhead, and in the following we show how to compute a minimum-score element with less comparisons and also in less than quadratic time. The idea of our algorithm is to iterate over all elements, and to keep a buffer of a relatively small number of elements that have a relatively small score with respect to the so-far inspected elements. To describe the algorithm, we will need the following terminology and observations. We refer to the largest possible minimum score that can be achieved with at most k faults as the *maxmin score*. An element e with $\text{score}(e) \leq$ maxmin is called a *candidate* for being a minimum-score element.

Theorem 8. *For all $k > 0$, maxmin $= \lfloor \frac{1}{2} \left(\sqrt{8k+1} + 1 \right) \rfloor$.*

Proof. Let j be the minimum score of an element. Each element with a smaller rank than j is involved in faults that increase its score: the rank 1 element in at least $j - 1$ faults, the rank 2 element in at least $j - 2$ faults, and so on. For each score increase, there needs to be at least one fault. Therefore, at least $\binom{j}{2}$ faults happen and maxmin $= \max_j \{ j \mid \binom{j}{2} \leq k \} = \lfloor \frac{1}{2} \left(\sqrt{8k+1} + 1 \right) \rfloor$. □

Theorem 9. *The number of candidates is at most $2 \cdot$ maxmin $- 1$.*

Proof. Let C be the set of all candidates. The comparison graph of C has $|C|$ vertices and $\binom{|C|}{2}$ directed arcs. Each arc points to one of the candidates and the score of each candidate is at most maxmin. Thus, each candidate has at most $(\text{maxmin} - 1)$ incoming arcs and the total number of arcs within C cannot exceed $|C| \cdot (\text{maxmin} - 1)$, i.e., $\binom{|C|}{2} \leq |C|(\text{maxmin} - 1)$. The claim follows trivially. □

We cannot compute the exact scores of all elements, without making all $\binom{n}{2}$ comparisons. Instead, we compute a score estimate, called the *scan score*, of every element e by simply comparing e to a specific subset of elements of S, and counting the comparisons in which e is supposedly larger. The scan score of e is then 1 plus this count. Obviously, such a score estimate is a lower bound on the actual score. We are now ready to present our *buffer scan algorithm*:

> Use a buffer of size $2 \cdot$ maxmin $- 1$ and iterate over all elements in any order. For each element e do the following: Compare e with all elements in the buffer and update the scan scores of e and the elements in the buffer. If e's scan score does not exceed maxmin, insert e into the buffer. If the buffer is already full, delete the element with the largest scan score to make room for the new element. After the iteration loop, compute the score (not scan score) of all elements in the buffer by comparing them to all elements in S and return an element with minimum score.

We need to ensure that throughout the scan each candidate is inserted in the buffer and never deleted from the buffer. Insertion is trivial, since the scan score is at most the score of an element. Showing the second requirement is slightly more involved.

Lemma 1. *The algorithm never removes a candidate from the buffer.*

Proof. Each element in the buffer has been compared with all other elements in the buffer and also with e. Since e is a candidate, its scan score is at most maxmin, i.e., in at most maxmin $- 1$ comparisons is e supposedly larger. In the comparison graph thus at least maxmin arcs point from e to an element in the buffer. Together with all the arcs within the buffer, at least $\binom{2 \cdot \text{maxmin}-1}{2} +$ maxmin arcs point towards a buffer element. By pigeon hole principle, one of these elements has at least maxmin incoming arcs, and thus a scan score of at least maxmin $+ 1$. The claim of the lemma trivially follows. $\square$

The space complexity of the algorithm is obviously $O(n + \text{maxmin})$. The time complexity is $O(n \cdot \text{maxmin})$. First, each element makes $O(n \cdot \text{maxmin})$ comparisons with buffer elements. Second, computing the scores of all elements in the buffer requires again $O(n \cdot \text{maxmin})$ comparisons. Furthermore, finding the maximum buffer score, replacing elements in the buffer and update the buffer scores needs constant time for every element in the buffer, i.e., $O(\text{maxmin})$ in total per one iteration.

Theorem 10. *The buffer-scan algorithm finds a minimum-score element making $O(n \cdot \text{maxmin})$ comparisons and in time $O(n \cdot \text{maxmin})$, where maxmin $= \lfloor \frac{1}{2}(\sqrt{8k + 1} + 1) \rfloor$.*

The algorithm using a smaller buffer cannot guarantee to find a minimum-score element any more. To see this, consider the following example. Let $k = 5$, and let faults happen between the rank pairs $(6, 1)$, $(5, 1)$, $(7, 2)$, $(8, 2)$, and $(6, 3)$. Consider a buffer of size four, and let the algorithm scan through the elements in the following rank order: $3, 4, 5, 6, 7, 8, 1, 2$. When arriving at 1, the elements $3, 4, 5, 6$ are in the buffer. After the comparisons with 1, all elements $1, 3, 4, 5, 6$ have buffer score maxmin – we cannot know which elements to keep in the buffer, although 1 is the only minimum-score element. This example is easily extendible for other values of k. Note that it would not help to consider the induced comparison graph of the current element and the elements in the buffer. In the above example, this graph would be totally symmetric.

We can generalize the algorithm to return all elements with a smaller score than a given threshold t. Naturally, the criterion for a considered element e to be inserted into the buffer is now whether its scan score is at most t. We also adapt the size of the buffer: the size needs to be as large as the maximum number of elements that can achieve a score of at most t.

Theorem 11. *Using the generalized buffer scan algorithm, we find all elements with a score of at most t in time $O((t + \text{maxmin}) \cdot n)$ and with a buffer of size $\min\{2t - 1, t + \text{maxmin} - 1\}$.*

Proof. The number of elements with a score of at most t is at most $2t + 1$. This is tight for $t \leq \text{maxmin}$. For larger t, the number of these elements is $t + \text{maxmin} - 1$. Hence, the claim about the size of the buffer follows. The runtime of the algorithm follows from the buffer size. $\square$

4 Conclusion

The value of k might or might not be known to the algorithm designer. Without knowledge of k, one cannot tell whether a permutation is a compatible solution, except for minimum compatible solutions. However, also for minimum compatible solutions the derived bounds on the quality measures are tight. It is easy to retrace the examples used in the respective tightness proofs to check that they involve minimum compatible solutions.

Computing a minimum compatible solution is NP-hard and in practice, one could opt to use a PTAS for the MINIMUMFEEDBACKARCSET problem for tournaments [11]. Note that our upper bounds on the quality of compatible solutions naturally extend. If $(1+\lambda)k$ is the size of the found feedback arc set, the Kemeny distance of a corresponding solution is at most $(2+\lambda)k$, the dislocation distance at most $(4+2\lambda)k$, and the maximum dislocation at most $(2+\lambda)k$.

We leave it as an open problem to provide a tight bound for the dislocation distance in score solutions.

References

1. Aigner, M.: Finding the maximum and minimum. Dis. Appl. Math. **74**(1), 1–12 (1997)
2. Ailon, N., Charikar, M., Newman, A.: Aggregating inconsistent information: Ranking and clustering. J. ACM 55(5) (2008)
3. Alon, N.: Ranking tournaments. SIAM J. Discrete Math. **20**(1), 137–142 (2006)
4. Alon, N., Lokshtanov, D., Saurabh, S.: Fast FAST. In: Automata, Languages and Programming, 36th International Colloquium, ICALP 2009, Proceedings, Part I, July 5–12, Rhodes, Greece, pp. 49–58 (2009)
5. Braverman, M., Mossel, E.: Noisy sorting without resampling. In: Proceedings of the Nineteenth Annual ACM-SIAM Symposium on Discrete Algorithms, SODA 2008, January 20–22, 2008, San Francisco, California, USA, pp. 268–276 (2008)
6. Cicalese, F.: Fault-tolerant search algorithms - reliable computation with unreliable information. Monographs in Theoretical Computer Science. An EATCS Series. Springer, Heidelberg (2013)
7. Feige, U.: Faster fast (feedback arc set in tournaments). CoRR, abs/0911.5094 (2009)
8. Feige, U., Peleg, D., Raghavan, P., Upfal, E.: Computing with unreliable information (preliminary version). In: Proceedings of the 22nd Annual ACM Symposium on Theory of Computing, May 13–17, 1990, Baltimore, Maryland, USA, pp. 128–137 (1990)
9. Garey, M.R., Johnson, D.S.: Computers and Intractability: A Guide to the Theory of NP-Completeness. W.H. Freeman, New York (1979)
10. Karp, R.M., Kleinberg, R.: Noisy binary search and its applications. In: Proceedings of the Eighteenth Annual ACM-SIAM Symposium on Discrete Algorithms, SODA 2007, January 7–9, 2007, New Orleans, Louisiana, USA, pp. 881–890 (2007)
11. Kenyon-Mathieu, C., Schudy, W.: How to rank with few errors. In: Proceedings of the 39th Annual ACM Symposium on Theory of Computing, June 11–13, 2007, San Diego, California, USA, pp. 95–103 (2007)

12. Klein, R., Penninger, R., Sohler, C., Woodruff, D.P.: Tolerant algorithms. In: Demetrescu, C., Halldórsson, M.M. (eds.) ESA 2011. LNCS, vol. 6942, pp. 736–747. Springer, Heidelberg (2011)
13. Raman, V., Saurabh, S.: Parameterized algorithms for feedback set problems and their duals in tournaments. Theor. Comput. Sci. **351**(3), 446–458 (2006)
14. Ravikumar, B., Ganesan, K., Lakshmanan, K.B.: On selecting the largest element in spite of erroneous information. In: STACS 1987, 4th Annual Symposium on Theoretical Aspects of Computer Science, Proceedings, February 19–21, 1987, Passau, Germany, pp. 88–99 (1987)

Graph Algorithms and Networking Applications

Minimal Disconnected Cuts in Planar Graphs

Marcin Kamiński[1], Daniël Paulusma[2],
Anthony Stewart[2(✉)], and Dimitrios M. Thilikos[3,4,5]

[1] Institute of Computer Science, University of Warsaw, Warszawa, Poland
mjk@mimuw.edu.pl
[2] School of Engineering and Computing Sciences, Durham University, Durham, UK
{daniel.paulusma,a.g.stewart}@durham.ac.uk
[3] Computer Technology Institute and Press "Diophantus", Patras, Greece
[4] Department of Mathematics, University of Athens, Athens, Greece
[5] AlGCo Project-team, CNRS, LIRMM, Montpellier, France
sedthilk@thilikos.info

Abstract. The problem of finding a disconnected cut in a graph is
NP-hard in general but polynomial-time solvable on planar graphs. The
problem of finding a minimal disconnected cut is also NP-hard but its
computational complexity is not known for planar graphs. We show that
it is polynomial-time solvable on 3-connected planar graphs but NP-
hard for 2-connected planar graphs. Our technique for the first result is
based on a structural characterization of minimal disconnected cuts in
3-connected $K_{3,3}$-free-minor graphs and on solving a topological minor
problem in the dual. We show that the latter problem can be solved in
polynomial-time even on general graphs. In addition we show that the
problem of finding a minimal connected cut of size at least 3 is NP-hard
for 2-connected apex graphs.

1 Introduction

A *cutset* or *cut* in a connected graph is a subset of its vertices whose removal
disconnects the graph. The problem STABLE CUT is that of testing whether a
connected graph has a cut that is an independent set. Le, Mosca, and Müller [12]
proved that this problem is NP-complete even for K_4-free planar graphs with
maximum degree 5. A connected graph $G = (V, E)$ is *k-connected* for some
integer k if $|V| \geq k+1$ and every cut of G has size at least k. It is not hard to see
that if one can solve STABLE CUT for 3-connected planar graphs in polynomial-
time then one can do so for all planar graphs (in particular the problem is trivial

The first author was supported by Foundation for Polish Science (HOMING
PLUS/2011-4/8) and National Science Center (SONATA 2012/07/D/ST6/02432).
The second author was supported by EPSRC Grant EP/K025090/1. The research of
the last author was co-financed by the European Union (European Social Fund ESF)
and Greek national funds through the Operational Program "Education and Life-
long Learning" of the National Strategic Reference Framework (NSRF) - Research
Funding Program: ARISTEIA II.

© Springer International Publishing Switzerland 2015
A. Kosowski and I. Walukiewicz (Eds.): FCT 2015, LNCS 9210, pp. 243–254, 2015.
DOI: 10.1007/978-3-319-22177-9_19

if the graph has a cut-vertex or a cut set of two vertices that are non-adjacent).
Hence, the problem is NP-complete for 3-connected planar graphs.

Due to the above it is a natural question whether one can relax the condition
on the cut to be an independent set. This leads to the following notion. For a
connected graph $G = (V, E)$, a subset $U \subseteq V$ is called a *disconnected* cut if U
disconnects the graph and the subgraph induced by U is disconnected as well,
that is, has at least two (connected) components. This problem is NP-compete
in general [13] but polynomial-time solvable on planar graphs [8]. However, the
property of the cut being disconnected can be viewed to be somewhat artificial
if one considers the 4-vertex path $P_4 = p_1 p_2 p_3 p_4$, which has two disconnected
cuts, namely $\{p_1, p_3\}$ and $\{p_2, p_4\}$. Both these cuts contain a vertex, namely
p_1 and p_4, respectively, such that putting this vertex out of the cut and back
into the graph keeps the graph disconnected. Therefore, Ito et al. [7] defined the
notion of a *minimal disconnected* cut of a connected graph $G = (V, E)$, that is, a
disconnected cut U so that $G[(V \setminus U) \cup \{u\}]$ is connected for every $u \in U$ (more
generally, we call a cut that satisfies the later condition a *minimal* cut). Here,
the graph $G[S]$ denotes the subgraph of G induced by $S \subseteq V(G)$. We note that
every vertex of a minimal cut U of a connected graph $G = (V, E)$ is adjacent to
every component of $G[V \setminus U]$. See Fig. 1 for an example of a planar graph with
a minimal disconnected cut.

The corresponding decision problem is defined as follows.

MINIMAL DISCONNECTED CUT
Instance: a connected graph G.
Question: does G have a minimal disconnected cut?

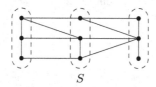

S

Fig. 1. An example of a planar graph with a minimal disconnected cut, namely the
set S.

Ito et al. [7] showed that MINIMAL DISCONNECTED CUT is NP-complete. However its computational complexity remains open for planar graphs. It can be
seen, via a straightforward reduction, that the problem of deciding whether a
graph has a minimal stable cut is NP-complete for any graph class (and thus
for the class of planar graphs) for which STABLE CUT is NP-complete. Moreover, the problem of deciding whether a graph has a minimal cut (that may be
connected or disconnected) is polynomial-time solvable: given a vertex cut U we
can remove vertices from U one by one until the remaining vertices in U form a
minimal cut.

Our Results. As a start we observe that MINIMAL DISCONNECTED CUT is
polynomial-time solvable for outerplanar graphs (as these graphs do not contain

$K_{2,3}$ as a minor, any minimal cut has size at most 2). In Sect. 2 we prove that MINIMAL DISCONNECTED CUT is also polynomial-time solvable on 3-connected planar graphs. The technique used by Ito et al. [8] for solving DISCONNECTED CUT in polynomial-time was based on the fact that a planar graph either has its treewidth bounded by some constant or else contains a large grid as a minor. However, grids (which are 3-connected planar graphs) do not have minimal disconnected cuts. Hence, we need to use a different approach, which we describe below.

We first provide a structural characterization of minimal disconnected cuts for the class of 3-connected $K_{3,3}$-free-minor graphs, which contains the class of planar graphs. In particular we show that any minimal disconnected cut of a 3-connected planar graph G has exactly two components and that these components are paths. In order to find such a cut we prove that it suffices to test whether G contains, for some fixed integer r, the biclique $K_{2,r}$ as a contraction. We show that G has such a contraction if and only if its dual contains for some fixed r the multigraph D_r, which is obtained from the r-vertex cycle by replacing each edge by two edges, as a subdivision (see also Fig. 2). We then present a characterization of any graph that contains such a subdivision. Next we use this characterization to prove that the corresponding decision problem is polynomial-time solvable even on general graphs.

In Sect. 3 we give our second result, namely that MINIMAL DISCONNECTED CUT stays NP-complete for the class of 2-connected planar graphs. This proof is based on a reduction from STABLE CUT and as such different from the NP-hardness proof for general graphs [7], the gadget of which contains large cliques. In the same section we show that the problem of finding a minimal *connected* cut of size at least 3 is NP-complete for 2-connected *apex graphs* (graphs that can be made planar by removing one vertex); to the best of our knowledge the computational complexity of this problem has not yet been determined even for general graphs. We note that the problem of finding whether a graph contains a (not necessarily minimal) connected cut of size at most k that separates two given vertices s and t is linear-time FPT when parameterized by k [14].

We finish our paper with some further observations on related problems in Sect. 4.

Related Work. Vertex cuts play an important role in graph connectivity. In the literature various kinds of vertex cuts, besides stable cuts, have been studied extensively and we briefly survey a number of results below.

A cut U of a graph $G = (V, E)$ is a clique cut if $G[U]$ is a clique, a k-clique cut if $G[U]$ has a spanning subgraph consisting of k complete graphs; a strict k-clique cut if $G[U]$ consists of k components that are complete graphs; and a matching cut if $E_{G[U]}$ is a matching. It follows from a classical result of Tarjan [17] that determining whether a graph has a clique cut is polynomial-time solvable. Whitesides [18] and Cameron et al. [3] proved that the problem of testing whether a graph has a k-clique cut is solvable in polynomial time for $k = 1$ and $k = 2$, respectively. Cameron et al. [3] also proved that testing whether a graph has a strict 2-clique cut can be solved in polynomial time. As mentioned the problem of testing whether a graph has a stable cut is NP-complete. This was first shown for general graphs by Chvátal [4]. Also the problem of testing whether a graph has a matching cut is NP-complete. This was shown by Brandstädt et al. [2]. Bonsma [1] proved that this problem is NP-complete even for planar graphs with girth 5 and for planar graphs with maximum degree 4.

The SKEW PARTITION problem is that of testing whether a graph $G = (V, E)$ has a disconnected cut U so that $V \setminus U$ induced a disconnected graph in the complement of G. De Figueiredo, Klein and Reed [5] proved that even the list version of this problem, where each vertex has been assigned a list of blocks in which it must be placed, is polynomial-time solvable. Afterwards, Kennedy and Reed [11] gave a faster polynomial-time algorithm for the non-list version.

Finally, for an integer $k \geq 1$, a cut U of a connected graph G is a k-cut of G if $G[U]$ contains exactly k components. For $k \geq 1$ and $\ell \geq 2$, a k-cut U is a (k, ℓ)-cut of a graph G if $G[V \setminus U]$ consists of exactly ℓ components. Ito et al. [8] proved that testing if a graph has a k-cut is solvable in polynomial time for $k = 1$ and NP-complete for every fixed $k \geq 2$. In addition they showed that testing if a graph has a (k, ℓ)-cut is polynomial-time solvable if $k = 1$, $\ell \geq 2$ and NP-complete otherwise [8]. The same authors show, by using the approach for solving DISCONNECTED CUT on planar graphs, that both problems are polynomial-time solvable on planar graphs.

Terminology. Let $G = (V, E)$ be a connected simple graph. A maximal connected subgraph of G is called a *component* of G. Recall that, for a subset $S \subseteq V(G)$, we let $G[S]$ denote the subgraph of G *induced* by S, which has vertex set S and edge set $\{uv \mid u, v \in S, uv \in E(G)\}$. A vertex $u \in V \setminus S$ is *adjacent* to a set $S \subseteq V \setminus \{u\}$ if u is adjacent to a vertex in S. We say that two disjoints sets $S \subset V$ and $T \subset V$ are *adjacent* if S contains a vertex adjacent to T, or equivalently if T contains a vertex adjacent to S.

Let G be a graph. We define the following operations. The *contraction* of an edge uv removes u and v from G, and replaces them by a new vertex made adjacent to precisely those vertices that were adjacent to u or v in G. Unless we explicitly say otherwise we remove all self-loops and multiple edges so that the resulting graph stays simple. The *subdivision* of an edge uv replaces uv by a new vertex w with edges uw and vw. Let $u \in V(G)$ be a vertex that has exactly two neighbours v, w, and moreover let v and w be non-adjacent. The *vertex dissolution* of u removes u and adds the edge vw.

A graph G contains a graph H as a *minor* if H can be obtained from G by a sequence of vertex deletions, edge deletions and edge contractions. We say that G contains H as a *contraction*, denoted by $H \leq_c G$, if H can be obtained from G by a sequence of edge contractions. Finally, G contains H as a *subdivision* if H can be obtained from G by a sequence of vertex deletions, edge deletions and vertex dissolutions, or equivalently, if G contains a subgraph H' that is a *subdivision* of H, that is, H can be obtained from H' after applying zero or more vertex dissolutions. We say that a vertex in H' is a *subdivision vertex* if we need to dissolve it in order to obtain H; otherwise it is called a *branch vertex* (that is, it corresponds to a vertex of H).

For some of our proofs the following global structure is useful. Let G and H be two graphs. An H-*witness structure* $\mathcal{W}$ is a vertex partition of a (not necessarily proper) subgraph of G into $|V(H)|$ nonempty sets $W(x)$ called (H-*witness) bags*, such that

(i) each $W(x)$ induces a connected subgraph of G,
(ii) for all $x, y \in V(H)$ with $x \neq y$, bags $W(x)$ and $W(y)$ are adjacent in G if x and y are adjacent in H.

In addition, we may require the following additional conditions:

(iii) for all $x, y \in V(H)$ with $x \neq y$, bags $W(x)$ and $W(y)$ are adjacent in G *only if* x and y are adjacent in H,
(iv) every vertex of G belongs to some bag.

By contracting all bags to singletons we observe that H is a minor or contraction of G if and only if G has an H-witness structure such that conditions (i)-(ii) or (i)-(iv) hold, respectively. We note that G may have more than one H-witness structure with respect to the same containment relation.

We denote the complete graph on k vertices by K_k and the complete bipartite graph with bipartition classes of size k and ℓ, respectively, by $K_{k,\ell}$.

2   The Algorithm

We first present a necessary and sufficient condition for a 3-connected $K_{3,3}$-minor-free graph to have a minimal disconnected cut.

Theorem 1. *A 3-connected $K_{3,3}$-minor-free graph G has a minimal disconnected cut if and only if $K_{2,r} \leq_c G$ for some $r \geq 2$.*

Proof. Let $G = (V, E)$ be a 3-connected graph that has no $K_{3,3}$ as a minor. First suppose that G has a minimal disconnected cut U. Let p and q be the number of components of $G[U]$ and $G[V \setminus U]$, respectively. Because U is a disconnected cut, $p \geq 2$ and $q \geq 2$. By definition, every vertex of every component of $G[U]$ is adjacent to all components in $G[V \setminus U]$. Hence, G contains $K_{p,q}$ as a contraction. Because G has no $K_{3,3}$ as a minor, G has no $K_{3,3}$ as a contraction. This means that $p \leq 2$ or $q \leq 2$. Because $p \geq 2$ and $q \geq 2$ holds as well, we find that $K_{2,r} \leq_c G$ for some $r \geq 2$.

Now suppose that $K_{2,r} \leq_c G$ for some $r \geq 2$. Throughout the remainder of the proof we denote the partition classes of $K_{k,\ell}$ by $X = \{x_1, \ldots, x_k\}$ and $Y = \{y_1, \ldots, y_\ell\}$. We refer to the bags in a $K_{k,\ell}$-witness structure of G corresponding to the vertices in X and Y as x-bags and y-bags, respectively. Because $K_{2,r} \leq_c G$, there exists a $K_{2,r}$-witness structure $\mathcal{W}$ of G that satisfies conditions (i)-(iv). Note that $W(x_1) \cup W(x_2)$ is a disconnected cut. However, it may not be minimal.

Suppose that $W(x_1)$ contains a vertex u that is adjacent to some but not all y-bags, i.e., the number of y-bags to which u is adjacent is h for some $1 \leq h < r$. Then we move u to a y-bag that contains one of its neighbors unless

$W(x_1) \cup W(x_2)$ no longer induce a disconnected graph (which will be the case if u is the only vertex in $W(x_1)$). We observe that $G[W(x_1) \setminus \{u\}]$ may be disconnected, namely when u is a cut vertex in $G[W(x_1)]$. We also observe that u together with its adjacent y-bags induces a connected subgraph of G. Hence, the resulting witness structure $\mathcal{W}'$ is a $K_{q,r'}$-witness structure of G with $q \geq 2$ (as the resulting vertices in $W(x_1) \cup W(x_2)$ still induce a disconnected graph) and $r' = r - (h-1)$. Because $1 \leq h < r$, we find that $2 \leq r' \leq r$. We repeat this rule as long as possible. During this process, $W(x_2)$ does not change, and afterwards, we do the same for $W(x_2)$. Let $\mathcal{W}^*$ denote the resulting witness structure that is a K_{q^*,r^*}-witness structure satisfying conditions (i)-(iv) for some $q^* \geq 2$ and $2 \leq r^* \leq r$.

We will now prove the following claim. Afterwards, we are done; due to this claim and because there are at least two x-bags and at least two y-bags in $\mathcal{W}^*$, the x-bags of $\mathcal{W}^*$ form a minimal disconnected cut U of G.

Claim 1. Every vertex of each x-bag of $\mathcal{W}^*$ is adjacent to all y-bags.

We prove Claim 1 as follows. First suppose that there exists an x-bag of $\mathcal{W}^*$, say $W^*(x_1)$, that contains a vertex u adjacent to some but not to all y-bags of $\mathcal{W}^*$, say u is not adjacent to $W^*(y_1)$. By our procedure we would have moved u to an adjacent y-bag unless that makes the disconnected cut connected. Hence we find that there are exactly two witness bags $W^*(x_1)$ and $W^*(x_2)$ and that $W^*(x_1) = \{u\}$. In our procedure we only moved vertices from x-bags to y-bags. This means that u belonged to an x-bag of the original witness structure $\mathcal{W}$. This x-bag was adjacent to all y-bags of $\mathcal{W}$ (as $\mathcal{W}$ was a $K_{2,r}$-witness structure). As we only moved vertices from x-bags to y-bags, this means that there must still exist a path from u to a vertex in $W^*(y_1)$ that does not use any vertex of $W^*(x_2)$; a contradiction. Hence every x-bag of $\mathcal{W}^*$ only contains vertices that are either adjacent to all y-bags or to none of them.

Now, in order to obtain a contradiction, suppose that an x-bag, say $W^*(x_1)$, contains a vertex u not adjacent to any y-bag. Because G is 3-connected, G contains three vertex-disjoints paths P_1, P_2, P_3 from u to a vertex in $W^*(y_1)$ (by Menger's Theorem). Each P_i contains a vertex v_i in $W^*(x_1)$ whose successor on P_i is outside $W^*(x_1)$. Hence, by our assumption, v_i has a neighbour in every y-bag (including $W^*(y_1)$). Recall that the number of y-bags is $r^* \geq 2$. Then the subgraph induced by the vertices from $W^*(y_1)$ and $W^*(y_2)$ together with the vertices on the three paths P_1, P_2, P_3 form a $K_{3,3}$-minor of G. This is not possible. Hence, every vertex of each x-bag of $\mathcal{W}^*$ is adjacent to all y-bags. This completes the proof of Claim 1 and thus the proof of Theorem 1.   □

By Theorem 1 we may restrict ourselves to finding a $K_{2,r}$-contraction for some $r \geq 2$ in a 3-connected planar graph. Below we state some additional terminology.

Recall that D_n is the multigraph obtained from the cycle on $n \geq 3$ vertices by doubling its edges. We let D_2 be the multigraph that has two vertices with four edges between them. The *dual* graph G_d of a plane graph G has a vertex for each face of G, and there exist k edges between two vertices u and v in G_d if and only if the two corresponding faces share k edges in G. Note that the dual

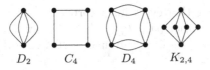

$D_2 \qquad C_4 \qquad D_4 \qquad K_{2,4}$

Fig. 2. The graphs D_2, C_4, D_2, $K_{2,4}$. Note that the dual of $C_4 = K_{2,2}$ is D_2, that D_4 is obtained from C_4 by duplicating each edge and that D_4 is the dual of $K_{2,4}$.

of a graph may be a multigraph. As 3-connected planar graphs have a unique embedding (see e.g. Lemma 2.5.1, p. 39 of [16]) we can speak of the dual of a 3-connected planar graph. We need the following lemma. Its proof, which we omit, follows from using a result from [9].

Lemma 1. *Let G be a 3-connected planar graph. Then G contains $K_{2,r}$ as a contraction for some $r \geq 2$ if and only if the dual of G contains D_r as a subdivision.*

By Lemma 1 it suffices to check if the dual of the 3-connected planar input graph contains D_r as a subdivision for some $r \geq 2$. We show how to solve this problem in polynomial time for general graphs. In order to do so we need the next lemma which gives a necessary condition for a graph G to be a yes-instance of this problem. In its proof we use the following notation. For a path $P = v_1 v_2 \ldots v_p$, we write $v_i P v_j$ to denote the subpath $v_i v_{i+1} \ldots v_j$ or $v_j P v_i$ if we want to emphasize that the subpath is to be traversed from v_j to v_i.

Lemma 2. *Let v, w be two distinct vertices of a multigraph G such that there exist four edge-disjoint v-w-paths in G. Then G contains a subdivision of D_r for some $r \geq 2$.*

Proof. We prove the lemma by induction on $|V(G)| + |E(G)|$. Then we can assume that G is the union of the four edge-disjoint v-w-paths. Let us call these paths P_1, P_2, P_3, and P_4. If these four paths are vertex-disjoint (apart from v and w) then they form a subdivision of D_2. Hence, we may assume that there exists at least one vertex of G not equal to v or w that belongs to more than one of the four paths.

First suppose that there exists a vertex u that belongs to all four paths P_1, P_2, P_3 and P_4. Let G' be the graph consisting of the vertices and edges of the four subpaths vP_1u, vP_2u, vP_3u and vP_4u. As G' does not contain w, it holds that $|V(G')| + |E(G')| < |V(G)| + |E(G)|$. By the induction hypothesis, G', and thus G, contains a subdivision of D_r for some $r \geq 2$.

Now suppose that there exists a vertex u that belong to only three of the four paths, say to P_1, P_2, and P_3. Let G' be the graph that consists of the vertices and edges of the four paths uP_1w, uP_2w, uP_3w and uP_1vP_4w. As G' does not contain an edge of vP_2u we find that $|V(G')| + |E(G')| < |V(G)| + |E(G)|$. By the induction hypothesis, G', and thus G, contains a subdivision of D_r for some $r \geq 2$.

From now on assume that every inner vertex of every path P_i $(i = 1, \ldots, 4)$ belongs to at most one other path P_j $(j \neq i)$. We say that two different paths P_i and P_j *cross* in a vertex u if u is an inner vertex of both P_i and P_j. Suppose P_i and P_j cross in some other vertex u' as well. Then we say that u is *crossed before* u' by P_i and P_j if u is an inner vertex of both vP_iu' and vP_ju'.

We now prove the following claim.

Claim 1. If P_i and P_j $(i \neq j)$ cross in both u and u' then we may assume without loss of generality that either u is crossed before u' or u' is crossed before u.

We prove Claim 1 as follows. Suppose that u is not crossed before u' by P_i and P_j and similarly that u' is not crossed before u by P_i and P_j. Then we may assume without loss of generality that u is an inner vertex of vP_iu' and that u' is an inner vertex of vP_ju. See Fig. 3 for an example of this situation. However, in that case we can replace P_i and P_j by the paths vP_iuP_jw and $vP_ju'P_iw$. These two paths together with the two unused original paths form a subgraph G' of G with fewer edges than G (as for instance no edge on uP_iu' belongs to G'). We apply the induction hypothesis on G'. This completes the proof of Claim 1.

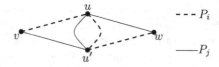

Fig. 3. The paths P_i and P_j where u is not crossed before u' by P_i and P_j and similarly u' is not crossed before u by P_j and P_i. Note that the paths P_i and P_j may have more common vertices, but for clarify this is not been shown.

We need Claim 1 to prove the following claim, which is crucial for our proof.

Claim 2. We may assume without loss of generality that there exists a vertex $u \notin \{v, w\}$ that is on two paths P_i and P_j $(i \neq j)$ so that every inner vertex of vP_iu and vP_ju has degree 2 in G.

We prove Claim 2 as follows. By our assumption there exists at least one vertex in G that is on two paths. Let $s \notin \{v, w\}$ be such a vertex, say s belongs to P_1 and P_2. Assume without los of generality that every inner vertex of vP_1s has degree 2. Then, by Claim 1, we find that P_1 and P_2 do not cross in an inner vertex of vP_2s.

If every inner vertex of vP_1s and vP_2s has degree 2 in G then the claim has been proven. Suppose otherwise, namely that there exists an inner vertex s' of vP_1s or vP_2s whose degree in G is larger than 2, say s' belongs to vP_2s. As P_1 does not cross vP_2s, we find that s' must belong to P_3 or to P_4. Choose s' in such a way that every inner vertex of vP_2s' has degree 2 in G. Assume without loss of generality that s' belongs to P_3.

If every inner vertex of vP_3s' has degree 2 then the claim has been proven (as every inner vertex of vP_2s' has degree 2 as well). Suppose otherwise, namely that there exists an inner vertex s'' of vP_3s' whose degree in G is larger than 2. Choose s'' in such a way that every inner vertex of vP_3s'' has degree 2 in G. By Claim 1, no inner vertex of vP_3s' belongs to P_2, so s'' does not lie on P_2. This means that s'' belongs either to P_1 or to P_4.

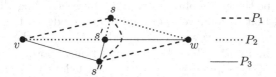

Fig. 4. The paths P_1, P_2 and P_3 where s belongs to P_1 and P_2, s' belongs to vP_2s and P_3 and s'' belongs to vP_3s' and P_1.

Suppose s'' belongs to P_1. See Fig. 4 for an example of this situation. As every inner vertex of vP_1s has degree 2, we find that s is an inner vertex of vP_1s''. However, we can now replace P_1, P_2 and P_3 by the three paths vP_1sP_2w, $vP_2s'P_3w$ and $vP_3s''P_1w$. These three paths form, together with P_4, a subgraph of G with fewer edges than G (for instance, no edge of sP_1s'' belongs to G'). We can apply the induction hypothesis on this subgraph. Hence we may assume that s'' does not belong to P_1.

From the above we conclude that s'' belongs to P_4. See Fig. 5 for an example of this situation. We consider the paths vP_3s'' and vP_4s''. If every inner vertex of vP_4s'' has degree 2 in G then we have proven Claim 2 (recall that every inner vertex of vP_3s'' has degree 2 in G as well). Suppose otherwise, namely that there exists an inner vertex t of vP_4s'' whose degree in G is larger than 2. Choose t in such a way that every inner vertex of vP_4t has degree 2 in G. By Claim 1 we find that t is not on P_3. If t is on P_2 we can use a similar replacement of three paths by three new paths as before that enables us to apply the induction hypothesis. Hence, we find that t belongs to P_1.

As every inner vertex of vP_1s has degree 2 in G we find that s is an inner vertex of vP_1t. Then we take the four paths vP_1sP_2w, $vP_2s'P_3w$, $vP_3s''P_4w$ and vP_4tP_1w. These four paths form a subgraph G' of G with fewer edges than G (as for instance G' contains no edge from sP_1t). We can apply the induction

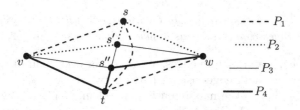

Fig. 5. The paths P_1, P_2, P_3 and P_4 where s belongs to P_1 and P_2, s' belongs to vP_2s and P_3, s'' belongs to vP_3s' and P_4 and t belongs to vP_4s'' and P_1.

hypothesis on G'. Hence we may assume that such a vertex t cannot exist. Thus we have found the desired vertex and subpaths, namely s'' with subpaths vP_3s'' and vP_4s''. This completes the proof of Claim 2.

By Claim 2 we may assume without loss of generality that there exists a vertex u that belongs to P_1 and P_2 such that every inner vertex of vP_1u and vP_2u has degree 2. Let G^* be the graph obtained from G by contracting all edges of vP_1u and vP_2u (recall that we remove loops and multiple edges). Let u^* be the new vertex to which all the edges were contracted. Notice that there are four edge-disjoint u^*-w-paths in G^*. Then, by the induction hypothesis, G^* contains a subdivision H of D_r for some $r \geq 2$. If u^* does not belong to H, then G contains H as well and we would have proven the lemma. Assume that u^* belongs to H.

First suppose that u^* is a subdivision vertex of H in G^*. Let u^* have neighbours s_1 and s_2 in H. Take a shortest path Q from s_1 to s_2 in the subgraph of G induced by s_1, s_2 and the vertices of vP_1u and vP_2u. This results in a graph H', which is a subgraph of G and which is a subdivision of D_r as well.

Now suppose that u^* is a branch vertex of H in G^*, say u^* corresponds to $z \in V(D_r)$. Note that any vertex in D_r has one neighbour if $r = 2$ and two neighbours if $r \geq 3$. We let s and t be the two branch vertices of H that correspond to the neighbours of z in D_r (note that $s = t$ if $r = 2$). Let s_1 and s_2 be the neighbours of u^* on the two paths from u^* to s, respectively, in H. Similarly, let t_1 and t_2 be the neighbours of u^* on the two paths from u^* to t, respectively, in H. Note that, as G is a multigraph, it is possible that $s_1 = s_2 = s$ and $t_1 = t_2 = t$.

Recall that every internal vertex on vP_1u and on vP_2u has degree 2 in G. As u is an inner vertex of P_1 and P_2 but not of P_3 and P_4, it has degree 4 in G. As G is the union of P_1, P_2, P_3 and P_4, we find that v has degree 4 as well. Then, after uncontracting u^*, we have without loss of generality one of the following two situations in G. First, u is adjacent to s_1 and s_2 and v is adjacent to t_1 and t_2. In that case u and v become branch vertices of a subdivision of D_{r+1} in G (to which the internal vertices on the paths uP_1v and uP_2v belong as well, namely as subdivision vertices). Second, u is adjacent to s_1 and t_1, whereas v is adjacent to s_2 and t_2. Then u and v become subdivision vertices of a subdivision of D_r in G (and we do not use the internal vertices on the paths uP_1v and uP_2v). This completes the proof of the lemma.   □

Lemma 2 gives us the following result.

Theorem 2. *It is possible to find in $O(mn^2)$ time whether a graph G with n vertices and m edges contains D_r as a subdivision for some $r \geq 2$.*

Proof. Let G be a graph with m edges. We check for every pair of vertices s and t whether G contains four edge-disjoint paths between them. We can do this via a standard reduction to the maximum flow problem. Replace each edge uv by the arcs (u, v) and (v, u). Give each arc capacity 1. Introduce a new vertex s' and an arc (s', s) with capacity 4. Also introduce a new vertex t' and an arc (t, t') with capacity 4. Check if there exists an (s', t')-flow of value 4 by using the

Ford-Fulkerson algorithm. As the maximum value of an (s', t')-flow is at most 4, this costs $O(m)$ time per pair, so $O(mn^2)$ time in total.

If there exists a pair s, t in G with four edge-disjoint paths between them then G has a subdivision of D_r, for some $r \geq 2$, by Lemma 2. If not then we find that G has no subdivision of any D_r ($r \geq 2$) as any subdivision of D_r immediately yields four edge-disjoint paths between two vertices and our algorithm would have detected this. $\qquad\square$

We are now ready to state our main result.

Theorem 3. MINIMAL DISCONNECTED CUT *can be solved in* $O(n^3)$ *time on 3-connected planar graphs with n vertices.*

Proof. Let G be a 3-connected planar graph with n vertices. By Theorem 1 it suffices to check whether $K_{2,r} \leq_c G$ for some $r \geq 2$. By Lemma 1, the latter is equivalent to checking whether the dual of G, which we denote by G^*, contains D_r as a subdivision for some $r \geq 2$. To find G^* we first embed G in the plane using the linear-time algorithm from Mohar [15]. As the number of edges in a planar graph is linear in the number of vertices, G^* has $O(n)$ vertices and $O(n)$ edges and can be constructed in $O(n)$ time. We are left to apply Theorem 2. $\square$

3 Hardness

We show the following result, which complements Theorem 3. We omit its proof.

Theorem 4. MINIMAL DISCONNECTED CUT *is* NP-*complete for the class of 2-connected planar graphs.*

A cut S in a graph G is a *minimal connected* cut if $G[S]$ is connected and for all $u \in S$ we have that $G[(V \setminus S) \cup \{u\}]$ is connected. We call the problem of testing whether a graph a minimal connected cut of size at least k the MINIMAL CONNECTED CUT(k) problem. By modifying the proof of Theorem 4 we obtain the following result (proof omitted).

Theorem 5. MINIMAL CONNECTED CUT(3) *is* NP-*complete even for the class of 2-connected apex graphs.*

We cannot use the reduction in the proof of Theorem 5 to get NP-hardness for MINIMAL CONNECTED CUT(1), the reason being that the gadget graph constructed in our omitted proof contains minimal disconnected cuts of size 2.

4 Conclusions

We proved that MINIMAL DISCONNECTED CUT is NP-complete for 2-connected planar graphs and polynomial-time solve for planar graphs that are 3-connected. Our proof technique for the latter result was based on translating the problem to a dual problem, namely the existence of a subdivision of D_r for *some* r, for which we obtained a polynomial-time algorithm even for general graphs. One can also solve the problem of determining whether a graph contains D_r as a subdivision for some *fixed* integer r by using the algorithm of Grohe, Kawarabayashi, Marx, and Wollan [6] which tests in cubic time, for any fixed graph H, whether a graph contains H as a subdivision. However, when r is part of the input we can show the following result via a reduction from HAMILTON CYCLE (proof omitted).

Theorem 6. *The problem of deciding whether a graph contains the graph D_r as a subdivision is* NP-*complete if r is part of the input.*

References

1. Bonsma, P.S.: The complexity of the matching-cut problem for planar graphs and other graph classes. J. Graph Theory **62**, 109–126 (2009)
2. Brandstädt, A., Dragan, F.F., Le, V.B., Szymczak, T.: On stable cutsets in graphs. Discrete Appl. Math. **105**, 39–50 (2000)
3. Cameron, K., Eschen, E.M., Hoáng, C.T., Sritharan, R.: The complexity of the list partition problem for graphs. SIAM J. Discrete Math. **21**, 900–929 (2007)
4. Chvátal, V.: Recognizing decomposable graphs. J. Graph Theory **8**, 51–53 (1984)
5. de Figueiredo, C.M.H., Klein, S., Reed, B.: Finding skew partitions efficiently. J. Algorithms **37**, 505–521 (2000)
6. Grohe, M., Kawarabayashi, K., Marx, D., Wollan, P.: Finding topological subgraphs is fixed-parameter tractable. In: Proceedings of the STOC 2011, pp. 479–488 (2011)
7. Ito, T., Kaminski, M., Paulusma, D., Thilikos, D.M.: On disconnected cuts and separators. Discrete Appl. Math. **159**, 1345–1351 (2011)
8. Ito, T., Kamiński, M., Paulusma, D., Thilikos, D.M.: Parameterizing cut sets in a graph by the number of their components. Theor. Comput. Sci. **412**, 6340–6350 (2011)
9. Kamiński, M., Paulusma, D., Thilikos, D.M.: Contractions of planar graphs in polynomial time. In: de Berg, M., Meyer, U. (eds.) ESA 2010, Part I. LNCS, vol. 6346, pp. 122–133. Springer, Heidelberg (2010)
10. Karp, R.M.: Reducibility among combinatorial problems. In: Complexity of Computer Computations, pp. 85–103. Plenum, New York (1972)
11. Kennedy, W.S., Reed, B.: Fast skew partition recognition. In: Ito, H., Kano, M., Katoh, N., Uno, Y. (eds.) KyotoCGGT 2007. LNCS, vol. 4535, pp. 101–107. Springer, Heidelberg (2008)
12. Le, V.B., Mosca, R., Müller, H.: On stable cutsets in claw-free graphs and planar graphs. J. Discrete Algorithms **6**, 256–276 (2008)
13. Martin, B., Paulusma, D.: The computational complexity of Disconnected Cut and 2K2-Partition. J. Comb. Theory, Series B **111**, 17–37 (2015)
14. Marx, D., O'Sullivan, B., Razgon, I.: Finding small separators in linear time via treewidth reduction. ACM Trans. Algorithms **9**, 30 (2013)
15. Mohar, B.: A linear time algorithm for embedding graphs in an arbitrary surface. SIAM J. Discrete Math. **12**, 6–26 (1999)
16. Mohar, B., Thomassen, C.: Graphs on Surfaces. The Johns Hopkins University Press, Baltimore (2001)
17. Tarjan, R.E.: Decomposition by clique separators. Discrete Math. **55**, 221–232 (1985)
18. Whitesides, S.H.: An algorithm for finding clique cut-sets. Inf. Proces. Letters **12**, 31–32 (1981)

ε-Almost Selectors and Their Applications

Annalisa De Bonis[(✉)] and Ugo Vaccaro

Dipartimento di Informatica, Università di Salerno,
Fisciano, SA, Italy
debonis@dia.unisa.it, uvaccaro@unisa.it

Abstract. Consider a group of stations connected through a multiple-access channel, with the constraint that if in a time instant exactly one station transmits a message, then the message is successfully received by any other station, whereas if two or more stations simultaneously transmit their messages then a conflict occurs and all messages are lost. Let us assume that n is the number of stations and that an (arbitrary) subset A of them, $|A| \leq k \leq n$, is *active*, that is, there are at most k stations that have a message to send over the channel. In the classical *Conflict Resolution Problem*, the issue is to schedule the transmissions of each station to let every active station use the channel alone (i.e., without conflict) at least once, and this requirement must be satisfied whatever might be the set of active stations A. The parameter to optimize is, usually, the worst case number of transmissions that any station has to attempt before all message transmissions are successful. In this paper we study the following question: is it possible to obtain a significant improvement on the protocols that solve the classical Conflict Resolution Problem if we allow the protocols to fail over a "small" fraction of all possible subsets of active stations? In other words, is it possible to significantly reduce the number of transmissions that must be attempted? In this paper we will show that this is indeed case. Our main technical tool is a generalization of the selectors introduced in [9]. As it turned out for selectors, we believe that our new combinatorial structures are likely to be useful also outside the present context in which they are introduced.

1 The Communication Model

Our scenario consists of a multiaccess system where n stations have access to the channel and at most a certain number k of stations might be *active* at the same time, i.e., might transmit simultaneously over the channel. An active station successfully transmits if and only if it transmits singly on the channel. We assume that time is divided into time slots and that transmissions occur during these time slots. We also assume that all stations have a global clock and that active stations start transmitting at the same time slot. A scheduling algorithm for such a multiaccess system is a protocol that schedules the transmissions of the n stations over a certain number t of time slots (*steps*) identified by integers $1, 2, \ldots, t$. In a distributed model, a scheduling algorithm can be represented by a set of n Boolean vectors of length t, identified by integers from 1 through n,

© Springer International Publishing Switzerland 2015
A. Kosowski and I. Walukiewicz (Eds.): FCT 2015, LNCS 9210, pp. 255–268, 2015.
DOI: 10.1007/978-3-319-22177-9_20

each of which corresponds to a distinct station, with the meaning that station j is scheduled to transmit at step i if and only if the i-th entry of its associated Boolean vector j is 1. In fact station j really transmits at step i if and only it is an active station and is scheduled to transmit at that step.

A *conflict resolution* algorithm for the above described multiaccess system is a scheduling protocol that allows active stations to transmit successfully. A *non adaptive* conflict resolution algorithm is a protocol that schedules all transmissions in advance, i.e., for each step $i = 1, \ldots, t$ establishes which stations should transmit at step i without looking at what happened over the channel at the previous steps. A non adaptive conflict resolution algorithm is conveniently represented by the Boolean matrix having as columns the n Boolean vectors associated with the scheduling of the transmissions of the n stations. Entry (i, j) of such a matrix is 1 if and only is station j is scheduled to transmit at step i. The parameter of interest to be minimized is the number of rows of the matrix which represents the number of time slots over which the conflict resolution algorithm schedules the transmissions of the n stations.

The Multiple-Access Channel Without Feedback. When stations receive no *feedback* from the channel then the conflict resolution algorithm must schedule transmissions in such a way that each active station transmits singly to the channel at some step, i.e., in such a way that no other active station is scheduled to transmit at that same step. In this case non adaptive algorithms are an obliged choice since at each given step the conflict resolution algorithm has to schedule transmissions without knowing which stations succeeded to transmit their messages in the previous steps. A conflict resolution algorithm for this model is represented by a Boolean matrix M with the property that for any k columns of M and for any column $\mathbf{c}$ chosen among these k columns, there exists a row in correspondence of which $\mathbf{c}$ has a 1-entry and the remaining $k - 1$ columns have 0-entries. In other words, for any choice of k out of n columns of M the submatrix formed by these k columns contains all rows of the identity matrix I_k. Matrices that satisfy this property have been very well studied in the literature where are known under different names, such as superimposed codes [14], $(k-1)$-cover free families [12], $(k-1)$-disjunct codes [10], and strongly selective families [6,7]. The best constructions for these combinatorial structures [19] imply the existence of conflict resolution algorithms that solve all conflicts among up to k active stations in $O(k^2 \log n)$ number of steps. Moreover, it is known that the number of rows of these matrices is lower bounded by $\Omega\left(\frac{k^2}{\log k} \log n\right)$ [11], and therefore any conflict resolution algorithm for this model should use at least this number of steps.

The authors of [9] introduced the following combinatorial structure that generalizes disjunct codes by introducing a parameter m which fixes the minimum number of distinct rows of the identity matrix I_k that should appear in any submatrix of k columns.

Definition 1 [9]. *Given integers k, m, and n, with $1 \leq m \leq k \leq n$, we say that a Boolean matrix M with t rows and n columns is a (k, m, n)-selector if* *for any choice of k out of n columns of M the submatrix formed by these k columns contains m rows of the identity matrix I_k. The integer t is the size of the (k, m, n)-selector. The minimum size of a (k, m, n)-selector is denoted by $t_s(k, m, n)$.*

A (k, m, n)-selector provides us with a non adaptive algorithm that allows at least m out of exactly k active stations to transmit successfully. Protocols based on (k, m, n)-selectors employ a number of steps which decreases with the maximum number of active stations that might not succeed in transmitting their messages, as shown by the following bound on the minimum number of rows of (k, m, n)-selectors [9].

$$t_s(k, m, n) = \frac{ek^2}{k - m + 1} \ln\left(\frac{n}{k}\right) + \frac{ek(2k - 1)}{k - m + 1}. \tag{1}$$

Notice that if there are less than k active stations then protocols based on (k, m, n)-selectors guarantee a number of successful transmission smaller than m since it might happen that some (eventually all) of the m out of k stations which are scheduled to transmit singly to the channel are not active. In particular, if there are $j \leq k$ active stations the protocol guarantees $m - (k - j)$ active stations to transmit successfully, and therefore, independently from the actual number of active stations, such a protocol schedules transmissions so that at most $k - m$ active stations do not succeed to transmit their messages.

The Multiple-Access Channel with Feedback. In addition to the situation when stations receive no feedback from the channel, we consider also a communication model in which any transmitting station receives feedback on whether its transmission has been successful or not. In such a model an active station has the capability to become *inactive* (i.e., to refrain from transmitting) after it has transmitted successfully. As in the previous model, a non adaptive conflict resolution algorithm should guarantee that for each active station there is a step at which it transmits singly. However, in this scenario an active station transmits singly to the channel also at time slots where it is scheduled to transmit simultaneously with some of the other k stations that were initially active, provided that these stations transmitted successfully at one of the previous steps. A Boolean matrix M represents a non adaptive conflict resolution algorithm for this more relaxed model *if and only if* any subset S of k columns of M satisfies the following property.

(*) *There are k row indices $i_1, i_2, \ldots, i_k$, with $i_1 \leq i_2 \leq \ldots \leq i_k$, and a permutation $[j_1, \ldots, j_k]$ of the indices of the columns in S, such that the submatrix of M formed by rows with indices $i_1, \ldots, i_k$, taken in this order, and columns with indices $j_1, \ldots, j_k$, taken in this order, form a $k \times k$ lower unitriangular matrix, i.e., a $k \times k$ matrix in which all entries in the diagonal are 1 and all those above the diagonal are 0.*

We will refer to a matrix in which all subsets of k column indices satisfy property (*) as a KG (k, n)-code and will denote the minimum length of KG

(k, n)-codes by $t_{KG}(k, n)$. The name KG (k, n)-code comes from the initials of Komlós and Greenberg [15] who were the first to prove the following upper bound on the minimum length of KG (k, n)-codes.

$$t_{KG}(k, n) = O(k \log(n/k)) \tag{2}$$

Interestingly, the above bound is tight with the $\Omega(k \log(n/k))$ lower bound proved by the authors of [7,8] for a combinatorial structure satisfying a weaker property than that of KG (k, n)-codes. The authors of [9,13] suggested a simpler construction which achieves the same asymptotic efficiency as the protocols in [15] and consists in concatenating $(2^i, 2^{i-1}, n)$-selectors starting from $i = \lceil \log k \rceil$ through $i = 1$, i.e., with the rows of the $(2^{\lceil \log k \rceil}, 2^{\lceil \log k \rceil - 1}, n)$-selector being placed at the top of the matrix and those of the $(2, 1, n)$-selector being placed at the bottom.

Our Results. In this paper we study non adaptive conflict resolution protocols for both the two above described multiple-access models. Our goal is to investigate what happens in terms of time efficiency if we allow the protocol to (*possibly*) fail over a "small" fraction of all possible subsets of active stations. In other words, we want to give an answer to the following question: is it possible to significantly reduce the number of steps used by the conflict resolution protocol if we tolerate that the protocol does not guarantee to behave correctly for a small fraction of all possible subsets of active stations? In order to study this question, we introduce two new combinatorial structures that consist in a generalization of selectors and KG codes, respectively. In these matrices only a certain ratio $(1 - \epsilon)$ of all k-column subsets is guaranteed to satisfy the desired properties, and therefore the corresponding conflict resolution protocols are guaranteed to work correctly only for $(1 - \epsilon)\binom{n}{k}$ subsets of active stations.

Our paper is organized as follows. In Sect. 2 we first introduce a new version of (k, m, n)-selectors that correspond to protocols that schedule transmissions so that, for at least a fraction $(1 - \epsilon)$ of all possible subsets of k active stations, one has that at least m out of k stations are scheduled to transmit singly to the channel. Then, we introduce a new version of KG (k, n)-codes that furnish scheduling protocols for the multiple-access channel with feedback that allow to solve all conflicts for at least a certain fraction $(1 - \epsilon)$ of all possible subsets of k active stations. In Sect. 3, we recall the basic notion of hypergraph along with the related concepts of cover and partial cover which are at the core of our constructions. Our main technical results are contained in Sect. 4 where we give our constructions for the combinatorial structures presented in Sect. 2 and derive upper bounds on the number of time slots used by the corresponding conflict resolution algorithms. Our main results are summarised by the following two theorems[1].

Theorem 1. *Let k, m, and n be integers such that $1 \le m \le k \le n$, and let ϵ be a real number such that $0 \le \epsilon \le 1$. There exists a conflict resolution algorithm for a multiple-access channel without feedback that schedules the transmissions of*

---

[1] However, see Sect. 4.1 for improvements.

*n stations in such a way that for at least a $(1 - \epsilon)$ ratio of all possible subsets of k
active stations, one has that at least m out k active stations transmit successfully.
The number of time slots used by the resolution algorithm is*

$$t \leq \frac{ek}{k - m + 1} \left(1 + \ln \frac{\binom{k}{k-m+1}}{\epsilon} \right).$$

Theorem 2. *Let k and n be integers such that $1 \leq k \leq n$, and let ϵ be a real
number such that $0 \leq \epsilon \leq 1$. There exists a conflict resolution algorithm for
a multiple-access channel with feedback that schedules the transmissions of n
stations in such a way that for at least a $(1 - \epsilon)$ ratio of all possible subsets of k
stations, one has that if the set of (up to k) active stations is entirely contained in
one of those k-subsets then all active stations transmits successfully. The number
of time slots used by the conflict resolution algorithm is*

$$t < 2e\lceil \log k \rceil \left(1 + \ln \left(\frac{\log k}{\epsilon} \right) \right) + O(k).$$

It should be remarked that our protocols, as well as those of [9,13], can
be made to work without any loss of efficiency also in the case in which the
parameter k is not known a priori. We will show this fact in the full version of
the paper.

2   Combinatorial Tools

For a positive integer n, we will denote by $[n]$ the set $\{1, \ldots, n\}$ and by $[n]_k$ the
family of all k element subsets of $[n]$. The following definition introduces a new
notion of selectors in which only a certain fraction $(1 - \epsilon)$ of all k-column subsets
is guaranteed to satisfy the condition described in Definition 1.

Definition 2. *Given integers k, m, and n, with $1 \leq m \leq k \leq n$, and a real
number $0 \leq \epsilon \leq 1$, we say that a Boolean matrix M with t rows and n columns
is an ϵ-almost (k, m, n)-selector if at least $(1 - \epsilon)\binom{n}{k}$ distinct subsets of k out
of n columns of M are such that the k columns in each of those subsets form a
submatrix that contains m rows of the identity matrix I_k. The integer t is the
size of the ϵ-almost (k, m, n)-selector.*

The ϵ-almost version of $(k + 1, k + 1, n)$-selectors are in fact the ϵ-almost version
of k-disjunct codes and correspond to the notion of type 2 $(k - 1, \epsilon)$-disjunct
codes introduced in [17]. In the context of conflict resolution in the presence of a
multiple-access channel without feedback, an ϵ-almost (k, m, n)-selector furnishes
a protocol that schedules transmissions so that, for at least a certain fraction
$(1 - \epsilon)$ of all possible subsets of exactly k active stations, one has that at least
m active stations transmit singly to the channel. By the same argument used is
Sect. 1, one can see that if the actual number of active stations is smaller than k,
then the protocol allows a number of active stations possibly smaller than m to

transmit singly to the channel. However, independently from the actual number of active stations, there are at most $k - m$ active stations that do not transmit successfully, provided that the active stations are contained in some of the k-station subsets corresponding to one of the k-column subsets that satisfy the property of (k, m, n)-selectors, i.e., form a submatrix that contains m rows of the identity matrix. We will use this observation later on in the proof of Theorem 5. One is expected to trade off the weaker selection capacity of ϵ-almost (k, m, n)-selectors for a better efficiency of conflict resolution algorithms. In fact, we will show that, for enough large values of ϵ, there are ϵ-almost (k, m, n)-selectors with a number of rows significantly smaller than that of their "exact" counterpart.

The following definition introduces the analogous notion of KG codes in which only a certain fraction $(1 - \epsilon)$ of all k-column subsets is guaranteed to satisfy property (*).

Definition 3. *Given integers k and n, with $1 \leq k \leq n$, and a real number $0 \leq \epsilon \leq 1$, we say that a Boolean matrix M with t rows and n columns is an ϵ-almost KG (k, n)-code if property (*) is satisfied by at least $(1 - \epsilon) \binom{n}{k}$ k-column subsets S of M. The integer t is the length of the code.*

In the context of conflict resolution for a multiple-access channel with feedback, ϵ-almost KG (k, n)-codes are equivalent to conflict resolution algorithms that guarantee all active stations to transmit successfully only in the presence of some subsets of active stations, i.e., if and only if the subset of active stations corresponds to a subset of one of the up to $(1 - \epsilon) \binom{n}{k}$ k-column subsets that satisfy property (*).

Our results rely on the fact that an ϵ-almost (k, m, n)-selector can be seen as a *partial cover* of a properly defined hypergraph.

In the following section, we briefly recall the definitions of hypergraph, cover and partial cover of an hypergraph, along with the related terminology.

3   Hypergraphs and Covers

Given a finite set X and a family $\mathcal{F}$ of subsets of X, a hypergraph is a pair $\mathcal{H} = (X, \mathcal{F})$. The set X will be denoted by $V(\mathcal{H})$ and its elements will be called vertices of $\mathcal{H}$, while the family $\mathcal{F}$ will be denoted by $E(\mathcal{H})$ and its elements will be called hyperedges of $\mathcal{H}$. A hypergraph is said to be *uniform* if all edges contain the same number of vertices and it is said to be *regular* if all vertices have the same degree, i.e., belong to the same number of edges. A vertex $v \in V(\mathcal{H})$ is said to *cover* an edge $E \in E(\mathcal{H})$ if $v \in E$. A *cover* of $\mathcal{H}$, also called *integral cover*, is a subset $T \subseteq V(\mathcal{H})$ such that for any hyperedge $E \in E(\mathcal{H})$ we have $T \cap E \neq \emptyset$, i.e., the vertices of T covers all edges in $E(\mathcal{H})$. A *fractional cover* of $\mathcal{H}$ is an assignment of vertex-weights $\{t_v \geq 0 : v \in V(\mathcal{H})\}$ such that the constraint $\sum_{v \in E} t_v \geq 1$ holds for all edges in $E(\mathcal{H})$. The minimum sizes of a cover and a fractional cover of $\mathcal{H}$ will be denoted by $\tau(\mathcal{H})$, and $\tau^*(\mathcal{H})$, respectively. Notice that the assignment of vertex-weights $\{t_v = \frac{1}{\min_{E \in \mathcal{F}} |E|} : v \in V(\mathcal{H})\}$ is a fractional cover of $\mathcal{H}$. Therefore, one has

$$\tau^*(\mathcal{H}) \leq \frac{|V(\mathcal{H})|}{\min_{E \in \mathcal{F}} |E|}. \tag{3}$$

Below, we recall the notions of partial cover $((1 - \epsilon)$-cover$)$ and fractional partial cover (fractional $(1 - \epsilon)$-cover). For $0 \leq \epsilon \leq 1$, a $(1 - \epsilon)$-cover T of a hypergraph $\mathcal{H} = (V(\mathcal{H}), E(\mathcal{H}))$ is a collection of vertices which cover at least $(1-\epsilon)|E(\mathcal{H})|$ edges, i.e., $|\{E \in E(\mathcal{H}) : T \cap E \neq \emptyset\}| \geq (1-\epsilon)|E(\mathcal{H})|$. A fractional $(1 - \epsilon)$-cover of a hypergraph $\mathcal{H} = (V(\mathcal{H}), E(\mathcal{H}))$ is an assignment of vertex-weights $\{t_v \geq 0 : v \in V(\mathcal{H})\}$ such that the constraint $\sum_{v \in E} t_v \geq 1$, holds for at least $(1 - \epsilon)|E(\mathcal{H})|$ edges $E \in E(\mathcal{H})$. The value of this partial cover is defined as $\sum_{v \in V(\mathcal{H})} t_v$. We denote by $\tau_\epsilon(\mathcal{H})$ and by $\tau_\epsilon^*(\mathcal{H})$ the minimum sizes of the integral $(1-\epsilon)$-cover and the fractional $(1-\epsilon)$-cover, respectively. The following theorem [18] allows to limit from above the minimum sizes of integral $(1 - \epsilon)$-covers of regular and uniform hypergraphs.

Theorem 3 [18]. *If $\mathcal{H}$ is a regular and uniform hypergraph then it holds that*

$$\tau_\epsilon(\mathcal{H}) \leq \tau_\epsilon^*(\mathcal{H}) \left(1 + \ln\left(\frac{1}{\epsilon}\right)\right).$$

4   Constructing ϵ-almost (k, m, n)-Selectors via Partial Coverings

We first give an upper bound on the length of ϵ-almost (k, m, n)-selectors and then exploit this result to derive an upper bound on the length of ϵ-almost KG (k, n)-codes. In the following, with a little abuse of notation, we will use the same capital letter to denote both a set of columns and the set of the stations associated with the columns in that set.

Theorem 4. *Given integers k, m, and n, with $1 \leq m \leq k \leq n$, and a real number $0 \leq \epsilon \leq 1$, there exists an ϵ-almost (k, m, n)-selector with size*

$$t \leq \frac{ek}{k - m + 1} \left(1 + \ln \frac{\binom{k}{k-m+1}}{\epsilon}\right).$$

Proof. We aim at constructing a hypergraph $\mathcal{H}$ in such a way that, for a given $0 \leq \epsilon' \leq 1$, any $(1 - \epsilon')$-partial cover of $\mathcal{H}$ is an ϵ-almost (k, m, n)-selector. Then, we will exploit Theorem 3 to derive the desired upper bound on the minimum selector size.

The hypergraph $\mathcal{H}$ is defined as in the proof of Theorem 1 of [9]. We denote by X the set of all binary vectors $\mathbf{x} = (x_1, \ldots, x_n)$ of length n containing n/k 1's For any integer i, $1 \leq i \leq k$, let $\mathbf{a}_i$ be the binary vector of length k having all components equal to zero but that in position i, that is, $\mathbf{a}_1 = (1, 0, \ldots, 0)$, $\mathbf{a}_2 = (0, 1, \ldots, 0)$, $\ldots$, $\mathbf{a}_k = (0, 0, \ldots, 1)$. Moreover, for any set of indices $S = \{i_1, \ldots, i_k\}$, with $1 \leq i_1 \leq i_2 < \ldots < i_k \leq n$, and for any binary vector $\mathbf{a} = (a_1, \ldots, a_k) \in \{\mathbf{a}_1, \ldots, \mathbf{a}_k\}$, let $E_{\mathbf{a},S}$ be the set of binary vectors $E_{\mathbf{a},S} = \{\mathbf{x} =$

$(x_1, \ldots, x_n) \in X : x_{i_1} = a_1, \ldots, x_{i_k} = a_k\}$. For any set $A \subseteq \{a_1, \ldots, a_k\}$ of size $r, r = 1, \ldots, k$, and any set $S \subseteq \{1, \ldots, n\}$, with $|S| = k$, we define the hyperedge $E_{A,S} = \bigcup_{a \in A} E_{a,S}$ and, for any $r = 1, \ldots, k$, we define the set of hyperedges $\mathbf{F}_r = \{E_{A,S} : A \subset \{a_1, \ldots, a_k\}, |A| = r, \text{ and } S \subseteq \{1, \ldots, n\}, |S| = k\}$ and the hypergraph $\mathcal{H}_r = (X, \mathbf{F}_r)$.

Let $\ell = \lceil (1 - \epsilon) \binom{n}{k} \rceil$ and let ϵ' be a real number such that $0 \leq \epsilon' \leq 1$ and

$$\left\lceil \binom{k}{k-m+1} \binom{n}{k} (1 - \epsilon') \right\rceil \geq \binom{n}{k} \left[\binom{k}{k-m+1} - 1 \right] + \ell. \tag{4}$$

We will prove that *any* $(1 - \epsilon')$-cover of $\mathcal{H}_{k-m+1}$ T is an ϵ-almost (k, m, n)-selector. Let T be *any* $(1-\epsilon')$-cover of $\mathcal{H}_{k-m+1}$. First notice that $|E(\mathcal{H}_{k-m+1})| = |\mathbf{F}_{k-m+1}| = \binom{k}{k-m+1} \binom{n}{k}$, therefore inequality (4) implies that any $(1-\epsilon')$-cover of $\mathcal{H}_{k-m+1}$ covers at least $\binom{n}{k} \left[\binom{k}{k-m+1} - 1 \right] + \ell$ edges. Moreover, by construction, for a fixed subset $S \in [n]_k$ there are exactly $\binom{k}{k-m+1}$ edges $E_{A,S}$ in $\mathcal{H}_{k-m+1}$, each for any of the possible subsets $A \subseteq \{a_1, \ldots, a_k\}$ of size $k - m + 1$. It follows that for at least ℓ subsets $S^* \in [n]_k$, T covers all $\binom{k}{k-m+1}$ edges E_{A,S^*}. Let us denote by $\mathcal{G} \subseteq [n]_k$ the set of these ℓ subsets. It is possible to prove that the submatrix of T formed by the columns in any subset $S^* \in \mathcal{G}$ contains m rows of the identity matrix I_k. Assume by contradiction that there exists a set of indices $S^* = \{i_1, \ldots, i_k\} \in \mathcal{G}$ such that the submatrix of T obtained by considering only the columns of T with indices $i_1, \ldots, i_k$ contains *at most* $m - 1$ distinct rows of I_k. It follows that there exists a subset A of $k - m + 1$ vectors of $\{a_1, \ldots, a_k\}$ that does not contain any of these $m - 1$ vectors, and consequently, E_{A,S^*} is not one of the edges covered by T, i.e., $T \cap E_{A,S^*} = \emptyset$. This contradicts the fact that for any $S^* \in \mathcal{G}$ and any subset A of $k - m + 1$ vectors of $\{a_1, \ldots, a_k\}$, T covers the edge E_{A,S^*}. Since $|\mathcal{G}| = \ell \geq (1 - \epsilon) \binom{n}{k}$ then T is an ϵ-almost (k, m, n)-selector. By Theorem 3 there exists a $(1 - \epsilon')$-cover of size

$$\tau_{\epsilon'}(\mathcal{H}_{k-m+1}) \leq \tau_{\epsilon'}^*(\mathcal{H}_{k-m+1}) \left(1 + \ln \left(\frac{1}{\epsilon'} \right) \right). \tag{5}$$

One can see that if we choose

$$\epsilon' = \frac{\epsilon}{\binom{k}{k-m+1}}, \tag{6}$$

then inequality (4) is satisfied with equality. To estimate the upper bound on $\tau_{\epsilon'}(\mathcal{H}_{k-m+1})$ in (5), we need to compute the minimum size $\tau_{\epsilon'}^*(\mathcal{H}_{k-m+1})$ of the fractional $(1 - \epsilon')$-cover of $\mathcal{H}_{k-m+1}$. Notice that it holds that $\tau_{\epsilon'}^*(\mathcal{H}_{k-m+1}) \leq \tau^*(\mathcal{H}_{k-m+1})$ and inequality (3) implies

$$\tau^*(\mathcal{H}_{k-m+1}) \leq \frac{|X|}{\min_{E \in \mathbf{F}_{k-m+1}} |E|} = \frac{\binom{n}{n/k}}{(k - m + 1) \binom{n-k}{n/k-1}}.$$

As shown in the proof of Theorem 1 of [9], inequality $\frac{\binom{n}{n/k}}{\binom{n-k}{n/k-1}} \leq ek$ holds, from which it follows

$$\tau_{\epsilon'}^*(\mathcal{H}_{k-m+1}) \leq \frac{ek}{k-m+1}. \tag{7}$$

The bound stated by the theorem follows from (5), (6) and (7).   □

By setting $m = k+1$ in the upper bound on the length of ϵ-almost $(k+1, k+1, n)$-selectors derived from Theorem 4, we obtain the following upper bound on the minimum length of ϵ-almost k-disjunct codes of size n.

Corollary 1. *Given integers k and n, with $1 \leq k < n$, and a real number $0 \leq \epsilon \leq 1$, there exists an ϵ-almost k-disjunct code with length $t \leq e(k+1)\left(1 + \ln \frac{(k+1)}{\epsilon}\right)$.*

Remark 1. In [17] it has been presented a construction for ϵ-almost k-disjunct codes with length $O\left((k-1)^{3/2} \ln n \frac{\sqrt{\frac{\ln(2k)}{\epsilon}}}{\ln(k-1) - \ln\ln \frac{2k}{\epsilon} + \ln(4a)}\right)$ for $\epsilon > 2ke^{-a(k-1)}$ and a being a constant larger than 1. This bound seems weak in several aspects. First of all, the constraint that $\epsilon > 2ke^{-a(k-1)}$ does not allow constructions of ϵ-almost k-disjunct codes in which the number $\epsilon\binom{n}{k}$ of "bad" k-column subsets is small enough, as in our constructions. In fact, $\epsilon > 2ke^{-a(k-1)}$ implies that $\epsilon\binom{n}{k} > 2ke^a \left(\frac{n}{ke^a}\right)^k$. Moreover, for $\epsilon > 2ke^{-a(k-1)}$ the bound of our Corollary 1 is $O(k^2)$ which is at least as good as the bound in [17] when $\epsilon \leq c \left(\frac{\ln n \sqrt{\ln(2k)}}{\sqrt{k-1}(\ln(k-1) - \ln\ln \frac{2k}{\epsilon} + \ln(4a))}\right)^2$, for any positive constant c, that is almost always the case.

Remark 2. It is interesting also to remark that for $\epsilon = \frac{1}{\binom{n}{k}} - \delta$, for any $\delta > 0$, the ϵ-almost disjunct codes reduce to the classical disjunct codes. Moreover, the bound one obtains from our Corollary 1 with that value of ϵ reduces to $t = O(k^2 \log \frac{n}{k})$, that corresponds to the best known upper bound on the length of classical disjunct codes.

We now turn our attention to ϵ-almost KG (k, n)-codes.

Theorem 5. *Given integers k and n, with $1 \leq k \leq n$, and a real number $0 \leq \epsilon \leq 1$, there exists an ϵ-almost KG (k, n)-code with length*

$$t < 2e\lceil \log k \rceil \left(1 + \ln\left(\frac{\lceil \log k \rceil}{\epsilon}\right)\right) + O(k).$$

Proof. We will prove that there exists a $n \times t$ matrix $\tilde{M}$, with t being upper bounded as in the statement of the theorem, such that $\tilde{M}$ contains at least $(1 - \epsilon)\binom{n}{k}$ subsets of k columns that satisfy property (*). This is equivalent to prove that there are at least $(1 - \epsilon)\binom{n}{k}$ subsets S of k stations such that the scheduling algorithm corresponding to $\tilde{M}$ allows all active stations in S to transmit successfully. Let us recall that Theorem 4 implies that for any integer

$1 \leq i \leq \lceil \log k \rceil$ and any real number ϵ_i with $0 \leq \epsilon_i \leq 1$, there exists an ϵ_i-almost $(2^i, 2^{i-1}, n)$-selector of size

$$t_i < 2e \left(1 + \ln \frac{\binom{2^i}{2^{i-1}+1}}{\epsilon_i} \right). \tag{8}$$

Let us denote by $\tilde{M}_{(\epsilon_i, 2^i, n)}$ such an ϵ_i-almost $(2^i, 2^{i-1}, n)$-selector, and let

$$\epsilon_i = \frac{\epsilon}{\binom{k}{2^i} \lceil \log k \rceil}. \tag{9}$$

We set $\tilde{M}$ to be the matrix obtained by concatenating the rows of matrices $\tilde{M}_{(\epsilon_i, 2^i, n)}$, for $i = \lceil \log k \rceil, \ldots, 1$, taken in this order, that is with the rows of $\tilde{M}_{(\epsilon_{\lceil \log k \rceil}, 2^{\lceil \log k \rceil}, n)}$ being placed at the top of the matrix and those of $\tilde{M}_{(\epsilon_1, 2, n)}$ being placed at the bottom, and by appending the all-1 row at the bottom. We will show that there are at least $(1 - \epsilon)\binom{n}{k}$ subsets S of k stations such that the conflict resolution algorithm based on $\tilde{M}$ allows any subset of active stations in S to transmit successfully, thus proving that $\tilde{M}$ is an ϵ-almost KG (k, n)-code.

For $i = 1, \ldots, \lceil \log k \rceil$, let us denote by $\mathcal{G}_i$ the family of the subsets of 2^i columns of $\tilde{M}_{(\epsilon_i, 2^i, n)}$ that satisfy the property of $(2^i, 2^{i-1}, n)$-selectors, i.e., the family of 2^i-column subsets for which it holds that the submatrix of $\tilde{M}_{(\epsilon_i, 2^i, n)}$ formed by the 2^i columns in any member of the family contains 2^{i-1} rows of the identity matrix I_{2^i}. We will prove that the following claim holds.

Claim. Let $S \in \mathcal{G}_{\lceil \log k \rceil}$. If, for $i = 1, \ldots, \lceil \log k \rceil$, it holds that each of the 2^i-column subsets of S are members of $\mathcal{G}_i$, then $\tilde{M}$ allows any subset A of initially active stations such that $A \subseteq S$ to transmit successfully.

Assume that the hypothesis of the claim is true. In Sect. 2 we observed that any protocol based on an ϵ-almost (k, m, n)-selector schedules transmissions in such a way that at the end there are at most $k - m$ active stations that did not succeed to transmit their messages, provided that all active stations are contained in one of the $(1 - \epsilon)\binom{n}{k}$ subsets of k stations corresponding to the k-column subsets that satisfy the property of (k, m, n)-selectors. Consequently, if the set A of active stations has size at most 2^i and A is subset of some set $S_i \in \mathcal{G}_i$, then the protocol based on $\tilde{M}_{(\epsilon_i, 2^i, n)}$ schedules transmissions so that at the end we are left with at most 2^{i-1} active stations. Recall, that an active station becomes inactive immediately after it has transmitted successfully. Let us consider the conflict resolution algorithm based on $\tilde{M}$. Since we are assuming that initially the number of active station is at most $k \leq 2^{\lceil \log k \rceil}$, then, by the above argument, one has that after the first $t_{\lceil \log k \rceil}$ steps there are at most $2^{\lceil \log k \rceil - 1}$ active stations. The hypothesis of the claim implies that these up to $2^{\lceil \log k \rceil - 1}$ active stations are contained in some set $S_{\lceil \log k \rceil - 1} \in \mathcal{G}_{\lceil \log k \rceil - 1}$ and therefore, there are at most $2^{\lceil \log k \rceil - 2}$ stations that are still active after the next $t_{\lceil \log k \rceil - 1}$ steps. Continuining in this way, one can see that, for each $i = 1, \ldots, \lceil \log k \rceil$, after the first $t_{\lceil \log k \rceil} + t_{\lceil \log k \rceil - 1} + \ldots + t_i$ steps, the algorithm is left with at most $2^{\lceil \log k \rceil - i}$ active stations. At the end, after $\sum_{i=0}^{\lceil \log k \rceil} t_i$ steps, there is at most a single station which is still active. Obviously, this last station can transmit without conflict with other active stations. The all 1-entry row at the bottom of the matrix takes care of this last active station. This concludes the proof of the claim.

In the following, we will prove that the hypothesis of the claim holds for at least $(1 - \epsilon)\binom{n}{k}$ members S of $\mathcal{G}_{\lceil \log k \rceil}$, thus showing that $\tilde{M}$ is an ϵ-almost KG (k, n)-code.

For $i = 1, \ldots, \lceil \log k \rceil - 1$, any of the $\epsilon_i \binom{n}{2^i}$ subsets of 2^i columns not belonging to $\mathcal{G}_i$ could be contained in up to $\binom{n-2^i}{k-2^i}$ members of $\mathcal{G}_{\lceil \log k \rceil}$. It follows that the number of k-column subsets in $\mathcal{G}_{\lceil \log k \rceil}$ that do not satisfy the condition of the claim is at most

$$\sum_{i=1}^{\lceil \log k \rceil - 1} \epsilon_i \binom{n}{2^i}\binom{n-2^i}{k-2^i} = \sum_{i=0}^{\lceil \log k \rceil - 1} \epsilon_i \binom{k}{2^i}\binom{n}{k}. \tag{10}$$

The total number of k-column subsets of $\mathcal{G}_{\lceil \log k \rceil}$ is at least $(1 - \epsilon_{\lceil \log k \rceil})\binom{n}{k}$ and inequality (10) implies that at most $\sum_{i=0}^{\lceil \log k \rceil - 1} \epsilon_i \binom{k}{2^i}\binom{n}{k}$ of these subsets do not satisfy the hypothesis of the claim. It follows that the number of members S of $\mathcal{G}_{\lceil \log k \rceil}$ for which the claim holds is at least

$$(1 - \epsilon_{\lceil \log k \rceil})\binom{n}{k} - \sum_{i=0}^{\lceil \log k \rceil - 1} \epsilon_i \binom{k}{2^i}\binom{n}{k} = \binom{n}{k} - \sum_{i=0}^{\lceil \log k \rceil - 1} \epsilon_i \binom{k}{2^i}\binom{n}{k} - \epsilon_{\lceil \log k \rceil}\binom{n}{k}.$$

Therefore, the total number of k-column subsets that do not satisfy property (*) is at most

$$\sum_{i=0}^{\lceil \log k \rceil} \epsilon_i \binom{k}{2^i}\binom{n}{k} + \epsilon_{\lceil \log k \rceil}\binom{n}{k}.$$

By setting ϵ_i as in (9), we obtain that the above bound is equal to $\epsilon\binom{n}{k}$ thus proving that $\tilde{M}$ is an ϵ-almost KG (k, n)-code.

Now we derive the claimed upper bound on the number of rows of $\tilde{M}$. Upper bound (8) and equation (9) imply that the length of $\tilde{M}$ is

$$t < 2e \sum_{i=1}^{\lceil \log k \rceil} \left(1 + \ln \frac{\binom{2^i}{2^{i-1}+1}}{\epsilon_i}\right) + 1 = 2e \sum_{i=1}^{\lceil \log k \rceil} \left(1 + \ln \frac{\binom{2^i}{2^{i-1}+1}\binom{k}{2^i}\lceil \log k \rceil}{\epsilon}\right) + 1. \tag{11}$$

We upper bound the righthand side of (11) by noticing that $\binom{2^i}{2^{i-1}+1} < \binom{2^i}{2^{i-1}}$ and applying the well known inequality $\binom{a}{c} \le (ea/c)^c$ to $\binom{2^i}{2^{i-1}}$ and $\binom{k}{2^i}$, thus obtaining

$$t < 2e\lceil \log k \rceil + 2e \sum_{i=1}^{\lceil \log k \rceil} \left(\ln \left(\frac{(2e)^{2^{i-1}}(ek/2^i)^{2^i}\lceil \log k \rceil}{\epsilon}\right)\right) + 1$$

$$= 2e\lceil \log k\rceil \left(1 + \ln\left(\frac{\lceil \log k\rceil}{\epsilon}\right)\right) + 2e \sum_{i=1}^{\lceil \log k\rceil} 2^i \left(\frac{1}{2}\ln(2e) + \ln e\right) + 2e \sum_{i=1}^{\lceil \log k\rceil} 2^i \ln(k/2^i) + 1$$

$$< 2e\lceil \log k\rceil \left(1 + \ln\left(\frac{\lceil \log k\rceil}{\epsilon}\right)\right) + 2e(4k-1)\ln(e\sqrt{2e}) + 4ek \sum_{i=0}^{\lceil \log k\rceil-1} \frac{1}{2^i}\ln 2^i + 1. \qquad (12)$$

By using the well known inequality $\sum_{i=0}^{\infty} ix^i \leq \frac{x}{(1-x)^2}$, holding for $|x| < 1$, to bound from above $\sum_{i=0}^{\lceil \log k\rceil-1} \frac{i}{2^i}$ in (12), it follows $t < 2e\lceil \log k\rceil \left(1 + \ln\left(\frac{\lceil \log k\rceil}{\epsilon}\right)\right) + 4e(2k-1)\ln(e\sqrt{2e}) + 8ek\ln 2 + 1$, that concludes the proof of the theorem.

4.1 Improvements

A possible drawback of our solutions is that our scheduling algorithms do not offer any guarantee for (at most) a fraction of ϵ of all $\binom{n}{k}$ k-subsets of conflicting stations. However, if one concatenates a (k, m, n)-selector to an ϵ-almost (k, k, n)-selector, then one obtains a scheduling algorithm that allows at least m active stations to transmit successfully even if the k-subset of active stations falls within these $\epsilon\binom{n}{k}$ subsets. The following theorem follows from setting $m = k$ in the upper bound of Theorem 4 and from upper bound (1) on the size of (k, m, n)-selectors.

Theorem 6. *Given integers k, m, and n, with $1 \leq m \leq k \leq n$, and a real number $0 \leq \epsilon \leq 1$, there exists an n-column matrix such that*

1. *each subset of k columns form a $k \times k$ submatrix that contains at least m rows of the identity matrix I_k and*
2. *for at least $(1 - \epsilon)\binom{n}{k}$ subsets of k columns one has that each of these subsets form a $k \times k$ submatrix that contains all rows of the identity matrix I_k*

and has a number of rows

$$t \leq ek\left(1 + \ln\frac{k}{\epsilon}\right) + \frac{ek^2}{k-m+1}\ln\left(\frac{n}{k}\right) + \frac{ek(2k-1)}{k-m+1}.$$

For $m = k+1-\frac{k\ln(n/k)}{c\ln(k/\epsilon)}$, where c is an arbitrary constant larger than or equal to 1, the scheduling algorithm of Theorem 6 uses the same asymptotic number of time slots as the algorithm based on ϵ-almost (k, k, n)-selectors. This is an immediate consequence of the following result obtained by setting $m = k + 1 - \frac{k\ln(n/k)}{c\ln(k/\epsilon)}$ in the statement of Theorem 6.

Corollary 2. *Let k and n be integers such that $1 \leq k \leq n$, and let ϵ be a real number such that $0 \leq \epsilon \leq 1$. There exists a conflict resolution algorithm for a multiple-access channel without feedback that schedules the transmissions of n stations in such a way that for at least a $(1 - \epsilon)$ ratio of all possible subsets of k active stations, the algorithm allows all k conflicting stations to transmit successfully, whereas for the remaining subsets of k stations it allows at least*

$k + 1 - \frac{k \ln(n/k)}{c \ln(k/\epsilon)}$ *stations to transmit successfully, where c is an arbitrary constant larger than or equal to 1. The number of time slots used by the conflict resolution algorithm is*

$$t \leq e(c+1)k \ln\left(\frac{k}{\epsilon}\right) + ec(2k-1)\frac{\ln(k/\epsilon)}{\ln(n/k)} + ek.$$

The above corollary implies that for any $\epsilon \leq \frac{k^2}{n}$ and for any arbitrary constant $c \geq 1$, there exists a scheduling algorithm such that

- it uses $t \leq e(c+1)k \ln\left(\frac{k}{\epsilon}\right) + ec(2k-1)\frac{\ln(k/\epsilon)}{\ln(n/k)} + ek$ time slots, i.e., the *same* asymptotic number of time slots as the scheduling algorithm based on ϵ-almost (k, k, n)-selectors, and
- in addition to solving *all conflicts* among a ratio $(1-\epsilon)$ of all possible subsets of k active stations, it allows at least $k\left(\frac{c-1}{c}\right)+1$ stations to transmit successfully *whichever* the subset of k active stations is.

Notice that for $\epsilon = \frac{k^2}{n}$, the above scheduling algorithm uses $t = O(k \log(n/k))$ time slots, i.e., the same asymptotic number of time slots used by the scheduling algorithm based on (classical) KG (k, n)-codes, which, however, solves conflicts only under the assumption that the stations receive feedback from the channel.

We can apply a similar idea to the multiple-access channel with feedback to obtain a scheduling algorithm that solves all conflicts among (up to) k active stations if the subset of active stations is contained in one of the $(1-\epsilon)\binom{n}{k}$ "good" k-subsets, and among a smaller subset of active stations otherwise. The idea is to concatenate ϵ-almost $(2^i, 2^{i-1}, n)$-selectors, as in the construction of Theorem 5, for values of i larger than an appropriately chosen value i^*, and (classical) $(2^i, 2^{i-1}, n)$-selectors for smaller i's. By a suitable choice of i^*, we obtain that for any $\epsilon \leq (\log k)\left(\frac{f(k)}{n}\right)^{\frac{cf(k)}{\log k}}$, where $f(k)$ is any non decreasing function such that $f(k) = O(k)$ and c is an arbitrary positive constant, there exists a scheduling algorithm for the multiple-access channel with feedback such that

- it uses $t = O\left((\log k)\ln\left(\frac{\log k}{\epsilon}\right) + k\right)$ time slots, i.e., the same asymptotic number of time slots of the scheduling algorithm based on ϵ-almost KG (k, n)-codes, and
- it allows *any* subset of up to $f(k)$ stations to transmit successfully, and for at least a ratio $(1 - \epsilon)$ of all subsets of k stations, it solves all conflicts among up to k active stations belonging to one of those k-subsets.

Notice that for $\epsilon = (\log k)\left(\frac{f(k)}{n}\right)^{\frac{cf(k)}{\log k}}$ and $f(k) = o(k)$, the above scheduling algorithm uses $t = O(f(k) \log(n/f(k)))$ time slots, i.e., an asymptotic number of time slots smaller than that used by the scheduling algorithm based on (classical) KG (k, n)-codes.

References

1. Alon, N., Fachini, E., Körner, J.: Locally thin set families. Comb., Prob. Comput. 9, 481–488 (2000)
2. Biglieri, E., Györfi, L.: Multiple Access Channels, NATO Security through Science Series - D: Information and Communication Security 10 (2007)
3. Chlebus, B.S.: Randomized communication in radio networks. In: Pardalos, M., Rajasekaran, S., Reif, J.H., Rolim, J.D.P. (eds.) Handbook on Randomized Computing, vol. I, pp. 401–456. Kluwer Academic Publishers, Dordrecht (2001)
4. Chlebus, B.S., Gasieniec, L., Kowalski, D.R., Shvartsman, A.: A robust randomized algorithm to perform independent tasks. J. Discrete Algorithms 6(4), 651–665 (2008)
5. Chlebus, B.S., Kowalski, D.R., Pelc, A., Rokicki, M.A.: Efficient distributed communication in ad-hoc radio networks. In: Aceto, L., Henzinger, M., Sgall, J. (eds.) ICALP 2011, Part II. LNCS, vol. 6756, pp. 613–624. Springer, Heidelberg (2011)
6. Chrobak, M., Gasieniec, L., Rytter, W.: Fast broadcasting and gossiping in radio networks, In: Proceedings of 42nd IEEE Annual Symposium on Found. of Computer Science (FOCS 2000), pp. 575–581 (2000)
7. Clementi, A.E.F., Monti, A., Silvestri, R.: Selective families, superimposed codes, and broadcasting on unknown radio networks, In: Proceedings of Symposium on Discrete Algorithms (SODA 2001), pp. 709–718 (2001)
8. Csűrös, M., Ruszinkó, M.: Single-user tracing and disjointly superimposed codes. IEEE Trans. Inf. Theory 51(4), 1606–1611 (2005)
9. De Bonis, A.: A, Gasieniec, U. Vaccaro: Optimal two-stage algorithms for group testing problems. SIAM J. Comput. 34(5), 1253–1270 (2005)
10. Du, D.Z., Hwang, F.K.: Combinatorial Group Testing and Its Applications. World Scientific, River Edge (2000)
11. D'yachkov, A.G., Rykov, V.V.: Bounds on the length of disjunct codes. Problemy Peredachi Informatsii 18(3), 7–13 (1982)
12. Erdös, P., Frankl, P., Füredi, Z.: Families of finite sets in which no set is covered by the union of r others. Israel J. of Math. 51, 75–89 (1985)
13. Kowalski, D.R.: On selection problem in radio networks. In: Proceedings of the Twenty-Fourth Annual ACM Symposium on Principles of Distributed Computing (PODC 2005), pp. 158–166. ACM Press (2005)
14. Kautz, W.H., Singleton, R.C.: Nonrandom binary superimposed codes. IEEE Trans Inf. Theory 10, 363–377 (1964)
15. Komlós, J., Greenberg, A.G.: An asymptotically fast non-adaptive algorithm for conflict resolution in multiple-access channels. IEEE Trans. Inform. Theory 31(2), 302–306 (1985)
16. Lovàsz, L.: On the ratio of optimal integral and fractional covers. Discrete Math. 13, 383–390 (1975)
17. Mazumdar, A.: On almost disjunct matrices for group testing. In: Chao, K.-M., Hsu, T., Lee, D.-T. (eds.) ISAAC 2012. LNCS, vol. 7676, pp. 649–658. Springer, Heidelberg (2012)
18. Sümer, Ö.: Partial covering of hypergraphs. In: SODA 2005, pp. 572–581 (2005)
19. Porat, E., Rothschild, A.: Explicit nonadaptive combinatorial group testing schemes. IEEE Trans. Inf. Theory 57(12), 7982–7989 (2011)

Derandomized Construction of Combinatorial Batch Codes

Srimanta Bhattacharya[(✉)]

Centre of Excellence in Cryptology, Indian Statistical Institute,
Kolkata 700108, India
mail.srimanta@gmail.com

Abstract. *Combinatorial Batch Codes* (CBCs), replication-based variant of *Batch Codes* introduced by Ishai *et al.* in [IKOS04], abstracts the following data distribution problem: n data items are to be replicated among m servers in such a way that any k of the n data items can be retrieved by reading at most one item from each server with the total amount of storage over m servers restricted to N. Given parameters m, c, and k, where c and k are constants, one of the challenging problems is to construct c-uniform CBCs (CBCs where each data item is replicated among exactly c servers) which maximizes the value of n.

In this work, we present explicit construction of c-uniform CBCs with $\Omega(m^{c-1+\frac{1}{k}})$ data items. The construction has the property that the servers are almost *regular*, i.e., number of data items stored in each server is in the range $[\frac{nc}{m} - \sqrt{\frac{n}{2}\ln(4m)}, \frac{nc}{m} + \sqrt{\frac{n}{2}\ln(4m)}]$. The construction is obtained through better analysis and derandomization of the randomized construction presented in [IKOS04]. Analysis reveals almost regularity of the servers, an aspect that so far has not been addressed in the literature. The derandomization leads to explicit construction for a wide range of values of c (for given m and k) where no other explicit construction with similar parameters, i.e., with $n = \Omega(m^{c-1+\frac{1}{k}})$, is known. Finally, we discuss possibility of parallel derandomization of the construction.

1 Introduction

1.1 Background

Batch Codes. An (n, N, k, m)-*batch code* (or $(n, N, k, m, t = 1)$-*batch code*) abstracts the following data distribution problem: n data items are to be distributed among m servers in such a way that any k of the n items can be retrieved by reading at most one item from each server and total amount of storage required for this distribution is bounded by N.[1] Batch codes were introduced

---

[1] Parameter t represents the maximum number of data items that can be read from each server during retrieval of any k items. In [IKOS04], batch codes were defined for general t. In [BT12], CBCs were studied for general t. However, in this work, we solely consider CBCs with $t = 1$ as it seems to capture the crux of the problem, and also this is more common in the literature.

© Springer International Publishing Switzerland 2015
A. Kosowski and I. Walukiewicz (Eds.): FCT 2015, LNCS 9210, pp. 269–282, 2015.
DOI: 10.1007/978-3-319-22177-9_21

in [IKOS04], and their primary motivation was to amortize computational work done by the servers during execution of *private information retrieval* protocol by batching several queries together while limiting total storage (see [IKOS04] for more details). It is also easy to see from the above description that these codes can have potential application in a distributed database scenario where the goal is to distribute load among the participating servers while optimizing total storage.

On the theoretical side batch codes closely resemble several combinatorial objects like *expanders, locally decodable codes,* etc., and there is also similarity with Rabin's *information dispersal.* However, there are fundamental differences of batch codes with these objects, especially as far as setting of parameter values are concerned, and it seems difficult to set up satisfactory correspondences with these objects (see [IKOS04] for a discussion on this). This dichotomy makes batch codes unique and very interesting objects.

Combinatorial Batch Codes (CBCs). These are replication based batch codes; each of the N stored data items is a copy of one of the n input data items. So, for CBCs, encoding is assignment (storage) of items to servers and decoding is retrieval (reading) of items from servers. This requirement makes CBCs purely combinatorial objects. As combinatorial objects CBCs are quite interesting, they have received considerable attention in recent literature ([PSW09, BKMS10, BT11b, BT11a, BT15, SG14, BRR12, BB14]). Before proceeding further, we introduce the formal framework for our study of CBCs.

We represent an (n, N, k, m)-CBC $\mathcal{C}$ as a bipartite graph $\mathcal{G}_\mathcal{C} = (\mathcal{L}, \mathcal{R}, \mathcal{E})$. Set of left vertices $\mathcal{L}$ represents $|\mathcal{L}| = n$ data items with vertex $u_i \in \mathcal{L}$ representing data item $x_i, 1 \le i \le n$, and set of right vertices $\mathcal{R}$ represents $|\mathcal{R}| = m$ servers with vertex $v_j \in \mathcal{R}$ representing server $s_j, 1 \le j \le m$. Hence, $(u_i, v_j) \in \mathcal{E}$ is an edge in $\mathcal{G}_\mathcal{C}$ if the data item x_i is stored in server s_j. Since the total storage is N, it follows that $\sum_{u \in \mathcal{L}} deg(u) = \sum_{v \in \mathcal{R}} deg(v) = |\mathcal{E}| = N$, where $deg(.)$ is the degree of a vertex in $\mathcal{G}_\mathcal{C}$. Now, it can be observed that any subset $\{x_{i_1}, \ldots, x_{i_k}\}$ of k distinct data items can be retrieved by reading one item from each of k distinct servers $s_{i_1}, \ldots, s_{i_k}$ iff there are distinct $v_{i_1}, \ldots, v_{i_k} \in \mathcal{R}$ such that $v_{i_j} \in \Gamma(u_{i_j})$ for all $1 \le j \le k$, where $\Gamma(u_r), r \in \{1, \ldots, n\}$, is the *neighbourhood* of the vertex $u_r \in \mathcal{L}$. [2] According to *Hall's theorem* (cf. [Bol86], pp. 6) this is equivalent to the condition that union of any j sets $\Gamma(u_{i_1}), \ldots \Gamma(u_{i_j})$, $\{u_{i_1} \ldots u_{i_j}\} \subset \mathcal{L}$, contains at least j elements for $1 \le j \le k$. These considerations lead naturally to the following theorem of [PSW09] which can also be thought as definition of a CBC.

Theorem 1 ([PSW09]). *A bipartite graph $\mathcal{G}_\mathcal{C} = (\mathcal{L}, \mathcal{R}, \mathcal{E})$ represents an (n, N, k, m)-CBC if and only if $|\mathcal{L}| = n, |\mathcal{R}| = m, |\mathcal{E}| = N$ and union of every collection of j sets $\Gamma(u_{i_1}), \ldots, \Gamma(u_{i_j})$, $\{u_{i_1}, \ldots, u_{i_j}\} \subset \mathcal{L}$ contains at least j elements for $1 \le j \le k$.*

---

[2] For a vertex $u \in \mathcal{L}$, its neighbourhood is $\Gamma(u) = \{v \in \mathcal{R}, (u, v) \in \mathcal{E}\}$. In the sequel, we will require extension of the definition of neighbourhood of a vertex to neighbourhood of a subset. More formally, given $S \subset \mathcal{L}$, we denote by $\Gamma(S)$ the set $\{v \in \mathcal{R} | \exists u \in S, (u, v) \in \mathcal{E}\}$.

Remark 1. Formal definition of general batch codes given in [IKOS04] also involves *decoding algorithm* of the code. Here we consider CBCs, a purely combinatorial subclass of general batch codes, with our only focus on their construction, and the above definition is sufficient for our purpose.

From now on, we will identify the graph $\mathcal{G}_C = (\mathcal{L}, \mathcal{R}, \mathcal{E})$ with an (n, N, k, m)-CBC, and omit the subscript C as it will not cause any trouble. A CBC $\mathcal{G} = (\mathcal{L}, \mathcal{R}, \mathcal{E})$ is called *c-uniform* if for each $u \in \mathcal{L}$, $deg(u) = c$, and it is called *l-regular* if for each $v \in R, deg(v) = l$.[3] Two optimization problems related to CBCs have been addressed in the literature: (i) given n, m, k find minimum value of N attained by a CBC (not necessarily uniform or regular), and provide explicit construction of corresponding extremal CBCs; (ii) given m, c, k, find maximum value of n, denoted as $n(m, c, k)$, attained by a c-uniform CBC (not necessarily regular), and provide explicit construction of corresponding extremal CBCs. In this work we will consider the latter problem in setting of parameters where c and k are constants while m is variable.

At this point, it may be observed that though the definition and the considered problem draws some similarity with those of bipartite expanders, especially the unbalanced expanders [GUV09], there are important differences between these two cases as well. On the similarity side, both are bipartite graphs with constant left-degree; in both the cases, it is required that every subset of vertices $\mathcal{L}$ of up to a specified size should have neighbourhood of specified sizes, and it is desirable that $|\mathcal{L}| >> |\mathcal{R}|$. However, in the case of unbalanced expanders, the goal is to stretch the expansion of subsets[4] (of specified sizes) of $\mathcal{L}$ as close to the left-degree as possible. Whereas, in case of CBCs, expansion of 1 is sufficient and it is more important to make $|\mathcal{L}|$ as large as possible with respect to $|\mathcal{R}|$. Also important is the fact that the parameter k is a constant in case of CBCs (within our setting of parameters), whereas for expanders k varies with n. These differences make the (desirable) parameters in these two cases essentially unrelated. So, it seems unlikely that the existing constructions of unbalanced expanders can be immediately used for construction of CBCs where c and k are constants.

Before discussing existing results and our contribution we briefly mention the notion of 'explicit' construction of a combinatorial object in general and CBCs in particular as that will be crucial to our discussion and result.

Explicit Construction. Construction of a combinatorial object with desirable properties is computation of a representation of the object and is tied with the resources used for the computation. In the literature, those constructions which require practically feasible amount of resources, such as polynomial time or logarithmic space are termed explicit. This can be contrasted with exhaustive search of a combinatorial object whose existence has been proven (e.g. by probabilistic argument); the search is done in the space of the object and requires infeasible

---

[3] Here the terminology is in keeping with the representation of a CBC as a *set-system* in some of the previous works, where the set $\mathcal{R}$ is treated as ground set and the multi-set $\{\Gamma(u_1) \dots \Gamma(u_n)\}, u_i \in \mathcal{L}$ is the collection of subsets.

[4] For a set S, expansion of S is $\frac{|\Gamma(S)|}{|S|}$.

amount of resources (e.g. exponential time). The notion of explicitness we will adhere to in this work is polynomial time constructibility, which requires that the time required for the construction be bounded by a polynomial in the size of the representation. Explicitness is further classified as following.[5]

Globally Explicit. In this case the whole object is constructed in time polynomial in the size of the object. For example, a globally explicit construction of CBC $(\mathcal{L}, \mathcal{R}, \mathcal{E})$ would list all the members of $\mathcal{E}$ in $poly(|\mathcal{E}|)$ time.

Locally Explicit. In this case the idea is to have quick local access to the object. More formally, for a desirable combinatorial object G, locally explicit construction of G is an algorithm which given an index of size $\log(|G|)$, outputs the member of G with the given index (or does some local computation on the member) in time $polylog(|G|)$. This is more specialized notion and depends on the context. For example, a locally explicit construction of a c-uniform CBC $(\mathcal{L}, \mathcal{R}, \mathcal{E})$ would list the neighbourhood of a vertex $v \in \mathcal{L}$ in time $poly(\log(|\mathcal{L}|), c)$ given the index of v (which is of size $|\mathcal{L}|$). It can be seen that locally explicit construction is a stronger notion than globally explicit construction and is always desirable as it is useful for algorithmic applications. In fact, common notion of construction of combinatorial objects (e.g. using various algebraic structures) falls in this category.

Now, we state relevant existing results for CBCs and subsequently discuss our contribution.

Existing Results

(i) In [IKOS04], the authors have shown, *inter alia*, that $n(m, c, k) = \Omega(m^{c-1})$; this bound was obtained using probabilistic method.

(ii) In [PSW09], the authors have refined the above estimate using the *method of deletion* (another probabilistic technique, see [AS92]) to $n(m, c, k) = \Omega(m^{\frac{ck}{k-1}-1})$. They have also shown through explicit construction that $n(m, k-1, k) = (k-1)\binom{m}{k-1}$, and $n(m, k-2, k) = \binom{m}{k-2}$.

(iii) In [BB14], it was shown that $n(m, c, k) = O(m^{c-\frac{1}{2^{c-1}}})$ for $7 \leq k$, and $3 \leq c \leq k - \lceil \log k \rceil - 1$. Also for $k - \lceil \log k \rceil \leq c \leq k-1$, it was shown through explicit construction that $n(m, c, k) = \Theta(m^c)$. For $c = 2$ case, the lower bound of (ii) from [PSW09] was improved (through explicit construction) to $n(m, 2, k) = \Omega(m^{\frac{k+1}{k-1}})$ for all $k \geq 8$ and infinitely many values of n.

(iv) In [BT15], the authors improved the general upper bound to show that
$$n(m, c, k) = O(m^{c-1+\frac{1}{\lfloor \frac{k}{c+1} \rfloor}}) \text{ for } c \leq \frac{k}{2} - 1.$$

All the constructions mentioned above are locally explicit.[6]

---

[5] Though the classification is with respect to polynomial time constructibility, it is applicable for other feasible resource bounds as well.

[6] In [SG14], constructions of CBCs are given for a setting of parameters where k and c vary with m. Since in our setting we require k and c to be constants, we do not discuss the results of [SG14].

Our Contribution. We address the question of explicit construction of uniform and regular CBCs. Here it is noteworthy that though uniform CBCs have been studied in the literature, aspects of regularity and uniformity have not been addressed together so far. While study of uniform and regular CBCs is theoretically interesting for its own sake, practical significance of regularity can be partly argued from the fact that for regular CBCs, number of data items stored in each server is same; this makes it simpler to allocate storage uniformly and optimally accross different servers, especially under dynamic conditions where the database (i.e., the set of distinct data items to be stored) changes with addition and deletion of data items.

We construct c-uniform (n, cn, k, m)-CBCs that are almost regular in the sense that for these CBCs number of data items stored in each server is $\frac{nc}{m} + o(n)$ (note that for regular CBCs with same parameters, this value is exactly $\frac{nc}{m}$); and our construction is globally explicit. More precisely, our result is the following.

Theorem 2. *Let c, k be positive constants. Then for all sufficiently large m, there exists c-uniform (n, cn, k, m)-CBC, where $n = \Omega(m^{c-1+\frac{1}{k}})$, and number of items in each server is in the range $\left[\frac{nc}{m} - \sqrt{\frac{n}{2}\ln(4m)}, \frac{nc}{m} + \sqrt{\frac{n}{2}\ln(4m)}\right]$. Moreover, the CBC can be globally constructed in $poly(m)$ time.*

We use the construction of [IKOS04] and derandomize it using the *method of conditional expectation* (see [AS92]), and analyze it in greater detail. The analysis shows almost regularity of the construction and also improves the exponent of the lower bound ($\Omega(m^{c-1+\frac{1}{k}})$ as opposed to $\Omega(m^{c-1})$ of [IKOS04]). Though the improved exponent is inferior to the one obtained in [PSW09] ($\Omega(m^{c-1+\frac{c}{k-1}})$), the importance of the construction lies in its almost regularity and *explicitness* which are not known for the construction of [PSW09]. In fact, apart from the range $k - \lceil \log k \rceil \leq c \leq k$, where $n = \Theta(m^c)$ is achieved (see [PSW09] for $c = k - 1$ and $k - 2$ and [BB14] for the remaining values), and for the 2-uniform case ([BB14]) there is no explicit construction in the literature. So, our construction serves to fill the void to some extent.

To describe our construction, we give an algorithm which, for given positive integers k, c, and sufficiently large m, runs in time $poly(m)$ and outputs the edges of a bipartite graph $(\mathcal{L}, \mathcal{R}, \mathcal{E})$, with $|\mathcal{R}| = m$, $|\mathcal{L}| = n = \frac{m^{c-1+\frac{1}{k}}}{4k^{c+1}}$, satisfying the following conditions.

(a) Each left vertex in $\mathcal{L}$ has degree c and each vertex in $\mathcal{R}$ has degree in the range $\left[\frac{nc}{m} - \sqrt{\frac{n}{2}\ln(4m)}, \frac{nc}{m} + \sqrt{\frac{n}{2}\ln(4m)}\right]$.
(b) Any subset of $i, 1 \leq i \leq k$, vertices in $\mathcal{L}$ has at least i neighbours in $\mathcal{R}$.

Note that there is a trivial non-explicit algorithm to construct the required graph which, given the input parameters, runs in time exponential in m; the algorithm searches the space of all possible bipartite graphs $(\mathcal{L}, \mathcal{R}, \mathcal{E})$, with $|\mathcal{R}| = m$, $|\mathcal{L}| = n = \frac{m^{c-1+\frac{1}{k}}}{4k^{c+1}}$, and outputs one that satisfies the above two conditions.

In the proof of Theorem 2, we will need the following version of Hoeffding's inequality.

Theorem 3 (Hoeffding's Inequality [Hoe63]**).** *Let* $X_1, X_2, \ldots, X_n$ *be independent random variables taking their values in the interval* $[0,1]$. *Let* $X = \sum_i X_i$. *Then for every real number* $a > 0$, $\mathbf{Pr}\{|X - \mathbf{E}[X]| \geq a\} \leq 2e^{\frac{-2a^2}{n}}$.

Also, given a set $\mathcal{S}$ and a positive integer $c(\leq |\mathcal{S}|)$, we will denote by $\binom{\mathcal{S}}{c}$, the set of all c element subsets of $\mathcal{S}$.

2   Proof of Theorem 2

Proof of Theorem 2 is split into two parts. In the first part, we give probabilistic proof of existence (which essentially is also a randomized algorithm) of the CBC. In the second part, we derandomize the construction using the method of conditional expectation. This is a commonly used technique having its genesis in [ES73] and was later on applied to prove many other derandomization results (e.g. [Rag88, Spe94]). Informally, the method systematically performs a binary (or more commonly a d-ary) search on the sample space from where the randomized algorithm makes its choices and finally finds a good point.

Proof of Existence. We construct a bipartite graph $\mathcal{G} = (\mathcal{L}, \mathcal{R}, \mathcal{E})$, where $\mathcal{L}$ is the set $\{u_1, \ldots, u_n\}$ of n left vertices, $\mathcal{R}$ is the set $\{v_1, \ldots, v_m\}$ of m right vertices, and $\mathcal{E}$ is the set of edges, in the following manner. For each vertex in $\mathcal{L}$ we choose its c distinct neighbours by picking randomly, uniformly, and independently a subset of c vertices from $\mathcal{R}$, i.e., its neighbourhood is an independently and uniformly chosen element of $\binom{\mathcal{R}}{c}$. So, for $u \in \mathcal{L}, S' \subseteq \mathcal{R}$,

$$\mathbf{Pr}\left\{\Gamma(u) \subseteq S'\right\} = \frac{\binom{|S'|}{c}}{\binom{m}{c}} \leq \left(\frac{|S'|}{m}\right)^c.$$

Next, for a subset $S \subset \mathcal{L}, |S| = i, c+1 \leq i \leq k$ and a subset $S' \subset \mathcal{R}, |S'| = i-1$, we say that event $Bad_{S,S'}$ has occured if $\Gamma(S) \subseteq S'$. So, we have

$$\mathbf{Pr}\{Bad_{S,S'}\} \leq \left(\frac{i-1}{m}\right)^{ic}, \tag{1}$$

using independence of the events $\Gamma(u) \subseteq S'$ for $u \in \mathcal{L}$. Now, our goal is to bound the probability of occurence of any $Bad_{S,S'}$, $S \subset \mathcal{L}$, $S' \subset \mathcal{R}$, and $c+1 \leq i \leq k$. To this end, we have

$$\sum_{\substack{c+1 \leq i \leq k}} \sum_{\substack{S \subset \mathcal{L}, \\ |S|=i}} \sum_{\substack{S' \subset \mathcal{R}, \\ |S'|=i-1}} \mathbf{Pr}\{Bad_{S,S'}\} \leq \sum_{c+1 \leq i \leq k} \binom{n}{i}\binom{m}{i-1}\left(\frac{i-1}{m}\right)^{ic} \tag{2}$$

$$\leq \sum_{1 \leq i \leq k} n^i m^{i-1} \left(\frac{i-1}{m}\right)^{ic}$$

$$\leq \sum_{1 \leq i \leq k} n^i m^{i-1} \left(\frac{k}{m}\right)^{ic}$$

$$\leq \sum_{1 \leq i \leq k} \left(\frac{1}{4k}\right)^i \quad \text{since } n = \frac{m^{c-1+\frac{1}{k}}}{4k^{c+1}}.$$

$$\leq \frac{1}{4}.$$

Next, for $u \in \mathcal{L}, v \in \mathcal{R}$ define the indicator random variable X_v^u such that

$$X_v^u = \begin{cases} 1 & (u,v) \in \mathcal{E} \\ 0 & \text{otherwise,} \end{cases}$$

and $X_v, v \in \mathcal{R}$, be a random variable denoting the degree of vertex v. Clearly $X_v = \sum_{u \in \mathcal{L}} X_v^u$. Now, $\mathbf{Pr}\{X_v^u = 1\} = \frac{c}{m}$. So, by linearity of expectation,

$$\mathbf{E}[X_v] = \mathbf{E}\left[\sum_{u \in \mathcal{L}} X_v^u\right] = \sum_{u \in \mathcal{L}} \mathbf{E}[X_v^u] = \frac{nc}{m}.$$

From the fact that neighbourhoods of vertices $u \in \mathcal{L}$ are chosen independently, it follows that the variables X_v^u for $u \in \mathcal{L}$, and a fixed $v \in \mathcal{R}$, are mutually independent. So, applying Theorem 3 with $a = \sqrt{\frac{n}{2}\ln(4m)}$ we have

$$\mathbf{Pr}\left\{|X_v - \mathbf{E}[X_v]| \geq \sqrt{\frac{n}{2}\ln(4m)}\right\} \leq 2(4m)^{-1} \leq \frac{1}{2m}. \tag{3}$$

So, by union bound, probability that the event $|X_v - \mathbf{E}[X_v]| \geq \sqrt{\frac{n}{2}\ln(4m)}$ occurs for some $v, v \in \mathcal{R}$ is bounded by

$$\sum_{v \in \mathcal{R}} \mathbf{Pr}\left\{|X_v - \mathbf{E}[X_v]| \geq \sqrt{\frac{n}{2}\ln(4m)}\right\} \leq \frac{1}{2}. \tag{4}$$

Hence, from Eqs. (2) and (4), with probability at least $1 - (\frac{1}{4} + \frac{1}{2}) = \frac{1}{4}$, none of the above events occur. □

Derandomization. Before presenting the algorithm we derive expressions for expected number of $Bad_{S,S'}$ events and expected number of vertices $v \in \mathcal{R}$ for which $|deg(v) - \frac{nc}{m}| > \sqrt{\frac{n}{2}\ln(4m)}$ conditional on fixed choices of $\Gamma(u_1), \ldots, \Gamma(u_t)$. Then we show that if at t-th stage, $1 \leq t \leq n$, (having fixed $\Gamma(u_1), \ldots, \Gamma(u_{t-1})$) choice of $\Gamma(u_t)$ is made in such a way to minimize the sum of these two expectations then in the final graph, which is no longer random since all the neighbourhoods are fixed, there are no events $Bad_{S,S'}$ and no vertices $v \in \mathcal{R}$ for which $|deg(v) - \frac{nc}{m}| > \sqrt{\frac{n}{2}\ln(4m)}$, i.e., no violations of conditions (a) and (b). So, the derandomization proceeds in n stages; the beginning of stage t, neighbourhood of vertices $u_1, \ldots, u_{t-1} \in \mathcal{L}$ are fixed, and neighbourhood of $u_t, \Gamma(u_t) \in \binom{\mathcal{R}}{c}$ is fixed in such a way that minimizes the expected number of violations of conditions (a) and (b). The algorithm (Algorithm 1) is immediate from these observations.

First, we introduce indicator random variables $Y_{S,S'}$ corresponding to each event $Bad_{S,S'}$, i.e.,

$$Y_{S,S'} = \begin{cases} 1 & \text{if } \Gamma(S) \subseteq S' \\ 0 & \text{otherwise.} \end{cases}$$

Also, we define $Y = \sum_{c+1 \leq i \leq k} \sum_{\substack{S \subseteq \mathcal{L}, \\ |S|=i}} \sum_{\substack{S' \subseteq \mathcal{R}, \\ |S'|=i-1}} Y_{S,S'}$. By linearity of expectation we have

$$\mathbf{E}[Y] = \mathbf{E}\left[\sum_{c+1 \leq i \leq k} \sum_{\substack{S \subseteq \mathcal{L}, \\ |S|=i}} \sum_{\substack{S' \subseteq \mathcal{R}, \\ |S'|=i-1}} Y_{S,S'} \right]$$

$$= \sum_{c+1 \leq i \leq k} \sum_{\substack{S \subseteq \mathcal{L}, \\ |S|=i}} \sum_{\substack{S' \subseteq \mathcal{R}, \\ |S'|=i-1}} \mathbf{E}[Y_{S,S'}]$$

$$= \sum_{c+1 \leq i \leq k} \sum_{\substack{S \subseteq \mathcal{L}, \\ |S|=i}} \sum_{\substack{S' \subseteq \mathcal{R}, \\ |S'|=i-1}} \mathbf{Pr}\{Y_{S,S'}\} \leq \frac{1}{4}, \text{ from}(2).$$

Let $\{u_1, u_2, \ldots, u_t\} \subseteq \mathcal{L}$, and $C_1, C_2, \ldots, C_t \in \binom{\mathcal{R}}{c}$ be fixed subsets such that $\Gamma(u_j) = C_j, 1 \leq j \leq t$, and for the remaining vertices in $\mathcal{L}$, their neighbourhoods are chosen independently, and uniformly at random from $\binom{\mathcal{R}}{c}$. Let $S \subseteq \mathcal{L}, S' \subseteq \mathcal{R}, |S| = i, |S'| = i - 1$ be fixed subsets for some $i, c + 1 \leq i \leq k$, also let $W = S \cap \{u_1, u_2, \ldots, u_t\}, |W| = w$, and $\Gamma(W) = \emptyset$ for $W = \emptyset$. Then we have

$$\mathbf{E}[Y_{S,S'}|\Gamma(u_1) = C_1, \ldots, \Gamma(u_t) = C_t]$$
$$= \mathbf{Pr}\{\Gamma(S) \subseteq S'|\Gamma(u_1) = C_1, \ldots, \ldots \Gamma(u_t) = C_t\}$$
$$= \begin{cases} 0 & \text{if } \Gamma(W) \not\subseteq S' \\ \left(\frac{\binom{i-1}{c}}{\binom{m}{c}}\right)^{i-w} & \text{otherwise.} \end{cases} \qquad (5)$$

So, by applying linearity of expectation and from above

$$\mathbf{E}[Y|\Gamma(u_1) = C_1, \ldots, \Gamma(u_t) = C_t]$$
$$= \sum_{c+1 \leq i \leq k} \sum_{\substack{S \subseteq \mathcal{L}, \\ |S|=i}} \sum_{\substack{S' \subseteq \mathcal{R}, \\ |S'|=i-1}} \mathbf{E}[Y_{S,S'}|\Gamma(u_1) = C_1, \ldots, \ldots \Gamma(u_t) = C_t]$$
$$= \sum_{c+1 \leq i \leq k} \sum_{\substack{S \subseteq \mathcal{L}, \\ |S|=i}} \sum_{\substack{S' \subseteq \mathcal{R}, \\ |S'|=i-1}} \mathbf{Pr}\{\Gamma(S) \subseteq S'|\Gamma(u_1) = C_1, \ldots, \ldots \Gamma(u_t) = C_t\}$$

$$= \sum_{c+1 \le i \le k} \sum_{\substack{S \subseteq \mathcal{L}, \\ |S|=i}} \sum_{\substack{S' \subseteq \mathcal{R}, \\ |S'|=i-1 \\ \Gamma(W) \subseteq S'}} \left(\frac{\binom{i-1}{c}}{\binom{m}{c}} \right)^{i-w}. \tag{6}$$

Next, corresponding to each vertex $v \in \mathcal{R}$ we introduce an indicator random variable Z_v such that

$$Z_v = \begin{cases} 1 & |deg(v) - \frac{nc}{m}| > \sqrt{\frac{n}{2} \ln(4m)} \\ 0 & \text{otherwise,} \end{cases}$$

and define $Z = \sum_{v \in \mathcal{R}} Z_v$. So, by linearity of expectation we have

$$\mathbf{E}[Z] = \mathbf{E}\left[\sum_{v \in \mathcal{R}} Z_v \right] = \sum_{v \in \mathcal{R}} \mathbf{E}[Z_v] = \sum_{v \in \mathcal{R}} \mathbf{Pr}\{Z_v = 1\} \le \frac{1}{2}, \text{ from (4).}$$

Like in the previous case, our goal is to estimate $\mathbf{E}[Z|\Gamma(u_1) = C_1, \dots, \Gamma(u_t) = C_t]$ by estimating $\mathbf{E}[Z_v|\Gamma(u_1) = C_1, \dots, \Gamma(u_t) = C_t]$ for each $v \in \mathcal{R}$. For a fixed $v \in \mathcal{R}$, let $l = |\{u_i | v \in \Gamma(u_i), 1 \le i \le t\}|$. Let $\alpha = \frac{nc}{m} - \sqrt{\frac{n}{2} \ln(4m)}$, and $\beta = \frac{nc}{m} + \sqrt{\frac{n}{2} \ln(4m)}$. Then we have

$$\mathbf{E}[Z|\Gamma(u_1) = C_1, \dots, \Gamma(u_t) = C_t]$$

$$= \sum_{v \in \mathcal{R}} \mathbf{E}[Z_v|\Gamma(u_1) = C_1, \dots, \Gamma(u_t) = C_t]$$

$$= \sum_{v \in \mathcal{R}} \mathbf{Pr}\{deg(v) < \alpha - l \text{ or } deg(v) > \beta - l | \Gamma(u_1) = C_1, \dots, \Gamma(u_t) = C_t\}$$

$$= \sum_{v \in \mathcal{R}} \Big(\sum_{i=0}^{i=\alpha-l-1} \binom{n-t}{i} \left(\frac{c}{m}\right)^i \left(1 - \frac{c}{m}\right)^{n-t-i} +$$

$$\sum_{i=\beta-l+1}^{n-t} \binom{n-t}{i} \left(\frac{c}{m}\right)^i \left(1 - \frac{c}{m}\right)^{n-t-i} \Big). \tag{7}$$

Finally, we show that if at j-th iteration (having fixed $\Gamma(u_1) = C_1, \dots, \Gamma(u_{j-1}) = C_{j-1}$ at the beginning) $\Gamma(u_j) = C_j$ is chosen so as to minimize $\mathbf{E}[Y + Z|\Gamma(u_1) = C_1, \dots, \Gamma(u_j) = C], C \in \binom{\mathcal{R}}{c}$, then in the final graph, which is no longer random, conditions (a) and (b) are met. To this end, we first observe that

$$\mathbf{E}[Y + Z|\Gamma(u_1) = C_1, \dots, \Gamma(u_{t-1}) = C_{t-1}]$$

$$= \frac{1}{\binom{m}{c}} \sum_{C \in \binom{\mathcal{R}}{c}} \mathbf{E}[Y + Z|\Gamma(u_1) = C_1, \dots, \Gamma(u_t) = C]$$

$$\ge \min_{C \in \binom{\mathcal{R}}{c}} \mathbf{E}[Y + Z|\Gamma(u_1) = C_1, \dots, \Gamma(u_t) = C]. \tag{8}$$

Algorithm 1. Algorithm to construct uniform and almost regular CBC

Input: Positive constants c, k, and sufficiently large m.
Output: A bipartite graph $(\mathcal{L}, \mathcal{R}, \mathcal{E})$, where

$\mathcal{L} = \{u_1, u_2, \ldots, u_n\}(n = \frac{m^{c-1+\frac{1}{k}}}{4k^{c+1}})$ and $\mathcal{R} = \{v_1, v_2, \ldots, v_m\}$
such that $\Gamma(u_j) = C_j \in \binom{\mathcal{R}}{c}, 1 \leq j \leq n$ meeting conditions (a)
and (b).

$\alpha = \frac{nc}{m} - \sqrt{\frac{n}{2} \ln(4m)}$, and $\beta = \frac{nc}{m} + \sqrt{\frac{n}{2} \ln(4m)}$;

for $j \leftarrow 1$ **to** n **do**

$\quad U_{j-1} = \{u_1, u_2, \ldots, u_{j-1}\}, min \leftarrow 1$

$\quad$ **for** $C \in \binom{\mathcal{R}}{c}$ **do**

$$Y' \leftarrow \sum_{\substack{c+1 \leq i \leq k \\ |S|=i}} \sum_{\substack{u_j \in S \subset \mathcal{L}, \\ |S|=i}} \sum_{\substack{S' \subset \mathcal{R}, \\ |S'|=i-1 \\ \Gamma(U_{j-1} \cap S) \cup C \subseteq S'}} \left(\frac{\binom{i-1}{c}}{\binom{m}{c}}\right)^{i-|U_{j-1} \cap S|-1}$$

$$Z \leftarrow \sum_{v \in \mathcal{R}} \Bigg(\sum_{i=0}^{\alpha - |U_{j-1} \cap \Gamma(v)| - |\{v\} \cap C| - 1} \binom{n-j}{i} \left(\frac{c}{m}\right)^i \left(1 - \frac{c}{m}\right)^{n-j-i}$$

$$+ \sum_{i=\beta - |U_{j-1} \cap \Gamma(v)| - |\{v\} \cap C| + 1}^{n-j} \binom{n-j}{i} \left(\frac{c}{m}\right)^i \left(1 - \frac{c}{m}\right)^{n-j-i} \Bigg)$$

$\quad$ **if** $min > Y' + Z$ **then**

$\quad\quad \Gamma(u_j) = C$

$\quad\quad min \leftarrow Y' + Z$

$\quad$ **end**

end

end

Hence, it follows that

$$\min_{C_1, \ldots, C_n \in \binom{\mathcal{R}}{c}} \mathbf{E}[Y + Z | \Gamma(u_1) = C_1, \ldots, \Gamma(u_n) = C_n]$$

$$\leq \min_{C_1, \ldots, C_{n-1} \in \binom{\mathcal{R}}{c}} \mathbf{E}[Y + Z | \Gamma(u_1) = C_1, \ldots, \Gamma(u_{n-1}) = C_{n-1}]$$

$$\vdots$$

$$\leq \min_{C_1 \in \binom{\mathcal{R}}{c}} \mathbf{E}[Y + Z | \Gamma(u_1) = C_1] \leq \mathbf{E}[Y + Z] \leq \frac{3}{4}. \tag{9}$$

Since Y and Z are integer valued random variables, the above essentially means that at the end, when all the neighbourhoods $\Gamma(u_1), \ldots, \Gamma(u_n)$ are fixed, $Y = 0$ and $Z = 0$. So, both the conditions (a) and (b) are met. Now, we have the following algorithm (Algorithm 1) to construct the bipartite graph.

Proof of Correctness of the Algorithm. At the beginning of j-th iteration, $\Gamma(u_1) = C_1, \ldots, \Gamma(u_{j-1}) = C_{j-1}$ are fixed and the algorithm selects $C = C_j$ which minimizes $Y' + Z$ for given $\Gamma(u_1) = C_1, \ldots, \Gamma(u_{j-1}) = C_{j-1}, \Gamma(u_j) = C$. Note that in (5), $\mathbf{E}[Y_{S,S'}|\Gamma(u_1) = C_1, \ldots, \Gamma(u_t) = C_t]$ is independent of particular choice of C_i if $u_i \notin S$. So, in j-th iteration of the algorithm, while computing Y', only those summands $\mathbf{E}[Y_{S,S'}|\Gamma(u_1) = C_1, \ldots, \Gamma(u_j) = C]$ are considered for which $u_j \in S$. Hence, $Y' \leq \mathbf{E}[Y_{S,S'}|\Gamma(u_1) = C_1, \ldots, \Gamma(u_j) = C]$, and particular choice of $C = C_j$ which minimizes $Y' + Z$ for given $\Gamma(u_1) = C_1, \ldots, \Gamma(u_{j-1}) = C_{j-1}, \Gamma(u_j) = C$ also minimizes $E[Y + Z|\Gamma(u_1) = C_1, \ldots, \Gamma(u_{j-1}) = C_{j-1}, \Gamma(u_j) = C]$; by (9), this also justifies setting min to 1 at the beginning of j-th iteration for $1 \leq j \leq n$. Hence the proof follows from the discussion preceeding Algorithm 1.

Runtime of the Algorithm. Now, we present a coarse analysis of the algorithm which is sufficient to indicate that the algorithm runs in time $poly(m)$. For that, we first estimate the time required by the algorithm to compute Y.[7] Note that the time reqired to compute $\binom{m}{c}$ and $\binom{i-1}{c}$ is $O(m^2)$ (through dynamic programming); the exponentiation takes time $O(\log k)$, and these operations are done $O(kn^{k-1}m^{k-1})$ times to get the summation. So, the time required by the algorithm to compute Y is $O(m^{(c+1)(k-1)+2})$. Similarly, in the case of computing Z, the binomial coefficients $\binom{n-i}{j}$ takes time $O(n^2)$ to be evaluated, exponentiations take time $O(\log n)$. So, the overall time requirement in this case is $O(mn^3 \log n) = O(m^{3c-1} \log m)$. The above two steps are done $O(nm^c) = O(m^{2c})$ times. So, the overall complexity of the algorithm is $O(m^{(k+1)(c+1)})$.

3   Concluding Remarks

Limitations of the Construction. It can be observed that the algorithm crucially depends on the fact that k is a constant, and this limits its applicability to wider setting where k is allowed to vary. Apart from being globally explicit, which, as discussed in the beginning, is a weaker notion of explicitness, the construction is on the slower side (even in terms of the number of edges which is $O(m^c)$), as indicated by the above analysis. One of the possible approaches to speed-up the algorithm is discussed below.

Towards Derandomization in NC. It can be observed from the first part of Theorem 2 that the construction is in RNC, i.e., the construction can be carried out on a probabilistic Parallel Random Access Machine (PRAM) (see [MR95] for definition) with $poly(m)$ many processors in constant time.[8] It is naturally interesting to investigate NC-derandomization of problems in RNC i.e., whether the same problem can be solved using a deterministic PRAM under same set of

---

[7] We consider RAM model (see [MR95]), so addition, multiplication, and division are atomic operations.

[8] In fact, it can be seen that the construction is in ZNC with the expected number of iterations at most 4.

restrictions on resources. Two of the most commonly used techniques employed for such derandomization are the method of conditional expectation and the *method of small sample spaces* (see [AS92]); sometimes they are used together [BR91, MNN94].

While the method of conditional expectation performs a binary search (or more commonly a d-ary search) on the sample space for a good point, method of small sample spaces takes advantage of small independence requirement of random variables involved in the algorithm and constructs a small sized (polynomial in the number of variables) sample space which ensures such independence, and searches the sample space for a good sample point. Since the sample space is polynomial sized the search can be done in polynomial time, and hence leads to polynomial time construction.

In the proof of Theorem 2 we used independence twice; in (1) we used k-wise independence and in (3) we used Hoeffding's inequality (Theorem 3) which requires the involved random variables $X_1, \ldots, X_n$ to be n-wise independent.[9] However, such independence comes at the cost of a large sample space. More precisely, in [ABI86] it was shown that in order to ensure k-independence among n random variables the sample space size have to be $\Omega(n^{\frac{k}{2}})$. So, in case of Theorem 2 requirement on the size of sample space is huge ($\Omega(m^m)$). However, we want to point out here that the requirement of n-wise independence in Theorem 2 can be brought down to $O(\ln(m))$-wise independence with the help of following limited independence Chernoff bound from [BR94]. First we state the bound.

Theorem 4 ([BR94]). *Let $t \geq 4$ be an even integer. Suppose $X_1, \ldots, X_n$ be t-wise independent random variables taking values in $[0, 1]$. Let $X = X_1 + \cdots + X_n$, and $a > 0$. Then*

$$\mathbf{Pr}\{|X - \mathbf{E}[X]| \geq a\} \leq C_t \left(\frac{nt}{a^2}\right)^{\frac{t}{2}},$$

where C_t is a constant depending on t, and $C_t < 1$ for $t \geq 6$.

Now, in inequality (3), we need $\frac{1}{2m}$ in the r.h.s. This can be achieved by setting $t = 2\ln(2m)$ in the above theorem (for simplicity we assume $2\ln(2m)$ to be even), and $a = \sqrt{2en\ln(2m)}$. Hence, $O(\ln(m))$-wise independence in choosing $\Gamma(u)$ for $u \in \mathcal{L}$ is sufficient for the randomized construction (with somewhat inferior bound on the deviation of the degrees from the average).

In [BR91, MNN94], the authors developed frameworks for NC-derandomization of certain algorithms (notably, the set discrepancy problem of Spencer [Spe94]) that require $O(\log^c(n))$-wise independence. One of the vital points of these frameworks is parallel computation of relevant conditional expectations for limited independence random variables in logarithmic time. In case of Algorithm 1, this means computation of Y' and Z by $poly(m)$ processors in $polylog(m)$ time under $O(\ln(m))$-wise independence among random choices of $\Gamma(u)$ for $u \in \mathcal{L}$. At present it is not clear to us how this can be achieved in the frameworks of [BR91, MNN94] and seems to require a more specialized technique.

---

[9] Here we again point out that though the events and random variables are different in two cases, k-wise independence in choices of $\Gamma(u), u \in \mathcal{L}$ induces k-wise independence among random variables $X_v^u, u \in \mathcal{L}$ for fixed $v \in \mathcal{R}$.

Acknowledgement. I would like to thank all the anonymous reviewers for their valuable remarks and suggestions. I would also like to thank Mr. Satrajit Ghosh and Mr. Subhabrata Samajder for commenting on an earlier version of the manuscript.

References

[ABI86] Alon, N., Babai, L., Itai, A.: A fast and simple randomized parallel algorithm for the maximal independent set problem. J. Algorithms **7**(4), 567–583 (1986)

[AS92] Alon, N., Spencer, J.H.: The Probabilistic Method. Wiley, New York (1992)

[BB14] Balachandran, N., Bhattacharya, S.: On an extremal hypergraph problem related to combinatorial batch codes. Discrete Appl. Math. **162**, 373–380 (2014)

[BKMS10] Brualdi, R.A., Kiernan, K., Meyer, S.A., Schroeder, M.W.: Combinatorial batch codes and transversal matroids. Adv. in Math. of Comm. **4**(3), 419–431 (2010)

[Bol86] Bollobás, B.: Combinatorics: Set Systems, Hypergraphs, Families of Vectors, and Combinatorial Probability. Cambridge University Press, New York (1986)

[BR91] Berger, B., Rompel, J.: Simulating $(\log^c n)$-wise independence in NC. J. ACM **38**(4), 1026–1046 (1991): Preliminary version. In: Proceedings of the 30th Annual Symposium on Foundations of Computer Science (FOCS 1989)

[BR94] Bellare, M., Rompel, J.: Randomness-efficient Oblivious Sampling. In: Proceedings of the 35th Annual Symposium on Foundations of Computer Science (FOCS 1994), Washington, DC, USA, pp. 276–287. IEEE Computer Society (1994)

[BRR12] Bhattacharya, S., Ruj, S., Roy, B.K.: Combinatorial batch codes: a lower bound and optimal constructions. Adv. in Math. of Comm. **6**(2), 165–174 (2012)

[BT11a] Bujtás, C., Tuza, Z.: Combinatorial batch codes: extremal problems under hall-type conditions. Electron. Notes Discrete Math. **38**, 201–206 (2011)

[BT11b] Bujtás, C., Tuza, Z.: Optimal batch codes: many items or low retrieval requirement. Adv. in Math. of Comm. **5**(3), 529–541 (2011)

[BT12] Bujtás, C., Tuza, Z.: Relaxations of hall's condition: optimal batch codes with multiple queries. Appl. Anal. Discrete Math. **6**(1), 72–81 (2012)

[BT15] Bujtás, C., Tuza, Z.: Turán numbers and batch codes. Discrete Appl. Math. **186**, 45–55 (2015)

[ES73] Erdős, P., Selfridge, J.: On a combinatorial game. J. Comb. Theory Ser. A **14**(3), 298–301 (1973)

[GUV09] Guruswami, V., Umans, C., Vadhan, S.: Unbalanced expanders and randomness extractors from parvaresh-vardy codes, J. ACM **56**(4), 20:1–20:34 (2009): Preliminary version. In: Proceedings of the Twenty-Second Annual IEEE Conference on Computational Complexity (CCC 2007)

[Hoe63] Hoeffding, W.: Probability inequalities for sums of bounded random variables. J. Am. Stat. Assoc. **58**(301), 13–30 (1963)

[IKOS04] Ishai, Y., Kushilevitz, E., Ostrovsky, R., Sahai, A.: Batch codes and their applications, In: Proceedings of the 36th Annual ACM Symposium on Theory of Computing (STOC 2004), Chicago, IL, USA, pp. 262–271. ACM (2004)

[MNN94] Motwani, R., Naor, J.S., Naor, M.: The probabilistic method yields deterministic parallel algorithms, J. Comput. Syst. Sci. 49(3), 478–516 (1994): Preliminary version in Proceedings of the 30th Annual Symposium on Foundations of Computer Science (FOCS 1989)

[MR95] Motwani, R., Raghavan, P.: Randomized Algorithms. Cambridge University Press, New York (1995)

[PSW09] Paterson, M.B., Stinson, D., Wei, R.: Combinatorial batch codes. Adv. Math. Commun. 3(1), 13–27 (2009)

[Rag88] Raghavan, P.: Probabilistic construction of deterministic algorithms: approximating packing integer programs, J. Comput. Syst. Sci. 37(2), 130–143 (1988). In: Proceedings of the 27th Annual Symposium on Foundations of Computer Science (FOCS'86)

[SG14] Silberstein, N., Gál, A.: Optimal combinatorial batch codes based on block designs, In: Designs, Codes and Cryptography, pp.1–16 (2014)

[Spe94] Spencer, J.: Ten Lectures on the Probabilistic Method, 2nd edn. Society for Industrial and Applied Mathematics, Philadelphia (1994)

On the Mathematics of Data Centre Network Topologies

Iain A. Stewart[(⊠)]

School of Engineering and Computing Sciences,
Durham University Science Labs, South Road,
Durham DH1 3LE, UK
i.a.stewart@durham.ac.uk

Abstract. In a recent paper, combinatorial designs were used to construct switch-centric data centre networks that compare favourably with the ubiquitous (enhanced) fat-tree data centre networks in terms of the number of servers within (given a fixed server-to-server diameter). Unfortunately there were flaws in some of the proofs in that paper. We correct these flaws here and extend the results so as to prove that the core combinatorial construction, namely the 3-step construction, results in data centre networks with optimal path diversity.

1 Introduction

Data centres are expanding both in terms of their physical size and their reach and importance as computational platforms for cloud computing, web search, social networking and so on. There is an increasing demand that data centres incorporate more and more servers but so that computational efficiency is not compromised. A key contributor as to the eventual performance of a data centre is the *data centre network* (*DCN*). New topologies are continually being developed so as to incorporate more servers and best utilize the additional computational power. It is with topological aspects of DCNs that concern us here.

The traditional design of a DCN is *switch-centric* whereby all routing intelligence resides amongst the switches. In such DCNs, there are no direct server-to-server links; only server-to-switch and switch-to-switch links. Switch-centric DCNs are traditionally tree-like with servers located at the 'leaves' of the tree-like structure, e.g., Fat-Tree [1], VL2 [3] and Portland [5]. Whilst it is generally acknowledged that tree-like, switch-centric DCNs have their limitations when it comes to, for example, scalability (with the core switches at the 'roots' quickly becoming bottlenecks), tree-like switch-centric DCNs remain popular and can usually be constructed from commodity hardware. A more recent paradigm, namely *server-centric* DCNs, has emerged so that deficiencies of tree-like, switch-centric DCNs might be ameliorated. In a server-centric DCN, routing intelligence resides within the servers with switches operating only as dumb crossbars; as

I.A. Stewart—Supported by EPSRC grant EP/K015680/1 'Interconnection Networks: Practice unites with Theory (INPUT)'.

© Springer International Publishing Switzerland 2015
A. Kosowski and I. Walukiewicz (Eds.): FCT 2015, LNCS 9210, pp. 283–295, 2015.
DOI: 10.1007/978-3-319-22177-9_22

such, there are only server-to-switch and server-to-server links. However, server-centric DCNs also suffer from deficiencies such as packet relay overheads caused by the need to route packets within the server (see [4] for the DCN state of the art). Both switch-centric and server-centric DCNs are abstracted as undirected graphs where the set of nodes is partitioned into a set of servers and a set of switches with edges depending upon the DCN type. It is with switch-centric DCNs that we are concerned here.

It is difficult to design computationally efficient DCNs so as to incorporate large numbers of servers as there are additional design considerations. For example, switches and (especially) servers have a limited number of ports; so, the more servers there are, the greater the average or worst-case link-count between two distinct servers and, consequently, there is a packet latency overhead to be borne. Also, so as to better support routing, fault-tolerance and load-balancing, we would prefer that there is path diversity in the form of numerous alternative (short) paths within the DCN joining any two distinct servers. There are many other design parameters to bear in mind (see, e.g., [7]).

A recent proposal in [6] advocated the use of *combinatorial design theory* in order to design switch-centric DCNs which incorporate more servers, have short server-to-server paths and possess path diversity. The use of combinatorial designs within the study of general interconnection networks is not new and originated in [2] where the targeted networks involved processors communicating via buses. A hypergraph framework was developed in [2] where the hypergraph nodes represented the processors and the hyperedges the buses, and likewise an analogous framework was developed in [6] where the hypergraph nodes represented the servers and the hyperedges represented the switches. However, some of the results derived in [6] are incorrect in that for some of the results there were errors in the proofs while for other results the actual claims are not true.

In this paper we provide correct proofs for some of the results from [6] and we also extend and improve the results from [6]. In particular, using the general construction for building switch-centric DCNs from bipartite graphs and transversal designs as adopted in [6], we prove that in the resulting switch-centric DCNs, there is the maximal number of internally disjoint paths joining any two distinct servers and provide a bound on the length of the longest such path. As can be seen from our proofs, the situation is far more subtle than was assumed in [6].

2   Basic Concepts

Hypergraphs provide the original framework for the 3-step construction as employed in [2] and [6]. A *hypergraph* $H = (V, E)$ consists of a finite set V of *nodes* together with a finite set E of *hyperedges* where each hyperedge is a non-empty set of nodes and each node appears in at least one hyperedge. The *degree* of a node is the number of hyperedges containing it and the *rank* of a hyperedge is its size as a subset of V. A hypergraph is *regular* (resp. *uniform*) if every node has the same degree (resp. every hyperedge has the same rank) with this degree (resp. rank) being the *degree* (resp. *rank*) of the hypergraph. Every graph $G = (V, E)$ has a natural representation as a hypergraph: the nodes of the hypergraph are V; and the hyperedges are E, where the hyperedge e consists of the pair of nodes incident with the edge e of G.

We can represent a hypergraph $H = (V, E)$ as a bipartite graph: the node set of the bipartite graph is $V \cup E$; and there is an edge (v, D), for $v \in V$ and $D \in E$, in the bipartite graph iff $v \in D$ in the hypergraph. It is clear that this yields a one-to-one correspondence between hypergraphs and bipartite graphs (without isolated nodes) that come complete with a partition of the elements into a 'left-hand side', which will correspond to the nodes of the hypergraph, and a 'right-hand side', which will correspond to the hyperedges of the hypergraph. We assume (henceforth) that every bipartite graph comes equipped with such a partition and for clarity we henceforth refer to the nodes on the left-hand side as *nodes* and the nodes on the right-hand side as *blocks*. Likewise, we refer to the degree of a node as its *degree* and the degree of a block as its *rank*. A bipartite graph corresponding to a regular, uniform hypergraph of degree d and rank Δ is called a (d, Δ)-*bipartite graph*. Every bipartite graph (and so every hypergraph) also describes its *dual hypergraph* where the roles of the nodes on the left-hand side and the blocks on the right-hand side of the partition are reversed in the definition of the hypergraph. With regard to our one-to-one correspondence between bipartite graphs and hypergraphs described above, if G is a bipartite graph then it corresponds to a hypergraph via this correspondence and it also corresponds to a (different) hypergraph via the natural representation highlighted in the previous paragraph.

A *path* in some hypergraph $H = (V, E)$ is an alternating sequence of nodes and hyperedges so that all nodes are distinct, all hyperedges are distinct and a node $v \in V$ follows or preceeds a hyperedge $D \in E$ in the sequence only if $v \in D$ in the hypergraph. The *length* of any path is its length in the bipartite graph corresponding to the hypergraph, and the *distance* between two distinct elements of $V \cup E$ is the length of a shortest path joining these two elements in the corresponding bipartite graph. The *diameter* of H is the maximum of the distances between every pair of distinct nodes of V, and the *line-diameter* of H is the maximum of the distances between every pair of distinct hyperedges of E.

We have two remarks. First, we have analogous notions of diameter and line-diameter in any bipartite graph. Note that our notion of diameter (which ignores node-to-block and block-to-block paths) is different from the usual graph-theoretic notion of diameter in a bipartite graph (and likewise for line-diameter). Second, our graph-theoretic notion of path length in a hypergraph differs from that in [6] where the focus is on the number of hyperedges in a hyperedge-to-hyperedge path in some hypergraph. We shall soon move to an exclusively graph-theoretic formulation in which our notion of length is the natural one.

We shall be interested in building sets of paths in some hypergraph H so that all paths have the same (distinct) source and destination nodes or hyperedges; moreover, we shall require that these paths do not 'interfere' with one another. We say that a set of paths in H joining two distinct elements of $V \cup E$ is: *pairwise internally disjoint* if every node and every hyperedge different from the source and destination lies on at most one path from this set; or *pairwise internally edge-disjoint* if every pair $(v, D) \in V \times E$ is such that v follows or precedes D on at most one path from this set. The reason we make the above differentiation as regards path disjointness is as follows. Given some hypergraph, our intention is to ultimately consider the nodes as servers and the hyperedges as switches (as it happens, we shall go on to compose such hypergraphs so that servers morph into switches but more later when we discuss composing DCNs). This intention is best appreciated by working with the corresponding bipartite graph where the nodes are to denote servers and the blocks switches. Consequently, we can regard a hypergraph as modelling a switch-centric DCN where there is one layer of switches. A set of pairwise internally disjoint (resp. edge-disjoint) paths is required if we want to enable simultaneous data transfer when the corresponding servers and switches are blocking (resp. non-blocking).

The notion of a transversal design is crucial to what follows.

Definition 1. *Let $k, \Delta \geq 2$. A $[\Delta, k]$-transversal design T is a triple (X, D, V) where: $|X| = \Delta k$; $D = (D_1, D_2, \ldots, D_\Delta)$ is a partition of X into Δ equal-sized groups (each of size k); and $V = \{V_j : j = 1, 2, \ldots, k^2\}$ is a family of k^2 subsets of X, each of size Δ and called a* block, *so that*

- $|D_i \cap V_j| = 1$, *for $i = 1, 2, \ldots, \Delta$, $j = 1, 2, \ldots, k^2$*
- *each pair of elements $\{x_i, x_j\}$, where $x_i \in D_i$, $x_j \in D_j$ and $i \neq j$, is contained in exactly 1 block (we say that the unique block containing x_i and x_j is the block generated by x_i and x_j).*

We adopt a graph-theoretic perspective on transversal designs as defined in Definition 1: we think of the $[\Delta, k]$-transversal design T as a bipartite graph where the elements of X (resp. V) lie on the left-hand side (resp. right-hand side) of the partition, and so are called nodes (resp. blocks) within the bipartite graph, and so that in this bipartite graph there is an edge (p, Q), for $p \in X$ and $Q \in V$, iff in the transversal design the element p is in the block Q. Note that the bipartite graph corresponding to the transversal design from Definition 1 is a (k, Δ)-bipartite graph. Henceforth, we regard both hypergraphs and transversals as bipartite graphs unless we state otherwise.

3   The 3-Step Construction and Its Extensions

We begin by describing the 3-*step construction* (originating in [2] and used in [6]) for building bipartite graphs by using a base bipartite graph and a transversal design. We'll then explain how one might iterate it and then compose bipartite graphs to obtain more complex DCNs (as was done in [6]).

Step 1: Let H_0 be a (d, Δ)-bipartite graph so that there are n nodes (on the left-hand side of the partition, each of degree d) and e blocks (on the right-hand side, each of rank Δ). Such an H_0 can be visualized as in Fig. 1(a).

Step 2: Let T be a $[\Delta, k]$-transversal design. In particular, there are Δ groups of k nodes (on the left-hand side) as well as k^2 blocks (on the right-hand side).

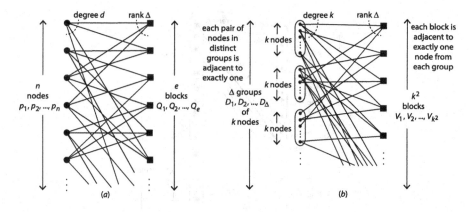

Fig. 1. A (d, Δ)-bipartite graph H_0 and a $[\Delta, k]$-transversal.

Such a T can be visualized as in Fig. 1(b). Build the bipartite graph H as follows. For every node p of H_0, introduce a group G_p of k nodes of H. For every block Q of H_0, adjacent to the nodes $p_1, p_2, \ldots, p_\Delta$ in H_0, introduce a copy of T, denoted T_Q, rooted on the Δ groups of nodes $G_{p_1}, G_{p_2}, \ldots, G_{p_\Delta}$ (so, corresponding to the block Q of H_0, we have introduced k^2 blocks in H). We refer to the Δ groups of nodes $G_{p_1}, G_{p_2}, \ldots, G_{p_\Delta}$ as the *roots* of the copy T_Q of T in H. Such a bipartite graph H can be visualized as in Fig. 2 where two of the copies of T are partially shown. The bipartite graph H_0 essentially provides a template as to where we introduce copies of T to form H.

Note that: each node of H can be indexed as $a_{p,j}$, where $p \in \{1, 2, \ldots, n\}$ and $j \in \{1, 2, \ldots, k\}$, so that p is the node of H_0 to which the group G_p in which $a_{p,j}$ sits corresponds and j is the index of $a_{p,j}$ in this group; and each block of H can be indexed as $B_{Q,V}$, where $Q \in \{1, 2, \ldots, e\}$ and $V \in \{1, 2, \ldots, k^2\}$, so that Q is the block of H_0 to which the set of blocks in which $B_{Q,V}$ sits corresponds and V is the block of T to which $B_{Q,V}$ corresponds. In addition, each node of T can be indexed $u_{i,j}$, where $i \in \{1, 2, \ldots, \Delta\}$ and $j \in \{1, 2, \ldots, k\}$, so that D_i is the group of nodes in which $u_{i,j}$ sits and j is the index of $u_{i,j}$ in that group.

Step 3: Let H^* be the bipartite graph obtained from the bipartite graph H by reversing the roles of nodes and blocks (so, H^* is the dual bipartite graph of H). Note that the bipartite graph H^* is regular of degree Δ and uniform of rank dk.

We refer to the (dk, Δ)-bipartite graph H (resp. the (Δ, dk)-bipartite graph H^*) constructed above as having been constructed by the 2-step (resp. 3-step) method using the (d, Δ)-bipartite graph H_0 and the $[\Delta, k]$-transversal T.

Our intention with our constructions is to ultimately design switch-centric DCNs with beneficial properties. Whilst there are many properties we would like our DCNs to have, it is important that DCNs can integrate a large number of servers so that the server-to-server distances are short and so that there is redundancy as to which short server-to-server routes we choose to use. In the parlance of bipartite graphs, this translates as building bipartite graphs with

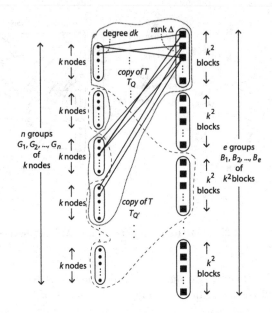

Fig. 2. Amalgamating H_0 and T to get H.

a large number of nodes and with redundant, short node-to-node paths. The following result was proven in [2] (it is actually derivable from the proofs of our upcoming results) and allows us control over the length of block-to-block paths in 2-step constructions (and so node-to-node paths in 3-step constructions).

Theorem 1 [2]. *Suppose that the (dk, Δ)-bipartite graph H has been constructed by the 2-step method using the (d, Δ)-bipartite graph H_0 and the $[\Delta, k]$-transversal T. If H_0 has line-diameter $\lambda \geq 2$ then H has line-diameter λ.*

We can iterate the 3-step construction (as was done in [6]). Note that if H_0 is a (d, Δ)-bipartite graph of line-diameter λ then the bipartite graph H_1 resulting from the 2-step construction (using H_0 and some $[\Delta, k]$-transversal design T) is a (dk, Δ)-bipartite graph of line-diameter λ. So, repeating the 2-step construction but with H_1 replacing H_0 (we keep the same T, though) yields a (dk^2, Δ)-bipartite graph H_2 of line-diameter λ. By iterating this construction, we can clearly obtain a (dk^i, Δ)-bipartite graph H_i of line-diameter λ. Converting H_i into H_i^* results in a bipartite graph with ek^{2i} nodes, with dk^i blocks, with diameter λ and that is regular of degree Δ and uniform of rank dk^i.

In [2], the 3-step construction was the focus as the application there was to build bus interconnection networks of large size but so as to limit the diameter of the resulting network. Similarly, in [6], the 3-step construction was the focus as the intention was to interpret nodes as servers and blocks as switches; were we to focus on the 2-step method and allow the server degree to grow (in H_i, above, the degree is dk^i), this would result in practically infeasible DCNs.

New methods of composing bipartite graphs (built using the 3-step construction) so as to obtain switch-centric DCNs were also derived in [6] where 4 such methods were given: Methods M_1, M_2 and M_3 are different cases of our Method A; and Method M_4 is our Method B. Let H be a (σ, ω)-bipartite graph which we think of as a DCN with the nodes as servers and the blocks as switches, and where $\sigma < \omega$.

<u>Method A</u>: We take c copies of H where $\omega - c\sigma > 0$ and $c \geq 1$. For each server (node) u of H: we remove the corresponding server in each of the c copies of H and introduce a new switch (this switch is common to all copies of H); we make all of the $c\sigma$ links incident with the c original servers incident with this new switch; and we attach $\omega - c\sigma$ pendant servers to the new switch (in [6], the new switches are termed *level-1 switches* with the original switches *level-2 switches*). So, the new DCN is such that: all switches have ω ports; there are links from (new) servers to level-1 switches; and links joining level-1 and level-2 switches. Note that there is some choice as regards the parameter c. The case where $c \geq 1$ corresponds to Method M_2 of [6]; the case when $c = 1$ corresponds to Method M_1; and the case when $c = \lfloor \frac{\lfloor \frac{\omega}{2} \rfloor}{\sigma} \rfloor$ corresponds to Method M_3 (here, the aim is to ensure that every level-1 switch is adjacent to roughly the same number of level-2 switches as it is nodes).

<u>Method B</u>: We now work with a switch-centric DCN as constructed by Method A. Let every level-1 switch have n_e adjacent servers. Suppose that there is an even number of level-1 switches. Partition the set of level-1 switches into pairs. For each pair of switches (S', S''): remove $\lfloor \frac{n_e}{2} \rfloor$ servers that are adjacent to S' and remove $\lceil \frac{n_e}{2} \rceil$ servers that are adjacent to S''; and make every server that is adjacent to the switch S' or the switch S'' also adjacent to the other switch.

In [6], various switch-centric DCNs were constructed using the 3-step construction allied with Methods A and B and were favourably compared with the ubiquitous 3-level fat-tree with regard to the number of servers therein when the diameter and the switch radix are held constant (see Tables 2–4 in [6]).

4   Constructions of Paths

We are now in a position to use transversal designs to build switch-centric DCNs, similarly to as was done in [6]. However, in [6] there were a number of problems with the proofs (so much so that some claimed results are false). We begin by highlighting these problems and then we provide not only correct proofs but also extend some of the claimed results in [6] with regard to path diversity.

In order to detail the difficulties in [6], we adopt the terminology of [6]. In Subcases (1.1) and (2.1) of the proof of Theorem 2 in [6], the situation when $r_j = s_j$, for some j where $p \neq t_j$, was not considered (although this is trivial to remedy). However, and more importantly, in Subcases (1.2) and (2.2) the construction does not work when $j = i$ as r_i, s_i, t_i all lie in the same group G_i^E and consequently we cannot infer the existence of R_i and S_i.

An attempt was also made in [6] to extend Theorem 2 of [6]; see Theorem 3 of [6]. Assumptions concerning the connectivity of H_0 are made and the existence of additional paths to those constructed in the proof of Theorem 2 are claimed in the situation when the two blocks $B_{Q,V}$ and $B_{Q',V'}$ are such that $Q \neq Q'$. However, there are serious flaws in the proof of Theorem 3 of [6], so much so that the theorem is untrue. In short, Theorem 3 of [6] claims that if there are ω pairwise internally disjoint paths in H_0 from Q to Q' then there are $\min\{\Delta\omega, k\omega\}$ pairwise internally disjoint paths in H from $B_{Q,V}$ to $B_{Q',V'}$. This does not make sense: the maximum number of pairwise internally disjoint paths in H from $B_{Q,V}$ to $B_{Q',V'}$ is Δ (as the bipartite graph H has rank Δ) and so we must have that $\min\{\Delta\omega, k\omega\} \leq \Delta$. For instance, in Example 1 of [6], where the bipartite graph H_0 is the cycle of length 10, so that $d = \Delta = 2$ and $n = e = 5$, and a $[2,3]$-transversal T is used, the bipartite graph H built by the 2-step method has degree 6 and rank 2. However, there are 2 paths from any block of H_0 to any other block of H_0 and so if Theorem 3 of [6] were true then there would be 4 pairwise disjoint paths from $B_{Q,V}$ to $B_{Q',V'}$ in H which clearly cannot be the case.

We now resurrect (some of) the proofs from [6] and extend the results claimed in that paper. We use the following easy-to-prove lemma repeatedly.

Lemma 1. *Let T be some $[\Delta, k]$-transversal with groups of nodes $\{D_1, D_2, \ldots, D_\Delta\}$. Let U be some block of T. For each $i \in \{1, 2, \ldots, \Delta\}$, let $r_i \in D_i$ be the unique node of D_i that is adjacent to U. Set $R = \{r_i : i = 1, 2, \ldots, \Delta\}$. Let P be a set of distinct pairs of nodes so that: exactly one node of any pair in P is in R and no node of R is in more than one pair of P; and no pair in P is such that both nodes lie in the same group. The blocks generated by the pairs in P are all distinct and different from U.*

Theorem 2. *Let $k, \Delta, d \geq 2$ but where $(k, \Delta) \notin \{(2,3), (2,5), (2,7)\}$. Let H be built by the 2-step method from the (d, Δ)-bipartite graph H_0 using the $[\Delta, k]$-transversal T.*

(a) *If Q and Q' are distinct blocks of H_0 so that there are $\lambda \geq 1$ pairwise internally disjoint paths in H_0 from Q to Q', each of length at most μ, then there are $\min\{\Delta, k\lambda\}$ pairwise internally disjoint paths from any block $B_{Q,V}$ of H to any other block $B_{Q',V'}$ of H, each of length at most $\mu + 4$.*

(b) *If $B_{Q,V}$ and $B_{Q,V'}$ are distinct blocks of H then there are Δ pairwise internally disjoint paths from $B_{Q,V}$ to $B_{Q,V'}$, each of length at most 6 and lying entirely within T_Q.*

Proof. (a) We may assume that $\lambda \leq \lceil \frac{\Delta}{k} \rceil$. Consider the λ pairwise internally disjoint paths from Q to Q' in H_0. We may clearly assume that either every path has length 2 or that every common neighbour of Q and Q' in H_0 lies on one of the λ paths (with each of these paths having length 2).

Suppose that $b + c = \lambda$, where $b \geq 1$ and $c \geq 0$, and that the nodes $p_1, p_2, \ldots, p_b$ are common neighbours in H_0 of Q and Q' (the case when there are no common neighbours is easy). As stated above, we may assume that either:

$b = \lambda$; or $c > 0$ and $\{p_1, p_2, \ldots, p_b\}$ consists of all common neighbours of Q and Q' in H_0. In the case when $c > 0$, let the nodes $q_1, q_2, \ldots, q_c$ be neighbours of Q but not of Q' in H_0, and let the nodes $q'_1, q'_2, \ldots, q'_c$ be neighbours of Q' but not of Q in H_0 so that the remaining c paths from Q to Q' in H_0 are of the form $Q, q_i, \ldots, q'_i, Q'$, for $i = 1, 2, \ldots, c$.

We begin with an involved construction. Set $k' = \Delta - k(\lceil \frac{\Delta}{k} \rceil - 1)$; so $1 \le k' \le k$. We can batch groups of nodes of T_Q and $T_{Q'}$ in H as follows:

- for $i \in \{1, 2, \ldots, b\}$, define $G_0^i = G_{p_i} = H_0^i$
- for $i \in \{1, 2, \ldots, c\}$ (where $c > 0$), define $G_0^{b+i} = G_{q_i}$ and $H_0^{b+i} = G_{q'_i}$
- for $i \in \{1, 2, \ldots, b + c - 1\}$, choose groups $G_1^i, G_2^i, \ldots, G_{k-1}^i$ within T_Q and groups $H_1^i, H_2^i, \ldots, H_{k-1}^i$ within $T_{Q'}$, and choose groups $G_1^{b+c}, G_2^{b+c}, \ldots, G_{k'-1}^{b+c}$ within T_Q and groups $H_1^{b+c}, H_2^{b+c}, \ldots, H_{k'-1}^{b+c}$ within $T_{Q'}$ so that:
  - all G_j^i, where $j > 0$, are distinct and different from $G_0^1, G_0^2, \ldots, G_0^{b+c}$
  - all H_j^i, where $j > 0$, are distinct and different from $H_0^1, H_0^2, \ldots, H_0^{b+c}$
  - any G_j^i, where $j > 0$, corresponds to some node p of H_0 that is adjacent to both Q and Q' iff the group H_j^i corresponds to the same node p of H_0, i.e., G_j^i and H_j^i are identical.

We have three remarks: each G_j^i, where $j \ge 0$, is in T_Q and each H_j^i, where $j \ge 0$, is in $T_{Q'}$, so that if $c > 0$ then the only groups common to both T_Q and $T_{Q'}$ are $G_{p_1}, G_{p_2}, \ldots, G_{p_b}$; the bound $b + c \le \lceil \frac{\Delta}{k} \rceil$ means that there are enough groups available in both T_Q and $T_{Q'}$ for us to be able to choose as above; and if some group of the form G_j^i, where $j > 0$, is identical to the group H_j^i then it must be the case that both are rooted at the same node p of H_0 that is a common neighbour of Q and Q' in H_0, and consequently that $c = 0$.

For each $i \in \{1, 2, \ldots, b + c\}$ and each $j \in \{0, 1, \ldots, k - 1\}$, if $i \ne b + c$, or each $j \in \{0, 1, \ldots, k' - 1\}$, if $i = b + c$, let $r_j^i \in G_j^i$ (resp. $s_j^i \in H_j^i$) be the unique node of G_j^i (resp. H_j^i) that is adjacent to $B_{Q,V}$ (resp. $B_{Q',V'}$) in H. Note that the pair r_j^i and s_j^i lie in the same group of H iff both G_j^i and H_j^i are rooted at the same node of H_0 and this node is adjacent to both Q and Q' in H_0.

For each $i \in \{1, 2, \ldots, b+c\}$, let $G_0^i = \{r_0^i, t_1^i, \ldots, t_{k-1}^i\}$ and $H_0^i = \{s_0^i, w_1^i, \ldots, w_{k-1}^i\}$ so that in the case when $G_0^i = H_0^i$: if $r_0^i = s_0^i$ then $t_j^i = w_j^i$, for $j \in \{1, 2, \ldots, k - 1\}$; and if $r_0^i \ne s_0^i$ then $r_0^i = w_1^i$ and $s_0^i = t_1^i$, with $t_j^i = w_j^i$, for $j \in \{2, 3, \ldots, k - 1\}$.

We are now ready to generate some blocks within T_Q and $T_{Q'}$ in H. For each $i \in \{1, 2, \ldots, b+c\}$ and each $j \in \{1, 2, \ldots, k - 1\}$, if $i \ne b + c$, or each $j \in \{1, 2, \ldots, k' - 1\}$, if $i = b + c$: let $B_{r_j^i, t_j^i}$ be the block of T_Q in H generated by $r_j^i \in G_j^i$ and $t_j^i \in G_0^i$; and let $B'_{s_j^i, w_j^i}$ be the block of $T_{Q'}$ in H generated by $s_j^i \in H_j^i$ and $w_j^i \in H_0^i$. So, we have generated $\Delta - \lambda$ blocks in T_Q and $\Delta - \lambda$ blocks in $T_{Q'}$. Note that any block of T_Q is necessarily distinct from any block of $T_{Q'}$. By Lemma 1 applied twice to both T_Q and $T_{Q'}$: all blocks of $\{B_{r_j^i, t_j^i} : i = 1, 2, \ldots, b+c-1 \text{ and } j = 1, 2, \ldots, k-1, \text{ or } i = b+c \text{ and } j = 1, 2, \ldots, k'-1\}$ are distinct and different from $B_{Q,V}$; and all blocks of $\{B'_{s_j^i, w_j^i} : i = 1, 2, \ldots, b+c-1$

and $j = 1, 2, \ldots, k - 1$ or $i = b + c$ and $j = 1, 2, \ldots, k' - 1$} are distinct and different from $B_{Q',V'}$; call these two sets of blocks our working sets of blocks.

Now we build some paths from $B_{Q,V}$ to $B_{Q',V'}$ in H. For each $i \in \{1, 2, \ldots, b\}$: if $r_0^i = s_0^i$ then define the paths:

- π_0^i as $B_{Q,V}, r_0^i, B_{Q',V'}$
- π_1^i as $B_{Q,V}, r_1^i, B_{Q',V'}$, if $r_1^i = s_1^i$, and as $B_{Q,V}, r_1^i, B_{r_1^i,t_1^i}, t_1^i, B'_{s_1^i,w_1^i}, s_1^i$, $B_{Q',V'}$, if $r_1^i \neq s_1^i$ (note that $t_1^i = w_1^i$)

and if $r_0^i \neq s_0^i$ then define the paths:

- π_0^i as $B_{Q,V}, r_0^i, B'_{s_1^i,w_1^i}, s_1^i, B_{Q',V'}$ (note that $w_1^i = r_0^i$)
- π_1^i as $B_{Q,V}, r_1^i, B_{r_1^i,t_1^i}, s_0^i, B_{Q',V'}$ (note that $t_1^i = s_0^i$).

The above definition of π_0^i and π_1^i presupposes that both paths exist; that is, that it is not the case that $\Delta = k(b-1) + 1$ and $r_0^b \neq s_0^b$ (as otherwise it is not clear how we build only π_0^b without having recourse to G_1^i; note that if $\Delta = k(b-1)+1$ and $r_0^b = s_0^b$ then π_0^b exists). We shall return to this special case later.

For each $i \in \{1, 2, \ldots, b\}$ and each $j \in \{2, 3, \ldots, k-1\}$, if $i < b + c$, or each $j \in \{2, 3, \ldots, k'-1\}$, if $i = b$ and $c = 0$: if $r_j^i \neq s_j^i$ then define the path π_j^i as $B_{Q,V}, r_j^i, B_{r_j^i,t_j^i}, t_j^i, B'_{s_j^i,w_j^i}, s_j^i, B_{Q',V'}$; and if $r_j^i = s_j^i$ then define the path π_j^i as $B_{Q,V}, r_j^i, B_{Q',V'}$.

Note that out of all the 'π-paths' constructed above, the only way that we can have that two of our paths are not internally disjoint is when $r_0^i \neq s_0^i$ but $r_1^i = s_1^i$, for some $i \in \{1, 2, \ldots, b\}$. In every such case, choose $x_1^i \in G_1^i \setminus \{r_1^i\}$. Let $B_{r_0^i,x_1^i}$ be the block of T_Q in H generated by $r_0^i \in G_0^i$ and $x_1^i \in G_1^i$, and let $B'_{s_0^i,x_1^i}$ be the block of $T_{Q'}$ in H generated by $s_0^i \in G_0^i$ and $x_1^i \in G_1^i$ (in essence, we have dispensed with the blocks $B_{r_1^i,t_1^i}$ and $B'_{s_1^i,w_1^i}$ and replaced them with the blocks $B_{r_0^i,x_1^i}$ and $B'_{s_0^i,x_1^i}$ in our working sets of blocks; we reiterate that we do this for every $i \in \{1, 2, \ldots, b\}$ for which $r_0^i \neq s_0^i$ and $r_1^i = s_1^i$). The conditions of Lemma 1 still hold and so the blocks in our working sets of blocks are all distinct and different from $B_{Q,V}$ and $B_{Q',V'}$. For each $i \in \{1, 2, \ldots, b\}$ for which $r_0^i \neq s_0^i$ and $r_1^i = s_1^i$, redefine the paths: π_0^i as $B_{Q,V}, r_1^i, B_{Q',V'}$; and π_1^i as $B_{Q,V}, r_0^i, B_{r_0^i,x_1^i}, x_1^i, B'_{s_0^i,x_1^i}, s_0^i, B_{Q',V'}$. The paths from the resulting set of π-paths are now pairwise internally disjoint.

Let us now return to the situation where $\Delta = k(b-1) + 1$ and $r_0^b \neq s_0^b$ (so, necessarily, $c = 0$). In this case, we proceed exactly as we did above but without building the path π_0^b. We need to build a path of the form $B_{Q,V}, r_0^b, \ldots, s_0^b, B_{Q',V'}$ (that is internally disjoint from all of the above $\Delta - 1$ π-paths). Suppose that $k \geq 3$; so, there is a node $x_0^{b-1} \in G_0^{b-1} \setminus \{r_0^{b-1}, s_0^{b-1}\}$. Generate the block $B_{r_0^b,x_0^{b-1}}$ of T_Q within H and the block $B_{s_0^b,x_0^{b-1}}$ of $T_{Q'}$ within H. By Lemma 1, these blocks are different from $B_{Q,V}$, $B_{Q',V'}$ and all other blocks so generated within T_Q and $T_{Q'}$. Define the path π_0^b as $B_{Q,V}, r_0^b, B_{r_0^b,x_0^{b-1}}, x_0^{b-1}, B_{s_0^b,x_0^{b-1}}, s_0^b, B_{Q',V'}$. This path is internally disjoint from all other π-paths. Suppose that $k = 2$. So, there are 4 blocks in T and consequently $\Delta \in \{3, 5, 7\}$ which yields a contradiction.

If $c > 0$ then we can define additional paths in H from $B_{Q,V}$ to nodes of $G_0^{b+1}, G_0^{b+2}, \ldots, G_0^{b+c}$ and from $B_{Q',V'}$ to nodes of $H_0^{b+1}, H_0^{b+2}, \ldots, H_0^{b+c}$ (note that in this scenario $G_j^i \neq H_j^i$ unless $i \in \{1, 2, \ldots, b\}$ and $j = 0$). For each $i \in \{b+1, b+2, \ldots, b+c\}$, define the paths: η_0^i as $B_{Q,V}, r_0^i$; and ν_0^i as $B_{Q',V'}, s_0^i$. For each $i \in \{b+1, b+2, \ldots, b+c\}$ and each $j \in \{1, 2, \ldots, k-1\}$, if $i \neq b+c$, or each $j \in \{1, 2, \ldots, k'-1\}$, if $i = b+c$, define the paths: η_j^i as $B_{Q,V}, r_j^i, B_{r_j^i, t_j^i}, t_j^i$; and ν_j^i as $B_{Q',V'}, s_j^i, B'_{s_j^i, w_j^i}, w_j^i$. Any 2 distinct paths from our collection of π-paths, η-paths and ν-paths clearly have no nodes in common and the only block in common is $B_{Q,V}$, $B_{Q',V'}$ or both.

If we can find a path in H from r_0^i or t_j^i to s_0^i or w_j^i, respectively, for each $i \in \{b+1, b_2, \ldots, b+c\}$ and each $j \in \{1, 2, \ldots, k-1\}$, if $i < b+c$, or each $j \in \{1, 2, \ldots, k'-1\}$, if $i = b+c$, so that no node or block of any of these paths, apart from the source and destination nodes, lies in T_Q or $T_{Q'}$ and so that the resulting paths are pairwise internally disjoint then we are done. Fix $i \in \{1, 2, \ldots, c\}$ and let $Q, q_i, Q_1, q_i^2, Q_2, q_i^3, \ldots, q_i^m, Q_m, q_i', Q'$, be one of our remaining c paths from Q to Q' in H_0; in particular, $m \in \{1, 2, \ldots, \frac{1}{2}(\mu - 2)\}$. In H: there are k paths of length 2, each path having a source in G_{q_i} and a destination in $G_{q_i^2}$ so that all sources are distinct as are all destinations and lying entirely within T_{Q_1}; there are k paths of length 2, each path having a source in $G_{q_i^2}$ and a destination in $G_{q_i^3}$ so that all sources are distinct as are all destinations and lying entirely within T_{Q_2}; $\ldots$; and there are k paths of length 2, each path having a source in $G_{q_i^m}$ and a destination in $G_{q_i'}$ so that all sources are distinct as are all destinations and lying entirely within T_{Q_m}. We are done.

Now return to the case ignored at the beginning of the proof, namely the case when $b = 0$ and $c = \lambda$. The above construction of paths from $B_{Q,V}$ to each node of G_{q_i}, concatenated with paths from each node of G_{q_i} to each node of $G_{q_i'}$, concatenated with paths from each node of $G_{q_i'}$ to $B_{Q',V'}$ still works.

(b) Consider the case when our two blocks are $B_{Q,V}$ and $B_{Q,V'}$. Suppose that the block Q of H_0 is adjacent to the nodes $p_1, p_2, \ldots, p_\Delta$. For each $i \in \{1, 2, \ldots, \Delta\}$, let $r_i \in G_{p_i}$ be adjacent to $B_{Q,V}$ in H and let $s_i \in G_{p_i}$ be adjacent to $B_{Q,V'}$ in H. W.l.o.g. suppose that $r_i \neq s_i$, for $i = 1, 2, \ldots, b$, and that $r_i = s_i$, for $i = b+1, b+2, \ldots, \Delta$.

Suppose that $b \geq 2$. For each $i \in \{1, 2, \ldots, b-1\}$, let $B_{r_i, s_{i+1}}$ be the unique block of T_Q that is generated by r_i and s_{i+1}, and let B_{r_b, s_1} be the unique block of T_Q that is generated by r_b and s_1. By Lemma 1, all blocks $B_{r_1, s_2}, B_{r_2, s_3}, \ldots, B_{r_{b-1}, s_b}, B_{r_b, s_1}$ are distinct and different from $B_{Q,V}$ and $B_{Q,V'}$. Hence, if π_i is the path $B_{Q,V}, r_i, B_{r_i, s_{i+1}}, s_{i+1}, B_{Q,V'}$, for $i \in \{1, 2, \ldots, b-1\}$, π_b is the path $B_{Q,V}, r_b, B_{r_b, s_1}, s_1, B_{Q,V'}$, and π_i is the path $B_{Q,V}, r_i, B_{Q,V'}$, for $i \in \{b+1, b+2, \ldots, \Delta\}$, then the set of paths are pairwise internally disjoint.

If $b = 0$ then the above construction trivially yields Δ paths of length 2 from $B_{Q,V}$ to $B_{Q,V'}$. Suppose that $b = 1$. Choose $x_2 \in G_{p_2} \setminus \{r_2\}$ and let B_{r_1, x_2} (resp. B_{s_1, x_2}) be the block of T_Q generated by r_1 and x_2 (resp. s_1 and x_2). By Lemma 1, $B_{r_1, x_2}, B_{s_1, x_2}, B_{Q,V}$ and $B_{Q,V'}$ are all distinct. So, if π_1 is the path $B_{Q,V}, r_1, B_{r_1, x_2}, x_2, B_{s_1, x_2}, s_1, B_{Q,V'}$ and π_i is the path $B_{Q,V}, r_i, B_{Q,V'}$, for $i \in \{2, 3, \ldots, \Delta\}$ then we obtain a pairwise internally disjoint set of paths. □

Note that Theorem 2 is optimal in the sense that if H_0 has blocks Q and Q' so that there are exactly λ pairwise internally disjoint paths from Q to Q' in H_0 then we can do no better than $\min\{\Delta, k\lambda\}$ pairwise internally disjoint paths from any block $B_{Q,V}$ to any block $B_{Q',V'}$ in H, as by Menger's Theorem we can remove λ nodes from H_0 so as to disconnect Q and Q', and so $k\lambda$ nodes from H so as to disconnect $B_{Q,V}$ and $B_{Q',V'}$. Note also that irrespective of the erroneous proofs in [6], Theorem 2 extends any claimed results in [6] by deriving $\min\{\Delta, k\lambda\}$ pairwise internally disjoint paths from any block $B_{Q,V}$ in H to any block $B_{Q',V'}$ where not only might we have $Q \neq Q'$ but also $Q = Q'$.

5   Conclusion

In this paper we have extended the use of mathematical techniques within the design of data centre networks. We feel that theoretical computer science has a lot to offer more practical areas such as data centre design and hope that this work provides some impetus to theoreticians. Naturally, our work provokes some directions for further research, both theoretical and applied. Whilst we have developed an optimal analysis of path diversity as regards using the 3-step construction, we have yet to use the additional path diversity so obtained. In order to do this we would need use bipartite graphs H_0 (with reference to the 3-step construction) with additional connectivity properties. In future we shall seek to build and use such bipartite graphs. Also, our constructions form part of a wider as yet untouched topic, analogous to the well-established study of Moore graphs, namely the analysis of (not graphs but) 'switch-server graphs' (the models of DCNs) as to the maximal number of servers that can be incorporated under 'degree and diameter' constraints. Finally, we would like to use techniques similar to those here so as to build not just switch-centric DCNs but also server-centric DCNs.

References

1. Al-Fares, M., Loukissas, A., Vahdat, A.: A scalable, commodity data center network architecture. SIGCOMM Comput. Commun. Rev. **38**(4), 63–74 (2008)
2. Bermond, J.C., Bond, J., Djelloul, S.: Dense bus networks of diameter 2. In: Proceedings of the Workshop on Interconnection Networks. DIMACS Series vol. 21, pp. 9–18 (1995)
3. Greenberg, A., Hamilton, J.R., Jain, N., Kandula, S., Kim, C., Lahiri, P., Maltz, D.A., Patel, P., Sengupta, S.: VL2: a scalable and flexible data center network. SIGCOMM Comput. Commun. Rev. **39**(4), 51–62 (2009)
4. Liu, Y., Muppala, J.K., Veeraraghavan, M., Lin, D., Katz, J.: Data Centre Networks: Topologies, Architectures and Fault-Tolerance Characteristics. Springer, Heidelberg (2013)
5. Mysore, R.N., Pamboris, A., Farrington, N., Huang, N., Miri, P., Radhakrishnan, S., Subramanya, V., Vahdat, A.: Portland: a scalable fault-tolerant layer 2 data center network fabric. SIGCOMM Comput. Commun. Rev. **39**(4), 39–50 (2009)

6. Qu, G., Fang, Z., Zhang, J., Zheng, S.-Q.: Switch-centric data center network structures based on hypergraphs and combinatorial block designs. IEEE Trans. on Par. Distrib. Sys. **26**(4), 154–1164 (2015)
7. Wu, K., Xiao, J., Ni, L.M.: Rethinking the architecture design of data center networks. Frontiers of Comput. Sci. **6**(5), 596–603 (2012)

Anonymity and Indistinguishability

Privacy in Elections: k-Anonymizing Preference Orders

Nimrod Talmon[✉]

Institut für Softwaretechnik und Theoretische Informatik,
TU Berlin, Berlin, Germany
nimrodtalmon77@gmail.com

Abstract. We study the (parameterized) complexity of a combinatorial problem, motivated by the desire to publish elections-related data, while preserving the privacy of the voters (humans or agents). In this problem, introduced and defined here, we are given an election, a voting rule, and a distance function over elections. The task is to find an election which is not too far away from the original election (with respect to the given distance function) while preserving the election winner (with respect to the given voting rule), and such that the resulting election is k-anonymized; an election is said to be k-anonymous if for each voter in it there are at least $k - 1$ other voters with the same preference order. We consider the problem of k-anonymizing elections for the Plurality rule and for the Condorcet rule, for the Discrete distance and for the Swap distance. We show that the parameterized complexity landscape of our problem is diverse, with cases ranging from being polynomial-time solvable to Para-NP-hard.

1 Introduction

We consider privacy issues when publishing preferences-related (or, election-related) data. Assume being given data consisting of a set of records, where each record (corresponding to a human or an agent) contains preferences-related information as well as some private (side) information. The task is to publish this data (for example, to let researchers analyze it) while preserving the privacy of the entities in it. Two of the most well-studied approaches for achieving privacy when publishing information is *differential privacy* (see, for example, Dwork and Roth [12]) and *k-anonymity* (and *l-diversity*; see, for example, Sweeney [25] and Machanavajjhala et al. [20]). Here we follow the k-anonymity framework (see Clifton and Tassa [9] for a recent comparison between these two approaches).

We say that an election is k-*anonymous* if each preference order in it appears at least k times (where a preference order is an order over a set of predefined alternatives). Given an input election, the goal is to generate an election which is k-anonymous but still preserves some properties of the original election.

It is natural to consider the distance between the original election and the resulting anonymized election (in a similar way as done when k-anonymizing

N. Talmon—Supported by DFG Research Training Group MDS (GRK 1408).

A. Kosowski and I. Walukiewicz (Eds.): FCT 2015, LNCS 9210, pp. 299–310, 2015.
DOI: 10.1007/978-3-319-22177-9_23

graphs; there, one can define the distance as the symmetric difference of the edge set, for example). Therefore, we consider distances over elections. We study the Discrete distance (where each preference order can be transformed into any other preference order at unit cost) and the Swap distance (where each two consecutive alternatives can be swapped at unit cost), as these are the most basic and well-studied distances defined on elections (see, for example, [14]). The idea is that if the distance is small, then the anonymized election does not differ too much from the original election; this is, arguably, more apparent in the Swap distance.

Besides requiring that the original election and the resulting k-anonymized election will be close (with respect to the considered distance), we would like to preserve some specific properties of the original election (in this, we follow ideas presented by Bredereck et al. [5], who considered preserving graph properties, such as the connectivity, the relative distances, and the diameter). Here, we require that the winner of the election will be preserved. For this, we need to fix a voting rule. We study two voting rules, the Plurality rule and the Condorcet rule, as these are the most basic and well-studied voting rules, which are also good representative rules (specifically, the Plurality rule, albeit simple, can be seen as a representative scoring rule, while the Condorcet rule can be seen as a representative tournament-based rule).

In what follows, we study the parameterized complexity of k-anonymizing elections, under the Plurality rule and under the Condorcet rule, for the Discrete distance and for the Swap distance. We consider two election-related parameters, specifically, the number of voters and the number of alternatives, and one anonymity-related parameter, the anonymity level k. We show that the parameterized complexity landscape of our problem is diverse, with cases ranging from being polynomial-time solvable to Para-NP-hard.

In a way, this paper can be seen as bringing the well-studied field of k-anonymity to the well-studied field of voting systems and social choice, with the hope of better understanding complexity issues of preserving privacy when publishing election-related data. We view our definition of k-anonymous elections as being a natural adaptation of the concept of k-anonymity to preferences-related (or, election-related) data.

1.1   Related Work

There is a big body of literature on security of elections and on preserving privacy of voters participating in (digital) elections. Chaum [7], Nurmi et al. [24], and Cuvelier et al. [10], among others, considered cryptographic mechanisms to encrypt the votes, while Chen et al. [8], among others, considered differential privacy. Ashur and Dunkelman [1] showed how to breach the privacy of voters for the Israeli parliament when an adversary can look at the publicly-available nation-wide election statistics. This work is of some relevance to us as it considers privacy with respect to (publicly-available) published data.

Sweeney [25] introduced the concept of k-anonymity as a way to preserve privacy over published data, after demonstrating how to identify many individuals by mixing publicly-available medical data with publicly-available voter lists (interestingly, Sweeney [25] already somewhat focuses on election-related data, specifically on the party affiliation; informally speaking, party affiliation corresponds to the Plurality rule, since only the first choice counts, while for the Condorcet rule we would need the complete preference orders publicly available). Much work has been done on k-anonymizing tables (for example, Meyerson and Williams [22], Bredereck et al. [4], and Bredereck et al. [6]; some of these concentrate on parameterized complexity), on k-anonymizing graphs (for example, Liu and Terzi [19], Hartung et al. [17], and Bredereck et al. [5]; some of these concentrate on parameterized complexity). Here, we consider neither general tables nor graphs, but instead we consider elections. Indeed, elections can be described as tables, but here we require to preserve the winner and allow different, election specific, operations. Specifically, while the Discrete distance can be natural also for general tables, this is not the case for the Swap distance.

For general information about social choice, elections, voting systems, and voting rules, we point the reader to any textbook on social choice, for example the book by Brandt et al. [3].

2 Preliminaries

Considering distances over elections, we follow some notation from Elkind et al. [14]. We assume familiarity with standard notions regarding algorithms, computational complexity, and graph theory. For a non-negative integer z, we denote the set $\{1, \ldots, z\}$ by $[z]$.

2.1 Elections and Distances

An election E is a pair (C, V) where $C = \{c_1, \ldots, c_m\}$ is the set of alternatives and $V = (v_1, \ldots, v_n)$ is the collection of voters. Each voter v_i is represented by a total order $\succ_{v_i}$ over C (her preference order; we use voters and preference orders interchangeably). A voter which prefers c to all other alternatives is called a c-voter. For reading clarity, we refer to the voters as females, while the alternatives are males. A voting rule $\mathcal{R}$ is a function that, given an election $E = (C, V)$, returns a set $\mathcal{R}(E) \subseteq C$ of election winners (indeed, we use the non-unique winner model). We consider the following voting rules:

1. **The Plurality Rule.** Each alternative receives one point for each voter which ranks him first, and the winners are the highest-scoring alternatives.
2. **The Condorcet Rule.** An alternative c is a (weak) Condorcet winner if for each other alternative $c' \in C \setminus \{c\}$, it holds that $|\{v \in V : c \succ_v c'\}| \geq |\{v \in V : c' \succ_v c\}|$, that is, if he beats (or ties with) all other alternatives in head-to-head contests. The Condorcet rule elects the Condorcet winners if some exist, returning an empty set otherwise.

Given a set V' of preference orders, we say that a function $d : V' \times V' \to \mathbb{N}$ is a *distance function* over preference orders if it is a metric over preference orders.

Given a distance function over preference orders d and two elections (over the same set of alternatives), $E = (C, (v_1, \ldots, v_n))$ and $E' = (C, (v'_1, \ldots, v'_n))$, the above definition can be naturally extended by fixing an arbitrary order for the voters of E, that is, $[v_1, \ldots, v_n]$, considering all the possible permutations for the voters of E', that is, $[v'_{\pi(1)}, \ldots, v'_{\pi(n)}]$, and defining the d-distance between E and E' to be $d(E, E') = \min_{\pi \in S_n} \sum_{i \in [n]} d(v_i, v'_{\pi(i)})$, where S_n is the set containing all the possible permutations of $[n]$. We define the following distance functions:

1. **Discrete Distance.** $d_{\mathrm{discr}}(v_1, v_2) = 0$ if and only if $v_1 = v_2$, while otherwise $d_{\mathrm{discr}}(v_1, v_2) = 1$. Indeed, for two elections, $E = (C, (v_1, \ldots, v_n))$ and $E' = (C, (v'_1, \ldots, v'_n))$, it holds that $d_{\mathrm{discr}}(E, E') = |\{i : v_i \neq v'_i\}|$ (hence, the Hamming distance).
2. **Swap Distance.** $d_{\mathrm{swap}}(v_1, v_2) = |\{(c, c') \in C \times C : c \succ_{v_1} c' \wedge c' \succ_{v_2} c\}|$. Indeed, the swap distance $d_{\mathrm{swap}}(v_1, v_2)$ (also called the Dodgson distance) is the minimum number of swaps of consecutive alternatives needed for transforming the preference order of v_1 to that of v_2.

Clearly, both the Discrete distance and the Swap distance are distance functions.

2.2 Anonymization

A group of voters with the same preference order is called a *block*. Using this notion, we have that an election is k-anonymous if and only if each block in it is of size at least k. We denote the number of voters in block B by $|B|$ and say that a block is *bad* if $0 < B < k$ (as it is not yet anonymized in this case). Since all voters in a block have the same preference order, it is valid to consider the preference order of the voters in the block. Specifically, a block of c-voters is called a *c-block*.

2.3 Parameterized Complexity

An instance (I, k) of a parameterized problem consists of the "classical" problem instance I and an integer k being the *parameter* [11, 15, 23]. A parameterized problem is called *fixed-parameter tractable* (FPT) if there is an algorithm solving it in $f(k) \cdot |I|^{O(1)}$ time, for an arbitrary computable function f only depending on the parameter k. In difference to that, algorithms running in $|I|^{f(k)}$ time prove membership in the class XP (clearly, FPT $\subseteq$ XP).

One can show that a parameterized problem L is (presumably) not fixed-parameter tractable by devising a *parameterized reduction* from a W[1]-hard problem (for example, the Clique problem, parameterized by the solution size) or a W[2]-hard problem (for example, the Set Cover problem, parameterized by the solution size) to L. A parameterized problem which is NP-hard even for instances for which the parameter is a constant is said to be Para-NP-hard.

Table 1. (parameterized) Complexity of EA.

	Discrete distance	Swap distance
Plurality rule	P (Theorem 1)	NP-h even when $n = k = 4$ (Theorem 3)
		FPT wrt. m (Theorem 5)
Condorcet rule	NP-h (Theorem 2) para. complexity wrt. k is open FPT wrt. n (Theorem 4) FPT wrt. m (Theorem 5)	NP-h even when $n = k = 4$ (Theorem 3) FPT wrt. m (Theorem 5)

2.4 Main Problem and Overview of Our Results

The main problem we consider in this paper is defined as follows.

$\mathcal{R}$-d-ELECTION ANONYMIZATION ($\mathcal{R}$-d-EA)
Input: An election $E = (C, V)$ where $C = \{c_1, \ldots, c_m\}$ is the set of alternatives and $V = (v_1, \ldots, v_n)$ is the collection of voters, anonymity level k, and a budget s.
Question: Is there a k-anonymous election E' such that $\mathcal{R}(E) = \mathcal{R}(E')$ and $d(E, E') \leq s$ (where $\mathcal{R}$ is a voting rule, d is a distance function over elections, and an election is said to be k-anonymous if for each voter in it there are at least $k - 1$ other voters with the same preference order)?

We study the (parameterized) complexity of $\mathcal{R}$-d-ELECTION ANONYMIZATION, where we consider both the Plurality rule and the Condorcet rule as the voting rule $\mathcal{R}$, and where we consider both the Discrete distance and the Swap distance as the distance d (that is, we consider the following four variants: Plurality-Discrete-EA, Condorcet-Discrete-EA, Plurality-Swap-EA, and Condorcet-Swap-EA). Our results are summarized in Table 1. Due to the lack of space, some proof details are deferred to the full version.

3 Results

Intuitively, from all variants considered in this paper, Plurality-Discrete-EA should be the most tractable, as the Plurality rule is conceptually simpler than the Condorcet rule and the Discrete distance is conceptually simpler than the Swap distance. This intuition is correct: it turns out that Plurality-Discrete-EA is polynomial-time solvable, while all other variants are NP-hard. We begin by describing a polynomial-time algorithm, based on dynamic programming, for Plurality-Discrete-EA.

Theorem 1. *Plurality-Discrete-EA is polynomial-time solvable.*

Proof. We describe an algorithm based on applying dynamic programming twice, in a nested way. To understand the general idea, consider an alternative c and its corresponding c-voters. We have two cases to consider with respect to the solution election: either (1) some of the c-voters are transformed to be c'-voters (for some, possibly several, other alternatives $c' \neq c$), or (2) some c'-voters (for some, possibly several, other alternatives $c' \neq c$) are transformed to be c-voters. The crucial observation is that, with respect to anonymizing the c-voters, we do not need to remember the specific c'-voters discussed above, but only their number. Therefore, we define a first (outer) dynamic program, iterating over the alternatives, and computing the most efficient way of anonymizing the c-voters, while considering all possible values for these numbers of c'-voters, and while making sure that the initial winner of the election stays the winner. In order to compute how to anonymize the c-voters we define a second (inner) dynamic program, considering the c-blocks one at a time. For each c-block, the inner dynamic program decides whether to make the respective c-block empty (with zero voters) or full (with at least k voters), by considering the possible ways of transforming other c-voters or other c'-voters, similarly in spirit to the first (outer) dynamic programming. The full proof is deferred to the full version. □

For the Condorcet rule, still considering the Discrete distance, we can show that EA is NP-hard, by a reduction from a restricted variant of the EXACT COVER BY 3-SETS problem.

Theorem 2. *Condorcet-Discrete-EA is* NP-*hard.*

Proof. We reduce from the following NP-hard problem [16], defined as follows.

RESTRICTED EXACT COVER BY 3-SETS
Input: Collection $\mathcal{S} = \{S_1, \ldots, S_n\}$ of sets of size 3 over a universe $X = \{x_1, \ldots, x_n\}$ such that each element appears in exactly three sets.
Question: Is there a subset $\mathcal{S}' \subseteq \mathcal{S}$ such that each element x_i occurs in exactly one member of $\mathcal{S}'$?

We assume, without loss of generality, that $\{\{x_1, x_2, x_3\}, \ldots, \{x_{n-2}, x_{n-1}, x_n\}\} = \{\{x_{3l-2}, x_{3l-1}, x_{3l}\} : l \in [n/3]\} \not\subseteq \mathcal{S}$, and that $n = 0 \pmod 3$. Given an instance for *Restricted Exact Cover by 3-Sets*, we create an instance for Condorcet-Discrete-EA, as follows.

We create two alternatives, p and d, and, also, for each element $x_i \in X$, we create an alternative x_i, such that the set of alternatives is $\{X, p, d\}$. We create a set of $n/3$ voters, called *jokers*, such that the ith joker (for $i \in [n/3]$) has preference order $x_{3i-2} \succ x_{3i-1} \succ x_{3i} \succ p \succ d \succ X \setminus \{x_{3i-2}, x_{3i-1}, x_{3i}\}$. For each set S_j, we create k voters, each with preference order $S_j \succ p \succ d \succ \overline{S_j}$. We refer to these voters as the *set voters*. We create another set of $(n-6)k + ((n/3) - 2)$ voters, each with preference order $d \succ X \succ p$, called the *init voters*. Finally, we set s to $n/3$ and k to $(n/3) + 2$. This finishes the construction.

Let us first compute the winner in the input election. Notice that the set voters and the jokers prefer p to d, while the init voters prefer d to p. As there are kn set voters, $n/3$ jokers, but only $(n-6)k + ((n/3) - 2)$ init voters, p defeats d. Consider an element x_i. There is exactly one joker which prefers x_i to p, and exactly three set voters which prefer x_i to p (as x_i appears in exactly three sets), and all the init voters prefer x_i to p. Other than these, all other voters prefer p to x_i. Therefore, there are $(n/3) - 1 + k(n - 3)$ voters which prefer p to x_i and $1 + 3k + (n - 6)k + ((n/3) - 2) = (n/3) - 1 + k(n - 3)$ voters which prefer x_i to p, so p and x_i are tied (we use the weak Condorcet criterion, but the reduction can be changed to work for the strong Condorcet criterion as well). Finally, it is not hard to see that d defeats x_i. Summarizing the above computations, we see that p is the winner in the input election. Thus, in an anonymized solution election, p shall be the winner as well. Given an exact cover $\mathcal{S}'$, we move all jokers to the set voters corresponding to the exact cover, that is, we move all jokers such that, for each set $S \in \mathcal{S}'$, we will have one joker in the block of set voters corresponding to S. Notice that initially the jokers form an exact cover. Notice further that they still form an exact cover in the solution, as we moved them to blocks of set voters corresponding to the sets of $\mathcal{S}'$. Specifically, considering only the jokers, the relative score of p and the x_i's did not change between the input and the solution. Therefore p is still the winner. Moreover, it is not hard to see that the election is anonymized. For the other direction, notice that the jokers are not anonymized (that is, their blocks are bad; specifically, each joker forms its own block) while all other blocks are anonymized. It is not possible to anonymize the jokers by moving other voters (or other jokers) to their blocks, as the budget is too small for that. Therefore, in a solution, all jokers should move to other blocks. There are two possibilities for each joker, either to move to the init voters or to move to some set voters. It is not a good idea to move some voters to the init voters, as the init voters prefer X to p. More formally, if in a solution, some jokers move to the init voters, then we can instead move them to an arbitrarily-chosen set voter. Therefore, we can assume that, in a solution, all jokers move to set voters. If the jokers move to set voters in a way that does not correspond to an exact cover, than at least one x_i would win over p (as at least one x_i would be covered twice, that is, at least two jokers would move to set voters preferring x_i to p; therefore, the relative score between x_i and p would change in such a way that x_i would win over p in a head-to-head contest). Therefore, a solution must correspond to an exact cover, and we are done. □

Moving further to the Swap distance, we show that EA is NP-hard for both voting rules considered, and even for elections with only four voters (and therefore, also for elections with anonymity level only four; indeed, any input with $k > n$ is a trivial no-instance). Technically, in the corresponding reduction, from KEMENY DISTANCE, we set both n and k to four. We mention that the next theorem actually holds for all voting rules which are *unanimous* (a voting rule is unanimous if for all elections where all voters prefer the same alternative c, it selects the preferred alternative c; see, for example, [14]).

Theorem 3. *For $\mathcal{R} \in \{Plurality, Condorcet\}$, $\mathcal{R}$-Swap-EA is NP-hard even if the number n of voters is four and the anonymity level k is four.*

Proof. We reduce from the KEMENY DISTANCE problem.

KEMENY DISTANCE
Input: An election $E' = (C', V')$ and a positive integer h.

Question: Is the Kemeny distance of $E' = (C', V')$ at most h (where the Kemeny distance is the minimum total number of swaps of neighboring alternatives needed to have all voters vote the same; such a similar vote is called a Kemeny vote)?

KEMENY DISTANCE is NP-hard already for four voters [2,13]. Given an input election for KEMENY DISTANCE, we create an instance for EA, as follows. We initialize our election $E = (C, V)$ with the election given for KEMENY DISTANCE and create another alternative c (that is, we set $C = C' \cup \{c\}$). For each voter in the election, we place c as the first choice of the voter, that is, for each voter $v' \in V'$, we create a voter $v \in V$ with the same preference order as v' while preferring c to all other alternatives. We set k to n and s to h. This finishes the construction.

It is clear that originally c is the winner of the election (both under the Plurality rule and under the Condorcet rule; indeed, for this theorem, we only require the rule to be *unanimous*; see, for example, [14]). The crucial observation is that there is no need to swap the new alternative c; this follows because all voters already agree on him, as they place him first in their preference orders. More formally, as k is set to n, it follows that if in a solution c is swapped in some voters, then he must be swapped in all voters. Therefore, we can simply "unswap" these swaps to get a cheaper solution.

Finally, since k is set to n, it follows that all voters should vote the same in the resulting k-anonymous solution election. Thus, the best way to anonymize the election is by finding a Kemeny vote, and transforming all voters to vote as this Kemeny vote. Therefore, the election can be k-anonymized by at most s swaps if and only if the Kemeny distance of the input election is at most h, and we are done. $\square$

With respect to the parameter number n of voters, the situation for the Discrete distance is different than the situation for the Swap distance. Specifically, it turns out that Condorcet-Discrete-EA is FPT wrt. n.

Theorem 4. *Condorcet-Discrete-*EA *is* FPT *wrt. the number n of voters.*

Proof. The crucial observation here is that there is no need to create new blocks, besides, perhaps, one arbitrarily-chosen p-block. To see this, consider a solution that adds a new block B which is not a p-block (recall that a p-block is a block of p-voters). Change the solution by moving the voters in B to a new arbitrarily-chosen p-block (instead of the block B). While using the same budget, the election is still anonymized and p is still a Condorcet winner. It follows that no new blocks, besides, perhaps, one additional arbitrarily chosen new p-block, are needed. We mention that an additional p-block might be needed when there are no p-voters (and, therefore, no p-blocks) in the input election.

The above observation suggests the following simple algorithm. We begin by guessing whether we need a new p-block. Then, for each voter, we guess whether (1) it will stay the same, (2) it will move to some other original block (out of the possible $n - 1$ other original blocks), or (3) it will move to the new p-block (if we guessed that such a new block p-block exists).

Correctness follows by the observation above and by the brute-force nature of the algorithm. Fixed-parameter tractability follows since we guess for each voter (out of the n original voters), where it will end up (out of n or $n + 1$ possibilities), resulting in running time $O(m \cdot n^n)$. □

We move on to consider the number m of alternatives. Similarly to numerous other problems in computational social choice, all variants of EA considered in this paper are FPT with respect to this parameter. This follows by applying the celebrated result of Lenstra [18] after formulating the problem as an integer linear program where the number of variables is upper-bounded by a function dependent only on the number m of alternatives.

Theorem 5. *For* $\mathcal{R} \in \{Plurality, Condorcet\}$ *and* $d \in \{Discrete, Swap\}$, $\mathcal{R}$-d-EA *is* FPT *wrt. the number* m *of alternatives.*

Proof. The crucial observation is that the number of different preference orders is $m!$, therefore upper-bounded by a function depending only on the parameter, m. We enumerate the set of $m!$ different preference orders and create a variable $x_{i,j}$, for each $i, j \in [m!]$, with the intended meaning that $x_{i,j}$ will represent the number of voters with preference order i in the input and preference order j in the solution.

We add a budget constraint:

$$\sum_{i,j \in [m!]} x_{i,j} \cdot d(i,j) \leq s.$$

(Note that we can precompute all the distances, in polynomial time.)

For preference order i, we denote the number of voters with preference order i in the input by start_i and the number of voters with preference order i in the solution by end_i. For each i, it holds that:

$$\text{end}_i = \text{start}_i + \sum_{j \in [m!]} x_{j,i} - \sum_{j \in [m!]} x_{i,j}.$$

We guess a set $Z \subseteq [m!]$ of preference orders with the intent that these will be the preference orders that will be present in the solution. For each preference order $i \in Z$ we add a k-anonymity constraint, and, similarly, for each preference order $i \notin Z$ we require that its end_i will be 0, as follows:

$$\forall i \in Z : \text{end}_i \geq k$$
$$\forall i \notin Z : \text{end}_i = 0$$

Differently for the Plurality rule and the Condorcet rule, we add more constraints to make sure that the winner will not change, as follows.

For the Plurality rule, we guess the highest score z in the solution (that is, the winning score), and we check that the initial winner p gets exactly z points, by adding the following constraint:

$$\sum_{\{i \,:\, p \text{ is at the first position of } i\}} \text{end}_i = z.$$

Similarly, for each non-winning alternative $c \neq p$, we add a non-winning constraint:

$$\sum_{\{i \,:\, c \text{ is at the first position of } i\}} \text{end}_i \leq z.$$

For the Condorcet rule, for the initial Condorcet winner p (if it exists), we check that he indeed beats all other alternatives in the solution:

$$\forall c \neq p : \sum_{i \in [m!] \text{ and } p \succ_i c} \text{end}_i \geq \sum_{i \in [m!] \text{ and } c \succ_i p} \text{end}_i.$$

Since the number of variables and the number of constraints is upper-bounded by the parameter, fixed-parameter tractability follows by applying the result of Lenstra [18].   □

4 Conclusion

Motivated by privacy issues when publishing election-related data, we initiated the study of k-anonymizing preference orders, investigating its computational complexity for the Plurality rule and the Condorcet rule, while considering the Discrete distance and the Swap distance. We showed a wide diversity of the (parameterized) complexity of EA, with respect to several natural parameters.

There are numerous opportunities for future research, some of which we briefly discuss next. Most immediately, there are still some open questions, as Table 1 suggests, and one might also consider other parameterizations as well as approximation algorithms for coping with the NP-hard cases. It is also natural to extend this line of research to other voting rules (for example, can the polynomial-time algorithm presented in Theorem 1 be extended to the Borda rule? what happens for Approval voting? what happens for multi-winner rules?). Similar in spirit, it is natural to consider other distances besides the Discrete distance and the Swap distance. While we considered a *minmax* approach, as we compute the distance between two elections as the sum of distances between their voters (and allow to permute the voters), it might also be interesting to study a *minmax* approach, where, roughly speaking, we would not allow any individual voter to change its vote by too much. This could be of particular interest as it might preserve the original election more closely. On a similar note, it is worth studying how to preserve more properties of the original election; while we only require to preserve the winner, one might require to preserve the full relative ranking of the alternatives (that is, fixing some scoring rule, and considering two alternatives c' and c'' such that c' is achieving a greater score than c'' in the original election, we might require c' to achieve greater score than c'' in the resulting election as well). Somehow related, one might consider some stronger notions of k-anonymity, for example l-diversity, which might be interesting to explore in the context of elections. Finally, one might experiment with real-world preference-data (for example, from PrefLib [21]) to evaluate the quality of the algorithms presented in this paper.

References

1. Ashur, T., Dunkelman, O.: Poster: on the anonymity of Israel's general elections. In: Proceedings of the 2013 ACM SIGSAC Conference on Computer & Communications Security, pp. 1399–1402. ACM (2013)
2. Biedl, T., Brandenburg, F.J., Deng, X.: On the complexity of crossings in permutations. Discrete Math. **309**(7), 1813–1823 (2009)
3. Brandt, F., Conitzer, V., Endriss, U.: Computational social choice. In: Weiss, G. (ed.) Multiagent Systems, 2nd edn., pp. 213–283. MIT Press, Cambridge (2013)
4. Bredereck, R., Nichterlein, A., Niedermeier, R.: Pattern-guided k-anonymity. Algorithms **6**(4), 678–701 (2013)
5. Bredereck, R., Froese, V., Hartung, S., Nichterlein, A., Niedermeier, R., Talmon, N.: The complexity of degree anonymization by vertex addition. In: Gu, Q., Hell, P., Yang, B. (eds.) AAIM 2014. LNCS, vol. 8546, pp. 44–55. Springer, Heidelberg (2014)
6. Bredereck, R., Nichterlein, A., Niedermeier, R., Philip, G.: The effect of homogeneity on the computational complexity of combinatorial data anonymization. Data Min. Knowl. Disc. **28**(1), 65–91 (2014b)
7. Chaum, D.L.: Untraceable electronic mail, return addresses, and digital pseudonyms. Commun. ACM **24**(2), 84–90 (1981)
8. Chen, Y., Chong, S., Kash, I.A., Moran, T., Vadhan, S.: Truthful mechanisms for agents that value privacy. In: Proceedings of the Fourteenth ACM Conference on Electronic Commerce, pp. 215–232. ACM (2013)
9. Clifton, C., Tassa, T.: On syntactic anonymity and differential privacy. In: 2013 IEEE 29th International Conference on Data Engineering Workshops (ICDEW), pp. 88–93. IEEE (2013)
10. Cuvelier, É., Pereira, O., Peters, T.: Election verifiability or ballot privacy: do we need to choose? In: Crampton, J., Jajodia, S., Mayes, K. (eds.) ESORICS 2013. LNCS, vol. 8134, pp. 481–498. Springer, Heidelberg (2013)
11. Downey, R.G., Fellows, M.R.: Fundamentals of Parameterized Complexity. Springer, Heidelberg (2013)
12. Dwork, C., Roth, A.: The algorithmic foundations of differential privacy. Theoret. Comput. Sci. **9**(3–4), 211–407 (2013)
13. Dwork, C., Kumar, R., Naor, M., Sivakumar, D.: Rank aggregation methods for the web. In: Proceedings of the 10th International Conference on World Wide Web, pp. 613–622. ACM (2001)
14. Elkind, E., Faliszewski, P., Slinko, A.: On the role of distances in defining voting rules. In: Proceedings of the 9th International Conference on Autonomous Agents and Multiagent Systems, vol. 1, pp. 375–382. International Foundation for Autonomous Agents and Multiagent Systems (2010)
15. Flum, J., Grohe, M.: Parameterized Complexity Theory. Springer, Heidelberg (2006)
16. Gonzalez, T.F.: Clustering to minimize the maximum intercluster distance. Theoret. Comput. Sci. **38**, 293–306 (1985)
17. Hartung, S., Nichterlein, A., Niedermeier, R., Suchý, O.: A refined complexity analysis of degree anonymization in graphs. In: Fomin, F.V., Freivalds, R., Kwiatkowska, M., Peleg, D. (eds.) ICALP 2013, Part II. LNCS, vol. 7966, pp. 594–606. Springer, Heidelberg (2013)
18. Lenstra, H.W.: Integer programming with a fixed number of variables. Math. Oper. Res. **8**, 538–548 (1983)

19. Liu, K., Terzi, E.: Towards identity anonymization on graphs. In: Proceedings of the 2008 ACM SIGMOD International Conference on Management of Data, pp. 93–106. ACM (2008)
20. Machanavajjhala, A., Kifer, D., Gehrke, J., Venkitasubramaniam, M.: l-diversity: privacy beyond k-anonymity. ACM Trans. Knowl. Disc. Data (TKDD) 1(1), 3 (2007)
21. Mattei, N., Walsh, T.: PREFLIB: a library for preferences http://www.preflib.org. In: Perny, P., Pirlot, M., Tsoukiàs, A. (eds.) ADT 2013. LNCS, vol. 8176, pp. 259–270. Springer, Heidelberg (2013)
22. Meyerson, A., Williams, R.: On the complexity of optimal k-anonymity. In: Proceedings of the Twenty-third ACM SIGMOD-SIGACT-SIGART Symposium on Principles of Database Systems, pp. 223–228. ACM (2004)
23. Niedermeier, R.: Invitation to Fixed-Parameter Algorithms. Oxford University Press, Oxford (2006)
24. Nurmi, H., Salomaa, A., Santean, L.: Secret ballot elections in computer networks. Comput. Secur. 10(6), 553–560 (1991)
25. Sweeney, L.: k-anonymity: a model for protecting privacy. Int. J. Uncertainty, Fuzziness Knowl. Based Syst. 10(05), 557–570 (2002)

On Equivalences, Metrics, and Polynomial Time

Alberto Cappai and Ugo Dal Lago[✉]

Università di Bologna and INRIA, Bologna, Italy
{alberto.cappai2,ugo.dallago}@unibo.it

Abstract. Interactive behaviors are ubiquitous in modern cryptography, but are also present in λ-calculi, in the form of higher-order constructions. Traditionally, however, typed λ-calculi simply do not fit well into cryptography, being both deterministic and too powerful as for the complexity of functions they can express. We study interaction in a λ-calculus for probabilistic polynomial time computable functions. In particular, we show how notions of context equivalence and context metric can both be characterized by way of traces when defined on linear contexts. We then give evidence on how this can be turned into a proof methodology for computational indistinguishability, a key notion in modern cryptography. We also hint at what happens if a more general notion of a context is used.

1 Introduction

Modern cryptography [14] is centered around the idea that security of cryptographic constructions needs to be defined precisely and, in particular, that crucial aspects are *how* an adversary interacts with the construction, and *when* he wins this game. The former is usually specified by way of an *experiment*, while the latter is often formulated stipulating that the probability of a favorable result for the adversary needs to be small, where being "small" usually means being *negligible* in a security parameter. This framework would however be vacuous if the adversary had access to an unlimited amount of resources, or if it were deterministic. As a consequence the adversary is usually assumed to work within probabilistic polynomial time (PPT in the following), this way giving rise to a robust definition. Summing up, there are three key concepts here, namely *interaction*, *probability* and *complexity*. Security as formulated above can often be spelled out semantically as the so-called *computational indistinguishability* between two distributions, the first one being the one produced by the construction and the second one modeling an idealized construction or a genuinely random object.

Typed λ-calculi as traditionally conceived, do not fit well into this picture. Higher-order types clearly allow a certain degree of interaction, but probability and complexity are usually absent: reduction is deterministic (or at least confluent), while the expressive power of λ-calculi tends to be very high. This picture

---

This work is partially supported by the ANR project 12IS02001 PACE.

A. Kosowski and I. Walukiewicz (Eds.): FCT 2015, LNCS 9210, pp. 311–323, 2015.
DOI: 10.1007/978-3-319-22177-9_24

has somehow changed in the last ten years: there have been some successful attempts at giving probabilistic λ-calculi whose representable functions coincide with the ones which can be computed by PPT algorithms [5,15,17]. These calculi invariably took the form of restrictions on Gödel's T, endowed with a form of binary probabilistic choice. All this has been facilitated by implicit computational complexity, which offers the right idioms to start from [11,12], themselves based on linearity and ramification. The emphasis in all these works were either the characterization of probabilistic complexity classes [5], or more often security [16,17]: one could see λ-calculi as a way to specify cryptographic constructions and adversaries for them. The crucial idea here is that computational indistinguishability can be formulated as a form of context equivalence. The real challenge, however, is whether all this can be characterized by handier notions, which would alleviate the inherently difficult task of dealing with all contexts when proving two terms to be equivalent.

The literature offers many proposals going precisely in this direction: this includes logical relations, context lemmas, or coinductive techniques. In applicative bisimulation [1], as an example, terms are modeled as interactive objects. This way, one focuses on how the interpreted program interacts with its environment, rather than on its internal evolution. None of them have so far been applied to calculi capturing probabilistic polynomial time, and relatively few among them handle probabilistic behavior.

In this paper, we study notions of equivalence and distance in one of these λ-calculi, called RSLR [5]. More precisely:

- After having briefly introduced RSLR and studied its basic metatheoretical properties (Sect. 2), we define *linear context equivalence*. We then show how the role of contexts can be made to play by *traces*. Finally, a coinductive notion of equivalence in the style of Abramsky's bisimulation is introduced. We also hint at how all this can be extended to metrics. This can be found in Sect. 4.

- We then introduce a notion of *parametrized context equivalence* for RSLR terms, showing that it coincides with computational indistinguishability when the compared programs are of base type. We then turn our attention to the problem of characterizing the obtained notion of equivalence by way of linear tests, giving a positive answer to that by way of a notion of parametrized trace metric. A brief discussion about the role of linear contexts in cryptography is also given. All this is in Sect. 5.

An extended version of this paper with more details is available [2].

2   Characterizing Probabilistic Polynomial Time

In this section we introduce RSLR [5], a λ-calculus for probabilistic polynomial time computation, obtained by extending Hofmann's SLR [12] with an operator for binary probabilistic choice. Compared to other presentations of the same calculus, we consider a call-by-value reduction and elide nonlinear function spaces and pairs. This has the advantage of making the whole theory less baroque, without any fundamental loss in expressiveness (see Sect. 5.1 below).

First of all, *types* are defined as follows:

$$A ::= \mathsf{Str} \mid \blacksquare A \to A \mid \Box A \to A.$$

The expression Str serves to type strings, and is the only *base* type. $\blacksquare A \to B$ is the type of functions (from A to B) which can be evaluated in constant time, while for $\Box A \to B$ the running time can be any polynomial. *Aspects* are the elements of $\{\Box, \blacksquare\}$ and are indeed fundamental to ensure polytime soundness. We denote them with metavariables like a or b. We define a partial order $<:$ between aspects simply as the relation $\{(\Box, \Box), (\Box, \blacksquare), (\blacksquare, \blacksquare)\}$, and subtyping by the rules below:

$$\frac{}{A <: A} \qquad \frac{A <: B \qquad B <: C}{A <: C} \qquad \frac{B <: A \qquad C <: D \qquad a <: b}{aA \to C <: bB \to D}$$

The syntactical categories of *terms* and *values* are defined by the following grammar:

$$t ::= x \mid v \mid 0(t) \mid 1(t) \mid \mathsf{tail}(t) \mid tt$$

$$\mid \mathsf{case}_A(t, t, t, t) \mid \mathsf{rec}_A(t, t, t, t) \mid \mathsf{rand};$$

$$v ::= \underline{m} \mid \lambda x : aA.t;$$

where m ranges over the set $\{0, 1\}^*$ of finite, binary strings, while x ranges over a denumerable set of variables X. We write T, V for the sets of terms and values, respectively. The operators 0 and 1 are constructors for binary strings, while tail is a destructor. The only nonstandard constant is rand, which returns $\underline{0}$ or $\underline{1}$, each with probability $\frac{1}{2}$, thus modeling uniform binary choice. The terms $\mathsf{case}_A(t, t_0, t_1, t_\epsilon)$ and $\mathsf{rec}_A(t, t_0, t_1, t_\epsilon)$ are terms for case distinction and recursion, in which first argument specifies the term (of base type) which guides the process. The expression $\underline{\epsilon}$ stands for the empty string and we set $\mathsf{tail}(\underline{\epsilon}) \to \underline{\epsilon}$.

As usual, a *typing context* Γ is a finite set of assignments of an aspect and a type to a variable, where as usual any variable occurs *at most once*. Any such assignment is indicated with $x : aA$. The expression Γ, Δ stands for the union of the two typing contexts Γ and Δ, which are assumed to be disjoint. The union Γ, Δ is indicated with $\Gamma; \Delta$ whenever we want to insist on Γ to only involve the base type Str. *Typing judgments* are in the form $\Gamma \vdash t : A$. Typing rules are in Fig. 1. The expression T_Γ^A (respectively, V_Γ^A) stands for the set of terms (respectively, values) of type A under the typing context Γ. Please observe how the type system we have just introduced enforces variables of higher-order type to occur free at most once and outside the scope of a recursion. Moreover, the type of terms which serve as step-functions in a recursion are assumed to be $\Box$-free, and this is precisely what allow this calculus to characterize polytime functions.

$$\frac{x : \mathsf{aA} \in \Gamma}{\Gamma \vdash x : \mathsf{A}} \qquad \frac{}{\Gamma \vdash \underline{m} : \mathsf{Str}} \qquad \frac{\Gamma \vdash t : \mathsf{Str}}{\Gamma \vdash \mathsf{0}(t), \mathsf{1}(t), \mathsf{tail}(t) : \mathsf{Str}} \qquad \frac{}{\vdash \mathsf{rand} : \mathsf{Str}}$$

$$\frac{\begin{array}{cc} \Gamma; \Delta_1 \vdash t : \mathsf{Str} & \Gamma; \Delta_3 \vdash t_1 : \mathsf{A} \\ \Gamma; \Delta_2 \vdash t_0 : \mathsf{A} & \Gamma; \Delta_4 \vdash t_\epsilon : \mathsf{A} \end{array}}{\Gamma; \Delta_1, \Delta_2, \Delta_3, \Delta_4 \vdash \mathsf{case}_{\mathsf{A}}(t, t_0, t_1, t_\epsilon) : \mathsf{A}} \qquad \frac{\Gamma, x : \mathsf{aA} \vdash t : \mathsf{B}}{\Gamma \vdash \lambda x : \mathsf{aA}.t : \mathsf{aA} \to \mathsf{B}} \qquad \frac{\Gamma \vdash t : \mathsf{A} \qquad \mathsf{A} <: \mathsf{B}}{\Gamma \vdash t : \mathsf{B}}$$

$$\frac{\begin{array}{cc} \Gamma_1; \Delta_1 \vdash t : \mathsf{Str} & \Gamma_1, \Gamma_2, \Gamma_3; \Delta_2 \vdash t_\epsilon : \mathsf{A} \\ \Gamma_1, \Gamma_2 \vdash t_0 : \square\mathsf{Str} \to \blacksquare\mathsf{A} \to \mathsf{A} & \Gamma_1, \Delta_1 <: \square \\ \Gamma_1, \Gamma_3 \vdash t_1 : \square\mathsf{Str} \to \blacksquare\mathsf{A} \to \mathsf{A} & \mathsf{A} \text{ is } \square\text{-free} \end{array}}{\Gamma_1, \Gamma_2, \Gamma_3; \Delta_1, \Delta_2 \vdash \mathsf{rec}_{\mathsf{A}}(t, t_0, t_1, t_\epsilon) : \mathsf{A}} \qquad \frac{\begin{array}{c} \Gamma; \Delta_1 \vdash t : \mathsf{aA} \to \mathsf{B} \\ \Gamma; \Delta_2 \vdash s : \mathsf{A} \qquad \Gamma, \Delta_2 <: \mathsf{a} \end{array}}{\Gamma; \Delta_1, \Delta_2 \vdash ts : \mathsf{B}}$$

Fig. 1. RSLR's typing rules

The operational semantics of RSLR is of course probabilistic: any closed term t evaluates not to a single value but to a *value distribution*, i.e., a function $\mathcal{D} : \mathrm{V} \to \mathbb{R}$ such that $\sum_{v \in \mathrm{V}} \mathcal{D}(v) = 1$. Judgments expressing this fact are in the form $t \Downarrow \mathcal{D}$, and are derived through a formal system (see [2] for more details). In the rest of this paper, we use some standard notation on distributions: the expression $\{v_1^{\alpha_1}, \ldots, v_n^{\alpha_n}\}$ stands for the distribution assigning probability α_i to v_i (for every $1 \leqslant i \leqslant n$), while the support of a distribution $\mathcal{D}$ is indicated with $\mathsf{S}(\mathcal{D})$. Given a set X, $\mathbb{P}_X$ is the set of all distributions over X. Noticeably:

Lemma 1. *For every term $t \in \mathrm{T}_{\varnothing}^{A}$ there is a unique distribution $\mathcal{D}$ such that $t \Downarrow \mathcal{D}$, which we denote as $[\![t]\!]$. Moreover, If $v \in \mathsf{S}(\mathcal{D})$, then $v \in \mathrm{V}_{\varnothing}^{A}$.*

This tells us both that $\Downarrow$ can be seen as a *function*, and that subject reduction holds.

A *probabilistic function* on $\{0,1\}^*$ is a function F from $\{0,1\}^*$ to $\mathbb{P}_{\{0,1\}^*}$. A term $t \in \mathrm{T}_{\varnothing}^{\mathsf{aStr} \to \mathsf{Str}}$ is said to *compute* F iff for every string $\underline{m} \in \{0,1\}^*$ it holds that $t\underline{m} \Downarrow \mathcal{D}$ where $\mathcal{D}(\underline{n}) = F(\underline{m})(\underline{n})$ for every $\underline{n} \in \{0,1\}^*$. What makes RSLR very interesting, however, is that it precisely captures those probabilistic functions which can be computed in polynomial time (see, e.g., [6] for a definition):

Theorem 1 (Polytime Completeness). *The set of probabilistic functions which can be computed by* RSLR *terms coincides with the polytime computable ones.*

This result is well-known [5,17], and can be proved in various ways, e.g. combinatorially or categorically.

3 Equivalences

Intuitively, we can say that two programs are equivalent if no one can distinguish them by observing their external, visible, behavior. A formalization of this intuition usually takes the form of context equivalence. A *context* is a term in which the *hole* $[\cdot]$ occurs at most once. Formally, contexts are defined by the following grammar:

$$C ::= t \mid [\cdot] \mid \lambda x.C \mid Ct \mid tC \mid \quad \mid 0(C) \mid 1(C) \mid \mathsf{tail}(C)$$

$$\mid \mathsf{case}_\mathsf{A}(C,t,t,t) \mid \mathsf{case}_\mathsf{A}(t,C,C,C) \mid \mathsf{rec}_\mathsf{A}(C,t,t,t).$$

What the above definition already tells us is that our emphasis in this paper will be on *linear* contexts, which are contexts whose holes lie outside the scope of any recursion operator. Given a term t we define $C[t]$ as the term obtained by substituting the occurrences of $[\cdot]$ in C (if any) with t. We only consider non-binding contexts here, i.e. contexts are meant to be filled with closed terms (this can be justified formally [2]). In other words, the type system from Sect. 2 can be turned into one for contexts whose judgments take the form $\Gamma \vdash C[\vdash \mathsf{A}] : \mathsf{B}$, which means that for every closed term t of type A, it holds that $\Gamma \vdash C[t] : \mathsf{B}$. See [2] for more details. Now that the notion of a context has been properly defined, one can finally give the central notion of equivalence in this paper.

Definition 1 (Context Equivalence). *Given two terms t, s such that $\vdash t, s :$ A, we say that t and s are* context equivalent *iff for every context C such that $\vdash C[\vdash \mathsf{A}] : Str$ we have that $[\![C[t]]\!](\underline{\epsilon}) = [\![C[s]]\!](\underline{\epsilon})$.*

The way we defined it means that context equivalence is a family of relations $\{\equiv_\mathsf{A}\}_{\mathsf{A} \in \mathsf{A}}$ indexed by types, which we denote as $\equiv$. Context equivalence is easily proved to be a congruence, i.e., a compatible equivalence relation.

3.1 Trace Equivalence

In this section we introduce a notion of *trace equivalence* for RSLR, and we show that it characterizes context equivalence.

We define a *trace* as a sequence of actions $l_1 \cdot l_2 \cdot \ldots \cdot l_n$ such that $l_i \in \{\mathsf{pass}(v), \mathsf{view}(\underline{m}) \mid v \in \mathsf{V}, \underline{m} \in \mathsf{V}^{\mathsf{Str}}\}$. Traces are indicated with metavariables like T, S. The *compatibility* of a trace T with a type A is defined inductively on the structure of A. If $\mathsf{A} = \mathsf{Str}$ then the only trace compatible with A is $\mathsf{T} = \mathsf{view}(\underline{m})$, with $\underline{m} \in \mathsf{V}^{\mathsf{Str}}$, otherwise, if $\mathsf{A} = a\mathsf{B} \to \mathsf{C}$ then traces compatible with A are in the form $\mathsf{T} = \mathsf{pass}(v) \cdot \mathsf{S}$ with $v \in \mathsf{V}^\mathsf{B}$ and S is itself compatible with C. With a slight abuse of notation, we often assume traces to be compatible with the underlying type.

Due to the probabilistic nature of our calculus, it is convenient to work with *term distributions*, i.e., distributions whose support is the set of closed terms of a certain type A, instead of plain terms. We denote term distributions with metavariables like $\mathcal{T}$ or $\mathcal{S}$. The effect traces have to distributions can be formalized by giving some binary relations:

- First of all, we need a binary relation on term distributions, called $\Rightarrow$. Intuitively, $\mathcal{T} \Rightarrow \mathcal{S}$ iff $\mathcal{T}$ evolves to $\mathcal{S}$ by performing internal moves, only. Furthermore, we use $\to$ to indicate a single internal move.

- We also need a binary relation $\Rightarrow^{\cdot}$ between term distributions, which is however labeled by a trace, and which models internal *and* external reduction.
- Finally, we need a labeled relation $\mapsto^{\cdot}$ between distributions and *real numbers*, which captures the probability that distributions accept traces.

The three relations are defined inductively by the rules in Fig. 2. The following gives basic, easy, results about the relations we have introduced:

$$\frac{}{\mathcal{T} \Rightarrow^{\epsilon} \mathcal{T}} \qquad \frac{\mathcal{T} \Rightarrow^{S} \{(\lambda x.t_i)^{p_i}\}}{\mathcal{T} \Rightarrow^{S \cdot \mathsf{pass}(v)} \{(t_i\{v/x\})^{p_i}\}} \qquad \frac{\mathcal{T} \Rightarrow^{S} \mathcal{S} \quad \mathcal{S} \Rightarrow \mathcal{U}}{\mathcal{T} \Rightarrow^{S} \mathcal{U}}$$

$$\frac{\mathcal{T} \Rightarrow^{S} \{(m_i)^{p_i}\}}{\mathcal{T} \mapsto^{S \cdot \mathsf{view}(\underline{m})} \sum_{m_i = \underline{m}} p_i} \qquad \frac{t \to \{(t_i)^{p_i}\}}{\mathcal{T} + \{(t)^p\} \Rightarrow \mathcal{T} + \{(t_i)^{p \cdot p_i}\}}$$

Fig. 2. Term distribution small-step rules

Lemma 2. *Let $\mathcal{T}$ be a term distribution for the type A. Then, there is a unique value distribution $\mathcal{D}$ such that $\mathcal{T} \Rightarrow^* \mathcal{D}$. As a consequence, for every trace T compatible for A there is a unique real number p such that $\mathcal{T} \mapsto^{\mathsf{T}} p$. This real number is denoted as $\Pr(\mathcal{T}, \mathsf{T})$.*

We are now ready to define what we mean by trace equivalence:

Definition 2. *Given two term distributions $\mathcal{T}, \mathcal{S}$ we say that they are* trace equivalent *(and we write $\mathcal{T} \simeq^{\mathsf{T}} \mathcal{S}$) if, for all traces T it holds that $\Pr(\mathcal{T}, \mathsf{T}) = \Pr(\mathcal{S}, \mathsf{T})$. In particular, then, two terms t, s are trace equivalent when $\{t^1\} \simeq^{\mathsf{T}} \{s^1\}$ and we write $t \simeq^{\mathsf{T}} s$ in that case.*

It is easy to prove that trace equivalence is an equivalence relation. The next step, then, is to prove that trace equivalence is compatible, thus paving the way to a proof of soundness w.r.t. context equivalence. Unfortunately, the direct proof of compatibility (i.e., an induction on the structure of contexts) simply does not work: the way the operational semantics is specified makes it impossible to track how a term behaves *in a context*. Following [7], we proceed by considering a refined semantics, defined not on terms but on pairs whose first component is a context and whose second component is a term distribution. Formally, a *context pair* has the form $(C, \mathcal{T})$, where C is a context and $\mathcal{T}$ is a term distribution. A *(context) pair distribution* is a distribution over context pairs. Such a pair distribution $\mathcal{P} = \{(C_i, \mathcal{T}_i)^{p_i}\}$ is said to be *normal* if for all i ad for all t in the support of $\mathcal{T}_i$ we have that $C_i[t]$ is a value. Similarly to terms, the internal and external evolution of traces can be defined by way of relations $\to$, $\Rightarrow$, $\Rightarrow^{\cdot}$ and $\mapsto^{\cdot}$ (see [2] for more details).

The following tells us that working with context pairs is the same as working with terms as far as traces are concerned:

Lemma 3. *Suppose given a context C, a term distribution $\mathcal{T}$, and a trace S. Then if $(C, \mathcal{T}) \Rightarrow^S \{(C_i, \mathcal{T}_i)^{p_i}\}$ then $C[\mathcal{T}] \Rightarrow^S \{(C_i[\mathcal{T}_i])^{p_i}\}$. Moreover, if $(C, \mathcal{T}) \mapsto^S p$, then $\mathrm{Pr}(C[\mathcal{T}], S) = p$.*

But how could we exploit context pairs for our purposes? The key idea can be informally explained as follows: there is a notion of "relatedness" for pair distributions which not only is stricter than trace equivalence, but can be proved to be preserved along reduction, even when interaction with the environment is taken into account.

Definition 3 (Trace Relatedness). *Let $\mathcal{P}, \mathcal{Q}$ be two pair distributions. We say that they are trace-related, and we write $\mathcal{P} \triangledown \mathcal{Q}$ if there exist families $\{C_i\}_{i \in I}$, $\{\mathcal{T}_i\}_{i \in I}$, $\{\mathcal{S}_i\}_{i \in I}$, and $\{p_i\}_{i \in I}$ such that $\mathcal{P} = \{(C_i, \mathcal{T}_i)^{p_i}\}, \mathcal{Q} = \{(C_i, \mathcal{S}_i)^{p_i}\}$ and for every $i \in I$, it holds that $\mathcal{T}_i \simeq^T \mathcal{S}_i$.*

Lemma 4 (Bisimulation, Externally). *Given two pair distributions $\mathcal{P}, \mathcal{Q}$ with $\mathcal{P} \triangledown \mathcal{Q}$, then for all traces S we have:*

1. *If $\mathcal{P} \Rightarrow^S \mathcal{M}$, with $\mathcal{M}$ normal distribution, then $\mathcal{Q} \Rightarrow^S \mathcal{N}$, where $\mathcal{M} \triangledown \mathcal{N}$ and $\mathcal{N}$ is a normal distribution too.*
2. *If $\mathcal{P} \mapsto^S p$ then $\mathcal{Q} \mapsto^S p$.*

We are now in a position to prove the main result of this section:

Theorem 2 (Full Abstraction). *Context equivalence and trace equivalence coincide.*

3.2 Some Words on Applicative Bisimulation

As we already discussed, the quantification over all contexts makes the task of proving two terms to be context equivalent burdensome, even if we restrict to linear contexts. And we cannot say that trace equivalence really overcomes this problem: there is a universal quantification anyway, even if contexts are replaced by objects (i.e. traces) having a simpler structure. It is thus natural to look for other techniques. The interactive view provided by traces suggests the possibility to go for coinductive techniques akin to Abramsky's applicative bisimulation, which has already been shown to be adaptable to probabilistic λ-calculi [3,4].

One could define (see [2] for more details) a notion of applicative bisimulation for RSLR by giving a labeled Markov chain whose states are closed terms and whose labels are either eval, which models evaluation, pass(v), which models parameter passing, and view($\underline{m}$), which models testing for equality with the string $\underline{m}$. With some effort, one can prove that a greatest applicative bisimulation exists, and that it consists of the union (at any type) of all bisimulation relations. This is denoted as $\sim$ and said to be (applicative) *bisimilarity*. One can then generalize $\sim$ to a relation $\sim_o$ on open terms by the usual open extension (see [2] for more details).

One way to show that bisimilarity is included in context equivalence consists in proving that $\sim_o$ is a congruence; to reach this goal one can go through the *Howe's method* [13], which works well but requires some care (see [2] for more details). Is there any hope to get full abstraction? The answer is negative: applicative bisimilarity is too strong to match context equivalence. A counterexample to that can be built easily following the analogous one from [4].

This, however, is not the end of the story on coinductive methodologies for context equivalence in RSLR. A different route, suggested by trace equivalence, consists in taking the naturally definable (deterministic) labeled transition system of term *distributions* and ordinary bisimilarity over it. What one obtains this way is a precise characterization of context equivalence. There is a price to pay however, since one is forced to reason on distributions rather than terms. For more details, see [2].

4   From Equivalences to Metrics

The notion of observation on top of which context equivalence is defined is the probability of evaluating to the empty string, and is thus quantitative in nature. This suggests the possibility of generalizing context equivalence into a notion of *distance* between terms:

Definition 4 (Context Distance). *For every type* A, *we define* $\delta_A^C : T_\varnothing^A \times T_\varnothing^A \to \mathbb{R}_{[0,1]}$ *as* $\delta_A^C(t,s) = \sup_{\vdash C[\vdash A]:Str} |[\![C[t]]\!](\epsilon) - [\![C[s]]\!](\epsilon)|$.

For every type A, the function δ_A^C is a *pseudometric*[1] on the space of closed terms. Obviously, $\delta_A^C(t,s) = 0$ iff t and s are context equivalent. As such, then, the context distance can be seen as a natural generalization of context equivalence, where a real number between 0 and 1 is assigned to each pair of terms and is meant to be a measure of how different the two terms are in terms of their behavior. δ^C refers to the family $\{\delta_A^C\}_{A \in A}$.

One may wonder whether δ^C, as we have defined it, can somehow be characterized by a trace-based notion of metric, similarly to what have been done in Sect. 3 for equivalences. First of all, let us *define* such a distance. Actually, the very notion of a trace needs to be slightly modified: in the action view($\cdot$), instead of observing a *single* string $\underline{m}$, we need to be able to observe the action on a *finite string set* M. The probability of accepting a trace in a term will be modified accordingly: $\Pr(t, \text{view}(M)) = [\![t]\!](M)$.

Definition 5 (Trace Distance). *For every type* A, *we define* $\delta_A^T : T_\varnothing^A \times T_\varnothing^A \to \mathbb{R}_{[0,1]}$ *as* $\delta_A^T(t,s) = \sup_T |\Pr(t,T) - \Pr(s,T)|$.

It is easy to realize that if $t \simeq^T s$ then $\delta_A^T(t,s) = 0$. Moreover, δ_A^T is itself a pseudometric. As usual, δ^T denotes the family $\{\delta_A^T\}_{A \in A}$.

But how should we proceed if we want to prove the two just introduced notions of distance to coincide? Could we proceed more or less like in Sect. 3.1?

---

[1] Following the literature on the subject, this stands for any function $\delta : A \times A \to \mathbb{R}$ such that $\delta(x,y) = \delta(y,x)$, $\delta(x,x) = 0$ and $\delta(x,y) + \delta(y,z) \geq \delta(x,z)$.

The answer is positive, but of course something can be found which plays the role of compatibility, since the latter is a property of *equivalences* and not of *metrics*. The way out is relatively simple: what corresponds to compatibility in metrics is non-expansiveness (see, e.g., [8]). A notion of distance δ is said to be *non-expansive* iff for every pair of terms t, s and for every context C, it holds that $\delta(C[t], C[s]) \leq \delta(t, s)$.

The proof of non-expansiveness for δ^{T} reflects the proof of compatibility for trace equivalence. What needs to be adapted, of course, is the notion of trace-relatedness, which should be made parametric on a real number ε, standing for the distance between the two context pair distributions at hand. Once we have non-expansiveness (see [2] for more details), full abstraction is within reach. As a corollary of non-expansiveness, one gets that:

Theorem 3 (Full Abstraction). *Context distance and trace distance coincide.*

One may wonder whether a coinductive notion of distance, sort of a metric analogue to applicative bisimilarity, can be defined. The answer is positive [8]. It however suffers from the same problems applicative bisimilarity has: in particular, it is not fully abstract.

5   Computational Indistinguishability

In this section we show how our notions of equivalence and distance relate to computational indistinguishability (CI in the following), a key notion in modern cryptography.

Definition 6. *Two distribution ensembles $\{D_n\}_{n \in \mathbb{N}}$ and $\{E_n\}_{n \in \mathbb{N}}$ (where both D_n and E_n are distributions on binary strings) are said to be* computationally indistinguishable *iff for every PPT algorithm $\mathcal{A}$ the following quantity is a negligible[2] function of $n \in \mathbb{N}$: $|\mathrm{Pr}_{x \leftarrow D_n}(\mathcal{A}(x, 1^n) = \epsilon) - \mathrm{Pr}_{x \leftarrow E_n}(\mathcal{A}(x, 1^n) = \epsilon)|$.*

It is a well-known fact in cryptography that in the definition above, $\mathcal{A}$ can be assumed to sample from x just *once* without altering the definition itself, provided the two involved ensembles are efficiently computable ([9], Theorem 3.2.6, page 108). This is in contrast to the case of arbitrary ensembles [10].

The careful reader should have already spotted the similarity between CI and the notion of context distance as given in Sect. 4. There are some key differences, though:

1. While context distance is an *absolute* notion of distance, CI depends on a parameter n, the so-called *security parameter*.
2. In computational indistinguishability, one can compare distributions over *strings*, while the context distance can evaluate how far terms of *arbitrary* types are.

---

[2] A negligible function is a function which tends to 0 faster than any inverse polynomial (see [9] for more details).

The discrepancy Point 1 puts in evidence, however, can be easily overcome by turning the context distance into something slightly more parametric.

Definition 7 (Parametric Context Equivalence). *Given two terms t, s such that $\vdash t, s : aStr \to A$, we say that t and s are* parametrically context equivalent *iff for every context C such that $\vdash C[\vdash A] : Str$ we have that $|[\![C[t1^n]]\!](\underline{\epsilon}) - [\![C[s1^n]]\!](\underline{\epsilon})|$ is negligible in* $\mathbf{n}$.

This way, we have obtained a characterization of CI:

Theorem 4. *Let t, s be two terms of type $aStr \to Str$. Then t, s are parametric context equivalent iff the distribution ensembles $\{[\![t1^n]\!]\}_{n \in \mathbb{N}}$ and $\{[\![s1^n]\!]\}_{n \in \mathbb{N}}$ are computationally indistinguishable.*

Please observe that Theorem 4 only deals with terms of type $aStr \to Str$. The significance of parametric context equivalence when instantiated to terms of type $aStr \to A$, where A is a higher-order type, will be discussed in Sect. 5.1 below. How could traces capture the peculiar way parametric context equivalence treats the security parameter? First of all, observe that, in Definition 7, the security parameter is passed to the term being tested *without* any intervention from the context. The most important difference, however, is that contexts are objects which test *families of terms* rather than terms. As a consequence, the action $view(\cdot)$ does not take strings or finite sets of strings as arguments (as in equivalences or metrics), but rather *distinguishers*, namely closed RSLR terms of type $aStr \to Str$ that we denote with the metavariable D. The probability that a term t of type Str satisfies one such action $view(D)$ is $\sum_m [\![t]\!](\underline{m}) \cdot [\![Dm]\!](\underline{\epsilon})$.

A trace T is said to be *parametrically compatible* for a type $aStr \to A$ if it is compatible for A. This is the starting point for the following definition:

Definition 8. *Two terms $t, s : aStr \to A$ are* parametrically trace equivalent, *and we write $t \simeq_n^T s$, iff for every trace T which is parametrically compatible with A, there is a negligible function $negl : \mathbb{N} \to \mathbb{R}_{[0,1]}$ such that $|\Pr(t, pass(1^n) \cdot T) - \Pr(s, pass(1^n) \cdot T)| \leq negl(\mathbf{n})$.*

The fact that parametric trace equivalence and parametric context equivalence are strongly related is quite intuitive: they are obtained by altering in a very similar way two notions which are already known to coincide (by Theorem 3). Indeed:

Theorem 5. *Parametric trace equivalence and parametric context equivalence coincide.*

The right-to-left inclusion is trivial, indeed every trace can be easily emulated by a context. The other one, as usual is more difficult, and requires a careful analysis of the behavior of terms depending on parameter, when put in a context. Overall, however, the structure of the proof is similar to the one we presented in Sect. 3.1 (see [2]).

5.1   Higher-Order Computational Indistinguishability?

Theorems 4 and 5 together tell us that two terms t, s of type aStr $\rightarrow$ Str are parametrically trace equivalent iff the distributions they denote are computationally indistinguishable. But what happens if the type of the two terms t, s is in the form aStr $\rightarrow$ A where A is an *higher-order* type? What do we obtain? Actually, the literature on cryptography does not offer a precise definition of "higher-order" computational indistinguishability, so a formal comparison with parametric context equivalence is not possible, yet.

Apparently, linear contexts do not capture equivalences as traditionally employed in cryptography, already when A is the first-order type aStr $\rightarrow$ Str. A central concept in cryptography, indeed, is *pseudorandomness*, which can be spelled out for strings, giving rise to the concept of a pseudorandom generator, but also for functions, giving rise to pseudorandom functions [14]. Formally, a function $F : \{0,1\}^* \rightarrow \{0,1\}^* \rightarrow \{0,1\}^*$ is said to be a pseudorandom function iff $F(s)$ is a function which is indistinguishable from a random function from $\{0,1\}^n$ to $\{0,1\}^n$ whenever s is drawn at random from n-bit strings. Indistinguishability, again, is defined in terms of PPT algorithms having *oracle* access to $F(s)$. Now, having access to an oracle for a function is of course different than having *linear* access to it. Indeed, building a *linear* pseudorandom function is very easy: $G(s)$ is defined to be the function which returns s independently on the value of its input. G is of course not pseudorandom in the classical sense, since testing the function multiple times a distinguisher immediately sees the difference with a truly random function. On the other hand, the RSLR term t_G implementing the function G above is such that $\lambda x.t_G s$ is trace *equivalent* to a term r where:

- s is a term which produces in output $|x|$ bits drawn at random;
- r is the term $\lambda x.q$ of type aStr $\rightarrow$ bStr $\rightarrow$ Str such that q returns a random function from $|x|$-bitstrings to $|x|$-bitstrings. Strictly speaking, r cannot be an RSLR term, but it can anyway be used as an idealized construction.

But this is not the end of the story. Sometime, enforcing linear access to primitives is necessary. Consider, as an example, the two terms

$$t = \lambda n.(\lambda k.\lambda x.\lambda y.ENC(x,k)) GEN(n);$$
$$s = \lambda n.(\lambda k.\lambda x.\lambda y.ENC(y,k)) GEN(n);$$

where ENC is meant to be an encryption function and GEN is a function generating a random key. t and s should be considered equivalent whenever ENC is a secure cryptoscheme. But if ENC is secure against passive attacks (but not against active attacks), the two terms can possibly be distinguished with high probability if copying is available. The two terms can indeed be proved to be parametrically context equivalent if ENC is the cryptoscheme induced by a pseudorandom generator (see [2] for a proof and for more details).

Summing up, parametrized context equivalence coincides with CI when instantiated on base types, has some interest also on higher-order types, but is different from the kind of equivalences cryptographers use when dealing with higher-order objects (e.g. when defining pseudorandom functions). This discrepancy is mainly due to the linearity of the contexts we consider here. It seems however very hard to overcome it by just considering arbitrary nonlinear contexts instead of linear ones. Indeed, it would be hard to encode any *arbitrary* PPT distinguisher accessing an oracle by an RSLR context: those adversaries are only required to be PPT for oracles implementing certain kinds of functions (e.g. n-bits to n-bits, as in the case of pseudorandomness), while filling a RSLR context with any PPT algorithm is guaranteed to result in a PPT algorithm. This is anyway a very interesting problem, which is outside the scope of this paper, and that we are currently investigating in the context of a different, more expressive, probabilistic λ-calculus.

6   Conclusions

We believe that the main contribution of this work is the new light it sheds on the relations between computational indistinguishability, linear contexts and traces. In particular, this approach, which is implicitly used in the literature on the subject [16,17], is shown to have some limitations, but also to suggest a notion of higher-order indistinguishability which could possibly be an object of study in itself. This is indeed the main direction for future work we foresee.

References

1. Abramsky, S.: The lazy lambda calculus. In: Turner, D. (ed.) Research Topics in Functional Programming, pp. 65–117. Addison Wesley, Boston (1990)
2. Cappai, A., Dal Lago, U.: On equivalences, metrics, and polynomial time (long version) (2015) http://arxiv.org/abs/1506.03710
3. Crubillé, R., Dal Lago, U.: On probabilistic applicative bisimulation and call-by-value λ-calculi. In: Shao, Z. (ed.) ESOP 2014 (ETAPS). LNCS, vol. 8410, pp. 209–228. Springer, Heidelberg (2014)
4. Dal Lago, U., Sangiorgi, D., Alberti, M.: On coinductive equivalences for higher-order probabilistic functional programs. In: POPL (2014)
5. Dal Lago, U., Parisen Toldin, P.: A higher-order characterization of probabilistic polynomial time. In: Peña, R., van Eekelen, M., Shkaravska, O. (eds.) FOPARA 2011. LNCS, vol. 7177, pp. 1–18. Springer, Heidelberg (2012)
6. Dal Lago, U., Zuppiroli, S., Gabbrielli, M.: Probabilistic recursion theory and implicit computational complexity. Sci. Ann. Comp. Sci. **24**(2), 177–216 (2014)
7. Deng, Y., Zhang, Y.: Program equivalence in linear contexts. CoRR, abs/1106.2872 (2011)
8. Desharnais, J., Gupta, V., Jagadeesan, R., Panangaden, P.: Metrics for labeled markov systems. In: Baeten, J.C.M., Mauw, S. (eds.) CONCUR 1999. LNCS, vol. 1664, pp. 258–273. Springer, Heidelberg (1999)
9. Goldreich, O.: The Foundations of Cryptography. Basic Techniques, vol. 1. Cambridge University Press, New York (2001)
10. Goldreich, O., Sudan, M.: Computational indistinguishability: a sample hierarchy. In: CCC, pp. 24–33 (1998)

11. Hofmann, M.: A mixed modal/linear lambda calculus with applications to bellantoni-cook safe recursion. In: Nielsen, M. (ed.) CSL 1997. LNCS, vol. 1414, pp. 275–294. Springer, Heidelberg (1997)
12. Hofmann, M.: Safe recursion with higher types and bck-algebra. Ann. Pure Appl. Logic **104**(1–3), 113–166 (2000)
13. Howe, D.J.: Proving congruence of bisimulation in functional programming languages. Inf. Comput. **124**(2), 103–112 (1996)
14. Katz, J., Lindell, Y.: Introduction to Modern Cryptography. Chapman and Hall/CRC Press, Boca Raton (2007)
15. Mitchell, J.C., Mitchell, M., Scedrov, A.: A linguistic characterization of bounded oracle computation and probabilistic polynomial time. In: FOCS, pp. 725–733 (1998)
16. Nowak, D., Zhang, Y.: A calculus for game-based security proofs. In: Heng, S.-H., Kurosawa, K. (eds.) ProvSec 2010. LNCS, vol. 6402, pp. 35–52. Springer, Heidelberg (2010)
17. Zhang, Y.: The computational SLR: a logic for reasoning about computational indistinguishability. Mathematical Structures in Computer Science **20**(5), 951–975 (2010)

Graphs, Automata, and Dynamics

Conjunctive Visibly-Pushdown Path Queries

Martin Lange[1]([✉]) and Etienne Lozes[2]

[1] School of Electrical Engineering and Computer Science,
University of Kassel, Kassel, Germany
`martin.lange@uni-kassel.de`
[2] LSV, ENS Cachan and CNRS, Cachan, France

Abstract. We investigate an extension of conjunctive regular path queries in which path properties and path relations are defined by visibly pushdown automata. We study the problem of query evaluation for extended conjunctive visibly pushdown path queries and their subclasses, and give a complete picture of their combined and data complexity. In particular, we introduce a weaker notion called extended conjunctive reachability queries for which query evaluation has a polynomial data complexity. We also show that query containment is decidable in 2-EXPTIME for (non-extended) conjunctive visibly pushdown path queries.

1 Introduction

Querying is the central mechanism for extracting information from a knowledge or data base. Queries have therefore been extensively studied in the fields of knowledge representation and database theory. Some of the most important foundational issues regarding queries and their decision problems like query evaluation or query containment concern the decidability, computational complexity and — related to that — expressive power of querying languages.

Graph-structured data [14] and their querying problems occur in many application areas such as semi-structured data the semantic web social networks transportation networks biological networks program analysis etc [2]. A well-established logical formalism for querying graph-structured data is the one of *conjunctive regular path queries* [5,7,8]. In this formalism, queries can express conditions on paths in the data-graph by regarding them as words over the alphabet of relation names; it is then possible to ask for instance whether or not there is a path whose associated word belongs to a certain regular language. This formalism has been extended, to context-free languages [10], regular relations [4], and later to rational ones [3], yielding *extended conjunctive regular path queries*. This extension allows to express rich queries, like the existence of two paths of the same length.

The European Research Council has provided financial support under the European Community's Seventh Framework Programme (FP7/2007-2013) / ERC grant agreement no 259267.

A. Kosowski and I. Walukiewicz (Eds.): FCT 2015, LNCS 9210, pp. 327–338, 2015.
DOI: 10.1007/978-3-319-22177-9_25

In the formalism used by Barceló et al. [4], no two atoms in a non-extended query may use the same path variable. Intuitively, this makes queries easier and less expressive: they merely express *reachability* properties on graphs. We therefore suggest a new 2-dimensional nomenclature for such queries, distinguishing *path* from *reachability* queries on one hand, and *extended* from *non-extended* queries on the other.

	Path query	Reachability query
Non-extended	CPQ	CRQ
Extended	ECPQ	ECRQ

Path queries may contain multiple occurrences of path variables, reachability queries may not. Extended queries speak about relations, non-extended ones about languages. In the notation of Barceló et al.'s [4], their conjunctive path queries correspond to CRQ in our setting, and their extended conjunctive path queries are the same as ours, i.e. ECPQ.

This paper deals with issues of decidability and computational complexity of extended conjunctive path queries over *visibly pushdown languages* (ECPQ[VPL]). Visibly pushdown languages [1,13] form an interesting class that lies between the regular and context-free ones because it basically has the same closure and decidability properties as the regular ones. Our contributions regarding the decidability and complexity of query evaluation and containment are the following.

1. We show that ECPQ[VPL] query evaluation is undecidable.
2. We show that, for CPQ[VPL] and ECRQ[VPL] queries, query evaluation is P-complete w.r.t. data complexity, thus only a bit more expensive than it is for regular queries (NLOGSPACE).
3. We give upper and lower bounds for the combined complexity of query evaluation for each subclass of queries.
4. We consider the query containment problem; this problem is already undecidable for extended queries in the regular case [9]. So we focus on CRQ[VPL] queries. We show that query containment is decidable among these queries, with a complexity upper bound of 2-EXPTIME, close to the EXPSPACE-complete [5] complexity of query containment for CRQ[REG] queries [5,8].

2   Preliminaries

Let $\mathbb{N}$ denote the set of non-negative integers. As usual, Σ denotes a finite alphabet, Σ^* is the set of all finite words over Σ, ε is the empty word. We assume a fixed alphabet Σ for the rest of this paper and note that the concepts introduced herein are to be understood with respect to this alphabet but do not depend on its actual content for as long as $|\Sigma| \geq 2$.

Visibly-Pushdown Relations. Let $\perp$ be a new symbol not occurring in Σ, and let $\Sigma_\perp := \Sigma \cup \{\perp\}$. Let $\overline{w} = (w_1, \ldots, w_k) \in (\Sigma^*)^k$, where $w_i = a_{i,1} \cdots a_{i,|w_i|}$ (and all $a_{i,j} \in \Sigma$). We define the string $[\overline{w}] \in (\Sigma_\perp^k)^*$ by $[\overline{w}] := b_1 \cdots b_n$, where n is the maximum of all $|w_i|$, and $b_j := (b_{j,1}, \ldots, b_{j,k})$, with $b_{j,i} = a_{i,j}$ if $j \leq |w_i|$, and $b_{j,i} = \perp$ if $j > |w_i|$. Intuitively, $[\overline{w}]$ is obtained by aligning all w_i to the left, and padding the unfilled space with $\perp$ symbols.

An alphabet Σ' is a visibly pushdown alphabet if it is partitioned into push, pop, and no-op symbols. A *visibly-pushdown automaton* [1,13] (VPA) is a non-deterministic pushdown automaton whose stack action (push, pop, no-op) is determined by the input letter they read, according to its type push, pop or no-op. A k-ary relation $R \subseteq (\Sigma^*)^k$ is called a *visibly pushdown relation* if $(\Sigma^k)_\perp$ is a visibly pushdown alphabet and $\{[\overline{w}] \mid \overline{w} \in R\}$ is recognised by a VPA.

DB-Graphs. A Σ-*labeled db-graph* (db-graph for short) is a directed graph $G = (V, E)$, where V is a finite set of nodes, and $E \subseteq V \times \Sigma \times V$ is a finite set of directed edges with labels from Σ. A *path* ρ between two nodes v_0 and v_n in G with $n \geq 0$ is a sequence $v_0 a_1 v_1 \ldots v_{n-1} a_n v_n$ with $(v_i, a_{i+1}, v_{i+1}) \in E$ for $0 \leq i < n$. We define the *label* $\lambda(\rho)$ of the path ρ by $\lambda(\rho) := a_1 \cdots a_n$.

Extended Conjunctive Path Queries. We generalise the definition of extended conjunctive regular path queries [4] to visibly pushdown relations. Fix a countable set of *node variables* and a countable set of *path variables*. Let $k \geq 1$. A k-dimensional *extended conjunctive visibly pushdown path query* Q is an expression of the form

$$\text{Ans}(\overline{z}, \overline{\chi}) \leftarrow \bigwedge_{1 \leq i \leq m} (x_i, \pi_i, y_i), \bigwedge_{1 \leq j \leq l} R_j(\overline{w}_j), \tag{1}$$

such that $m \geq 1$, $l \geq 0$, and

1. there is a fixed partition of the alphabet $(\Sigma_\perp)^k$ into push, pop, and no-op symbols
2. each R_j is a k-dimensional visibly pushdown relation
3. $\overline{x} = (x_1, \ldots, x_m)$ and $\overline{y} = (y_1, \ldots, y_m)$ are tuples of (not necessarily distinct) node variables,
4. $\overline{\pi} = (\pi_1, \ldots, \pi_m)$ is a tuple of distinct path variables,
5. $\overline{w}_1, \ldots, \overline{w}_l$ are tuples of path variables, such that each $\overline{w}_j$ is a tuple of variables from $\overline{\pi}$, of the same arity as R_j,
6. $\overline{z}$ is a tuple of node variables among $\overline{x}, \overline{y}$, and
7. $\overline{\chi}$ is a tuple of path variables among those in $\overline{\pi}$.

The expression $\text{Ans}(\overline{z}, \overline{\chi})$ is the *head*, and the expression to the right of $\leftarrow$ is the *body* of Q. If $\overline{z}$ and $\overline{\chi}$ are the empty tuple (i.e., the head is of the form $\text{Ans}()$), Q is a *Boolean query*. The *relational* part of Q is $\bigwedge_{1 \leq i \leq m}(x_i, \pi_i, y_i)$, and the *labeling* part is $\bigwedge_{1 \leq j \leq l} R_j(\overline{w}_j)$. We denote the set of node variables in Q by $\text{nvar}(Q)$. The size of the query is defined as $m + \sum_{1 \leq i \leq n} |R_i|$, where $|R_i|$ denotes the size (number of states and transitions) of the VPA representing the relation R_i.

ECPQ[VPL] will denote the class of all extended conjunctive visibly pushdown path queries, and CPQ[VPL] will denote the class of queries of dimension $k = 1$. A query is called an extended *reachability* query if $\overline{w}_i \cap \overline{w}_j = \emptyset$ for $i \neq j$, in other words if every path variable occurs at most in one relation constraint. In that case, we abbreviate $R(\boldsymbol{x}, \boldsymbol{y})$ for $(\boldsymbol{x}, \boldsymbol{\omega}, \boldsymbol{y}) \wedge R(\boldsymbol{\omega})$. ECRQ[VPL] will denote the class of extended visibly pushdown reachability queries. Finally, we write ECPQ[REG] for the class of extended conjunctive regular path queries, i.e. extended path queries where all relations are regular.

Example 1. Let $L_0 := \{a^n b^n \in \Sigma^* \mid n \geq 0\}$, $L = L_0^*$, and $R = \{(w_1, w_2) \in L \times L \mid w_1 \neq w_2\}$. The query $\mathrm{Ans}(x, y) \leftarrow (x, \pi, y) \wedge L^*(\pi)$ is a CRQ[VPL] query. The query $\mathrm{Ans}(x, y) \leftarrow (x, \pi_1, y) \wedge (x, \pi_2, y) \wedge R(\pi_1, \pi_2)$ is an ECRQ[VPL] query that would be hard to express as a CRQ[VPL] query or an ECPQ[REG] query. Finally, the query $\mathrm{Ans}(x, y) \leftarrow (x, \pi_1, y) \wedge (x, \pi_2, y) \wedge (x, \pi_3, y_3) \wedge R(\pi_1, \pi_2) \wedge R(\pi_1, \pi_3) \wedge R(\pi_2, \pi_3)$ is an ECPQ[VPL] query.

Remark 1. Since visibly pushdown languages are closed under intersection, every CPQ[VPL] query is equivalent to a CRQ[VPL] query.

Query Evaluation and Query Containment. The evaluation $Q(G)$ of a query Q over a db-graph G is intuitively obtained by interpreting all variables as quantified existentially, and path constraints as constraints on the words formed by the labels along the paths. Formally, for every db-graph G, every ECPQ[VPL] query Q (of the form described in (1)), every mapping σ from the node variables of Q to nodes in G, and every mapping μ from the path variables of Q to paths in G, we write $G, \sigma, \mu \models Q$ if

1. $\mu(\pi_i)$ is a path from $\sigma(x_i)$ to $\sigma(y_i)$ for every $1 \leq i \leq m$,
2. for each $\overline{w}_j = (\pi_{j_1}, \ldots, \pi_{j_k})$, $1 \leq j \leq l$, the tuple $(\lambda(\mu(\pi_{j_1})), \ldots, \lambda(\mu(\pi_{j_k})))$ belongs to the relation R_j.

We define the output of Q on G as

$$Q(G) := \{ (\sigma(\overline{z}), \mu(\overline{\chi})) \mid G, \sigma, \mu \models Q \}.$$

The query Q is contained in the query Q' if $Q(G) \subseteq Q'(G)$ for all db-graphs G.

The problem of query evaluation is: given a Boolean query Q and a db-graph G over the same underlying alphabet, decide whether or not $Q(G) = \{()\}$, or in other words, whether there are σ, μ such that $G, \sigma, \mu \models Q$. We distinguish the *combined complexity* from the *data complexity*. The former considers both Q and G to be the input, the latter considers the query to be fixed and measures the complexity of query evaluation only in terms of the size of the db-graph G. For complexity considerations we focus on the decision problem of query evaluation for Boolean queries, but in general, when this problem is decidable, it is possible to compute a representation of the output of a query.

3   The Complexity of Evaluating CPQ[VPL] Queries

We address the problem of evaluation of both CPQ[VPL] and CRQ[VPL] queries w.r.t. combined and data complexity. First note that, since every CPQ[VPL] query can be converted into a CRQ[VPL] query, the data complexity of query evaluation is the same for these two formalisms. It can also be noticed that, for a given db-graph G, a CRQ[VPL] query Q, and a mapping σ, the model-checking problem $G, \sigma \models Q$ can be solved in polynomial time. Indeed, for two variables x, y, the language $L(\sigma(x), \sigma(y))$ of all paths from $\sigma(x)$ to $\sigma(y)$ is recognized by a NFA of size $|G|$, and G satisfies the atom (x, L, y) iff $L \cap L(\sigma(x), \sigma(y))$ is not empty. Since $G, \sigma \models Q$ can be decided in polynomial time, the data complexity of query evaluation is PTIME – the enumeration of all possible σ takes time $O(|G|^{|\mathsf{Invars}(Q)|})$. This upper bound is actually tight.

Theorem 1. *Evaluating CRQ[VPL] queries is P-complete (data complexity).*

Proof. For a natural number n let $bin(n)$ denote its binary representation using a special symbol to mark the end of the code, e.g. $bin(13) = 1101\ \$$. Let $\overleftarrow{bin}(n)$ be its representation in reverse order using different symbols, i.e. $\overleftarrow{bin}(13) = \$'1'0'1'1'$. Finally, let $\#$ be an extra symbol. It is not hard to see that the language L_0 described recursively by

$$L_0' = \mathsf{set} + \mathsf{or}\,\{bin(n)\,L_0'\,\overleftarrow{bin}(n) \mid n \in \mathbb{N}\} +$$
$$\left(\mathsf{and}\,\{bin(n)\,L_0'\,\overleftarrow{bin}(n) \mid n \in \mathbb{N}\}\right)^2$$

as well as $L_0 := \#L_0'$ is a VPL over the visibly pushdown alphabet with push-symbols \$, 1, 0, pop-symbols \$', 1' 0' and no-op symbols $\#$, set, or, and.

We present a logspace reduction from the Circuit Value Problem, known to be P-complete [12], to the problem of evaluating the query $\mathrm{Ans}() \leftarrow (x, \pi, x) \wedge L_0(\pi)$. We informally describe how to turn a circuit C into a db-graph G_C. W.l.o.g. we can assume that every internal gate in C has fan-in exactly 2, and is identified by a unique number i.

Every input gate that is set to 0 becomes a single node ●. Every input gate that is set to 1 becomes a node ●⊃ The OR-gate with number i and the AND-gate with number i are translated as the filled nodes below

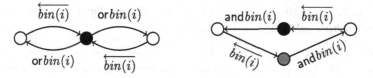

assuming the non-filled nodes represent the nodes associated to the inputs of these gates, and the gray node is a local auxiliary node. Additionally, we append a loop labeled with $\#$ to the output gate of the circuit. Note that this is a logspace reduction.

Now we have $Q_0(G_C) = \mathtt{true}$ iff there is a path starting and ending in the output gate's node that traverses through the graph such that in every OR-gate node it continues to one successor, and in every AND-gate node it traverses first through one input gate and then, after coming back to it goes through the other successor. The path can only traverse through input gates that are 1. Thus, this is the case iff the circuit evaluates to 1. □

We now consider the combined complexity of query evaluation. Note that even if a CPQ[VPL] query can be converted into an equivalent CRQ[VPL] query, the original query might be exponentially more concise, so the two families of queries might have different combined complexities – and indeed, they do. Consider first evaluating a CRQ[VPL] query Q. Since $G, \sigma \models Q$ can be decided in polynomial time and σ can be guessed in polynomial time, query evaluation is in NP. Moreover, query evaluation is NP-hard for basic conjunctive queries [6]. As a consequence, CRQ[VPL] query evaluation is NP-complete wrt combined complexity. The situation is rather different for CPQ[VPL] queries.

Theorem 2. *Evaluating CPQ[VPL] queries is EXPTIME-complete (combined complexity).*

Proof. The emptiness of the intersection of a family of VPAs is EXPTIME-complete due to the EXPTIME-completeness of the emptiness problem of a familiy of tree automata. We show that this is also equivalent to CPQ[VPL] query evaluation upto logspace reductions. Assume first a fixed query Q in the form of a conjunct of k queries $Q_1, \ldots, Q_k$ with Q_i of the form $(x_i, \pi_i, y_i) \wedge L_{i,1}(\pi_i) \wedge \cdots \wedge L_{i,n}(\pi_i)$. Then, for every db-graph G and mapping σ, $G, \sigma \models Q$ iff $L(\sigma x, \sigma y) \cap L_{i,1} \cap \ldots L_{i,n} \neq \emptyset$ for all $i = 1, \ldots, k$, which shows the upper bound. Conversely, if $L_1, \ldots, L_n$ is a family of visibly pushdown languages, and if G is the graph restricted to a single node with a self loop labeled with Σ, then $G \models (x, \pi, x) \wedge L_1(\pi) \wedge \cdots \wedge L_n(\pi)$ iff $L_1 \cap \cdots \cap L_n \neq \emptyset$.   □

4   The Complexity of Evaluating ECPQ[VPL] Queries

We now address the complexity of evaluating *extended* queries. We just saw that reachability and path queries are equally expressive in the non-extended case. In the extended setting, however, reachability and path queries have important differences. Let us first consider the richest language of extended path queries.

Theorem 3. *Evaluating ECPQ[VPL] is undecidable for dimension 2 and two path constraints (data and combined complexity).*

In other words, there is a query $Q \in$ ECPQ[VPL] containing two two-dimensional relations $R_1, R_2 \subseteq (\Sigma_\perp^2)^*$, such that query evaluation for Q is undecidable. Note that according to the definition of ECPQ[VPL], $\Sigma_\perp^2$ is partitioned into the same visibly pushdown alphabet for R_1 and R_2, and note moreover that the query Q is fixed.

Proof. We claim that there are two visibly pushdown languages L_1, L_2 over two visibly pushdown alphabet Σ_1, Σ_2 containing the same symbols, but with different associated operations, such that the word problem for $L_0 := \{w \mid$ there is v with $wv \in L_1 \cap L_2\}$ is undecidable. Indeed, it is folklore that for every Turing machine M, there are pushdown languages L_1, L_2 such that $L_1 \cap L_2$ is the set of runs of M coded as words, and L_0 can be defined as the set of inputs accepted by a universal Turing machine M.

Now we reduce this problem to the problem of evaluating a fixed ECPQ[VPL] query. Let $\Sigma' := \Sigma \cup \{\$, \sharp\}$ with the two symbols not occurring in Σ. A symbol $(\$, a)$ of $(\Sigma'_\perp)^2$ is a push (resp. pop, remain) symbol if a is a push (resp. pop, remain) symbol in L_1. Similarly, $(\#, a)$ is a push (resp. pop, remain) symbol if a is a push (resp. pop, remain) symbol in L_2. $w = a_1 \ldots a_n \in \Sigma^*$. The fixed query we consider is $Q =$

$$\mathrm{Ans}() \leftarrow (x, \pi, y) \wedge (x, \pi_\$, x) \wedge (y, \pi_\#, y) \wedge (L_1 \times \$^*)(\pi, \pi_\$) \wedge (L_2 \times \#^*)(\pi, \pi_\#).$$

Now, for every db-graph G of the form

it is not hard to see that $w \in L_0$ iff $Q(G) = \mathbf{true}$.   □

We now turn our attention to extended reachability queries. Evaluating an ECRQ[VPL] query of dimension k over a db-graph G is equivalent to evaluating a CRQ[VPL] query over G^k, which shows by Theorem 1 that the combined complexity of evaluating ECRQ[VPL] queries also is in EXPTIME. The same remark shows that the problem is in PTIME w.r.t. data complexity, and even P-complete by Theorem 1. For the combined complexity, we have get EXPTIME-hardness. The proof relies on EXPTIME-hardness of the following decision problem VPADFAISECT.

- INPUT: a VPA $\mathcal{A}$, and arbitrarily many DFAs $\mathcal{A}_1, \ldots, \mathcal{A}_n$
- QUESTION: is $L(\mathcal{A}) \cap L(\mathcal{A}_1) \cap \cdots \cap L(\mathcal{A}_n)$ empty?

Lemma 1. VPADFAISECT *is EXPTIME-hard.*

Proof. By a reducton the problem to decide whether or not the intersection of the languages of given top-down tree automata (TA) $\mathcal{A}_1, \ldots, \mathcal{A}_n$ over some alphabet Σ is empty. This is known to be EXPTIME-hard [15]. For the sake of simplicity we assume that all the TA share a common state space Q, transition function δ and acceptance set $F \subseteq Q$. The n different TA are then simply given by n different initial states $q_1^I, \ldots, q_n^I$. W.l.o.g. we furthermore assume that the underyling alphabet Σ has binary symbols a, b and a unique leaf symbol c, and that trees have height (number of nodes) at least 2.

Next we consider a particular encoding of trees t with node labels of the form $\Sigma \times Q^n$ as a word $\mathsf{rep}(t)$ over the alphabet $\widehat{\Sigma} = \{\overrightarrow{(x,y)}, \overleftarrow{(x,y)} \mid x, y \in \Sigma$ or $x, y \in Q\}$. This can easily be done by induction on the structure of t. If t is the single-leaf tree with Σ-label c and any label from Q^n then $\mathsf{rep}(t)$ is the empty word. If t is of the form and t' and t'' are the left and right subtrees starting with y and z respectively, then we have

$$\mathsf{rep}(t) = \overrightarrow{\binom{x}{y}}\overrightarrow{\binom{q_1}{p_1}}\cdots\overrightarrow{\binom{q_n}{p_n}}\,\mathsf{rep}(t')\,\overleftarrow{\binom{q_n}{p_n}}\cdots\overleftarrow{\binom{q_1}{p_1}}\overleftarrow{\binom{x}{y}}\overrightarrow{\binom{x}{z}}\overrightarrow{\binom{q_1}{r_1}}\cdots\overrightarrow{\binom{q_n}{r_n}}\,\mathsf{rep}(t'')\,\overleftarrow{\binom{q_n}{r_n}}\cdots\overleftarrow{\binom{q_1}{r_1}}\overleftarrow{\binom{x}{z}}$$

It should be clear that the language $\{\mathsf{rep}(t) \mid t \text{ is a } \Sigma \times Q^n\text{-labelled tree}\}$ is recognisable by a VPA $\mathcal{B}$ over the visibly pushdown alphabet with push-symbols of the form $\overrightarrow{(\cdot,\cdot)}$, pop-symbols of the form $\overleftarrow{(\cdot,\cdot)}$ and no internal symbols. It uses $\widehat{\Sigma}$ as the stack alphabet and can then easily check the symmetries in the given word. It needs $n+1$ states to check that the length of each segment encoding a node label is $n+1$ while it pushes such an encoding onto the stack, and one more state in which it pops such an encoding and compares it to the $\overleftarrow{(\cdot,\cdot)}$-parts of the word.

It remains to be seen that on such words a DFA $\mathcal{B}_i$ can check whether or not the labelling in the i-th component of the state tuple forms an accepting run of the TA $\mathcal{A}_i$ on the tree given by the Σ-labels. We describe its behaviour informally, using the term *segment* to denote each part of length $n+1$ in such a word that enocdes the labels of two consecutive nodes in the tree.

1. It discards if the symbol at position $i+1$ of the entire word is not of the form $\overrightarrow{(q_i^I,\cdot)}$, i.e. the run does not start in the initial state. Otherwise it continues.

2. On a $\overrightarrow{\cdot}$-segment it remembers the symbol $\overrightarrow{(q,p)}$ at position $i+1$. If this segment is followed by another $\overrightarrow{\cdot}$-segment then it discards this information and continues. Otherwise the second components in this segment encode the label of a leaf node, and $\mathcal{B}_i$ discards if $p \notin F$. Otherwise it continues.

3. On a $\overleftarrow{\cdot}$-segment it remembers the symbols $\overleftarrow{(q,p)}$ and $\overleftarrow{(a,b)}$ at positions $n+1-i$ and $n+1$. If this segment is followed by another $\overleftarrow{\cdot}$-segment then it discards this information because this segment encodes the labels on a node and its right child. Otherwise this segment is followed by a $\overrightarrow{\cdot}$-segment, and these two encode the labels on a node with Σ-label a and its two children. $\mathcal{B}_i$ then also reads the symbol $\overrightarrow{(q',r)}$ at position i of this successor segment and discards if $q \neq q'$ or $(p,r) \notin \delta(q,a)$. Otherwise it continues reading the next segment.

It should be clear that each $\mathcal{B}_i$ only needs polynomial size in $\mathcal{A}_i$ and n. Moreover, we get that $L(\mathcal{B}) \cap \bigcap_{i=1}^n L(\mathcal{B}_i) \neq \emptyset$ iff there is a word $\mathsf{rep}(t)$ in this intersection which encodes a $\Sigma \times Q^n$-labelled tree such that the $(i+1)$-st components of the labels form an accepting run of the TA $\mathcal{A}_i$ on the tree given by the Σ-labels. This is the case iff $\bigcap_{i=1}^n L(\mathcal{A}_i) \neq \emptyset$. $\square$

Theorem 4. *Evaluating ECRQ[VPL] queries is EXPTIME-hard (combined complexity).*

Proof. By a reduction from VpaDfaISect. Let $\mathcal{A}$ be a VPA and $\mathcal{A}_1,\ldots,\mathcal{A}_n$ be n DFAs. We then construct a db-graph $G_{\mathcal{A}_1,\ldots,\mathcal{A}_n}$ as their disjoint union and add a self-loop labeled with a new symbol start_i to every starting state in $\mathcal{A}_i$ as well as a self-loop labeled with a new symbol end to each of its final states. Then consider the n-ary relation

$$R_n = \left\{ \begin{pmatrix} \text{start}_1 \\ \vdots \\ \text{start}_n \end{pmatrix} \begin{pmatrix} w \\ \vdots \\ w \end{pmatrix} \begin{pmatrix} \text{end} \\ \vdots \\ \text{end} \end{pmatrix} \mid w \in L(\mathcal{A}) \right\}.$$

Note that it can be recognised by a VPA $\mathcal{A}'$ of linear size in the size of $\mathcal{A}$. Then $L(\mathcal{A}) \cap \bigcap_{i=1}^n L(\mathcal{A}_i) \neq \emptyset$ iff $G_{\mathcal{A}_1,\ldots,\mathcal{A}_n} \models \bigwedge_{i=1}^n (x_i, \pi_i, y_i) \wedge R_n(\pi_1, \ldots, \pi_n)$.   □

5   Query Containment

We now consider the problem of query containment. For ECRQ[REG] (extended conjunctive regular reachabilty queries), Freydenberger and Schweikardt showed that this problem is undecidable. In the remainder, we focus on query containment for CRQ[VPL] queries where no path variables occur in the head. That is, we consider the problem $Q_1 \subseteq Q_2$ for two CRQ[VPL] queries Q_1, Q_2 of the form $Q_h =$

$$\text{Ans}(z_1, \ldots, z_n) \leftarrow \bigwedge_{i=1\ldots m_h} (x_{h,i}, L_{h,i}, y_{h,i}).$$

Note that the distinguished variables $z_1, \ldots, z_n$ are the same for the two queries. We assume that these are the only variables shared by the queries – the non-distinguished variables are assumed properly renamed. We also assume that the languages $L_{h,i}$ are recognised by VPAs working with a fixed visibly pushdown alphabet Σ. The set $\mathcal{V}_i := \text{nvars}(Q_i)$ is the set of all variables of Q_i. A tuple $G = (V, E, \sigma)$ is a *canonical candidate* of Q_1 if $\sigma : \mathcal{V}_1 \to V$ and G is the union of m_1 cycle-free paths $\pi_1, \ldots, \pi_n$, one for each conjunct of Q_1, such that π_i goes from $\sigma(x_i)$ to $\sigma(y_i)$. A canonical candidate of Q_1 is a *canonical model* of Q_1 if moreover $\lambda(\pi_i) \in L_{1,i}$. In particular, if G is a canonical model of Q_1, then $\sigma(z) \in Q_1(G)$. The following result has been proved in several places, see for instance [11].

Lemma 2. $Q_1 \subseteq Q_2$ *iff for all canonical model* $G = (V, E, \sigma)$ *of* Q_1*, there is* $\sigma' : \mathcal{V}_2 \to V$ *such that* $G, \sigma' \models Q_2$ *and* $\sigma(z) = \sigma'(z)$ *for all distinguished variables* z.

We now follow the automata-based approach [5] for deciding query containment. The main idea of this approach is that canonical db-graphs can be represented by means of words. To every canonical db-graph $G = (V, E, \sigma)$, we associate the word $w_G := \$d_1 w_1 d'_1 \$d_2 w_2 d'_2 \$ \ldots \$d_{m_1} w_{m_1} d'_{m_1} \$$ where the new symbol $\$ \notin \Sigma$ acts as a separator and $d_1, d'_1, \ldots, d_{m_1}, d'_{m_1}$ range over $\mathcal{D}_1 := 2^{\mathcal{V}_1}$. The i-th block of w_G contains $\$d_i w_i d'_i \$$ if $d_i = \sigma^{-1}(\sigma(x_{1,i}))$, $d'_i = \sigma^{-1}(\sigma(y_{1,i}))$, and w_i labels the path associated to the atom $(x_{1,i}, L_{1,i}, y_{1,i})$. In the extended visibly pushdown alphabet $\Sigma \cup \{\$\} \cup \mathcal{D}_1$, the symbols $\$$ and $\mathcal{D}_1$ are considered no-op symbols.

Lemma 3. *Let* Q_1 *be a query of size* n. *Then* $L(Q_1) := \{w_G \mid G$ *is a canonical model of* $Q_1\}$ *is recognised by a VPA with* $2^{O(n \log n)}$ *states.*

Proof. Let L_{part} be the language of words w over $\Sigma \cup \{\$\} \cup \mathcal{D}_1$ such that the set of $\mathcal{D}_1$ symbols of w define a partition of $\mathcal{D}_1$. Then L_{part} is recognised by an NFA that guesses the partition P of $\mathcal{V}_1$ and then checks that the set of $\mathcal{D}_1$ symbols of w is P. Since there are $2^{O(|\mathcal{V}_1|\log(|\mathcal{V}_1|))}$ different partitions of $\mathcal{V}_1$, the automaton is of size $2^{O(n\log n)}$.

Let $\{x\}^{\uparrow}$ denote the set of $\mathcal{D}_1$ symbols d such that $x \in d$. The language $L_{\mathsf{query}} := \${x_1}^{\uparrow}L_{1,1}\{y_1\}^{\uparrow}\$ \ldots \${x_{m_1}}^{\uparrow}L_{1,m_1}\{y_{m_1}\}^{\uparrow}\$$ is recognised by a VPA with $O(|Q_1|)$ states. Since $L(Q_1) = L_{\mathsf{part}} \cap L_{\mathsf{query}}$, this shows the result.

Let $G = (V, E, \sigma)$ be a canonical db-graph of Q_1, and let $\sigma' : \mathcal{V}_1 \cup \mathcal{V}_2 \to V$ be an extension of σ. The pair (G, σ') can be represented as a word $w_{G,\sigma'}$ over the alphabet $\Sigma \cup \$ \cup \mathcal{D}_2$ with $\mathcal{D}_2 := 2^{\mathcal{V}_1 \cup \mathcal{V}_2}$ following the same idea as before, except that now symbols $d \in \mathcal{D}_{2\setminus 1} := 2^{\mathcal{V}_2 \setminus \mathcal{V}_1}$ can occur anywhere – a variable $x \in \mathcal{V}_2 \setminus \mathcal{V}_1$ can be mapped to a node $\sigma'(x)$ that is not in the image of σ. Let $\pi_{2\to1}$ be the substitution that sends $d \in \mathcal{D}_2$ to $d \cap \mathcal{V}_1$ if $d \cap \mathcal{V}_1 \neq \emptyset$, ϵ otherwise. Note that if L is definable by a VPA, then $\pi_{2\to1}(L)$ is definable by a VPA, because $\pi_{2\to1}$ only changes or erases no-op symbols.

Let $L(Q_2|Q_1)$ be the set of words $w_{G,\sigma'}$ such that $G, \sigma' \models Q_2$ and G is a canonical candidate for Q_1. Then, by virtue of Lemma 3, $Q_1 \subseteq Q_2$ iff $L(Q_1) \subseteq \pi_{2\to1}(L(Q_2|Q_1))$.

Example 2. Consider the query Q_1 whose unique canonical model is the graph G such that $w_G = \${x_1}aa\{y_1\}\${x_2}cc\{y_2\}\${y_1}b\{x_2\}\$$. Let Q_2 be the query (x, abc, z). Then $G, \sigma' \models Q_2$ for σ' such that $w_{G,\sigma'} =$

$$\${x_1}a\{x\}a\{y_1\}\${x_2}c\{z\}c\{y_2\}\${y_1}b\{x_2\}\$,$$

so $Q_1 \subseteq Q_2$ (assuming G is the unique canonical model of Q_1).

As seen in the example above, it is not straightforward to recognise $L(Q_2|Q_1)$ with a VPA, because checking the atom (x, L, y) requires an L-path to be guessed that might be coded in different blocks, in any order, and the extremities of this path can be at intermediate positions inside some blocks. This motivates the following definition.

Definition 1. *A* jumping visibly pushdown automaton (JPA) *over a visibly pushdown alphabet Σ is a tuple $\mathcal{A} = (Q, \Gamma, \delta, q_I, F, J)$ such that $(Q, \Gamma, \delta, q_I, F)$ is a VPA and $J \subseteq Q$ is a set of* jumping states.

Intuitively, a JPA reads a word on a tape from left to right managing the stack as usual, but when it enters a jumping state, the head of the tape can instantaneously move to any position, while erasing the content of the stack. Formally, a configuration of the run of a jumping VPA over a nested word $w \in \Sigma^*$ is a tuple $(q, s, i) \in Q \times \Gamma^* \times \{1, \ldots, |w|\}$; it is accepting if $q \in F$. The configuration (q, s, i) leads to the configuration (q', s', i'), $(q, s, i) \vdash_w (q', s', i')$, if there is a transition from $(q, a, \alpha, q') \in \delta$ such that a is the i-th symbol of w, α is some stack action that transforms s into $\alpha(s)$, and either $i' = i + 1$ and $s' = \alpha(s)$, or $q' \in J$ and $s' = \bot$. A nested word w is accepted if $(q_I, \bot, 1) \vdash_w^* (q, s, i)$ for some accepting configuration (q, s, i).

Lemma 4. $L(Q_2|Q_1)$ *is recognised by a JPA with* $O(|Q_1| + |Q_2|)$ *states.*

Proof. We first assume that Q_2 contains only one atom (x, L, y). Let $\mathcal{A} = (Q, \Sigma, \delta, q_I, F)$ be the VPA representing L. We informally define a JPA $\mathcal{B}$ that accepts $w_{G,\sigma'}$ iff $G, \sigma' \models (x, L, y)$. In the first step, the automaton jumps anywhere in the word. Then it checks that the symbol it reads is a $\mathcal{D}_2$ symbol d that contains x, and start running $\mathcal{A}$. When it meets a $\mathcal{D}_2$ symbol d, it may jump anywhere else provided the next symbol it will read is a $\mathcal{D}_2$ symbol d' such that $d \cap d' \cap \mathcal{V}_1 \neq \emptyset$. It accepts when it reads a symbol d that contains y.

Clearly, if $w_{G,\sigma'}$ is accepted by $\mathcal{B}$, the piece of w_G over which $\mathcal{B}$ ran corresponds to a path satisfying the atom (x, L, y). Conversely, assume that $G, \sigma' \models (x, L, y)$. Then there is a word $w_1 w_2 \ldots w_n$ in L and blocks indices $i_1, \ldots, i_n$ such that w_1 is a suffix of the word of i_1th block of $w_{G,\sigma'}$, $w_2, \ldots, w_{n-1}$ are the words of the blocks $i_2, \ldots, i_{n-1}$, and w_n is a prefix of the word of the i_nth block. To ensure that an accepting run of $\mathcal{A}$ over $w_1 w_2 \ldots w_n$ can be mimicked by an accepting run of $\mathcal{B}$, it must be checked that the w_i are well nested, so that it does not harm to erase the stack when jumping. For $k = 2, \ldots n - 1$, w_k is well-nested because it belongs to the visibly pushdown language L_{1,i_k}, and w_1 and w_n are also well-nested because they are prefixes of well-nested words of L and L_{1,i_n} respectively.

Let us assume now that Q_2 is $\bigwedge_{i=1}^{n}(x_i, L_i, y_i)$, and let $\mathcal{B}_i$ be the JPA associated to the ith atom following the previous construction. Then we define $\mathcal{B}$ as the automaton that executes B_1, then B_2, ..., then B_n. Clearly $|\mathcal{B}| = O(|\mathcal{B}_1| + \cdots + |\mathcal{B}_n|)$ and $L(\mathcal{B}) = \bigcap_{i=1}^{n} L(\mathcal{B}_i)$, which shows the result. □

Observe now that if a JPA has an accepting run over a word w, then it has an accepting run over w with at most $|J|$ jumps, where J is its set of jumping states. This leads to the following result.

Lemma 5. *For every JPA $\mathcal{A}$ with n states, there is a VPA $\mathcal{B}$ such that $L(\mathcal{A}) = L(\mathcal{B})$ and $|\mathcal{B}| = 2^{O(n \log n)}$.*

Proof. Let $\mathcal{A}$ be a fixed JPA. Observe that if $\mathcal{A}$ has an accepting run over a word w, then it has an accepting run with at most $|J|$ jumps, where J is the set of jumping states. Indeed, if during a run $\mathcal{A}$ jumps in jumping state q from position i_1 to position j_1 and later from i_2 to j_2 in the same jumping state q, then a shorter run is obtained by jumping from i_1 to j_2 directly.

Consider a sequence $S = (q_1, q_1')(q_2, q_2') \ldots (q_{|S|}, q_{|S|}')$ of $|S| \leq n$ pairs of set of states of $\mathcal{A}$. We say that S is accepting if q_1 is the initial state, $q_{|S|}'$ is accepting, and $q_i' = q_{i+1} \in J$ for each $i = 1, \ldots, |S| - 1$. There are thus $2^{O(n \log n)}$ accepting sequences. For a fixed sequence S, and for each $i = 1, \ldots, |S|$, there is a VPA $\mathcal{A}_{S,i}$ with $O(n)$ states that accepts a word w iff w contains a factor w' and there is a run of $\mathcal{A}$ from q_i to q_i' over w'. There is also a VPA $\mathcal{A}_S$ that accepts $\bigcap_{i=1}^{n} L(\mathcal{A}_{S,i})$ and such that $|\mathcal{A}_S| = 2^{O(n \log n)}$. Let $\mathcal{B}$ be the automaton that accepts $\bigcup \{L(\mathcal{A}_S) \mid S \text{ is an accepting sequence}\}$. Then $L(\mathcal{B}) = L(\mathcal{A})$ and $|\mathcal{B}| = 2^{O(n \log n)}$.

To sum up, Lemmas 4 and 5 show that $L(Q_2|Q_1)$ can be recognised by a VPA of exponential size, and so can $L(Q_1)$ (Lemma 3). Language inclusion between two VPAs can be decided in exponential time, so the inclusion $L(Q_1) \subseteq L(Q_2|Q_1)$ can be decided in 2-EXPTIME.

Theorem 5. *Containment for CRQ[VPL] queries is decidable in 2-EXPTIME.*

Acknowledgments. We would like to thank Nicole Schweikardt for introducing the world of conjunctive path queries to us and for some interesting discussions on that topic.

References

1. Alur, R., Madhusudan, P.: Visibly pushdown languages. In: Proceedings of the STOC 2004, pp. 202–211. ACM (2004)
2. Angles, R., Gutiérrez, C.: Survey of graph database models. ACM Comput. Surv. **40**(1), 1–39 (2008)
3. Barceló, P., Figueira, D., Libkin, L.: Graph logics with rational relations. Logical Methods Comput. Sci. **9**(3:1), 1–44 (2013)
4. Barceló, P., Libkin, L., Lin, A.W., Wood, P.T.: Expressive languages for path queries over graph-structured data. ACM Trans. Database Syst. **37**(4), 31 (2012)
5. Calvanese, D., De Giacomo, G., Lenzerini, M., Vardi, M.Y.: Containment of conjunctive regular path queries with inverse. In: Proceedings of the KR 2000, pp. 176–185. Morgan Kaufmann (2000)
6. Chandra, A.K., Merlin, P.M.: Optimal implementation of conjunctive queries in relational databases. In: Proceedings of the STOC 1977, pp. 77–90 (1977)
7. Deutsch, A., Tannen, V.: Optimization properties for classes of conjunctive regular path queries. In: Ghelli, G., Grahne, G. (eds.) DBPL 2001. LNCS, vol. 2397, pp. 21–39. Springer, Heidelberg (2002)
8. Florescu, D., Levy, A.Y., Suciu, D.: Query containment for conjunctive queries with regular expressions. In: Proceedings of the PODS1998, pp. 139–148 (1998)
9. Freydenberger, D., Schweikardt, N.: Expressiveness and static analysis of extended conjunctive regular path queries. J. Comp. Syst. Sci. **79**(6), 892–909 (2013)
10. Hellings, J.: Conjunctive context-free path queries. In: Proceedings of the 17th International Conference on Database Theory (ICDT), pp. 119–130. Athens, Greece, 24–28 March 2014
11. Kolaitis, P.G., Vardi, M.Y.: Conjunctive-query containment and constraint satisfaction. In: Proceedings of the PODS 1998, pp. 205–213. ACM (1998)
12. Ladner, R.E.: The circuit value problem is log-space complete for P. SIGACT News **6**(2), 18–20 (1975)
13. Mehlhorn, K.: Pebbling mountain ranges and its application to DCFL-recognition. In: Proceedings of the ICALP 1980, vol. 85 of LNCS, pp. 422–435. Springer (1980)
14. Mendelzon, A.O., Wood, P.T.: Finding regular simple paths in graph databases. In: Proceedings of the VLDB 1989, pp. 185–193. Morgan Kaufmann (1989)
15. Seidl, H.: Deciding equivalence of finite tree automata. SIAM J. Comput. **19**(3), 424–437 (1990)

On the Power of Color Refinement

V. Arvind[1], Johannes Köbler[2(✉)], Gaurav Rattan[1], and Oleg Verbitsky[2,3]

[1] The Institute of Mathematical Sciences, Chennai 600 113, India
{arvind,grattan}@imsc.res.in

[2] Institut für Informatik, Humboldt Universität zu Berlin, Berlin, Germany
{koebler,verbitsk}@informatik.hu-berlin.de

[3] On leave from the Institute for Applied Problems of Mechanics
and Mathematics, Lviv, Ukraine

Abstract. *Color refinement* is a classical technique used to show that two given graphs G and H are non-isomorphic; it is very efficient, although it does not succeed on all graphs. We call a graph G *amenable* to color refinement if the color-refinement procedure succeeds in distinguishing G from any non-isomorphic graph H. Babai, Erdős, and Selkow (1982) have shown that random graphs are amenable with high probability. We determine the exact range of applicability of color refinement by showing that amenable graphs are recognizable in time $O((n+m)\log n)$, where n and m denote the number of vertices and the number of edges in the input graph.

1 Introduction

The well-known *color refinement* (also known as *naive vertex classification*) procedure for Graph Isomorphism works as follows: it begins with a uniform coloring of the vertices of two graphs G and H and refines the vertex coloring step by step. In a refinement step, if two vertices have identical colors but differently colored neighborhoods (with the multiplicities of colors counted), then these vertices get new different colors. The procedure terminates when no further refinement of the vertex color classes is possible. Upon termination, if the multisets of vertex colors in G and H are different, we can correctly conclude that they are not isomorphic. However, color refinement sometimes fails to distinguish non-isomorphic graphs. The simplest example is given by any two non-isomorphic regular graphs of the same degree with the same number of vertices. Nevertheless, color refinement turns out to be a useful tool not only in isomorphism testing but also in a number of other areas; see [9,12,17] and references there.

For which pairs of graphs G and H does the color refinement procedure succeed in solving Graph Isomorphism? Mainly this question has motivated the study of color refinement from different perspectives.

This work was supported by the Alexander von Humboldt Foundation in its research group linkage program. The second author and the fourth author were supported by DFG grants KO 1053/7-2 and VE 652/1-2, respectively.

© Springer International Publishing Switzerland 2015
A. Kosowski and I. Walukiewicz (Eds.): FCT 2015, LNCS 9210, pp. 339–350, 2015.
DOI: 10.1007/978-3-319-22177-9_26

Immerman and Lander [10], in their highly influential paper, established a close connection between color refinement and 2-variable first-order logic with counting quantifiers. They show that color refinement distinguishes G and H if and only if these graphs are distinguishable by a sentence in this logic.

A well-known approach to tackling intractable optimization problems is to consider an appropriate linear programming relaxation. A similar approach to isomorphism testing, based on the notion of a fractional isomorphism (introduced by Tinhofer [18] using the term doubly stochastic isomorphism), turns out to be equivalent to color refinement. Building on Tinhofer's work [18], it is shown by Ramana, Scheinerman and Ullman [16] (see also Godsil [8]) that two graphs are indistinguishable by color refinement if and only if they are fractionally isomorphic.

We say that color refinement *applies* to a graph G if it succeeds in distinguishing G from any non-isomorphic H. A graph to which color refinement applies is called *amenable*. There are interesting classes of amenable graphs:

1. An obvious class of graphs to which color refinement is applicable is the class of *unigraphs*. Unigraphs are graphs that are determined up to isomorphism by their degree sequences; see, e.g., [5,19].
2. Trees are amenable (Edmonds [6,20]).
3. It is easy to see that all graphs for which the color refinement procedure terminates with all singleton color classes (i.e. the color classes form the discrete partition) are amenable. Babai, Erdös, and Selkow [2] have shown that a random graph $G_{n,1/2}$ has this property with high probability. Moreover, the discrete partition of $G_{n,1/2}$ is reached within at most two refinement steps. This implies that graph isomorphism is solvable very efficiently in the average case (see also [3]).

Our Contribution. What is the class of graphs to which color refinement applies? The logical and linear programming based characterizations of color refinement do not provide any efficient criterion answering this question.

We aim at determining the exact range of applicability of color refinement. We find an efficient characterization of the entire class of amenable graphs, which allows for a quasilinear-time test whether or not color refinement applies to a given graph. This result is shown in Sect. 5, after we unravel the structure of amenable graphs in Sects. 3 and 4. We note that a weak *a priori* upper bound for the complexity of recognizing amenable graphs is $coNP^{GI[1]}$, where the superscript means the one-query access to an oracle solving the graph isomorphism problem. To the best of our knowledge, no better upper bound was known before.

Combined with the Immerman-Lander result [10] mentioned above, it follows that the class of graphs definable by first-order sentences with 2 variables and counting quantifiers is recognizable in polynomial time.

Related Work. In an accompanying paper [1], we use our characterization of amenable graphs to prove that the polytope of fractional automorphisms of an amenable graph is integral. A characterization of amenable graphs similar to that in the present paper has been suggested independently by Kiefer, Schweitzer, and Selman [13]. Moreover, they obtain a generalization of this result to arbitrary relational structures (which includes, in particular, directed graphs).

Notation. The vertex set of a graph G is denoted by $V(G)$. The vertices adjacent to a vertex $u \in V(G)$ form its neighborhood $N(u)$. A set of vertices $X \subseteq V(G)$ induces a subgraph of G, that is denoted by $G[X]$. For two disjoint sets X and Y, $G[X, Y]$ is the bipartite graph with vertex classes X and Y formed by all edges of G connecting a vertex in X with a vertex in Y. The vertex-disjoint union of graphs G and H will be denoted by $G + H$. Furthermore, we write mG for the disjoint union of m copies of G. The *bipartite complement* of a bipartite graph G with vertex classes X and Y is the bipartite graph G' with the same vertex classes such that $\{x, y\}$ with $x \in X$ and $y \in Y$ is an edge in G' if and only if it is not an edge in G. We use the standard notation K_n for the complete graph on n vertices, $K_{s,t}$ for the complete bipartite graph whose vertex classes have s and t vertices, and C_n for the cycle on n vertices.

2   Basic Definitions and Facts

Throughout the paper, we consider vertex-colored graphs. A *vertex-colored graph* is an undirected simple graph G endowed with a vertex coloring $c : V(G) \to \{1, \ldots, k\}$. Isomorphisms between vertex-colored graphs are required to preserve vertex colors. We get usual graphs when c is constant.

Given a graph G, the *color-refinement* algorithm (to be abbreviated as *CR*) iteratively computes a sequence of colorings C^i of $V(G)$. The initial coloring C^0 is the vertex coloring of G, i.e., $C^0(u) = c(u)$. Then,

$$C^{i+1}(u) = \left(C^i(u), \{\!\{\, C^i(a) : a \in N(u) \}\!\} \right), \tag{1}$$

where $\{\!\{\ldots\}\!\}$ denotes a multiset.

The partition $\mathcal{P}^{i+1}$ of $V(G)$ into the color classes of C^{i+1} is a refinement of the partition $\mathcal{P}^i$ corresponding to C^i. It follows that, eventually, $\mathcal{P}^{s+1} = \mathcal{P}^s$ for some s; hence, $\mathcal{P}^i = \mathcal{P}^s$ for all $i \geq s$. The partition $\mathcal{P}^s$ is called the *stable partition* of G and denoted by $\mathcal{P}_G$.

Given a partition $\mathcal{P}$ of the vertex set of a graph G, we call its elements *cells*. We call $\mathcal{P}$ *equitable* if:

(i)   Each cell $X \in \mathcal{P}$ is monochromatic, i.e., all vertices $u, v \in X$ have the same color $c(u) = c(v)$.
(ii) For any cell $X \in \mathcal{P}$ the graph $G[X]$ induced by X is *regular*, that is, all vertices in $G[X]$ have equal degrees.
(iii) For any two cells $X, Y \in \mathcal{P}$ the bipartite graph $G[X, Y]$ induced by X and Y is *biregular*, that is, all vertices in X have equally many neighbors in Y and vice versa.

It is easy to see that the stable partition of G is equitable; our analysis in the next section will make use of this fact.

A straightforward inductive argument shows that the colorings C^i are preserved under isomorphisms.

Lemma 1. *If ϕ is an isomorphism from G to H, then $C^i(u) = C^i(\phi(u))$ for any vertex u of G.*

Lemma 1 readily implies that, if graphs G and H are isomorphic, then

$$\{\!\{ C^i(u) : u \in V(G) \}\!\} = \{\!\{ C^i(v) : v \in V(H) \}\!\} \tag{2}$$

for all $i \geq 0$. When used for isomorphism testing, the CR algorithm accepts two graphs G and H as isomorphic exactly when the above condition is met on input $G + H$. Note that this condition is actually finitary: If Equality (2) is false for some i, it must be false for some $i < 2n$, where n denotes the number of vertices in each of the graphs. This follows from the observation that the partition $\mathcal{P}^{2n-1}$ induced by the coloring C^{2n-1} must be the stable partition of the disjoint union of G and H. In fact, Equality (2) holds true for all i if it is true for $i = n$; see, e.g., [15]. Thus, it is enough that CR verifies (2) for $i = n$.

Note that computing the vertex colors literally according to (1) would lead to an exponential growth of the lengths of color names. This can be avoided by renaming the colors after each refinement step. Then CR never needs more than n color names (appearance of more than n colors is an indication that the graphs are non-isomorphic).

Definition 2. *We call a graph G amenable if for every graph H, procedure CR works correctly on the input pair G and H. That is, Equality (2) is false for $i = n$ whenever $H \not\cong G$.*

3   Local Structure of Amenable Graphs

Consider the stable partition $\mathcal{P}_G$ of an amenable graph G. The following lemma gives a list of all possible regular and biregular graphs that can occur, respectively, as $G[X]$ and $G[X, Y]$ for cells X, Y of $\mathcal{P}_G$.

Lemma 3. *The stable partition $\mathcal{P}_G$ of an amenable graph G fulfills the following properties:*

(A) *For any cell $X \in \mathcal{P}_G$, $G[X]$ is an empty graph, a complete graph, a matching graph mK_2, the complement of a matching graph, or the 5-cycle;*

(B) *For any two cells $X, Y \in \mathcal{P}_G$, $G[X, Y]$ is an empty graph, a complete bipartite graph, a disjoint union of stars $sK_{1,t}$ where X and Y are the set of s central vertices and the set of st leaves, or the bipartite complement of the last graph.*

The proof of Lemma 3 is based on the following facts.

Lemma 4 (Johnson [11]). *A regular graph of degree d with n vertices is a unigraph if and only if $d \in \{0, 1, n-2, n-1\}$ or $d = 2$ and $n = 5$.*[1]

---

[1] The last case, in which the graph is the 5-cycle, is missing from the statement of this result in [11, Theorem 2.12]. The proof in [11] tacitly considers only graphs with at least 6 vertices.

Lemma 5 (Koren [14]). *A bipartite graph is determined up to isomorphism by the conditions that every of the m vertices in one part has degree c and every of the n vertices in the other part has degree d if and only if $c \in \{0, 1, n-1, n\}$ or $d \in \{0, 1, m-1, m\}$.*

If G contains a subgraph $G[X]$ or $G[X, Y]$ that is induced by some $X, Y \in \mathcal{P}_G$ but not listed in Lemma 3, then Lemmas 4 and 5 imply that this subgraph can be replaced by a non-isomorphic regular or biregular graph with the same parameters. Hence, in order to prove Lemma 3 it suffices to show that the resulting graph H is indistinguishable from G by color refinement. The graphs G and H in the following lemma have the same vertex set. Given a vertex u, we distinguish its colors $C_G^i(u)$ and $C_H^i(u)$ in the two graphs.

Lemma 6. *Let X and Y be cells of the stable partition of a graph G.*

(i) If H is obtained from G by replacing the edges of the subgraph $G[X]$ with the edges of an arbitrary regular graph of the same degree on the same vertex set X, then $C_G^i(u) = C_H^i(u)$ for any $u \in V(G)$ and any i.

(ii) If H is obtained from G by replacing the edges of the subgraph $G[X, Y]$ with the edges of an arbitrary biregular graph with the same vertex partition such that the vertex degrees remain unchanged, then $C_G^i(u) = C_H^i(u)$ for any $u \in V(G)$ and any i.

Proof of Lemma 3. **(A)** If $G[X]$ is a graph not from the list, by Lemma 4, it is not a unigraph. Hence, we can modify G locally on X by replacing $G[X]$ with a non-isomorphic regular graph with the same parameters. Part (i) of Lemma 6 implies that the resulting graph H satisfies Equality (2) for any i, implying that CR does not distinguish between G and H. The graphs G and H are non-isomorphic because, by Part (i) of Lemma 6 and by Lemma 1, an isomorphism from G to H would induce an isomorphism from $G[X]$ to $H[X]$. This shows that G is not amenable.

(B) This condition follows, similarly to Condition **A**, from Lemma 5 and Part (ii) of Lemma 6.   □

4   Global Structure of Amenable Graphs

Recall that $\mathcal{P}_G$ is the stable partition of the vertex set of a graph G, and that elements of $\mathcal{P}_G$ are called cells. We define the auxiliary *cell graph* $C(G)$ of G to be the complete graph on the vertex set $\mathcal{P}_G$ with the following labeling of vertices and edges. A vertex X of $C(G)$ is called *homogeneous* if the graph $G[X]$ is complete or empty and *heterogeneous* otherwise. An edge $\{X, Y\}$ of $C(G)$ is called *isotropic* if the bipartite graph $G[X, Y]$ is either complete or empty and *anisotropic* otherwise. A path $X_1 X_2 \ldots X_l$ in $C(G)$ where every edge $\{X_i, X_{i+1}\}$ is anisotropic will be referred to as an *anisotropic path*. If also $\{X_l, X_1\}$ is an anisotropic edge, we speak of an *anisotropic cycle*. In the case that $|X_1| = |X_2| = \ldots = |X_l|$, such a path (or cycle) is called *uniform*.

For graphs fulfilling Conditions **A** and **B** of Lemma 3 we refine the labeling of the vertices and edges of $C(G)$ as follows. A heterogeneous cell $X \in \mathcal{P}_G$ is called *matching, co-matching,* or *pentagonal* depending on the type of $G[X]$. Note that a matching or co-matching cell X always consists of at least 4 vertices. Further, an anisotropic edge $\{X, Y\}$ is called *constellation* if $G[X, Y]$ is a disjoint union of stars, and *co-constellation* otherwise (in the latter case, the bipartite complement of $G[X, Y]$ is a disjoint union of stars). Likewise, homogeneous cells X (and isotropic edges $\{X, Y\}$) are called *empty* if the graph $G[X]$ (resp. $G[X, Y]$) is empty, and *complete* otherwise.

Note that if an edge $\{X, Y\}$ of a uniform path or cycle is a constellation, then $G[X, Y]$ is a matching graph.

Lemma 7. *The cell graph $C(G)$ of an amenable graph G has the following properties:*

(C) $C(G)$ *contains no uniform anisotropic path connecting two heterogeneous cells;*

(D) $C(G)$ *contains no uniform anisotropic cycle;*

(E) $C(G)$ *contains neither an anisotropic path $XY_1 \ldots Y_l Z$ such that $|X| < |Y_1| = \ldots = |Y_l| > |Z|$ nor an anistropic cycle $XY_1 \ldots Y_l X$ such that $|X| < |Y_1| = \ldots = |Y_l|$;*

(F) $C(G)$ *contains no anisotropic path $XY_1 \ldots Y_l$ such that $|X| < |Y_1| = \ldots = |Y_l|$ and the cell Y_l is heterogeneous.*

Proof. **(C)** Suppose that P is a uniform anisotropic path in $C(G)$ connecting two heterogeneous cells X and Y. Let $k = |X| = |Y|$. Complementing $G[A, B]$ for each co-constellation edge $\{A, B\}$ of P, in G we obtain k vertex-disjoint paths connecting X and Y. These paths determine a one-to-one correspondence between X and Y. Given $v \in X$, denote its mate in Y by v^*. Call P *conducting* if this correspondence is an isomorphism between $G[X]$ and $G[Y]$, that is, two vertices u and v in X are adjacent exactly when their mates u^* and v^* are adjacent. In the case that one of X and Y is matching and the other is co-matching, we call P *conducting* also if the correspondence is an isomorphism between $G[X]$ and the complement of $G[Y]$.

We construct a non-isomorphic graph H such that CR does not distinguish between G and H. Since X and Y are heterogeneous, we can replace the edges of the subgraph $G[X]$ with the edges of an isomorphic but different graph on the same vertex set X such that P is a conducting path in the resulting graph H if and only if P is a non-conducting path in G. Now, Part (i) of Lemma 6 implies that CR computes the same coloring for G and H and does not distinguish between them. On the other hand, Lemma 1 implies that any isomorphism ϕ between G and H must map each cell to itself. Since $\phi(v^*) = \phi(v)^*$, ϕ must also preserve the conducting property along the path P. It follows that G and H are not isomorphic. Hence, G is not amenable.

(D) Suppose that $C(G)$ contains a uniform anisotropic cycle Q of length m. All cells in Q have the same cardinality; denote it by k. Complementing $G[A, B]$

for each co-constellation edge $\{A, B\}$ of Q, in G we obtain the vertex-disjoint union of cycles whose lengths are multiples of m. As two extreme cases, we can have k cycles of length m each or we can have a single cycle of length km. Denote the isomorphism type of this union of cycles by $\tau(Q)$. Note that this type is isomorphism invariant: For an isomorphism ϕ from G to another graph H, $\tau(\phi'(Q)) = \tau(Q)$ for the induced isomorphism ϕ' from $C(G)$ to $C(H)$.

Let X and Y be two consecutive cells in Q. We can replace the subgraph $G[X, Y]$ with an isomorphic but different bipartite graph so that in the resulting graph H, $\tau(Q)$ becomes either kC_m or C_{km}, whatever we wish. In particular, we can replace the subgraph $G[X, Y]$ in such a way that $\tau(Q)$ is changed.

Similarly as for Condition **C**, we use Part (ii) of Lemma 6 to argue that CR does not distinguish between G and H. Furthermore, $G \not\cong H$ because the types $\tau(Q)$ in G and H are different. Therefore, G is not amenable.

(E) Suppose that $C(G)$ contains an anisotropic path $P = XY_1 \ldots Y_l Z$ such that $|X| < |Y_1| = \ldots = |Y_l| > |Z|$ (for the case of a cycle, where $Z = X$, the argument is virtually the same). Let $G[X, Y_1] = sK_{1,t}$ and $G[Z, Y_l] = aK_{1,b}$, where $s, a, t, b \geq 2$ (if any of these subgraphs is a co-constellation, we consider its complement). Thus, $|X| = s$, $|Z| = a$, and $|Y_1| = |Y_l| = st = ab$.

Like in the proof of Condition**C**, the uniform anisotropic path $Y_1 \ldots Y_l$ determines a one-to-one correspondence between the cells Y_1 and Y_l that can be used to make the identification $Y_1 = Y_l = \{1, 2, \ldots, st\} = Y$. For each $x \in X$, let Y_x denote the set of vertices in Y adjacent to x. The set Y_z is defined similarly for each $z \in Z$. Note that for any $x \neq x'$ in X and $z \neq z'$ in Z,

$$|Y_x| = t, \quad |Y_z| = b, \quad Y_x \cap Y_{x'} = \emptyset, \text{ and} Y_z \cap Y_{z'} = \emptyset.$$

We regard $\mathcal{Y}_G = \{Y_x\}_{x \in X} \cup \{Y_z\}_{z \in Z}$ as a hypergraph on the vertex set Y. Note that $\mathcal{Y}_G$ has multiple hyperedges if $Y_x = Y_z$ for some x and z. Without loss of generality, we can assume that the hyperedges Y_z, $z \in Z$, form consecutive intervals in Y. We call the anisotropic path P *flat*, if there exists no pair $(x, z) \in X \times Z$ such that one of the two hyperedges Y_x and Y_z is contained in the other.

We construct a non-isomorphic graph H such that CR does not distinguish between G and H. If P is flat in G, we replace the edges of the subgraph $G[X, Y_1]$ by the edges of an isomorphic but different biregular graph such that P becomes non-flat in the resulting graph H. More precisely, we replace the edges in such a way that all hyperedges of $\mathcal{Y}_H$ form consecutive intervals in Y by letting $\mathcal{Y}_H = \{Y_i\}_{i \in [s]} \cup \{Y_z\}_{z \in Z}$, where $Y_i = \{(i-1)t + 1, \ldots, it\}$. Likewise, if P is non-flat in G, we replace the edges of $G[X, Y_1]$ such that P becomes flat in H by letting $Y_i = \{i, i + s, \ldots, i + (t-1)s\}$.

Now, Part (i) of Lemma 6 implies that CR computes the same coloring for G and H and does not distinguish between them. On the other hand, Lemma 1 implies that any isomorphism ϕ between G and H must map each cell to itself. As ϕ must also preserve the flatness property of the path P, it follows that G and H are not isomorphic. Hence, G is not amenable.

(F) Suppose that $C(G)$ contains an anisotropic path $XY_1 \ldots Y_l$ where $|X| < |Y_1| = \ldots = |Y_l|$ and Y_l is heterogeneous. Let $G[X, Y_1] = sK_{1,t}$ (in the case of a co-constellation, we consider the complement). Since $s, t \geq 2$ and $|Y_1| = st$, the cell Y_l cannot be pentagonal. Considering the complement if needed, we can assume without loss of generality that Y_l is matching. Like in the proof of Condition **E**, the uniform anisotropic path $Y_1 \ldots Y_l$ determines a one-to-one correspondence between the cells Y_1 and Y_l that can be used to make the identification $Y_1 = Y_l = \{1, 2, \ldots, st\} = Y$. Consider the hypergraph $\mathcal{Y}_G = \{Y_x\}_{x \in X} \cup \mathcal{E}$, where $Y_x = N(x) \cap Y_1$ and $\mathcal{E}$ consists of the pairs of adjacent vertices in $G[Y_l]$. Now, exactly as in the proof of Condition **E**, we can change the isomorphism type of $\mathcal{Y}_G$ by replacing the edges of the subgraph $G[X, Y_1]$ by the edges of an isomorphic biregular graph. This yields a non-isomorphic graph H that is indistinguishable from G by CR. □

It turns out that Conditions **A–F** are not only necessary for amenability (as shown in Lemmas 3 and 7) but also sufficient. As a preparation we first prove the following Lemma 8 that reveals a tree-like structure of amenable graphs. By an *anisotropic component* of the cell graph $C(G)$ we mean a maximal connected subgraph of $C(G)$ whose edges are all anisotropic. Note that if a vertex of $C(G)$ has no incident anisotropic edges, it forms a single-vertex anisotropic component.

Lemma 8. *Suppose that a graph G satisfies Conditions **A–F**. Then for any anisotropic component A of $C(G)$, the following is true.*

(G) *A is a tree with the following monotonicity property. Let R be a cell in A of minimum cardinality and let A_R be the rooted directed tree obtained from A by rooting A at R. Then $|X| \leq |Y|$ for any directed edge (X, Y) of A_R.*

(H) *A contains at most one heterogeneous vertex. If R is such a vertex, it has minimum cardinality among the cells of A.*

Proof. **(G)** A cannot contain any uniform cycle by Condition **D** and any other cycle by Condition **E**. The monotonicity property follows from Condition **E**.

(H) Assume that A contains more than one heterogeneous cell. Consider two such cells S and T. Let $S = Z_1, Z_2, \ldots, Z_l = T$ be the path from S to T in A. The monotonicity property stated in Condition **G** implies that there is j (possibly $j = 1, l$) such that $|Z_1| \geq \ldots \geq |Z_j| \leq \ldots \leq |Z_l|$. Since the path cannot be uniform by Condition **C**, at least one of the inequalities is strict. However, this contradicts Condition **F**.

Suppose that R is a heterogeneous cell in A. Consider now a path $R = Z_1, Z_2, \ldots, Z_l = S$ in A where S is a cell with the smallest cardinality. By the monotonicity property and Condition **F**, this path must be uniform, proving that $|R| = |S|$. □

In combination with Conditions **A** and **B**, Conditions **G** and **H** on anisotropic components give a very stringent characterization of amenability.

Theorem 9. *For a graph G the following conditions are equivalent:*

(i) G is amenable.
*(ii) G satisfies Conditions **A–F**.*

(iii) G satisfies Conditions **A, B, G** *and* **H**.

Proof. It only remains to show that any graph G fulfilling the Conditions **A, B, G** and **H** is amenable. Let H be a graph indistinguishable from G by CR. Then we have to show that G and H are isomorphic.

Consider the coloring C^s corresponding to the stable partition $\mathcal{P}^s$ of the disjoint union $G + H$. Since G and H satisfy Equality (2) for $i = s$, there is a bijection $f : \mathcal{P}_G \to \mathcal{P}_H$ matching each cell X of the stable partition of G to the cell $f(X) \in \mathcal{P}_H$ such that the vertices in X and $f(X)$ have the same C^s-color. Moreover, Equality (2) implies that $|X| = |f(X)|$. We claim that for any cells X and Y of G,

(a) $G[X] \cong H[f(X)]$ and
(b) $G[X, Y] \cong H[f(X), f(Y)]$,

implying that f is an isomorphism from $C(G)$ to $C(H)$.

Indeed, since X and $f(X)$ are cells of the stable partitions $\mathcal{P}_G$ and $\mathcal{P}_H$, both $G[X]$ and $H[f(X)]$ are regular. Since $X \cup f(X)$ is a cell of the stable partition $\mathcal{P}^s$ of $G + H$, the graphs $G[X]$ and $H[f(X)]$ have the same degree. By Condition **A**, $G[X]$ is a unigraph, implying Property (a). Property (b) follows from Condition **B** by a similar argument.

We now construct an isomorphism ϕ from G to H. By Lemma 1, we should have $\phi(X) = f(X)$ for each cell X. Therefore, we have to define the map $\phi : X \to f(X)$ on each X.

By Condition **H**, an anisotropic component A of the cell graph $C(G)$ contains at most one heterogeneous cell. Denote it by R_A if it exists. Otherwise fix R_A to be an arbitrary cell of the minimum cardinality in A.

For each A, define ϕ on $R = R_A$ to be an arbitrary isomorphism from $G[R]$ to $H[f(R)]$, which exists according to (a). After this, propagate ϕ to any other cell in A as follows. By Condition **G**, A is a tree. Let A_R be the directed rooted tree obtained from A by rooting it at R. Suppose that ϕ is already defined on X and (X, Y) is an edge in A. By the monotonicity property in Condition **G** and our choice of R, we can assume that $|X| \leq |Y|$. Then ϕ can be extended to Y so that this is an isomorphism from $G[X, Y]$ to $H[f(X), f(Y)]$. This is possible by (b) due to the fact that all vertices in Y have degree 1 in $G[X, Y]$ or its bipartite complement (and the same holds for all vertices in $f(Y)$ in the graph $H[f(X), f(Y)]$).

It remains to argue that the map ϕ obtained in this way is indeed an isomorphism from G to H. It suffices to show that ϕ is an isomorphism between $G[X]$ and $H[f(X)]$ for each cell X of G and between $G[X, Y]$ and $H[f(X), f(Y)]$ for each pair of cells X and Y.

If X is homogeneous, $f(X)$ is homogeneous of the same type, complete or empty, according to (a). In this case, any ϕ is an isomorphism from $G[X]$ to $H[f(X)]$. If X is heterogeneous, the assumption of the lemma says that it belongs to a unique anisotropic component A (and $X = R_A$). Then ϕ is an isomorphism from $G[X]$ to $H[f(X)]$ by construction.

If $\{X, Y\}$ is an isotropic edge of $C(G)$, then (b) implies that $\{f(X), f(Y)\}$ is an isotropic edge of $C(H)$ of the same type, complete or empty. In this case, ϕ is an isomorphism from $G[X, Y]$ to $H[f(X), f(Y)]$, no matter how it is defined. If $\{X, Y\}$ is anisotropic, it belongs to some anisotropic component A, and ϕ is an isomorphism from $G[X, Y]$ to $H[f(X), f(Y)]$ by construction. $\square$

5   Examples and Applications

Theorem 9 is a convenient tool for verifying amenability. For example, amenability of discrete graphs is a well-known fact. Recall that those are graphs whose stable partitions consist of singletons. Since the cell graph has no anisotropic edge in this case, any anisotropic component of a discrete graph consists of a single cell. Hence, Conditions **A** and **B** as well as Conditions **G** and **H** on anisotropic components are fulfilled by trivial reasons.

Checking these four conditions, we can also reprove the amenability of trees. Moreover, our argument extends to the class of forests. Note in this respect that the class of amenable graphs is not closed under disjoint unions. For example, $C_3 + C_4$ is indistinguishable by CR from C_7 and, hence, is not amenable.

Corollary 10. *All forests are amenable.*

Proof. A regular acyclic graph is either an empty or a matching graph. This implies Condition **A**. Condition **B** follows from the observation that biregular acyclic graphs are either empty or disjoint unions of stars.

Let $C^*(G)$ be the version of the cell graph $C(G)$ where all empty edges are removed. If $C^*(G)$ contains a cycle, G must contain a cycle as well. Therefore, if G is acyclic, then $C^*(G)$ is acyclic too, and any anisotropic component of $C(G)$ must be a tree. To prove the monotonicity property in Condition **G**, it suffices to show that $C(G)$ cannot contain an anisotropic path $XY_1 \ldots Y_l Z$ with $|X| < |Y_1| = \cdots = |Y_l| > |Z|$. But this easily follows since in this case each vertex of the induced subgraph $G[X \cup Y_1 \cup \ldots \cup Y_l \cup Z]$ has degree at least 2 in G, contradicting the acyclicity of G.

To prove Condition **H**, suppose that $C(G)$ contains an anisotropic path $X_0, X_1, \ldots, X_l$ connecting two heterogeneous cells X_0 and X_l. Then each vertex of the induced subgraph $G[X_0 \cup X_1 \cup \ldots \cup X_{l-1} \cup X_l]$ has degree at least 2 in G, a contradiction. The same contradiction arises if such a path connects a heterogeneous cell X_0 with an arbitrary cell X_l, where $|X_l| < |X_{l-1}|$. Hence, X_0 must have minimum cardinality among all cells belonging to the same anisotropic component. $\square$

Our characterization of amenable graphs via Conditions **A**, **B**, **G** and **H** leads to an efficient test for amenability of a given graph, that has the same time complexity as CR. It is known (Cardon and Crochemore [7]; see also [4]) that the stable partition of a given graph G can be computed in time $O((n+m) \log n)$. It is supposed that G is presented by its adjacency list.

Corollary 11. *The class of amenable graphs is recognizable in time* $O((n + m) \log n)$, *where n and m denote the number of vertices and edges of the input graph.*

Proof. Using known algorithms, we first compute the stable partition $\mathcal{P}_G = \{X_1, \ldots, X_k\}$ of the input graph G. Let $C^*(G)$ be the version of the cell graph $C(G)$ where all empty edges are removed. We can compute the adjacency list of each vertex X_i of $C^*(G)$ by traversing the adjacency list of an arbitrary vertex $u \in X_i$ and listing all cells X_j that contain a vertex v adjacent to u. Simultaneously, we compute for each pair (i, j) such that $i = j$ or $\{X_i, X_j\}$ is an edge of $C^*(G)$ the number d_{ij} of neighbors in X_j of any vertex in X_i. Knowing the numbers $|X_i|$, $|X_j|$ and d_{ij} allows us to determine whether all the subgraphs $G[X_i]$ and $G[X_i, X_j]$ fulfill Conditions **A** and **B** of Lemma 3.

To check Conditions **G** and **H** we use breadth-first search in the graph $C^*(G)$ to find all anisotropic components A of $C(G)$ and, simultaneously, to check that each component A is a tree containing at most one heterogeneous cell. If we restart the search from an arbitrary cell in A having minimum cardinality, we can also check for each forward edge of the resulting search tree whether the monotonicity property of Condition **G** is fulfilled. $\square$

We conclude by considering logical aspects of our result. A *counting quantifier* $\exists^m$ opens a sentence saying that there are at least m elements satisfying some property. Immerman and Lander [10] discovered an intimate connection between color refinement and 2-variable first-order logic with counting quantifiers. This connection implies that amenability of a graph is equivalent to its definability in this logic. Thus, Corollary 11 asserts that the class of graphs definable by a first-order sentence with counting quantifiers and occurrences of just 2 variables is recognizable in polynomial time.

References

1. Arvind, V., Köbler, J., Rattan, G., Verbitsky, O.: On Tinhofer's linear programming approach to isomorphism testing. In: In: Proceedings of the 40th International Symposium on Mathematical Foundations of Computer Science (MFCS), Lecture Notes in Computer Science. Springer (2015) (to appear)
2. Babai, L., Erdős, P., Selkow, S.M.: Random graph isomorphism. SIAM J. Comput. **9**(3), 628–635 (1980)
3. Babai, L., Kucera, L.: Canonical labelling of graphs in linear average time. In: Proceedings of the 20th Annual Symposium on Foundations of Computer Science, pp. 39–46, (1979)
4. Berkholz, C., Bonsma, P., Grohe, M.: Tight lower and upper bounds for the complexity of canonical colour refinement. In: Bodlaender, H.L., Italiano, G.F. (eds.) ESA 2013. LNCS, vol. 8125, pp. 145–156. Springer, Heidelberg (2013)
5. Borri, A., Calamoneri, T., Petreschi, R.: Recognition of unigraphs through superposition of graphs. J. Graph Algorithms Appl. **15**(3), 323–343 (2011)
6. Busacker, R., Saaty, T.: Finite Graphs and Networks: An Introduction with Applications. International Series in Pure and Applied Mathematics. McGraw-Hill Book Company, New York (1965)

7. Cardon, A., Crochemore, M.: Partitioning a graph in $O(|A| \log_2 |V|)$. Theor. Comput. Sci. **19**, 85–98 (1982)
8. Godsil, C.: Compact graphs and equitable partitions. Linear Algebra Appl. **255**(13), 259–266 (1997)
9. Grohe, M., Kersting, K., Mladenov, M., Selman, E.: Dimension reduction via colour refinement. In: Schulz, A.S., Wagner, D. (eds.) ESA 2014. LNCS, vol. 8737, pp. 505–516. Springer, Heidelberg (2014)
10. Immerman, N., Lander, E.: Describing graphs: a first-order approach to graph canonization. In: Selman, A.L. (ed.) Complexity Theory Retrospective, pp. 59–81. Springer, Heidelberg (1990)
11. Johnson, R.: Simple separable graphs. Pac. J. Math. **56**, 143–158 (1975)
12. Kersting, K., Mladenov, M., Garnett, R., Grohe, M.: Power iterated color refinement. In: Proceedings of the Twenty-Eighth AAAI Conference on Artificial Intelligence, pp. 1904–1910. AAAI Press (2014)
13. Kiefer, S., Schweitzer, P., Selman, E.: Graphs identified by logics with counting. In:Graphs identified by logics with counting. In: Proceedings of the 40th International Symposium on Mathematical Foundations of Computer Science (MFCS), Lecture Notes in Computer Science. Springer (2015) (to appear)
14. Koren, M.: Pairs of sequences with a unique realization by bipartite graphs. J. Comb. Theor. Series B **21**(3), 224–234 (1976)
15. Krebs, A., Verbitsky, O.: Universal covers, color refinement, and two-variable logic with counting quantifiers: lower bounds for the depth. In: Proceedings of the 30th ACM/IEEE Annual Symposium on Logic in Computer Science (LICS), IEEE Computer Society (2015) (to appear)
16. Ramana, M.V., Scheinerman, E.R., Ullman, D.: Fractional isomorphism of graphs. Discrete Math. **132**(1–3), 247–265 (1994)
17. Shervashidze, N., Schweitzer, P., van Leeuwen, E.J., Mehlhorn, K., Borgwardt, K.M.: Weisfeiler-Lehman graph kernels. J. Mach. Learn. Res. **12**, 2539–2561 (2011)
18. Tinhofer, G.: Graph isomorphism and theorems of Birkhoff type. Computing **36**, 285–300 (1986)
19. Tyshkevich, R.: Decomposition of graphical sequences and unigraphs. Discrete Math. **220**(1–3), 201–238 (2000)
20. Valiente, G.: Algorithms on Trees and Graphs. Springer, Heidelberg (2002)

Block Representation of Reversible Causal Graph Dynamics

Pablo Arrighi[1], Simon Martiel[2]([✉]), and Simon Perdrix[3]

[1] Aix-Marseille University, LIF, 13288 Marseille Cedex 9, France
pablo.arrighi@univ-amu.fr
[2] University Nice-Sophia Antipolis, I3S, 06900 Sophia Antipolis, France
martiel@i3s.unice.fr
[3] CNRS, LORIA, Inria Project Team CARTE,
University de Lorraine, Nancy, France
simon.perdrix@loria.fr

Abstract. Causal Graph Dynamics extend Cellular Automata to arbitrary, bounded-degree, time-varying graphs. The whole graph evolves in discrete time steps, and this global evolution is required to have a number of physics-like symmetries: shift-invariance (it acts everywhere the same) and causality (information has a bounded speed of propagation). We study a further physics-like symmetry, namely reversibility. More precisely, we show that Reversible Causal Graph Dynamics can be represented as finite-depth circuits of local reversible gates.

Keywords: Bijective · Invertible · Locality · Cayley graphs · Reversible cellular automata

1 Introduction

Cellular Automata (CA) consist in a $\mathbb{Z}^n$ grid of identical cells, each of which may take a state among a finite set Σ. Thus the configurations are in $\Sigma^{\mathbb{Z}^n}$. The state of each cell at time $t + 1$ is given by applying a fixed local rule f to the cell and its neighbours, synchronously and homogeneously across space. CA constitute the most established model of computation that accounts for euclidean space. They are widely used to model spatially distributed computation (self-replicating machines, synchronization problems...), as well as a great variety of multi-agents phenomena (traffic jams, demographics...). But their origin lies in Physics, where they are commonly used to model waves or particles. And since small scale physics is understood to be reversible, it was natural to endow them with this further, physics-like symmetry: reversibility. The study of Reversible CA (RCA) was further motivated by the promise of lower energy consumption, according to Landauer's principle. RCA have turned out to have a beautiful mathematical theory, which relies on topological and algebraic characterizations [9] in order to prove that any RCA can be expressed as a finite-depth circuits of local reversible permutations or 'blocks' [7,10,11].

© Springer International Publishing Switzerland 2015
A. Kosowski and I. Walukiewicz (Eds.): FCT 2015, LNCS 9210, pp. 351–363, 2015.
DOI: 10.1007/978-3-319-22177-9_27

Causal Graph Dynamics (CGD) [1,3], on the other hand, deal with a twofold extension of CA. First, the underlying grid is extended to being an arbitrary – possibly infinite – bounded degree graph G. Informally, this means for that each vertex of the graph may take a state among a finite set Σ, so a configuration is an element of $\Sigma^{V(G)}$, and the edges of the graph stand for the locality of the evolution: the next state of a vertex depends only on the states of the vertices which are at distance at most k, i.e. in a disk of radius k, for some fixed integer k. Second, the graph itself is allowed to evolve over time. Informally, this means having configurations in $\bigcup_G \Sigma^{V(G)}$. This has led to a model where the local rule f is applied synchronously and homogeneously on every possible subdisk of the input graph, thereby producing small patches of the output graphs, whose union constitutes the output graph. Figure 1 illustrates the concept of these CA over graphs.

CGD are motivated by the countless situations in which some agents interact with their neighbours, leading to a global dynamics in which the notion of who is next to whom also varies in time (e.g. agents become physically connected, get to exchange contact details, move around...). Indeed, several existing models (of physical systems, computer processes, biochemical agents, economical agents, social networks...) feature such neighbour-to-neighbour interactions with time-varying neighbourhood, thereby generalizing CA for their specific sake (e.g. self-reproduction as [17], discrete general relativity à la Regge calculus [14],

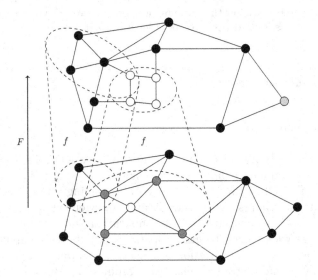

Fig. 1. *Informal illustration of Causal Graph Dynamics.* The entire graph evolves into another according to a global function F. But this evolution is causal (information propagates at a bounded speed) and homogeneous (same causes lead to same effects). This global function approach has been proved equivalent [3] to applying a local rule f to every subdisk of the input graphs, leading to small output graphs whose union make up the output graph. But this local rule f is not reversible. In this paper, we take the global approach as the starting point, and show that it can be implemented by reversible local mechanisms.

etc.). CGD provide a theoretical framework, for these models. Some graph rewriting models, such as Amalgamated Graph Transformations [6] and Parallel Graph Transformations [8,15,16], also work out rigorous ways to apply a local rewriting rule synchronously throughout a graph, albeit with a different, category-theory-based perspective. In particular the topological approach we follow and the reversibility question that we address have not been considered in these works.

This paper studies CGD in the reversible regime. From a theoretical Computer Science perspective, the point is therefore to generalize RCA theory to arbitrary, bounded-degree, time-varying graphs. From this perspective, our main result is the proof that Reversible CGD admit a block representation, i.e. an implementation as a finite-depth circuit of local gates. This is a non-trivial problem: the [11] construction seems unapplicable with dynamical graphs. We manage to apply, after some work, a proof scheme which comes from Quantum CA theory [5].

From a theoretical physics perspective, the question whether the reversibility of small scale physics can be reconciled with the time-varying topology of large scale physics (relativity), is a topic of debate and constant investigation. This paper provides all by itself a first toy, discrete, classical model where reversibility and time-varying topology coexist and interact. But ultimately, this deep question would need to be addressed in a quantum mechanical setting. This paper is indeed part of a long standing program to approach these issues in the framework of CA, and more precisely through the "axiomatic definition of a global evolution versus existence of an implementation for it, via local mechanisms" question.

Can CA be implemented by local mechanisms?			
CA type...	Classical	Reversible	Quantum
Grid	Yes [9]	Yes [10]	Yes [5]
Graphs	Yes [1,3]	This paper	Future

2 Pointed Graph Modulo, Paths, and Operations

Pointed Graph Modulo. There are two main approaches to CA. The one with a local rule, usually denoted f, is the constructive one, but CA can also be defined in a more topological way as being exactly the shift-invariant continuous functions from Σ^{Z^n} to itself, with respect to a certain metric. Through a compactness argument, the two approaches are equivalent. This topological approach carries through to CA over graphs. But for this purpose, one has to make the set of graphs into an appropriate compact metric space, which can only be done for certain pointed graph modulo isomorphism – referred to as generalized Cayley graphs in [3]. This is worth the trouble, as the topological characterization is one of the crucial ingredients to prove that the inverse of a CGD is a CGD.

Basically, the pointed graphs modulo isomorphism (or pointed graphs modulo, for short) are the usual, connected, undirected, possibly infinite, bounded-degree graphs, but with a few added twists:

- Each vertex has *ports* in a finite set π. A vertex and its port are written $u : a$.
- An *edge* is an unordered pair $\{u : a, v : b\}$. I.e. edges are between ports of vertices, rather than vertices themselves. Because the port of a vertex can only appear in one edge, the degree of the graphs is bounded by $|\pi|$. We shall consider connected graphs only.
- The graphs are rooted i.e., there is a privileged pointed vertex playing the role of an origin, so that any vertex can be referred to relative to the origin, via a sequence of ports that lead to it.
- The graphs are considered modulo isomorphism, so that only the relative position of the vertices can matter.
- The vertices and edges are given labels taken from finite sets Σ and Δ, so that they may carry an internal state just like the cells of a cellular automaton.
- The labelling functions are partial, so that we may express our partial knowledge about part of a graph. For instance it is common that a local function may yield a vertex, its internal state, its neighbours, and yet have no opinion about the internal state of those neighbours.

The set of all pointed graphs modulo (see Fig. 2(c)) of ports π, vertex labels Σ and edge labels Δ is denoted $\mathcal{X}_{\Sigma,\Delta,\pi}$. A thorough formalization of pointed graphs modulo can be found in [3]. For the sake of this paper, Fig. 2 summarizes the construction of pointed graphs modulo from pointed graphs whose vertex names are dropped.

Paths and Vertices. Since we are considering pointed graphs modulo isomorphism, vertices no longer have a unique identifier, which may seem impractical when it comes to designating a vertex. Two elements come to our rescue. First, these graphs are pointed, thereby providing an origin. Second, the vertices are connected through ports, so that each vertex can tell between its different neighbours. It follows that any vertex of the graph can be designated by a sequence of ports in $(\pi^2)^*$ that lead from the origin to this vertex. The origin is designated by ε. For instance, say two vertices designated by a path u and a path v, respectively. Suppose there is an edge $e = \{u : a, v : b\}$. Then, v can be designated by

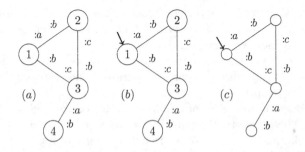

Fig. 2. *The different types of graphs.* (a) A graph G. (b) A pointed graph $(G, 1)$. (c) A pointed graph modulo isomorphism. These are anonymous: vertices have no names and can only be distinguished using the graph structure.

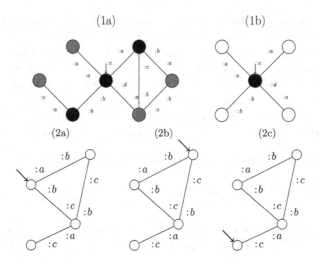

Fig. 3. *Operations over pointed graphs modulo.* (1) From X to X^0: taking the *subdisk of radius* 0. In general the neighbours of radius r are just those vertices which can be reached in r steps starting from the origin, whereas the disk of radius r, written X^r, is the subgraph induced by the neighbours of radius $r+1$, with labellings restricted to the neighbours of radius r and the edges between them. (*2a*) A pointed graph modulo X. (*2b*) X_{ab} the pointed graph modulo X *shifted* by ab. (*2c*) $X_{bc.ac}$ the pointed graph modulo X *shifted* by $bc.ac$, which also corresponds to the graph X_{ab} *shifted* by $cb.ac$. *Shifting* this last graph by $\overline{cb.ac} = ca.bc$ produces the graph (*2b*) again.

the path $u.ab$, where "." stands for the word concatenation. A thorough formalization of pointed graphs modulo and naming conventions can be found in [3].

Operations. Given a pointed graph modulo X, X^r denotes the subdisk of radius r around the pointer. The pointer of X can be moved along a path u, leading to $Y = X_u$. The pointer can be moved back where it was before, leading to $X = Y_{\bar{u}}$. We use the notation X_u^r for $(X_u)^r$ i.e., first the pointer is moved along u, then the subdisk of radius r is taken. A thorough formalization of these operations on pointed graph modulo can be found in [3]. For the sake of this paper, Fig. 3 illustrates the operations.

3   Reversible Causal Graph Dynamics

We will now recall the definition of CGD. We provide a topological definition in terms of shift-invariant continuous functions, rather than a constructive definition based on a local rule f applied synchronously across space (Fig. 1). The two were proved equivalent in [3].

A crucial point in the topological characterization of CGD is the correspondence between the vertices of a pointed graph modulo X, and those of its image $F(X)$. Indeed, on the one hand it is important to know that a given vertex $u \in X$ has become $u' \in F(X)$, e.g. in order to express shift-invariance $F(X_u) = F(X)_{u'}$, or to express continuity. But on the other hand since u' is named relative to ε, its determination requires some knowledge of X.

The following analogy provides a useful way of tackling this issue. Say that we were able to place a white stone on the vertex $u \in X$ that we wish to follow across evolution F. Later, by observing that the white stone is found at $u' \in F(X)$, we would be able to conclude that u has become u'. This way of grasping the correspondence between an image vertex and its antecedent vertex is a local, operational notion of an observer moving across the dynamics.

Definition 1 (Dynamics). *A dynamics* $(F, R_\bullet)$ *is given by*

- *a function* $F \colon \mathcal{X}_{\Sigma,\Delta,\pi} \to \mathcal{X}_{\Sigma,\Delta,\pi}$;
- *a map* $R_\bullet$, *with* $R_\bullet \colon X \mapsto R_X$ *and* $R_X \colon V(X) \to V(F(X))$.

For all X, *the function* R_X *can be pointwise extended to sets of vertices i.e.,* $R_X \colon \mathcal{P}(V(X)) \to \mathcal{P}(V(F(X)))$ *maps* S *to* $R_X(S) = \{R_X(u) \mid u \in S\}$.

The intuition is that R_X indicates which vertices $\{u', v', \dots\} = R_X(\{u, v, \dots\}) \subseteq V(F(X))$ will end up being marked as a consequence of $\{u, v, \dots\} \subseteq V(X)$ being marked. Now, clearly, the set $\{(X, S) \mid X \in \mathcal{X}_{\Sigma,\Delta,\pi}, S \subseteq V(X)\}$ is isomorphic to $\mathcal{X}_{\Sigma',\Delta,\pi}$ with $\Sigma' = \Sigma \times \{0, 1\}$. Hence, we can define the function F' that maps $(X, S) \cong X' \in \mathcal{X}_{\Sigma',\Delta,\pi}$ to $(F(X), R_X(S)) \cong F'(X') \in \mathcal{X}_{\Sigma',\Delta,\pi}$, and think of a dynamics as just this function $F' \colon \mathcal{X}_{\Sigma',\Delta,\pi} \to \mathcal{X}_{\Sigma',\Delta,\pi}$.

Since continuity and uniform continuity are equivalent over compact spaces, we give directly the definition of uniform continuity:

Definition 2 (Uniform Continuity). *A dynamics* $(F, R_\bullet)$ *is said to be continuous if for any* $m \geq 0$, *there exists* $n \geq 0$ *such that for every* X, Y, $X^n = Y^n$ *implies both*

- $F(X)^m = F(Y)^m$.
- $\operatorname{dom} R_X^m \subseteq V(X^n)$, $\operatorname{dom} R_Y^m \subseteq V(Y^n)$, *and* $R_X^m = R_Y^m$.

where R_X^m *denotes the partial map obtained as the restriction of* R_X *to the codomain* $F(X)^m$, *using the natural inclusion of* $F(X)^m$ *into* $F(X)$.

In the $F' \colon \mathcal{X}_{\Sigma',\Delta,\pi} \to \mathcal{X}_{\Sigma',\Delta,\pi}$ formalism, the two above conditions are equivalent to just one: F' continuous.

A dynamics can be be shifted to act over the region surrounding some vertex u.

Definition 3 (Shifted Dynamics). *Given a dynamics* $(F, R_\bullet)$ *over* $\mathcal{X}_{\Sigma,\Delta,\pi}$ *and* $u \in \pi^*$, $(F, R_\bullet)$ *shifted at* u *is the dynamics* $(F_u, R_{u,\bullet})$ *where*

$$F_u = X \mapsto \begin{cases} (F(X_u))_{R_{X_u}(\overline{u})} & \text{if } u \in X \\ X & \text{otherwise} \end{cases} \quad \text{and } R_{u,X} = v \mapsto \begin{cases} \overline{R_{X_u}(\overline{u})}.R_{X_u}(\overline{u}.v) & \text{if } u \in X \\ v & \text{otherwise} \end{cases}$$

The dynamics $(F, R_\bullet)$ which are shift invariant – i.e. $\forall u, (F_u, R_{u,\bullet}) = (F, R_\bullet)$ – satisfy the following property:

Lemma 1 (Shift-invariance). *A dynamics* $(F, R_\bullet)$ *is shift-invariant if and only if for every* X, $u \in X$, *and* $v \in X_u$,

- $F(X_u) = F(X)_{R_X(u)}$
- $R_X(u.v) = R_X(u).R_{X_u}(v)$.

The second condition expresses the shift-invariance of $R_\bullet$ itself. Notice that $R_X(\varepsilon) = R_X(\varepsilon).R_X(\varepsilon)$; hence $R_X(\varepsilon) = \varepsilon$.

Definition 4 (Boundedness). *A dynamics* $(F, R_\bullet)$ *is said to be bounded if there exists a bound* b *such that for any* X *and any* $w' \in F(X)$, *there exist* $u' \in \operatorname{im} R_X$ *and* $v' \in F(X)^b_{u'}$, *such that* $w' = u'.v'$.

The following is the topological definition of CGD:

Definition 5 (Causal Graph Dynamics). *A CGD is a shift-invariant, continuous, bounded dynamics.*

A reversible causal graph dynamics (RCGD) is an invertible CDG which inverse is a CGD itself:

Definition 6 (Reversible). *A CGD* $(F, R_\bullet)$ *is reversible if there exists* $S_\bullet$ *such that* $(F^{-1}, S_\bullet)$ *is a CGD.*

Actually it can be proven that if a CGD has an inverse, this inverse is necessarily a CGD [4]. This works through a compactness argument just like for CA, but goes beyond the scope of this paper. It can also be shown that:

Theorem 1 (Reversible Implies Almost-Vertex-Preserving). *Let* $(F, R_\bullet)$ *be a Reversible CGD over* $\mathcal{X}_{\Sigma, \Delta, \pi}$. *Then there exists a bound* p, *such that for any graph* X, *if* $|X| > p$ *then* R_X *is bijective.*

Moving Head. Figure 4 is an example of invertible CGD. In this example, a vertex, representing the head of an automaton, is moving along a line graph, representing a tape. The line graph is built using $ab-$edges, while the head is attached using either a $cc-$edge if it is travelling forward along the $ab-$edges, or $dd-$edges if it is travelling backwards. The transformation can be completed into a bijection over the entire set of graphs with $\pi = \{a, b, c, d\}$. It then accounts for several heads, etc. The resulting transformation is continuous, as the moving heads travel at speed one along the tape, and shift-invariant as it is possible to build a $R_\bullet$ operator verifying the right commutation properties.

4   Locality

Causal Graph Dynamics change the entire graph in one go. The word causal there refers to the fact that information does not propagate too fast. Local operations, on the other hand, act just in one bounded region of the graph, leaving the rest unchanged. We introduce the following locality definition:

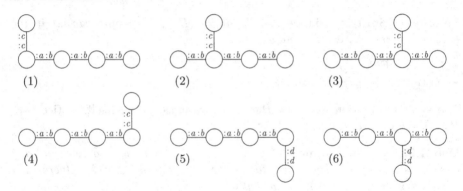

Fig. 4. *Moving head dynamics.* In this example, a moving head is running along a "tape" formed by a linear graph of alternating ab edges. When reaching the end of the line, the head starts moving backwards and changes the ports on its attaching edge to dd. (1) to (6) represent 6 consecutive configurations.

Definition 7 (Local Dynamics). *A dynamics $(L, S_\bullet)$ is r-local if it is continuous and bounded, and for any X and any $v \in L(X)$ with $|v| > r$, there exists $u \in X$ such that $L(X)_v^0 = X_u^0$ and $\forall w \in X_u^0$, $S_X(u.w) = v.w$.*

A local dynamics acts around the pointer of the graph modulo. To act around another position u, one can shift the local dynamics at u. Moreover, we may wish to apply a series of local operations at several positions u_i i.e., a circuit. However, applying a local operation may change the graph and hence vertex names, hence some care must be taken.

Definition 8 (Product). *Consider a local dynamics $(L, S_\bullet)$ and X a pointed graph modulo in its domain we define the product $\prod(L, S)$ as the limit when r goes to infinity of $(L^r, S_\bullet^r)$:*

$$L^r(X) = \prod_{i \in [1, \ldots, |V(X^r)|]} L_{u_i'}(X)$$

$$S_X^r = \prod_{i \in [1, \ldots, |V(X^r)|]} S_{u_i', \prod_{k \in [1, \ldots, i-1]} L_{u_k'}(X)}$$

where $\{u_1, u_2, \ldots\} = V(X)$ such that $i < j \Rightarrow |u_i| \le |u_j|$, and $u_1' = u_1$, $u_2' = S_{u_1', X}(u_2)$, $u_3' = S_{u_2', L_{u_1'}(X)}(u_3), \ldots$

Soundness of the Definition. For infinite graphs, the image of a graph X through the application of $\prod(L, S)$ needs to be defined as the limit of the sequence of graphs $(L^r(X))$ obtained by applying L to every node of the disk X^r. As $\mathcal{X}_{\Sigma, \Delta, \pi}$ is compact (see [3] for details), this sequence of graphs converges toward a limit graph X'. Moreover, for all radii r' there exists a radius r such that $X'^r = L^{r'}(X)$. Thus, X' corresponds to the graph X where the local dynamics $(L, S_\bullet)$ has been applied on every vertex.

5 Block Representation

A famous result on RCA [10], is that these admit a finite-depth, reversible circuit form, with gates acting only locally. The result carries through, with a different proof, to Quantum CA [5]. It is the Quantum CA proof scheme that we managed to adapt to RCGD. First, we show that conjugating a local operation with an RCGD still yields a local operation.

Proposition 1. *If $(F, R_\bullet)$ is an RCGD and $(L, S_\bullet)$ is a local dynamics, then $(L', T_\bullet)$ is a local dynamics, with $L' = F^{-1} \circ L \circ F$ and $T_X(u) = R'_{F^{-1}(L(F(X)))}(S_{F(X)}(R_X(u)))$, where the function $R'_\bullet$ is such that $(F^{-1}, R'_\bullet)$ is a CDG.*

Proof Outline. To say that 'L is local' is a formal way of saying that far away from ε, it acts like the identity. Hence, far away from ε, the expression for L' becomes $F^{-1} \circ F$ which is just the identity, and so L' is itself local. The actual proof is much more technical, see [4]. Notice that the function $R'_\bullet$ iwhich makes $(F^{-1}, R'_\bullet)$ a CGD is not unique in general.

Second, we give ourselves a little more space so as to mark which parts of the graph have been updated, or not.

Definition 9 (Marked Pointed Graphs Modulo). *Consider the set of pointed graph modulo $\mathcal{X}_{\Sigma,\Delta,\pi}$ with labels in Σ, and ports in π. Let $\Sigma' = \Sigma \times \{0,1\}$ and $\pi' = \pi \times \{0,1\}$. We define the set of marked pointed graph modulo $\overline{\mathcal{X}}_{\Sigma',\Delta,\pi'}$ to be the subset of $\mathcal{X}_{\Sigma',\Delta,\pi'}$ such that:*

(1) $\forall u \in X$, if u is labelled with (x,a) and $\{u{:}(i,b), v{:}(j,c)\} \in X$, then $a = c$.
(2) $\forall v \in X$, if $\{u{:}(i,b), v{:}(j,c)\} \in X$ and $\{u'{:}(i',b'), v{:}(j,c')\} \in X$, then $u = u'$.

Condition (1) states that a marked vertex is connected to the rest of the marked graph through marked ports only. Condition (2) states that if a vertex has both the marked instance and unmarked instance of a same port affected to an edge, then both edges lead to the same vertex.

Definition 10 (Mark Operation). *We define the mark operation μ as the following local dynamics $(L^\mu, S_\bullet^\mu)$ over $\mathcal{X}_{\Sigma',\Delta,\pi'}$. For any X in $\mathcal{X}_{\Sigma',\Delta,\pi'}$:*

- *if the label of ε is (x,a) in X then its label is $(x, 1{-}a)$ in $L^\mu(X)$.*
- *if $\{\varepsilon{:}(x,a), \varepsilon{:}(y,b)\} \in X$ then $\{\varepsilon{:}(x, 1{-}a), \varepsilon{:}(y, 1{-}b)\} \in L^\mu(X)$.*
- *if $\{\varepsilon{:}(x,a), v{:}(y,b)\} \in X$ with $v \neq \varepsilon$ then $\{\varepsilon{:}(x,a), v{:}(y, 1{-}b)\} \in L^\mu(X)$,*
- $S_X^\mu(u) = \begin{cases} \varepsilon & \text{if } u = \varepsilon \\ (x,a)(y, 1{-}b) & \text{if } \{\varepsilon{:}(x,a), v{:}(y,b)\} \in X \text{ with } v \neq \varepsilon \\ S_X^\mu(v).pq & if u = v.pq \, with \, p, q \in \pi' \end{cases}$

and leaving the rest of the graph X unchanged.

Notice that the set of marked graphs $\overline{\mathcal{X}}_{\Sigma',\Delta,\pi'}$ is nothing but the subset of $\mathcal{X}_{\Sigma',\Delta,\pi'}$ obtained by as closure of μ, and shifts, upon $\mathcal{X}_{\Sigma\times\{0\},\Delta,\pi\times\{0\}}$. Notice also that $\overline{\mathcal{X}}_{\Sigma',\Delta,\pi'}$ is a compact subset of $\mathcal{X}_{\Sigma',\Delta,\pi'}$.

It turns out that any RCGD admits an extension that allows for these marks.

Definition 11 (Reversible Extension). *Let* $(F, R_\bullet)$ *be an RCGD over* $\mathcal{X}_{\Sigma,\Delta,\pi}$. *We say that* $(F', R'_\bullet)$ *is a reversible extension of* $(F, R_\bullet)$ *if it is an RCGD over* $\overline{\mathcal{X}}_{\Sigma',\Delta,\pi'}$ *such that,*

- *For any* $X \in \mathcal{X}_{\Sigma,\Delta,\pi}$ *and* $u \in X$:

$$F'(X\times\{0\}) = F(X)\times\{0\} \qquad R'_{X\times\{0\}}(u\times\{0\}) = R_X(u)\times\{0\}$$
$$F'(X\times\{1\}) = X\times\{1\} \qquad R'_{X\times\{1\}}(u\times\{1\}) = u\times\{1\}$$

- *For any* $X \in \overline{\mathcal{X}}_{\Sigma',\Delta,\pi'}$ *such that* $|X| \leq p$ *and* $X \notin \mathcal{X}_{\Sigma\times\{0\},\Delta,\pi\times\{0\}}$:

$$F'(X) = X \qquad R'_X = u \mapsto u$$

where p *is that of Theorem 1;* $X\times\{0\}$ *consists in pairing with* 0 *all the vertex-states and edges of* X; *and* $u\times\{0\}$ *is defined as* $\varepsilon\times\{0\} = \varepsilon$ *and* $u.ab\times\{0\} = (u\times\{0\}){:}(a,0)(b,0)$.

Proposition 2 (Reversible Extension). *Any RCGD* $(F, R_\bullet)$ *over* $\mathcal{X}_{\Sigma,\Delta,\pi}$ *admits a reversible extension* $(F', R'_\bullet)$ *over* $\overline{\mathcal{X}}_{\Sigma',\Delta,\pi'}$.

Proof Outline. Since $(F, R_\bullet)$ is be induced by a local rule f (see [3] for details), one can construct another local rule f' acting over graphs in the set $\mathcal{X}_{\Sigma',\Delta,\pi'}$ as follow:

- When applied on an unmarked vertex u, f' creates the same subgraph $f(X_u^r)$ while preserving edges on marked ports.
- When applied to a marked vertex u, f' creates the subgraph X_u^0, preserving the connectivity of u.

Basically, the induced CGD $(F', R'_\bullet)$ performs the identity around every marked vertex, and performs $(F, R_\bullet)$ around unmarked vertices. More details are provided in [4].

In order to obtain our circuit-like form for RCGD, we will proceed by reversible, local updates.

Definition 12 (Conjugate Mark). *Given a reversible extension* $(F', R'_\bullet)$ *over* $\overline{\mathcal{X}}_{\Sigma',\Delta,\pi'}$, *we define the conjugate mark* K *as a dynamics* $(L^K, S_\bullet^K)$ *over* $\overline{\mathcal{X}}_{\Sigma',\Delta,\pi'}$ *as follows:*

$$L^K = F'^{-1} \circ L^\mu \circ F' \quad and \quad S_X^K(u) = T_{F'^{-1}(L^\mu(F'(X)))}(S_{F(X)}^\mu(R'_X(u)))$$

where the function $T_\bullet$ *is such that* $(F'^{-1}, T_\bullet)$ *is a CGD.*

Notice that by Proposition 1, the local update blocks are local operations. Moreover, since they are defined as a composition of invertible dynamics, they are invertible. In order to represent the whole of an RCGD, it suffices to apply these local update blocks at every vertex.

Theorem 2 (Reversible Localizability). *For any RCGD* $(F, R_\bullet)$ *over* $\mathcal{X}_{\Sigma,\Delta,\pi}$, $(F, R_\bullet)$ *and* $(\prod \mu)(\prod K)$ *act the same on all but a finite number of graphs, where* K *is the conjugate of* μ *with respect to* F' *a reversible extension of* F *i.e.,* $K = F' \circ \mu \circ F'^{-1}$.

Proof Outline. By Theorem 1, there exists $p > 0$ s.t. if $|X| > p$, R_X is invertible. These are the graphs we consider, i.e. all but a finite number. On these graphs, the action of $(\prod \mu)(\prod K)$ is equivalent to $(\prod \mu)(F'^{-1}, R_{\bullet}^{-1})(\prod \mu)(F', R_{\bullet})$. Therefore, given X s.t. $|X| > p$, we have $X \times \{0\} \xmapsto{(F', R_{\bullet})} F(X) \times \{0\} \xmapsto{\prod \mu}$ $F(X) \times \{1\} \xmapsto{(F'^{-1}, R_{\bullet}^{-1})} F(X) \times \{1\} \xmapsto{\prod \mu} F(X) \times \{0\}$. Full details are provided in [4].

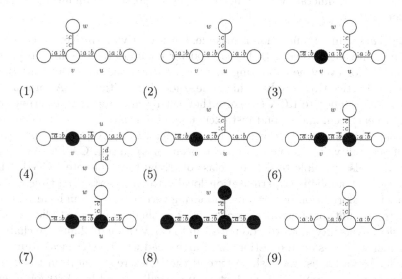

Fig. 5. Block representation of the moving head dynamics. (1) Initially, no vertices are marked. (2) to (4) Application of K_v. First F is applied, then v is marked, followed by the application of F^{-1}. (5) to (7) Application of K_u. (8) The graph once every K have been applied. The vertices just need to be unmarked by the μ's. (9) Altogether this implements one time step of F.

Notice that for the finite number of graphs when the decomposition of Theorem 2 does not apply, F is bijective. Therefore it just permutes those cases. Thus, this theorem generalizes the block decomposition of reversible cellular automata, which represents any reversible cellular automata as a circuit of finite depth of local permutations. Here, the mark μ and its conjugate K are the local permutations. The circuit is again of finite depth, a vertex u will be attained by all those K that act over $X_u^{r'}$, where r' is the locality radius of K. Therefore, the depth is less than $|\pi|^{r'}$. An example of such a decomposition is described in Fig. 5.

6   Conclusion

Summary of Results. The pointed graphs modulo are arbitrary bounded-degree networks, with a pointed vertex serving as the origin, and modulo renaming of vertices. We have studied Reversible Causal Graph Dynamics (RCGD), extending Reversible Cellular Automata results to time-varying, pointed graphs modulo. We have shown that RCGD admit a Block representation according to the formula $(\prod \mu)(\prod K)$, where the K's are commuting permutations, each of them acts locally on the neighbourhood of a vertex which gets marked, whereas μ is a local permutation that just unmarks a vertex. The result entails that RCGD can be implemented by reversible local mechanisms. The result also entails that RCGD can be enumerated: by enumerating every possible commuting, marking, local permutation K.

Future Work. This enumeration is feasible, but non-trivial, since every candidate local permutation K needs to be checked for commutation against its shifted version K_u for every neighbouring u. In CA theory, the notion of Partitioned CA, a.k.a Lattice-Gas, resolves this issue. Indeed, Partitioned CA form a class of trivial-to-enumerate RCA (because they act by first applying a permutation locally to each cell, and second just exchanging information between neighbouring cells). Yet, they are intrinsically universal for RCA [7,13], so this is without loss of generality. The same has been proven for Quantum CA [2]. As regards CGD, it is also possible to define a class of trivial-to-enumerate RCGD, which works by first applying a permutation locally to each vertex, and second just exchanging information between neighbouring vertices — although care must be taken to account for dynamical connectivity in this second step. The proofs of intrinsic universality, however, do not seem to carry through. Thus, we challenge the reader with this open question: are Partitioned CGD instrinsically universal for RCGD? As for us, we wish to extend the Block representation theorem of this paper to quantum CGD. Such structure results would be of strong interest to theoretical physics, where quantum, discrete spacetime dynamics are being studied in relation to Quantum Gravity [12].

Acknowledgements. This work has been funded by the ANR-12-BS02-007-01 TARMAC grant, the ANR-10-JCJC-0208 CausaQ grant, and the John Templeton Foundation, grant ID 15619. The authors acknowledge enlightening discussions with Bruno Martin and Emmanuel Jeandel. This work has been partially done when PA was delegated at Inria Nancy Grand Est, in the project team Carte.

References

1. Arrighi, P., Dowek, G.: Causal graph dynamics. In: Czumaj, A., Mehlhorn, K., Pitts, A., Wattenhofer, R. (eds.) ICALP 2012, Part II. LNCS, vol. 7392, pp. 54–66. Springer, Heidelberg (2012)
2. Arrighi, P., Grattage, J.: Partitioned quantum cellular automata are intrinsically universal. Nat. Comput. **11**, 13–22 (2012)
3. Arrighi, P., Martiel, S., Nesme, V.: Generalized Cayley graphs and cellular automata over them. submitted (long version) (2013). Pre-print arXiv:1212.0027
4. Arrighi, P., Martiel, S., Perdrix, P.: Reversible Causal Graph Dynamics (2015). Pre-print arXiv:1502.04368

5. Arrighi, P., Nesme, V., Werner, R.: Unitarity plus causality implies localizability. J. Comput. Syst. Sci. **77**, 372–378 (2010). QIP 2010 (long talk)
6. Boehm, P., Fonio, H.R., Habel, A.: Amalgamation of graph transformations: a synchronization mechanism. J. Comput. Syst. Sci. **34**(2–3), 377–408 (1987)
7. Durand-Lose, J.O.: Representing reversible cellular automata with reversible block cellular automata. Discrete Math. Theor. Comput. Sci. **145**, 154 (2001)
8. Ehrig, H., Lowe, M.: Parallel and distributed derivations in the single-pushout approach. Theor. Comput. Sci. **109**(1–2), 123–143 (1993)
9. Hedlund, G.A.: Endomorphisms and automorphisms of the shift dynamical system. Math. Syst. Theory **3**, 320–375 (1969)
10. Kari, J.: Representation of reversible cellular automata with block permutations. Theory Comput. Syst. **29**(1), 47–61 (1996)
11. Kari, J.: On the circuit depth of structurally reversible cellular automata. Fundamenta Informaticae **38**(1–2), 93–107 (1999)
12. Konopka, T., Markopoulou, F., Smolin, L.: Quantum graphity. Arxiv preprint (2006). Pre-print arXiv:hep-th/0611197
13. Morita, K.: Computation-universality of one-dimensional one-way reversible cellular automata. Inf. Process. Lett. **42**(6), 325–329 (1992)
14. Sorkin, R.: Time-evolution problem in Regge calculus. Phys. Rev. D. **12**(2), 385–396 (1975)
15. Taentzer, G.: Parallel and distributed graph transformation: Formal description and application to communication-based systems. Ph.D. thesis, Technische Universitat Berlin (1996)
16. Taentzer, G.: Parallel high-level replacement systems. Theor. comput. sci. **186**(1–2), 43–81 (1997)
17. Tomita, K., Kurokawa, H., Murata, S.: Graph automata: natural expression of self-reproduction. Phys. D: Nonlin. Phenom. **171**(4), 197–210 (2002)

Logic and Games

Reasoning with Global Assumptions in Arithmetic Modal Logics

Clemens Kupke[1], Dirk Pattinson[2], and Lutz Schröder[3]([✉])

[1] University of Strathclyde, Glasgow, UK
[2] Australian National University, Canberra, Australia
[3] Friedrich-Alexander Universität Erlangen-Nürnberg, Erlangen, Germany
lutz.schroeder@fau.de

Abstract. We establish a generic upper bound ExpTime for reasoning with global assumptions in coalgebraic modal logics. Unlike earlier results of this kind, we do not require a tractable set of tableau rules for the instance logics, so that the result applies to wider classes of logics. Examples are Presburger modal logic, which extends graded modal logic with linear inequalities over numbers of successors, and probabilistic modal logic with polynomial inequalities over probabilities. We establish the theoretical upper bound using a type elimination algorithm. We also provide a global caching algorithm that offers potential for practical reasoning.

Arithmetic modal logics feature arithmetical constraints on the number or total weight of successors. The simplest logics of this type compare weights to constants, such as graded modal logic [12] or some variants of probabilistic modal logic [17,19]. More involved examples are *Presburger modal logic* [7], which allows Presburger constraints on numbers of successors, and probabilistic modal logic with polynomial inequalities over probabilities. The former logic allows for statements like 'the majority of university students are female', or 'dance classes have an even number of participants', while probabilistic modal logic with polynomial inequalities can assert, for example, independence of events.

These logics are the main examples we address in a more general coalgebraic framework in this paper. Our main observation is that satisfiability for coalgebraic logics can be decided in a step-by-step fashion, peeling off one layer of operators at a time. We thus reduce the overall satisfiability problem to instances of a *one-step* satisfiability problem involving only immediate successor states, and hence no nesting of modalities [21,26]. We define a *strict* variant of this problem, distinguished by a judicious redefinition of its input size; if strict one-step satisfiability is in ExpTime, we obtain a (typically optimal) ExpTime upper bound for satisfiability under global assumptions in the full logic. For our two main examples, the requisite complexity bounds (in fact, even PSpace) on strict one-step satisfiability follow in essence directly from known complexity results in integer programming and the existential theory of reals, respectively; in other words, even in fairly complex examples the complexity bound for the full logic is obtained with comparatively little effort once the generic result is in place.

© Springer International Publishing Switzerland 2015
A. Kosowski and I. Walukiewicz (Eds.): FCT 2015, LNCS 9210, pp. 367–380, 2015.
DOI: 10.1007/978-3-319-22177-9_28

Applied to Presburger constraints, our results complement a recent result [6,7] showing that the complexity of Presburger modal logic without global assumptions is PSPACE, the same as for the modal logic K (or equivalently the description logic $\mathcal{ALC}$). For polynomial inequalities on probabilities, our syntax generalizes propositional *polynomial weight* formulae [11] to a full modal logic allowing nesting of weights (and global assumptions).

In more detail, our first contribution is to show via a type elimination algorithm [24] that also in presence of global assumptions (and, hence, in presence of the universal modality [13]), the satisfiability problem for coalgebraic modal logics is no harder than for K, i.e. in EXPTIME, provided strict one-step satisfiability is in EXPTIME. We then refine the algorithm to use global caching in the spirit of Goré and Nguyen [15], i.e. bottom-up expansion of a tableau-like graph and propagation of satisfiability and unsatisfiability through the graph. We thus potentially avoid constructing the whole exponential-sized tableau, and provide maneuvering space for heuristic optimization. Global caching algorithms have been demonstrated to perform well in practice [16].

Related Work. Our algorithms use a semantic method, and as such complement earlier results on global caching in coalgebraic description logics that rely on tractable sets of tableau rules [14], which are not currently available for our leading examples. (In fact, [18] gives tableau-style axiomatizations of various logics of linear inequalities over the reals and over the integers; however, over the integers the rules appear to be incomplete: if $\sharp(p)$ denotes the integer weight of successors satisfying p, the formula $2\sharp(p) < 1 \sqcup -2\sharp(p) < -1$ is clearly valid, but cannot be derived.)

Work related to XML query languages has shown that reasoning in Presburger fixpoint logic is EXPTIME complete [30], and that a logic with Presburger constraints and nominals is in EXPTIME [3], when these logics are interpreted *over finite trees*, thus not subsuming our EXPTIME upper bound for Presburger modal logic with global assumptions. It will likely be possible to obtain this bound via looping tree automata like for graded modal logic [31]. However, this would mean translating the target formula into an exponential-sized automaton, making exponential runtime the typical rather than the worst case; contrastingly, the main goal of our global caching algorithm is to avoid building the full exponential-sized set of types. Description logics with explicit quantification over integer variables and number restrictions mentioning integer variables [2] appear to be incomparable to Presburger modal logic: they do not support general linear inequalities, but on the other hand allow integer variables to be used at different modal depths. Reasoning with polynomial inequalities over probabilities has been studied in propositional logics [11] and in many-dimensional modal logics [20], which work with a single distribution on worlds rather than with world-dependent probability distributions as in [10,17,19].

1   Coalgebraic Logic

We briefly describe the key concepts of coalgebraic logic, a general framework that allows us to treat structurally different modal logics, such as Presburger and probabilistic modal logics, in a uniform way. We parametrize modal logics in terms of their *syntax* and their coalgebraic *semantics*. Syntactically, we work with a modal similarity type Λ of modal operators with given finite arities. The set $\mathcal{F}(\Lambda)$ of Λ-*formulas* is then given by the grammar

$$\mathcal{F}(\Lambda) \ni \phi, \psi ::= \bot \mid \phi \wedge \psi \mid \neg \phi \mid \heartsuit(\phi_1, \ldots, \phi_n) \qquad (\heartsuit \in \Lambda n\text{-ary}).$$

We omit explicit propositional atoms; these can be regarded as nullary modalities. The operators $\top$, $\rightarrow$, $\vee$, $\leftrightarrow$ are assumed to be defined in the standard way.

The semantics of formulas is then parametrized over the choice of a Λ-*structure* consisting of a set functor $T : \mathsf{Set} \rightarrow \mathsf{Set}$ and the assignment of an n-ary predicate lifting $[\![\heartsuit]\!]$ to each modality $\heartsuit \in \Lambda$, of arity n; we briefly refer to this structure just by T. We recall that an n-*ary predicate lifting* for T is a natural transformation $\lambda : Q^n \rightarrow Q \circ T^{op}$ where $Q : \mathsf{Set}^{op} \rightarrow \mathsf{Set}$ is the contravariant powerset functor, $T^{op} : \mathsf{Set}^{op} \rightarrow \mathsf{Set}^{op}$ acts like T, and Q^n denotes the pointwise n-th Cartesian power, i.e. $Q^n(X) = Q(X)^n$. Naturality of λ then amounts to commutation with preimage, i.e. $\lambda_X(f^{-1}[A_1], \ldots, f^{-1}[A_n]) = Tf^{-1}[\lambda_Y(A_1, \ldots, A_n)]$ for $f : X \rightarrow Y$.

The idea here is that T determines the type of systems underlying the semantics, as the coalgebras of T: Recall that a T-coalgebra $C = (X, \gamma)$ consists of a set X of *states* and a map $\gamma : X \rightarrow TX$, which should be thought of as assigning to each state s a structured collection $\gamma(x)$ of successors. The basic example has $T = \mathcal{P}$, the powerset functor; in this case, $\gamma(x)$ is just a set of successors, so a $\mathcal{P}$-coalgebra is a Kripke frame. The predicate liftings then turn predicates on the set X of states into predicates on the set TX of structured collections of successors. A basic example is the predicate lifting for the usual diamond modality $\Diamond$, given by $[\![\Diamond]\!]_X(A) = \{B \in \mathcal{P}(X) \mid B \cap A \neq \emptyset\}$.

Satisfaction $x \models_C \phi$ of formulas $\phi \in \mathcal{F}(\Lambda)$ in states x of a coalgebra $C = (X, \gamma)$ is defined inductively by the expected clauses for Boolean operators, and

$$x \models_C \heartsuit(\phi_1, \ldots, \phi_n) \quad \text{iff} \quad \gamma(x) \models \heartsuit([\![\phi_1]\!]_C, \ldots, [\![\phi_n]\!]_C)$$

where we write $[\![\phi]\!]_C = \{x \in X \mid x \models_C \phi\}$, and for $t \in TX$, $t \models \heartsuit(A_1, \ldots, A_n)$ is short for $t \in [\![\heartsuit]\!]_X(A_1, \ldots, A_n)$. Continuing the above example, the predicate lifting $[\![\Diamond]\!]$ thus induces exactly the usual semantics of $\Diamond$, i.e. $x \models_C \Diamond\phi$ iff the set of successors of x intersects with $[\![\phi]\!]_C$.

We will be interested in satisfiability under global assumptions, or, in description logic terminology, reasoning with general TBoxes [1]: Given a formula ψ, the *global assumption* (or *TBox*), a coalgebra $C = (X, \gamma)$ is a ψ-*model* if $[\![\psi]\!]_C = X$; and a formula ϕ is ψ -*satisfiable* if there exists a ψ-model C such that $[\![\phi]\!]_C \neq \emptyset$. The *satisfiability problem with global assumptions* is to decide, given ψ and ϕ, whether ϕ is ψ-satisfiable. For the complexity analysis of these problems, we assume a suitable encoding of the modal operators in Λ that enters into the calculation of the *size* $|\phi|$ of formulas ϕ; in particular, we assume that numbers occurring in the description of modal operators are coded in binary.

Previous generic algorithms in coalgebraic logic did for the most part rely on complete rule sets for the given operators [27]. Our interest in the present paper is in cases for which suitable rule sets are not (currently) available. We proceed to present our leading examples of this kind, Presburger modal logic and a probabilistic modal logic with polynomial inequalities. For the sake of readability, we focus on the case with a single (weighted) transition relation, and omit propositional atoms. Both features are easily added, e.g. using compositionality results in coalgebraic logic [28], and in fact we use them freely in the examples.

1.1   Presburger Modal Logic

Presburger modal logic [7] admits statements in Presburger arithmetic over numbers $\sharp\phi$ of successors satisfying a formula ϕ, called *cardinalities*. Throughout, we let Rels denote the set $\{<,>,=\} \cup \{\equiv_k | k \in \mathbb{N}\}$ of *arithmetic relations*, with $\equiv_k$ read as congruence modulo k. Syntactically, Presburger modal logic is then defined by taking Λ to contain all modal operators of the form

$$L_{a_1,\dots,a_n;\sim b} = \textstyle\sum_{i=1}^n a_i \sharp(\cdot)_i \sim b$$

where $(\cdot)_i$ denotes the i-th argument of the operator, $\sim\ \in$ Rels, and $a_1,\dots,a_n, b \in \mathbb{Z}$. Weak inequalities can be coded as strict ones, replacing, e.g., $\geq k$ with $> k-1$. The numbers a_i and b, as well as the modulus k in $\equiv_k$, are referred to as the *coefficients* of a Presburger constraint. We also apply these terms to constraints $\sum_{i=1}^n a_i x_i \sim b$ in general, interpreted over the integers.

The semantics of Presburger modal logic was originally defined over standard Kripke frames; in order to make sense of sums with arbitrary integer coefficients, one clearly needs to restrict to finitely branching frames. We consider an alternative more general semantics in terms of *multigraphs*, which have some key technical advantages [5]. Informally, a multigraph is a Kripke frame but with every transition edge annotated with an integer-valued multiplicity; ordinary finitely branching Kripke frames can be viewed as multigraphs by just taking edges to be transitions with multiplicity 1. Formally, a multigraph can be seen as a coalgebra for the *finite multiset functor* $\mathcal{B}$: For a set X, $\mathcal{B}(X)$ consists of the *finite multisets over* X, which are maps $\mu : X \to \mathbb{N}$ with finite support, i.e. $\mu(x) > 0$ for only finitely many x. We view μ as an $\mathbb{N}$-valued measure, and write $\mu(Y) = \sum_{x\in Y} \mu(x)$ for $Y \subseteq X$. Then, $\mathcal{B}(f)$, for maps f, acts as image measure formation. A coalgebra $\gamma : X \to \mathcal{B}(X)$ assigns to each state x a multiset $\gamma(x)$ of successor states, i.e. each successor state is assigned a transition multiplicity.

The semantics of the operators is then given by the predicate liftings

$$[\![L_{a_1,\dots,a_n;\sim b}]\!]_X(A_1,\dots,A_n) = \{\mu \in \mathcal{B}(X) \mid \textstyle\sum_{i=1}^n a_i \cdot \mu(A_i) \sim b\},$$

that is, a state x in a $\mathcal{B}$-coalgebra $C = (X,\gamma)$ satisfies $\sum_{i=1}^n a_i \cdot \sharp\phi_i \sim b$ iff $\sum_{i=1}^n a_i \cdot \gamma(x)([\![\phi_i]\!]_C) \sim b$. This setup generalises effortlessly to multiple (weighted) transition relations: If $\mathcal{R}$ is a set of roles, we take the modal operators to be

$$L_{a_1^{r_1},\dots,a_n^{r_n};\sim b} = \sum_{i=1}^n a_i \sharp_{r_i}(\cdot)_i \sim b$$

where $r_i \in \mathcal{R}$ for every $1 \leq i \leq n$ and $\sharp_r(\cdot)$ is the number of successors along the (weighted) transition relation r. Logics with operators of this kind are then intrpreted by assigning $\mathcal{R}$-many multisets of successors to each world, i.e. as coalgebras of type $X \to \mathcal{B}(X)^{\mathcal{R}}$.

We note that satisfiability is the same over Kripke and over multigraphs:

Lemma 1 *[25]. A formula ϕ is ψ-satisfiable over multigraphs iff ϕ is ψ-satisfiable over Kripke frames.*

(The proof of the non-trivial direction is by making copies of states according to their multiplicity.)

Expressiveness and Examples. Presburger modal logic subsumes graded modal logic [12]: the graded formula $\Diamond_k \phi$, read 'more than k successors satisfy ϕ,' becomes $\sharp(\phi) > k$ in Presburger modal logic. Moreover, Presburger modal logic subsumes majority logic [22]: The *weak majority* formula $W\phi$ ('at least half the successors satisfy ϕ') is expressed in Presburger modal logic as $\sharp(\phi) - \sharp(\neg\phi) \geq 0$. Using propositional atoms as indicated above, we express the examples given in the abstract by the formulas

$$\text{University} \to \sharp_{\text{hasStudent}}(\text{Female}) - \sharp_{\text{hasStudent}}(\text{Male}) > 0$$
$$\text{DanceCourse} \to \sharp_{\text{hasParticipant}}(\top) \equiv_2 0$$

where indices informally indicate the understanding of the successor relation, and the formulae are sensibly understood as global assumptions. As an example involving non-unit coefficients, a chamber of parliament in which a motion requiring a 2/3 majority has sufficient support is described by the concept

$$\sharp_{\text{hasMember}}(\text{SupportsMotion}) - 2\sharp_{\text{hasMember}}(\neg\text{SupportsMotion}) \geq 0.$$

1.2 Probabilistic Modal Logic with Polynomial Inequalities

Probabilistic logics of various forms have been studied in different contexts such as reactive systems [19] and uncertain knowledge [10,17]. A typical feature of such logics is that they talk about probabilities $w(\phi)$ of formulas ϕ holding for the successors of a state; the concrete syntax then variously includes only inequalities of the form $w(\phi) \sim p$ for $\sim \in \{>, \geq, =, <, \leq\}$ and $p \in \mathbb{Q} \cap [0,1]$ [17,19], linear inequalities over terms $w(\phi)$ [10], or polynomial inequalities, with the latter so far treated only in either purely propositional settings [11] or in many-dimensional logics such as the probabilistic description logic Prob-$\mathcal{ALC}$ [20], which use a single global distribution over worlds. An important use of polynomial inequalities over probabilities is to express independence constraints [20]; e.g. two properties ϕ and ψ (of successors) are independent if $w(\phi \wedge \psi) = w(\phi)w(\psi)$.

We thus define the following *probabilistic modal logic with polynomial inequalities*: the system type is given by the *distribution functor* $\mathcal{D}$ that assigns to a set X the set $\mathcal{D}(X)$ of discrete probability distributions on X; again, for a map

f, $\mathcal{D}(f)$ takes image measures. Then, a $\mathcal{D}$-coalgebra $\gamma : X \to \mathcal{D}(X)$ assigns to each state x a distribution $\gamma(x)$ over successor states. We can thus view γ as a Markov chain (interpreting $\gamma(x)$ as a distribution over possible future evolutions of the system), or as a (single-agent) type space in the sense of epistemic logic [17] (interpreting $\gamma(x)$ as the subjective probabilities assigned by the agent to possible alternative worlds in world x). We let the modal similarity type Λ consist of modalities L_p indexed over polynomials $p \in \mathbb{Q}[X_1, \ldots, X_n]$, $n \geq 0$; L_p then has arity n. We denote the application of L_p to formulas $\phi_1, \ldots, \phi_n$ by substituting each variable X_i in p with $w(\phi_i)$ and postulating the result to be non-negative; e.g., the formula $w(\phi \wedge \psi) - w(\phi)w(\psi) \geq 0$ denotes one half of the above-mentioned independence constraint. We correspondingly interpret L_p by the predicate lifting

$$[\![L_p]\!]_X(A_1, \ldots, A_n) = \{\mu \in \mathcal{D}(X) \mid p(\mu(A_1), \ldots, \mu(A_n)) \geq 0\}.$$

2 One-Step Satisfiability

The key to our approach is to deal with modalities level by level; the core concepts of the arising notion of one-step satisfiability checking go back to [21,25,26]. From now on, we mostly restrict the notation to unary operators although our central examples all have operators with higher arities, to avoid cumbersome notation; a fully general treatment requires no more than additional indexing. We fix a Λ-structure T throughout. Peeling off one level of modalities and abstracting from their arguments leads to the following notions.

Definition 2 (One-Step Pairs, One-Step Satisfiability). We assume a set $\mathcal{V}$ of (propositional) variables. We denote the set of Boolean formulas over a set Z of atoms by $\mathsf{Prop}(Z)$, and by $\Lambda(Z) = \{\heartsuit a \mid \heartsuit \in \Lambda, a \in Z\}$ the set of *modal atoms* over Z. As usual, a *literal* over Z is an element $z \in Z$ or a negation thereof, written ϵz where ϵ is either nothing or negation. A *modal literal* over Z is a literal over $\Lambda(Z)$. A *conjunctive clause* over Z is a finite set of literals over Z, read as a conjunction. A *one-step pair* (ϕ, η) over $V \subseteq \mathcal{V}$ consists of

- a conjunctive clause ϕ over $\Lambda(V)$ mentioning each variable at most once, and
- a Boolean formula $\eta \in \mathsf{Prop}(V)$ mentioning only variables occurring in ϕ.

A *one-step model* $M = (X, \tau, t)$ over V consists of

- a set X together with a $\mathcal{P}(X)$-valuation $\tau : V \to \mathcal{P}(X)$; and
- an element $t \in TX$, thought of as the successor structure of an anonymous state.

For $\eta \in \mathsf{Prop}(V)$, $\tau(\eta)$ is the interpretation of η in the Boolean algebra $\mathcal{P}(X)$ under the valuation τ. For a modal atom $\heartsuit a \in \Lambda(V)$, we put $\tau(\heartsuit a) = [\![\heartsuit]\!]_X(\tau(a)) \subseteq TX$. Via the Boolean algebra structure on $\mathcal{P}(TX)$, this extends to an assignment of $\tau(\phi) \in \mathcal{P}(TX)$ to each $\phi \in \mathsf{Prop}(\Lambda(V))$. We say that the one-step model $M = (X, \tau, t)$ *satisfies* the one step pair (ϕ, η), and write $M \models (\phi, \eta)$, if

$$\tau(\eta) = X \qquad \text{and} \qquad t \in \tau(\phi).$$

Then, (ϕ, η) is *(one-step) satisfiable* if there exists a one-step model M such that $M \models (\phi, \eta)$. The *one-step satisfiability problem* is to decide whether a given (ϕ, η) is one-step satisfiable, with η given as a DNF consisting of conjunctive clauses each mentioning every variable occurring in ϕ. The *strict one-step satisfiability problem* is the same problem, but with the *input size* defined to be just the size of ϕ; the representation of η is, then, irrelevant. We say that Λ has the *one-step small model property* if there is a polynomial p such that every one-step satisfiable (ϕ, η) has a one-step model (X, τ, t) with $|X| \leq p(|\phi|)$ (no bound is assumed on the representation of t).

The intuition behind these definitions is that propositional variables are placeholders for argument concepts; their valuation τ in a one-step model represents the extensions of these argument concepts; and the second component η of a one-step pair captures the Boolean constraints on the argument concepts that are globally satisfied in a given model. One-step satisfiability is precisely what will allow us to construct satisfying models later on. Note that most of a one-step pair (ϕ, η) is disregarded for purposes of determining the input size of the *strict* one-step satisfiability problem, as η, a propositional formula, can be exponentially larger than the conjunctive clause ϕ.

Example 3. In Presburger modal logic, let $\phi = \sharp(a) \geq 1 \wedge \sharp(b) \geq 1$. Then (ϕ, η) is one-step satisfiable as long as η does not force the interpretation of either a or b to be empty, i.e. both $\eta \wedge a$ and $\eta \wedge b$ need to be (propositionally) satisfiable. Thus, the strongest possible η are $a \wedge b$ and $(a \wedge \neg b) \vee (\neg a \wedge b)$.

Lemma 4. *A one-step pair (ϕ, η) over V is satisfiable iff it is satisfiable by a one-step model of the form (X, τ, t) where X is the set of valuations $V \to 2$ satisfying η (where $2 = \{\top, \bot\}$ is the set of Booleans) and $\tau(a) = \{\kappa \in X \mid \kappa(a) = \top\}$ for $a \in V$.*

Under the one-step small model property, the two versions of the one-step satisfiability problem coincide for our purposes:

Lemma 5. *Let T have the one-step small model property. Then for any complexity class C containing PSPACE, strict one-step satisfiability is in C iff one-step satisfiability is in C.*

Although not phrased in these terms, the complexity analysis of (TBox-free) Presburger modal logic by Demri and Lugiez [7] is based on showing that the strict one-step satisfiability problem is in PSPACE [26], without using the one-step small model property – in fact, the latter is based on more recent results from integer programming:

Lemma 6 *[8]. Every system of d linear inequalities over the integers with coefficients of binary length at most s has a solution with at most polynomially many non-zero components in d and s.*

The corresponding statement over the rationals (where in fact one has at most d non-zero components) is well-known, and features centrally in the analysis of probabilistic logics [11]. From these observations, we obtain sufficient tractability of one-step satisfiability in our key examples:

Example 7. 1. Presburger modal logic has the one-step small model property. To see this, let (ϕ, η) be satisfied by $M = (X, \tau, \mu)$, where by Lemma 4 we can assume that X consists of satisfying valuations of η, hence of at most exponential size in $|\phi|$. Let $V = \{a_1, \ldots, a_n\}$, and put $q_i = \mu(\tau(a_i))$. By standard estimates in integer programming [23] we can assume that the $\mu(x)$ and, hence, the q_i (being sums of at most exponentially many $\mu(x)$) have polynomial binary length in $|\phi|$. Now all we need to know about τ to guarantee that M satisfies ϕ is that

$$\sum_{x \in \tau(a_i)} \mu(x) = q_i.$$

We can see this as a system of linear constraints on the $\mu(x)$, which by Lemma 6 has a solution with only m nonzero components where m is polynomially bounded in n and the binary length s of the largest q_i, and hence in $|\phi|$; from this solution, we immediately obtain a one-step model of (ϕ, η) with m states.

Moreover, again using Lemma 4, one-step satisfiability in Presburger modal logic easily reduces to checking solvability of Presburger constraints over the integers, which can be done in NP and hence in PSPACE; by Lemma 5, we obtain that *strict one-step satisfiability in Presburger modal logic is in* PSPACE.

2. By a completely analogous (slightly easier) argument as for Presburger modal logic, probabilistic modal logic with polynomial inequalities has the one-step small model property. In this case, one-step satisfiability reduces to solvability of systems of polynomial inequalities over the reals, which can be checked in PSPACE [4] (this argument can essentially be found in [11]). Again, we obtain that *strict one-step satisfiability in probabilistic modal logic with polynomial inequalities is in* PSPACE.

By [26], these observations imply decidability in PSPACE of the *plain* satisfiability problem. We show below that one obtains an optimal upper bound EXPTIME for satisfiability under global assumptions. One should note that the proof of the one-step small model property will in both cases work for any coalgebraic modal logic over integer- or real-weighted systems whose modalities depend only on the measures of their arguments.

Remark 8. Most previous generic complexity results in coalgebraic logic have relied on tractable sets of tableau rules, e.g. [14,27,29]. These rules are of the shape ϕ/η where ϕ is a conjunctive clause over $\Lambda(V)$ and $\eta \in \mathsf{Prop}(V)$, to be read, within a system including also the standard propositional rules, as 'in order to establish that ψ is satisfiable, show that the conclusions of all rule matches to ψ are satisfiable'. E.g. PSPACE-tractability [27] of a rule set essentially amounts to the rules being codable in such a way that it suffices, for each ψ, to consider only rules with polynomial-sized codes. In terms of one-step pairs, this means essentially that there are certificates (in the shape of rule codes) for *unsatisfiability* of a one-step pair (ϕ, η) that are of polynomial size in $|\phi|$ and can be checked in polynomial space in $|\phi|$ (in particular comparing η with the conclusion of the encoded rule), so that the strict one-step satisfiability problem is in PSPACE. Summing up, complexity bounds obtained by our current semantic approach subsume earlier tableau-based ones.

3   Type Elimination

We now describe a type elimination algorithm that realizes an EXPTIME upper bound for reasoning with global assumptions in coalgebraic logics. Like all type elimination algorithms, it is not suited for practical use, as it begins by constructing the full exponential-sized set of types. We therefore refine the algorithm to a global caching algorithm in Sect. 4.

As usual, we rely on defining a scope of relevant concepts:

Definition 9. A set Σ of concepts is *closed* if Σ is closed under subconcepts and single negations.

We fix from now on a global assumption ψ and a formula ϕ_0 to be checked for ψ-satisfiability. We denote the closure of $\{\psi, \phi_0\}$ in the above sense by Σ.

Definition 10. A ψ-*type* is a subset $T \subseteq \Sigma$ such that

- $\psi \in T \not\ni \bot$;
- whenever $\neg\phi \in \Sigma$, then $\neg\phi \in T$ iff $\phi \notin T$;
- whenever $\phi \wedge \chi \in \Sigma$, then $\phi \sqcap \chi \in T$ iff $\phi, \chi \in T$.

The design of the algorithm relies on one-step satisfiability as an abstraction: We denote the set of all ψ-types by S_0. We take V to be the set of propositional variables $a_{\heartsuit\phi}$ for all modal atoms $\heartsuit\phi \in \Sigma$; we then define a substitution σ by $\sigma(a_{\heartsuit\phi}) = \phi$ for all $a_{\heartsuit\phi} \in V$. For $S \subseteq S_0$ and $T \in S$, we construct a one-step pair (ϕ_T, η_S) over V by taking ϕ_T to be the set of all modal literals $\epsilon\heartsuit a$ over V such that $\epsilon\heartsuit\sigma(a) \in T$, and η_S a DNF consisting of all conjunctive clauses ϑ (seen as sets of literals L) over V such that $\{L\sigma \mid L \in \vartheta\} \subseteq T$ for some $T \in S$. Then we define a functional $\mathcal{E} : \mathcal{P}(S_0) \to \mathcal{P}(S_0)$ by $S \mapsto \{T \in S_0 \mid (\phi_T, \eta_S) \text{ one-step satisfiable}\}$.

Lemma 11. $\mathcal{E}$ *is monotone w.r.t. set inclusion.*

We can thus compute the greatest postfixpoint $\nu\mathcal{E}$ of $\mathcal{E}$ by just iterating $\mathcal{E}$:

Algorithm 12. (Decide by type elimination whether ϕ_0 is satisfiable over ψ)

1. Set $S := S_0$.
2. Compute $S' = \mathcal{E}(S)$; if $S' \neq S$ then put $S := S'$ and repeat.
3. Return 'yes' if $\phi_0 \in T$ for some $T \in S$, and 'no' otherwise.

If strict (!) one-step satisfiability is in ExpTime, then this algorithm has at most exponential run time. We analyse correctness:

Definition 13. A type $\mathcal{T}$ is *realized* in a ψ-model $C = (X, \gamma)$ if there exists $x \in X$ such that $x \models \phi$ for all $\phi \in \mathcal{T}$.

Lemma 14. *The set of types realized in a given ψ-model is a postfixpoint of $\mathcal{E}$.*

By Lemma 14, all ψ-satisfiable types are in $\nu\mathcal{E}$. Thus, the algorithm is sound, i.e. answers 'yes' on ψ-satisfiable concepts. To see completeness, we show

Lemma 15. *Let S be a postfixpoint of $\mathcal{E}$. Then there exists a T-coalgebra $C = (S, \gamma)$ such that for each $\phi \in \Sigma$, $[\![\phi]\!]_C = \{\mathcal{T} \in S \mid \phi \in \mathcal{T}\}$.*

An interpretation as in Lemma 15 is clearly a ψ-model, so that Algorithm 12 is complete, i.e. answers 'yes' *only* on ψ-satisfiable concepts.

Theorem 16. *If strict one-step satisfiability in T is in ExpTime, then satisfiability with global assumptions is in ExpTime.*

Example 17. By the results of the previous section and by inheriting lower bounds from reasoning with global assumptions in K, we obtain that reasoning with global assumptions in Presburger modal logic and in probabilistic modal logic with polynomial inequalities is ExpTime-complete.

4   Global Caching

We now develop the type elimination algorithm from the preceding section into a global caching algorithm. Existing global caching algorithms work with systems of tableau rules (satisfiability is guaranteed if every applicable rule has at least one satisfiable conclusion) [14]. The fact that we work with a semantics-based decision procedure impacts on the design of the algorithm in two ways:

- In a tableaux setting, node generation is driven by the tableau rules, and a global caching algorithm generates successor nodes by applying tableau rules. In principle, however, successor nodes can be generated at will, with the rules just pointing to relevant nodes. In our setting, we make the relevant nodes explicit using the concept of *children*.
- The rules govern the propagation of satisfiability and unsatisfiability among the nodes. Semantic propagation of satisfiability is straightforward, but propagation of unsatisfiability again needs the concept of children: a node can only be marked as unsatisfiable once all its children have been generated (and too many of them are unsatisfiable).

We continue to work with a closed set Σ as in Sect. 3 (generated by the global assumption ψ and the target formula ϕ_0) but replace types with *(tableau) sequents*, i.e. arbitrary subsets $\Gamma, \Theta \subseteq \Sigma$, understood conjunctively; in particular, a sequent need not mention every formula in Σ. We write $\mathsf{Seqs} = \mathcal{P}(\Sigma)$.

A *state* is a sequent consisting of modal literals only (recall that we take atomic propositions as nullary operators). We denote the set of states by States.

To convert sequents into states, we emply the usual *propositional rules*

$$\frac{\Gamma,\phi_1 \sqcap \phi_2}{\Gamma,\phi_1,\phi_2} \quad \frac{\Gamma,\neg(\phi_1 \sqcap \phi_2)}{\Gamma,\neg\phi_1 \mid \Gamma,\neg\phi_2} \quad \frac{\Gamma,\neg\neg\phi}{\Gamma,\phi} \quad \frac{\Gamma,\bot}{}$$

where | denotes alternative conclusions. (As usual, a rule $\Gamma,\phi,\neg\phi/$ with no conclusions is admissible.)

Definition 18. The *children* of a state Γ are the sequents consisting of ψ and, for each modal literal $\epsilon\heartsuit\phi \in \Gamma$, a choice of either ϕ or $\neg\phi$. The *children* of a non-state sequent are its conclusions under the propositional rules.

We modify the functional $\mathcal{E}$ defined in the previous section to work also with sequents and depend on a set $G \subseteq$ Seqs of sequents already generated: we define $\mathcal{E}_G : \mathcal{P}(G) \to \mathcal{P}(G)$ by taking $\mathcal{E}_G(S)$ to contain

- a non-state sequent $\Gamma \in G -$ States iff every propositional rule that applies to Γ has a satisfiable conclusion that is contained S, and
- a state $\Gamma \in G \cap$ States iff, for C the set of children of Γ, the one-step pair $(\phi_\Gamma, \eta_{S \cap C})$ over V_Γ is one-step satisfiable where V_Γ contains a variable $a_{\epsilon\heartsuit\phi}$ for each modal literal $\epsilon\heartsuit\phi \in \Gamma$, and ϕ_Γ, $\eta_{S \cap C}$ are defined like ϕ_T, η_S in the previous section, using the substitution $\sigma_\Gamma(a_{\epsilon\heartsuit\phi}) = \phi$ in place of σ.

To propagate unsatisfiability, we introduce a second functional $\mathcal{A}_G : \mathcal{P}(G) \to \mathcal{P}(G)$, where we take $\mathcal{A}_G(S)$ to contain

- a non-state sequent $\Gamma \in G -$ States iff there is a propositional rule applying to Γ all whose conclusions are in S, and
- a state $\Gamma \in G \cap$ States iff, for C the set of children of Γ, we have $C \subseteq G$ and the one-step pair $(\phi_\Gamma, \eta_{C \setminus S})$ is one-step unsatisfiable.

The global caching algorithm maintains, as global variables, a set G of sequents with subsets E and A of sequents already decided as satisfiable or unsatisfiable, respectively.

Algorithm 19. (Decide $\mathcal{T}$-satisfiability of ϕ_0 by global caching.)

1. Initialize $G = \{\Gamma_0\}$ with $\Gamma_0 = \{\phi_0, \psi\}$, and $E = A = \emptyset$.
2. (Expand) Select a sequent $\Gamma \in G$ that has children that are not in G, and add any number of these children to G. If no sequents with missing children are found, go to Step 5
3. (Propagate) Optionally recalculate E as the greatest fixed point $\nu S. \mathcal{E}_G(S \cup E)$, and A as $\mu S. \mathcal{A}_G(S \cup A)$. If $\Gamma_0 \in E$, return 'yes'; if $\Gamma_0 \in A$, return 'no'.
4. Go to Step 2.
5. Recalculate E as $\nu S. \mathcal{E}_G(S \cup E)$; return 'yes' if $\Gamma_0 \in E$, and 'no' otherwise.

Theorem 20. *If the strict one-step satisfiability problem of T is in* EXPTIME *then the global caching algorithm decides satisfiability under global assumptions in* EXPTIME.

The key feature of the algorithm is that it avoids generating the full set of types by detecting satisfiability or unsatisfiability on the fly in the intermediate propagation step. The non-determinism in the formulation of the algorithm can be resolved arbitrarily, i.e. any choice (e.g. of which sequents to add in the expansion step and whether or not to trigger propagation) leads to correct results; thus, it affords room for heuristic optimization. Detecting *unsatisfiability* (but not satisfiability) in Step 19 requires previous generation of all, in principle exponentially many, children of a sequent. This is presumably not necessarily prohibitive in practice, as the exponential dependence is only in the number of *top-level* modalities in a sequent. As an extreme example, if we encode $\Diamond\phi$ as $\sharp(\phi) > 0$, then the sequent $\{\Diamond^n\top\}$ (n successive diamonds) induces 2^n types but has only two children, $\{\Diamond^{n-1}\top\}$ and $\{\neg\Diamond^{n-1}\top\}$.

5   Conclusions

We have provided a generic upper bound EXPTIME for reasoning with global assumptions in coalgebraic modal logics, based on a generic semantic approach centered around *one-step satisfiability checking*. This approach is particularly suitable for logics for which no tractable sets of modal tableau rules are known; our core examples of this type are Presburger modal logic and probabilistic modal logic with polynomial inequalities. (Another example is Elgesem's logic of agency [9], which also satisfies the conditions of our generic result [26].) The upper complexity bounds we obtain for these logics by instantiating our generic results appear to be new. The upper bound is based on a type elimination algorithm; additionally, we have designed a more practical global caching algorithm that offers a perspective for efficient reasoning in practice.

Acknowledgements. We wish to thank Erwin R. Catesbeiana for remarks on unsatisfiability. Work of the first author forms part of the DFG project GenMod2 (SCHR 1118/5-2).

References

1. Baader, F., Calvanese, D., McGuinness, D., Nardi, D., Patel-Schneider, P. (eds.): The Description Logic Handbook. Cambridge University Press, Cambridge (2003)
2. Baader, F., Sattler, U.: Description logics with symbolic number restrictions. In: European Conference on Artificial Intelligence, ECAI 1996, pp. 283–287. Wiley (1996)
3. Bárcenas, E., Lavalle, J.: Expressive reasoning on tree structures: recursion, inverse programs, presburger constraints and nominals. In: Castro, F., Gelbukh, A., González, M. (eds.) MICAI 2013, Part I. LNCS, vol. 8265, pp. 80–91. Springer, Heidelberg (2013)

4. Canny, J.: Some algebraic and geometric computations in PSPACE. In: Symposium on Theory of Computing, STOC 1988, pp. 460–467. ACM (1988)
5. D'Agostino, G., Visser, A.: Finality regained: A coalgebraic study of Scott-sets and multisets. Arch. Math. Logic 41, 267–298 (2002)
6. Demri, S., Lugiez, D.: Presburger modal logic is PSPACE-complete. In: Furbach, U., Shankar, N. (eds.) IJCAR 2006. LNCS (LNAI), vol. 4130, pp. 541–556. Springer, Heidelberg (2006)
7. Demri, S., Lugiez, D.: Complexity of modal logics with Presburger constraints. J. Appl. Logic 8, 233–252 (2010)
8. Eisenbrand, F., Shmonin, G.: Carathéodory bounds for integer cones. Oper. Res. Lett. 34(5), 564–568 (2006)
9. Elgesem, D.: The modal logic of agency. Nordic J. Philos. Logic 2, 1–46 (1997)
10. Fagin, R., Halpern, J.: Reasoning about knowledge and probability. J. ACM 41, 340–367 (1994)
11. Fagin, R., Halpern, J., Megiddo, N.: A logic for reasoning about probabilities. Inform. Comput. 87, 78–128 (1990)
12. Fine, K.: In so many possible worlds. Notre Dame J. Form. Log. 13, 516–520 (1972)
13. Goranko, V., Passy, S.: Using the universal modality: Gains and questions. J. Log. Comput. 2, 5–30 (1992)
14. Goré, R., Kupke, C., Pattinson, D.: Optimal tableau algorithms for coalgebraic logics. In: Esparza, J., Majumdar, R. (eds.) TACAS 2010. LNCS, vol. 6015, pp. 114–128. Springer, Heidelberg (2010)
15. Goré, R., Nguyen, L.: EXPTIME tableaux for ALC using sound global caching. In: Description Logics, DL 2007, CEUR Workshop Proceedings, vol. 250 (2007)
16. Goré, R.P., Postniece, L.: An experimental evaluation of global caching for ALC (System description). In: Armando, A., Baumgartner, P., Dowek, G. (eds.) IJCAR 2008. LNCS (LNAI), vol. 5195, pp. 299–305. Springer, Heidelberg (2008)
17. Heifetz, A., Mongin, P.: Probabilistic logic for type spaces. Games Econ. Behav. 35, 31–53 (2001)
18. Kupke, C., Pattinson, D.: On modal logics of linear inequalities. In: Advances in Modal Logic, AiML 2010, pp. 235–255. College Publications (2010)
19. Larsen, K., Skou, A.: Bisimulation through probabilistic testing. Inf. Comput. 94, 1–28 (1991)
20. Lutz, C., Schröder, L.: Probabilistic description logics for subjective uncertainty. In: Principles of Knowledge Representation and Reasoning, KR 2010, pp. 393–403. AAAI (2010)
21. Myers, R., Pattinson, D., Schröder, L.: Coalgebraic hybrid logic. In: de Alfaro, L. (ed.) FOSSACS 2009. LNCS, vol. 5504, pp. 137–151. Springer, Heidelberg (2009)
22. Pacuit, E., Salame, S.: Majority logic. In: Principles of Knowledge Representation and Reasoning, KR 2004, pp. 598–605. AAAI Press (2004)
23. Papadimitriou, C.: On the complexity of integer programming. J. ACM 28, 765–768 (1981)
24. Pratt, V.: Models of program logics. In: Foundations of Computer Science, FOCS 1979, pp. 115–122. IEEE Comp. Soc. (1979)
25. Schröder, L.: A finite model construction for coalgebraic modal logic. J. Log. Algebr. Prog. 73, 97–110 (2007)
26. Schröder, L., Pattinson, D.: Shallow models for non-iterative modal logics. In: Dengel, A.R., Berns, K., Breuel, T.M., Bomarius, F., Roth-Berghofer, T.R. (eds.) KI 2008. LNCS (LNAI), vol. 5243, pp. 324–331. Springer, Heidelberg (2008)
27. Schröder, L., Pattinson, D.: PSPACE bounds for rank-1 modal logics. ACM Trans. Comput. Log. 10,13:1–13:33 (2009)

28. Schröder, L., Pattinson, D.: Modular algorithms for heterogeneous modal logics via multi-sorted coalgebra. Math. Struct. Comput. Sci. **21**, 235–266 (2011)
29. Schröder, L., Pattinson, D., Kupke, C.: Nominals for everyone. In: International Joint Conference on Artificial Intelligence, IJCAI 2009, pp. 917–922 (2009)
30. Seidl, H., Schwentick, T., Muscholl, A.: Counting in trees. In: Logic and Automata: History and Perspectives (in Honor of Wolfgang Thomas), pp. 575–612. Amsterdam Univ. Press (2008)
31. Tobies, S.: Complexity results and practical algorithms for logics in knowledge representation. Ph.D. thesis, RWTH Aachen (2001)

Nearest Fixed Points and Concurrent Priority Games

Bruno Karelovic and Wiesław Zielonka[✉]

LIAFA, Université Paris Diderot, Paris 7, France
bruno.karelovic@gmail.com, zielonka@liafa.univ-paris-diderot.fr

Abstract. As it is known the values of different states in parity games (deterministic parity games, or stochastic perfect information parity games or concurrent parity games) can be expressed by formulas of μ-calculus – a fixed point calculus alternating the greatest and the least fixed points of monotone mappings on complete lattices.

In this paper we examine concurrent priority games that generalize parity games and we relate the value of such games to a new form of fixed point calculus – the nearest fixed point calculus.

1 Introduction

As it is well known parity games are closely related to μ-calculus. This fact was first observed in the context of turn based deterministic games [1,2], next for perfect information stochastic games [3,4] and for concurrent stochastic games [5].

Intuitively, parity games capture a situation where we meet two types of properties, desirable ones and undesirable ones. Moreover properties are ordered by a priority relation. This leads to a classification of infinite runs of a system, a run is desirable iff the property associated with the maximal priority encountered infinitely often during the run is desirable.

We can try however a finer classification of properties by quantifying them by real numbers from a closed bounded interval $I = [p_1, p_2]$ of real numbers. To this end we associate with the most preferable properties the reward p_2 and with the most undesirable the reward p_1. However in general we can have also a whole spectrum of intermediate properties with rewards between p_1 and p_2. As in parity games the properties can be ordered by a priority relation, the priority over properties has nothing to do with the natural reward order on I, given two properties a and b with rewards $r(a), r(b) \in I$, it is possible to have $r(a) < r(b)$ (b gives a better payoff than a) with the priority of a greater than the priority of b, i.e. property a whenever happens then it "invalidates" property b. As in parity games, given an infinite run we take the property a of maximal priority encountered infinitely often during the run and define the reward of the run as the reward associated with this property.

We obtain in this way a class of games that we call priority games. Deterministic priority games can be reduced to parity games, in particular solving

© Springer International Publishing Switzerland 2015
A. Kosowski and I. Walukiewicz (Eds.): FCT 2015, LNCS 9210, pp. 381–393, 2015.
DOI: 10.1007/978-3-319-22177-9_29

a sequence of parity games we can find the values of all states in the priority games and optimal memoryless strategies for both players. Perfect information stochastic priority games also admit optimal memoryless strategies, however we do not know if they can be reduced to parity games. In this paper we examine concurrent stochastic priority games. As it turns out the values of such games can be obtained by a new kind of μ-calculus. The traditional μ-calculus alternates the greatest and the least fixed points, the μ-calculus in this paper defines for each $r \in I = [p_1, p_2]$ the nearest fixed point of a monotone function (nearest to r). The greatest and the least fixed points are just special cases of the nearest fixed points (they are nearest to p_1 and to p_2 respectively).

Even if priority games just extend parity games we think that our approach contributes also to a better comprehension of parity games. Indeed it is notoriously difficult to comprehend the μ-calculus formulas that give solutions to parity games. This follows from the fact that it is difficult to grasp the meaning of a μ-calculus formula alternating several greatest and least fixed points.

Our approach has advantage because it provides a natural interpretation in terms of games of a partially evaluated μ-calculus formula, where fixed points are applied only to some variables while other variables are left free. In our approach each variable in the formula corresponds to a state of the game and free variables correspond to absorbing states. Then adding a new fixed point over a free variable has the following interpretation in term of games – the state corresponding to this variable changes its nature from absorbing to non-absorbing. And the usual method for approximating a new fixed point giving the value of the state turns out to be nothing else but the natural algorithm for calculating the value the new non-absorbing state. At the end, when all fixed points are applied, then this corresponds to the final situation where all states are transformed from absorbing to non-absorbing.

2   Concurrent Stochastic Priority Games

A two-player arena $G = (\mathbf{S}, \mathbf{A}, \mathbf{B}, p)$ is composed of a finite set of states $\mathbf{S} = \{1, 2, \ldots, n\} \subset \mathbb{N}$ (we assume without loss of generality that $\mathbf{S}$ is a subset of positive integers) and finite sets $\mathbf{A}$ and $\mathbf{B}$ of actions of players Max and Min. For each state s, $\mathbf{A}(s) \subseteq \mathbf{A}$ and $\mathbf{B}(s) \subseteq \mathbf{B}$, are the set of actions that players Max and Min have at their disposal at s. We assume that $\mathbf{A}$ and $\mathbf{B}$ are disjoint and $(\mathbf{A}(s))_{s \in \mathbf{S}}$, $(\mathbf{B}(s))_{s \in \mathbf{S}}$ are partitions of $\mathbf{A}$ and $\mathbf{B}$.

For $s, s' \in \mathbf{S}, a \in \mathbf{A}(s), b \in \mathbf{B}(s)$, $p(s'|s, a, b)$ is the probability to move to s' if players Max and Min execute respectively actions a and b at s. The couple (a, b) is called a joint action.

An infinite play is played by players Max and Min. At each stage, given the current state s, the players choose simultaneously and independently actions $a \in \mathbf{A}(s)$ and $b \in \mathbf{B}(s)$ and the game moves to a new state s' with probability $p(s'|s, a, b)$. The couple (a, b) is called a joint action.

A finite history is a sequence $h = (s_1, a_1, b_1, s_2, a_2, b_2, s_3 \ldots, s_n)$ alternating states and joint actions and beginning and ending with a state. The length of h is the number of joint actions in h, in particular a history of length 0 consists of just one state and no actions. The set of finite histories is denoted H.

A strategy of player Max is a mapping $\sigma : H \to \Delta(\mathbf{A})$, where $\Delta(\mathbf{A})$ is the set of probability distributions on $\mathbf{A}$. We require that $\mathrm{supp}(\sigma(h)) \subseteq \mathbf{A}(s)$, where s is the last state of h and $\mathrm{supp}(\sigma(h)) := \{a \in \mathbf{A} \mid \sigma(h)(a) > 0\}$ is the support of the measure $\sigma(h)$.

A strategy σ is *stationary* if $\sigma(h)$ depends only on the last state of h. Thus stationary strategies of player Max can be identified with mappings from $\mathbf{S}$ to $\Delta(\mathbf{A})$ such that $\mathrm{supp}(\sigma(s)) \subseteq \mathbf{A}(s)$ for each $s \in \mathbf{S}$.

A strategy σ is *pure* if $\mathrm{supp}(\sigma(h))$ is a singleton for each h. Pure stationary strategies of player Max are identified with mappings $\sigma : \mathbf{S} \to \mathbf{A}$ such that $\sigma(s) \in \mathbf{A}(s)$.

Strategies for player Min are defined in a similar way.

We write $\Sigma(\mathbf{G})$ and $\mathcal{T}(\mathbf{G})$ to denote the sets of all strategies for player Max and Min respectively.

We omit $\mathbf{G}$ and write $\Sigma, \mathcal{T}$ if $\mathbf{G}$ is clear from the context. We use σ and τ (with subscripts or superscripts) to denote strategies of players Max and Min respectively.

An infinite history or a play is an infinite sequence $h = (s_1, a_1, b_1, s_2, a_2, b_2, s_3, a_3, b_3, \ldots)$ alternating states and joint actions. The set of infinite histories is denoted H^∞. For a finite history h by h^+ we denote the cylinder generated by h consisting of all infinite histories with prefix h. We assume that H^∞ is endowed with the σ-algebra $\mathcal{B}(H^\infty)$ generated by the set of cylinders.

Strategies σ, τ of players Max and Min and the initial state s determine in the usual way a probability measure $\mathbb{P}_s^{\sigma,\tau}$ on $(H^\infty, \mathcal{B}(H^\infty))$.

A concurrent stochastic priority game is obtained by adding to $\mathbf{G}$ a reward mapping

$$\rho : \mathbf{S} \to I$$

associating with each state s a reward $\rho(s)$ belonging to a closed interval $I \subset \mathbb{R}$.

The payoff $u_\rho(h)$ of an infinite history $h = (s_1, a_1, b_1, s_2, a_2, b_2, s_3, a_3, b_3, \ldots)$ in the priority game is defined as

$$u_\rho(h) = \rho(\limsup_n s_n). \tag{1}$$

Thus the payoff is equal to the reward of the greatest (in the usual integer order) state visited infinitely often.

The aim of player Max (player Min) is to maximize (resp. minimize) the expected payoff

$$\mathbb{E}_s^{\sigma,\tau}[u_\rho] = \int_{H^\infty} u_\rho(h) \mathbb{P}_s^{\sigma,\tau}(dh).$$

Concurrent priority games contain two well known classes of games.

(1) Concurrent parity games [6] correspond to concurrent priority games with the reward mapping having rewards in the two element set $\{0, 1\}$ rather than arbitrary rewards in the interval I.

(2) The second subclass of concurrent priority games is the class of Everett's recursive games [7]. Everett games are concurrent priority games such that

all non-absorbing states have reward 0 (a state s is absorbing if $p(s|s,a,b) = 1$ for all joint actions (a,b)).

Thus in Everett games players receive the payoff 0 if the play remains forever in non-absorbing states, otherwise, for plays ending in an absorbing state, the payoff is equal to the reward associated with this state. Note that Everett games contain as a subclass the class of reachability games which correspond to Everett games such that all absorbing states have reward 1.

From the determinacy of Blackwell games proved by Martin [8] it follows that concurrent priority games have values, i.e. for each state s, $\sup_\sigma \inf_\tau \mathbb{E}_s^{\sigma,\tau}[u_\rho] = \inf_\tau \sup_\sigma \mathbb{E}_s^{\sigma,\tau}[u_\rho]$. (Blackwell games do not have states but the result of Martin extends immediately to games with states as shown by Maitra and Sudderth [9].)

For two subclasses of concurrent priority games mentioned earlier we have more precise results. A proof of determinacy of concurrent stochastic parity games using fixed points was given by de Alfaro and Majumdar [5]. And for Everett's games Everett proved not only that such games have values but also that both players have ε-optimal stationary strategies [7].

Notation: For an arena G by $G(\rho)$ we will denote the priority game obtained by endowing G with the reward mapping ρ. Another notation used frequently in the paper is $G(x_1, \ldots, x_n)$ which denotes the priority game with n states $\{1, \ldots, n\}$ having rewards $x_1, \ldots, x_n$ respectively.

3   Interval Lattice and Nearest Fixed Point

Let us recall that a complete lattice is a partially ordered set $(E, \leq)$ such that each subset X of E has the least upper bound $\bigvee X$ and the greatest lower bound $\bigwedge X$. A mapping $f : E \to F$ from a lattice E to a lattice F is monotone if for all $x, y \in E$, $x \leq y$ implies $f(x) \leq f(y)$. The set of such monotone mappings is denoted $\mathrm{Mon}(E, F)$. The greatest and the least element of a complete lattice will be denoted respectively $\top$ and $\bot$.

Theorem 1 (Tarski [10]). *For each complete lattice $(E, \leq)$ and a monotone mapping $f : E \to E$ the set P of fixed points of f is non-empty and is a complete lattice, in particular P has the greatest and the least element.*

In this paper we are interested in the complete lattice $I \subseteq \mathbb{R}$ of real numbers from a closed interval I and in the product lattice I^n endowed with the componentwise order. In the sequel we will fix $I = [\bot, \top]$ and $\bot, \top$ will always denote the minimal and maximal elements of I.

Let $f : x \mapsto f(x)$ be a monotone mapping from I to itself. For a monotone mapping $f : I \to I$ and $r \in I$ we define *the nearest fixed point* $\mu_r x.f(x)$:

(1) if $f(r) = r$ then $\mu_r x.f(x) = r$,
(2) if $f(r) < r$ then f maps the interval $[\bot, r]$ into itself and by Tarski's fixed point theorem there exists the greatest fixed-point of f in $[\bot, r]$ and $\mu_r x.f(x)$ denotes this fixed point (in other words, $\mu_r x.f(x)$ is the greatest fixed point of f in $[\bot, r]$),

(3) if $f(r) > r$ then f maps the interval $[r, \top]$ into itself and by Tarski's fixed point theorem, there exists the least fixed-point of f in $[r, \top]$ and $\mu_r x. f(x)$ denotes this fixed point (in other words, $\mu_r x. f(x)$ is the least fixed point of f in $[r, \top]$).

A mapping $f : I \to I$ is *nonexpansive* if for all $x, y \in I$, $|f(x) - f(y)| \leq |x - y|$.

It is known that in general complete lattices a transfinite induction can be necessary in order to calculate the least and the greatest fixed points of monotone mappings. The following lemma shows that for monotone nonexpansive mappings from I to I the situation is much simpler:

Lemma 2. *Let $f : I \to I$ be monotone and nonexpansive.*

(i) *Let $r \in I$ and let $(r_n)_{n \geq 0}$ be the sequence of real numbers such that $r_0 = r$ and $r_{n+1} = f(r_n)$. The sequence (r_n) is monotone and it converges (in the usual sense of convergence in $\mathbb{R}$ with the euclidean metric) to $\mu_r x. f(x)$.*

(ii) *The set of fixed points of f is a closed subinterval of I.*

(iii) *Let e_1, e_2 be respectively the least and the greatest fixed points of f. Then $\mu_r x. f(x) = e_1$ if $r < e_1$, $\mu_r x. f(x) = r$ if $r \in [e_1, e_2]$ and $\mu_r x. f(x) = e_2$ if $r > e_2$.*

Note that (iii) shows that $\mu_r x. f(x)$ is indeed the fixed point which is closest (in the sense of the euclidean distance) to r.

Proof. (i) If $f(r) \propto r$ then by monotonicity $f^{n+1}(r) \propto f^n(r)$ for all $n \geq 0$, where $\propto$ is either $\leq$ or $\geq$. But a bounded monotone sequence of real numbers converges to some $r_\infty \in I$. Since f is nonexpansive $|f(r_\infty) - f^{n+1}(r)| \leq |r_\infty - f^n(r)|$. The left-hand side of this inequality converges to $|f(r_\infty) - r_\infty|$ while the right-hand side converges to 0.

(ii) Suppose that $x < y$ are two fixed points of f and $z \in [x, y]$. Then $x = f(x) \leq f(z) \leq f(y) = y$ and $|x - f(z)| = |f(x) - f(z)| \leq |x - z|$. Similarly $|f(z) - y| \leq |z - y|$. However both these inequalities can hold simultaneously only if $f(z) = z$. Thus if e_1 and e_2 are the least and the greatest fixed points of f then all elements of $[e_1, e_2]$ are fixed points of f.

(iii) Direct consequence of (ii).

3.1 Nested Nearest Fixed Points

For any set D we endow the set I^D of mappings from D to I with the order relation: for $f, g \in I^D$, $f \leq g$ if $f(x) \leq g(x)$ for all $x \in D$. Then $(I^D, \leq)$ is also a complete lattice, for a set $F \subseteq I^D$, ΘF is a mapping f such that $f(x) = \Theta_{h \in F} h(x)$, $\Theta \in \{\bigvee, \bigwedge\}$. Note that the product lattice I^n can be seen as a mapping from $\{1, \ldots, n\}$ into I, i.e. is covered by this definition.

The lattice I^I contains the lattice $\mathrm{Mon}(I, I)$ of all monotone mappings from I to I. Note that for any set $F \subseteq \mathrm{Mon}(I, I)$, ΘF calculated in the lattice I^I or in the lattice $\mathrm{Mon}(I, I)$ gives the same result, $\Theta \in \{\bigwedge, \bigvee\}$.

Lemma 3. *Let $f, g \in \mathrm{Mon}(I, I)$ and $r \in I$. If $f \leq g$ then for each $r \in I$, $\mu_r x. f(x) \leq \mu_r x. g(x)$.*

Proof. If $f(r) \leq r \leq g(r)$ then $\mu_r x.f(x) \leq f(r) \leq r \leq g(r) \leq \mu_r x.g(x)$.

If $r < f(r) \leq g(r)$ then f and g are monotone mappings from $[r, \top]$ to $[r, \top]$ and then $\mu_r x.f$ and $\mu_r x.g$ are the least fixed points of f and g considered as mappings from the lattice $[r, \top]$ to $[r, \top]$. However, if $f \leq g$, where f and g monotone, then the least (greatest) fixed point of f is $\leq$ than the least (greatest) fixed point of g (Proposition 1.2.18 in [11]).

The case $f(r) \leq g(r) < r$ is symmetric to the previous one. $\qquad \square$

We endow $\mathbb{R}^n$ with the norm $\|(x_1, \ldots, x_n)\| = \max_i |x_i|$.

We say that a mapping $f \in \mathrm{Mon}(I^n, I^m)$ is monotone nonexpansive if for all $x = (x_1, \ldots, x_n), y = (y_1, \ldots, y_n) \in I^n$, $\|f(x) - f(y)\| \leq \|x - y\|$.

By $\mathrm{Mon}_e(I^n, I^m)$ we denote the set of monotone nonexpansive mappings from I^n to I^m.

Given $f \in \mathrm{Mon}(I^n, I)$ by

$$\mu_r z_i.f(z_1, \ldots, z_{i-1}, z_i, z_{i+1}, \ldots, z_n) \tag{2}$$

we denote the mapping from I^{n-1} to I which maps $(z_1, \ldots, z_{i-1}, z_{i+1}, \ldots, z_n) \in I^{n-1}$ to the nearest fixed point of the mappinng $z_i \mapsto f(z_1, \ldots, z_{i-1}, z_i, z_{i+1}, \ldots, z_n)$.

Lemma 4. *Let us fix* $r \in I$.

If $f \in \mathrm{Mon}(I^n, I)$ *then the mapping* (2) *belongs to* $\mathrm{Mon}(I^{n-1}, I)$.

If $f \in \mathrm{Mon}_e(I^n, I)$ *then the mapping* (2) *belongs to* $\mathrm{Mon}_e(I^{n-1}, I)$.

Proof. The first assertion follows immediately from Lemma 3.

For $(x_1, \ldots, x_{i-1}, x_{i+1}, \ldots, x_n) \in I^{n-1}$ we define inductively a sequence of mappings:

$$g^0(x_1, \ldots, x_{i-1}, x_{i+1}, \ldots, x_n) = f(x_1, \ldots, x_{i-1}, r, x_{i+1}, \ldots, x_n)$$

$$g^{k+1}(x_1, \ldots, x_{i-1}, x_{i+1}, \ldots, x_n) =$$
$$f(x_1, \ldots, x_{i-1}, g^k(x_1, \ldots, x_i, x_{i+1}, \ldots, x_n), x_{i+1}, \ldots, x_n).$$

We shall prove that for all k and all $x = (x_1, \ldots, x_{i-1}, x_{i+1}, \ldots, x_n)$ and $y = (y_1, \ldots, y_{i-1}, y_{i+1}, \ldots, y_n)$ in I^{n-1},

$$|g^k(x) - g^k(y)| \leq \|x - y\|. \tag{3}$$

For $k = 0$ this follows directly from the fact that f is monotone nonexpansive. Suppose that (3) holds for k. Then

$$|g^{k+1}(x) - g^{k+1}(y)|$$
$$= |f(x_1, \ldots, x_{i-1}, g^k(x), x_{i+1}, \ldots, x_n) - f(y_1, \ldots, y_{i-1}, g^k(y), y_{i+1}, \ldots, y_n)|$$
$$\leq \max\{|x_1 - y_1|, \ldots, |x_{i-1} - y_{i-1}|, |g^k(x) - g^k(y)|, |x_{i+1} - y_{i+1}|, \ldots |x_n - y_n|\}$$
$$\leq \max\{|x_1 - y_1|, \ldots, |x_{i-1} - y_{i-1}|, \|x - y\|, |x_{i+1} - y_{i+1}|, \ldots |x_n - y_n|\} \leq \|x - y\|.$$

Now it suffices to note that, by Lemma 2, $g^k(x)$ and $g^k(y)$ converge respectively to $\mu_r x_i.f(x_1, \ldots, x_i, \ldots, x_n)$ and to $\mu_r y_i.f(y_1, \ldots, y_i, \ldots, y_n)$ when $k \uparrow \infty$. $\qquad \square$

Let $f \in \mathrm{Mon}_e(I^n, I^n)$. Thus f is a vector of mappings $f = (f_1, \ldots, f_n)$ where, for each i, $f_i \in \mathrm{Mon}_e(I^n, I)$. Let $(r_1, \ldots, r_n) \in I^n$.

For each k, $1 \le k \le n$, we define a monotone nonexpansive mapping $F^{(k)}$: $I^{n-k} \to I^k$:

$$I^{n-k} \ni (x_{k+1}, \ldots, x_n) \mapsto F^{(k)}(x_{k+1}, \ldots, x_n) \in I^k.$$

(for $k = n$, $F^{(n)}$ will be just a constant from I^n not depending on any variable).

Since $F^{(k)}$ is a mapping into I^k, it is composed of k mappings into I, $F^{(k)} = (F_1^{(k)}, \ldots, F_k^{(k)})$.

For $k = 1$, $F^{(1)}$ is mapping into I and we will identify it with $F_1^{(1)}$.

We define $F^{(k)}$ by induction. For $k = 1$,

$$F^{(1)}(x_2, \ldots, x_n) = F_1^{(1)}(x_2, \ldots, x_n) = \mu_{x_1} r_1 . f_1(x_1, x_2, \ldots, x_n).$$

Suppose that $F^{(k-1)}(x_k, \ldots, x_n) = (F_1^{(k-1)}(x_k, \ldots, x_n), \ldots, F_{k-1}^{(k-1)}(x_k, \ldots, x_n))$ is already defined.

Intuitively, given $F^{(k-1)}$ as above to obtain $F^{(k)}$ we should eliminate the variable x_k. To this end we use the kth component mapping f_k of f.

First we define

$$F_k^{(k)}(x_{k+1}, \ldots, x_n) =$$
$$\mu_{r_k} x_k . f_k(F_1^{(k-1)}(x_k, \ldots, x_n), \ldots, F_{k-1}^{(k-1)}(x_k, \ldots, x_n), x_k, x_{k+1}, \ldots, x_n), \quad (4)$$

and subsequently we put

$$F_i^{(k)}(x_{k+1}, \ldots, x_n) = F_i^{(k-1)}(F_k^{(k)}(x_{k+1}, \ldots, x_n), x_{k+1}, \ldots, x_n), \quad \text{for } 1 \le i < k. \tag{5}$$

By Lemma 4 and since the composition of monotone nonexpansive mappings is monotone nonexpansive we can see that all mappings $F^{(k)}$ are monotone nonexpansive.

We shall write

$$\mu_{r_k} x_k . \ldots . \mu_{r_1} x_1 . f(x_1, \ldots, x_n)$$

to denote the mapping $F^{(k)}$ defined above and we call it the k-th nested fixed point of f. For $k = n$ we will speak about the nested fixed point without mentioning k.

4 Value of the Concurrent Priority Game as the Nested Nearest Fixed Point

4.1 Auxiliary One-Shot Game

In this section we define auxiliary matrix games.

Let $x = (x_1, \ldots, x_n) \in I^n$ and let $G(x)$ be a priority game with n states.

A one shot game $\Gamma_k(x_1, \ldots, x_n)$ is the game played on G in the following way. The game starts at state k, players Max and Min choose independently and simultaneously actions $a \in \mathbf{A}(k)$ and $b \in \mathbf{B}(k)$. Suppose that the next state is m. Then player Max receives from player Min the payoff x_m and the game $\Gamma_k(x_1, \ldots, x_n)$ ends.

Note that $\Gamma_k(x_1, \ldots, x_n)$ can be seen a zero-sum matrix game where the payoff obtained by player Max from player Min when players play actions a, b respectively is equal to $\sum_{m \in \mathbf{S}} x_m \cdot p(m|k, a, b)$. The value of the game $\Gamma_k(x_1, \ldots, x_n)$ will be denoted by

$$\Phi_k(x_1, \ldots, x_n) := \mathrm{val}(\Gamma_k(x_1, \ldots, x_n)). \tag{6}$$

We will be interested in $\Phi_k(x_1, \ldots, x_n)$ as a function of the reward vector $x = (x_1, \ldots, x_n)$.

Since all entries in the matrix game $\Gamma_k(x_1, \ldots, x_n)$ belong to I, $\mathrm{val}(\Gamma_k(x_1, \ldots, x_n)) \in I$, i.e. Φ_k is a mapping from I^n into I.

The following well known properties of matrix games are essential (see for example [12]):

Theorem 5. *Let M_1 and M_2 be two matrix games of the same size. Then*

- *If $M_1 \leq M_2$ (where the inequality holds componentwise) then $\mathrm{val}[M_1] \leq \mathrm{val}[M_2]$.*
- *$|\mathrm{val}[M_1] - \mathrm{val}[M_2]| \leq \|M_1 - M_2\|$, where $\|M\| = \max_{i,j} |M(i,j)|$.*

From Theorem 5 it follows that

Proposition 6. *The mapping Φ_k defined in (6) is monotone and nonexpansive.*

4.2   Priority Games with One Non-absorbing State

In this section we will study concurrent priority games with one non-absorbing state. Let us recall that a state i is absorbing if for all $(a, b) \in \mathbf{A}(i) \times \mathbf{B}(i)$, $p(i|i, a, b) = 1$.

We shall write $G_k(x_1, \ldots, x_n)$ to denote a priority game G with n states having rewards $x_1, \ldots, x_n$ and such that all states, except state k, are absorbing. We shall call such a game absorbing. A game starting in an absorbing state i, $i \neq k$, is trivial, the game remains forever in i and the payoff is equal to the reward x_i associated with state i. For plays starting in the non-absorbing state k either at some moment we hit an absorbing state i and the payoff obtained for such plays is x_i (and it is irrelevant what players play once an absorbing state is attained) or we remain forever in k and the payoff for such a play is x_k. Such games are equivalent to Everett games with one non-absorbing state. Thus only the value and players' strategies in the non-absorbing state k are of interest in $G_k(x_1, \ldots, x_n)$.

In the sequel we will use the following notation. For $x = (x_1, \ldots, x_n) \in I^n$ and $e \in I$ we write (x_{-k}, e) to denote the element $(x_1, \ldots, x_{k-1}, e, x_{k+1}, \ldots, x_n)$ of I^n.

Moreover if σ, τ are strategies of players Max and Min in the one shot game Γ (i.e. for each state $s \in \mathbf{S}$, $\sigma(s) \in \Delta(\mathbf{A}(s))$ and $\tau(s) \in \Delta(\mathbf{B}(s))$ are mappings from states to distributions over actions) then σ^∞ and τ^∞ will denote the stationary strategies in the priority game G such that at each stage players select actions independently of the past history with probabilities given by σ and τ.

Lemma 7. *Let $G_k(x)$ be an absorbing priority game and $r \in I$. Then*

$$\mathrm{val}(G_k(x_{-k}, r)) = \mu_r x_k.\Phi_k(x_1, \ldots, x_k, \ldots, x_n).$$

(i) *If $\mu_r x_k.\Phi_k(x) \geq r$ then player Min has an optimal stationary strategy while, for each $\varepsilon > 0$, player Max has an ε-optimal stationary strategy.*

(ii) *If $\mu_r x_k.\Phi_k(x) \leq r$ then player Max has an optimal stationary strategy in $G_k(x_{-k}, r)$ while, for each $\varepsilon > 0$, player Min has an ε-optimal stationary strategy.*

Note that from this lemma it follows that if $\mu_r x_k.\Phi_k(x) = r$ then both players have optimal stationary strategies in the absorbing game $G_k(x_{-k}, r)$.

4.3   General Priority Games

Let G be a priority game.

Note that

$$(x_1, \ldots, x_n) \mapsto \Phi(x_1, \ldots, x_n) = (\Phi_1(x_1, \ldots, x_n), \ldots, \Phi_n(x_1, \ldots, x_n))$$

where Φ_i defined in (6) are monotone and nonexpansive mappings from I^n to I.

By $G_{\leq k}$ we will denote the priority game obtained from G by transforming all states i, such that $i > k$ into absorbing states. On the other hand, all states $j \leq k$ have the same available actions and transition probabilities as they have in G.

Of course, the value of each absorbing state j of the game $G_{\leq k}(r_1, \ldots, r_k, x_{k+1}, \ldots, x_n)$, $k < j \leq n$, is x_j thus only the values of non-absorbing states $1, \ldots, k$ are of interest.

It turns out that these values are obtained as nested fixed points:

Theorem 8. *Let $(r_1, \ldots, r_k) \in I^k$. Then the nested fixed point*

$$(F_1^{(k)}(x_{k+1}, \ldots, x_n), \ldots, F_k^{(k)}(x_{k+1}, \ldots, x_n)) := \mu_{r_k} x_k. \cdots \mu_{r_1} x_1.\Phi(x)$$

is the vector of values of non-absorbing states $(1, \ldots, k)$ of the game $G_{\leq k}(r_1, \ldots, r_k, x_{k+1}, \ldots, x_n)$.

As it is known for parity games, which form a special subclass of priority games, the winning regions (in the deterministic case [2]) or the values (for concurrent stochastic parity games [5]) can be described by an appropriate formulas of μ-calculus – a fixed point calculus over an appropriate complete lattice where we alternate the greatest and the least fixed points. From this point of view Theorem 8 looks just as an extension of known results to a wider framework of priority games. However there is one ingredient of Theorem 8 that seems to be new.

It is notoriously difficult to comprehend a μ-calculus formula alternating several greatest and least fixed point.

Theorem 8 provides a natural interpretation in the term of games of a formula where only some initial fixed points are applied.

Let

$$(v_1, \ldots, v_k) = \mu_{r_k} x_k . \mu_{r_{k-1}} x_{k-1} . \cdots \mu_{r_1} x_1 . \boldsymbol{\Phi}(x_1, \ldots, x_k, r_{k+1}, \ldots, r_n) \quad (7)$$

Then $(v_1, \ldots, v_k)$ are the values of states $1, \ldots, k$ in the priority game

$$G_{\leq k}(r_1, \ldots, r_k, r_{k+1}, \ldots, r_n) \quad (8)$$

which differs from the original priority game $G(r_1, \ldots, r_k, r_{k+1}, \ldots, r_n)$ in that the states $k+1, k+2, \ldots, n$ are absorbing in the game (8).

Now when we add another fixed point to (7) to obtain

$$(v'_1, \ldots, v'_k, v'_{k+1}) =$$
$$\mu_{r_{k+1}} x_{k+1} . \mu_{r_k} x_k . \mu_{r_{k-1}} x_{k-1} . \cdots \mu_{r_1} x_1 . \boldsymbol{\Phi}(x_1, \ldots, x_k, x_{k+1}, r_{k+2}, \ldots, r_n)$$

then this can be interpreted as an operation transforming state $k+1$ form a non-absorbing in the game (8) into a non-absorbing in the game $G_{\leq k+1}(r_1, \ldots, r_k, r_{k+1}, \ldots, r_n)$.

5   Algorithmic Issues

One can wonder if the recursive formulas for the nested nearest point cannot be used to approximate the values of fixed points, i.e. the values of the states in the priority game. Unfortunately in general this seems to be difficult, if at levels $1, \ldots, k$ we stop iterations before attaining the fixed points then the resulting errors can even change the direction of iterations at level $k+1$. Moreover it is difficult to see when we can stop iterations (there is no criterion to estimate the distance between the value obtained at some iteration and the limit fixed point).

However there is one case when the recursive formulas developed in this paper can be used to solve the priority game, this is the case of perfect information stochastic priority games where for each state only one of the two players chooses actions to play (the other player can be seen as having only one action at this state). First we have the following counterpart of Lemma 7:

Lemma 9. *Let $G_k(x)$ be a priority game with the unique non-absorbing state k, $r \in I$. Let k be controlled by one player (either Max or Min) who chooses an action to play and the probability distribution over next states depends uniquely upon the action chosen by the controlling player. Then the value*

$$\mathrm{val}(G_k(x_{-k}, r)) = \mu_r x_k . \Phi_k(x_1, \ldots, x_k, \ldots, x_n).$$

of state k can be calculated in polynomial time and the controlling player has an optimal pure strategy.

Proof. For each action a of the player controlling k we have the following formula for the expected reward after playing a once:

$$\mathbb{E}_k^a[R] = \mathbb{E}_k^a[R|A] \cdot p(A|k, a) + r \cdot p(k|k, a)$$

where R is the expected reward after playing a once, A is the event that the next state is absorbing, $p(A|k, a) = \sum_{j \in A} p(j|k, a)$ is the probability that the next state is absorbing when a is executed, $r \cdot p(k|k, a)$ is the probability that we remain in k when the player plays a in k. From this formula we can calculate $\mathbb{E}_k^a[R|A]$ i.e. the expected reward under condition that the next state is absorbing.

The expected payoff obtained when we play the strategy a^∞ (play only a as long as the current state is k) is equal

$$\mathbb{E}_k^{a^\infty}[u] = \begin{cases} \mathbb{E}_k^a[R|A] & \text{if } p(A|k, a) > 0, \\ r & \text{otherwise.} \end{cases} \tag{9}$$

Thus if k is controlled by player Max (Min) then he should play action a that maximizes (minimizes) (9) as long as we are in k. $\qquad\square$

Thus Lemma 9 shows how to solve one state perfect information stochastic priority game. To solve a perfect information priority game with any number of non-absorbing states we use the induction. However instead of value iteration algorithm (which can be non-terminating) we use the strategy iteration (which always terminate as the number of pure strategies is finite). In fact this algorithm just tries to accelerate and optimize the procedure calculating the nested nearest fixed points.

Algorithm 1 on page 12 implements a recursive procedure SolveGame $(k, (r_1, \ldots, r_k, x_{k+1}, \ldots, x_n))$ that calculates the vector $(a_1, \ldots, a_k)$ of actions played in non-absorbing states $1, \ldots, k$ by optimal pure stationary strategies in the perfect information stochastic priority game $G_{\leq k}(r_1, \ldots, r_k, x_{k+1}, \ldots, x_n)$.

Algorithm 1. Calculate optimal pure stationary strategies in a perfect information stochastic priority game with k non-absorbing states. We assume that $\texttt{OneState}(k, c_1, \ldots, c_n)$ is the procedure described in Lemma 9 returning the optimal action for the game with one non-absorbing state k and reward vector $(c_1, \ldots, c_n)$ and $\texttt{value}(k, (a_1, \ldots, a_k))$ is the value of state k when players play actions $(a_1, \ldots, a_k)$ in states $1, \ldots, k$ respectively.

```
1  SolveGame(k, (r₁,...,rₖ, xₖ₊₁,...,xₙ))  Result: a vector (a₁,...,aₖ) of
                              actions, aᵢ action played in state i by the optimal strategy of the player
                              controlling i.
2  if k = 1 then
3  │  return OneState(1, r₁, x₂,...,xₙ);
4  end
5  w ← rₖ;
6  while true do
7  │  (a₁,...,aₖ₋₁) ← SolveGame(k − 1, r₁,...,rₖ₋₁, w, xₖ₊₁,...,xₙ);
8  │  aₖ ← OneState(k, a₁,...,aₖ₋₁, w, xₖ₊₁,...,xₙ);
9  │  z ← value(k, (a₁,...,aₖ));
10 │  if z = w then
11 │  │  return (a₁,...,aₖ);
12 │  end
13 │  w ← z;
14 end
```

References

1. Emerson, E., Jutla, C.: Tree automata, mu-calculus and determinacy. In: FOCS 1991, pp. 368–377. IEEE Computer Society Press (1991)
2. Walukiewicz, I.: Monadic second-order logic on tree-like structures. Theoret. Comput. Sci. **275**, 311–346 (2002)
3. McIver, A., Morgan, C.: Games, probability and the quantitative μ-calculus qmu. In: Baaz, M., Voronkov, A. (eds.) LPAR 2002. LNCS (LNAI), vol. 2514, pp. 292–310. Springer, Heidelberg (2002)
4. McIver, A., Morgan, C.: A novel stochastic game via the quantitative mu-calculus. In: Cerone, A., Wiklicky, H., (eds.) Proceedings of the Third Workshop on Quantitative Aspects of Programming Languages (QAPL 2005), ENTCS, vol. 153(2), pp. 195–212. Elsevier (2005)
5. de Alfaro, L., Majumdar, R.: Quantitative solution to omega-regular games. J. Comput. Syst. Sci. **68**, 374–397 (2004)
6. Chatterjee, K., de Alfaro, L., Henzinger, T.: Qualitative concurrent parity games. ACM Trans. Comput. Logic **12**, 28:1–28:51 (2011)
7. Everett, H.: Recursive games. In: Contributions to the Theory of Games, vol. III, pp. 47–78. Princeton University Press (1957)
8. Martin, D.: The determinacy of Blackwell games. J. Symbolic Logic **63**(4), 1565–1581 (1998)
9. Maitra, A., Sudderth, W.: Stochastic games with Borel payoffs. In: Neyman, A., Sorin, S. (eds.) Stochastic Games and Applications, NATO Science Series C, Mathematical and Physical Sciences, vol. 570, pp. 367–373. Kluwer Academic Publishers (2004)

10. Tarski, A.: A lattice-theoretical fixpoint theoem and its aplications. Pacific J. Math. **5**, 285–309 (1955)
11. Arnold, A., Niwiński, D.: Rudiments of mu-calculus. Studies in Logic and the Foundations of Mathematics, vol. 146. Elsevier (2001)
12. Parthasarathy, T., Raghavan, T.: Some Topics in Two-Person Games. Elsevier, New York (1971)

Author Index

Printed in the United States
By Bookmasters